PLASMA PROTEIN METABOLISM

Regulation of Synthesis, Distribution, and Degradation

CONTRIBUTORS

CHESTER A. ALPER
G. BENDIXEN
R. BIANCHI
GUNNAR BIRKE
R. MICHAEL BLAESE
G. BONORRIS
E. CARSON
A. Y. CHEN
L. DONATO
EDWARD C. FRANKLIN
TRISTRAM FREEMAN
M. CLELIA GANOZA
S. GOLDEN
A. H. GORDON
EVA HEDFORS
R. HOFFENBERG
S. JARNUM
K. B. JENSEN
DAVID W. JOHN
E. A. JONES
J. KATZ
TERRY LANGER
ROBERT I. LEVY
JERRY LYNN
A. S. McFARLANE
P. MANCINI
CL. MARCHE
G. MARIANI
LEON L. MILLER
HAMISH N. MUNRO
RENEE NORBERG
MURRAY ORATZ
A. PILO
L.-O. PLANTIN
E. B. REEVE
ERWIN REGOECZI
VICTOR M. ROSENOER
MARCUS A. ROTHSCHILD
E. J. SARCIONE
SIDNEY S. SCHREIBER
A. L. SELLERS
HERSCHEL SIDRANSKY
J. SOLTOFT
HANS L. SPIEGELBERG
KENNETH STERLING
WARREN STROBER
Y. TAKEDA
P. VESIN
F. VITEK
THOMAS A. WALDMANN
B. WEEKE
H. WESTERGAARD
CURTIS A. WILLIAMS

PLASMA PROTEIN METABOLISM
Regulation of Synthesis, Distribution, and Degradation

Edited by

MARCUS A. ROTHSCHILD

Veterans Administration Hospital and New York University School of Medicine
New York, New York

AND

THOMAS WALDMANN

Immunophysiology Section of the Metabolism Branch, National Cancer Institute,
National Institutes of Health, Bethesda, Maryland

1970

ACADEMIC PRESS New York and London

ACADEMIC PRESS, INC.
111 Fifth Avenue, New York, New York 10003

United Kingdom Edition published by
ACADEMIC PRESS, INC. (LONDON) LTD.
Berkeley Square House, London W1X 6BA

LIBRARY OF CONGRESS CATALOG CARD NUMBER: 77-117634

PRINTED IN THE UNITED STATES OF AMERICA

CONTENTS

Part V Immunoglobulin Metabolism

16. Genetic Regulation of Immunoglobulin Levels in Man

Edward C. Franklin

17. Physiological Factors Controlling Immunoglobulin Metabolism

T. A. Waldmann, R. M. Blaese, and W. Strober

18. Abnormalities of Immunoglobulin Metabolism

Warren Strober, R. Michael Blaese, and Thomas A. Waldmann

Part VI Acute Phase Proteins: Stress–Trauma

Part IX Hormonal Transport

32. Regulation of Hormones by Binding Proteins With Special Reference to the Transport of Thyroxine in the Circulation

Kenneth Sterling

LIST OF CONTRIBUTORS

Numbers in parentheses indicate the pages on which the authors' contributions begin.

CHESTER A. ALPER (393), Department of Pediatrics, Harvard Medical School; Department of Medicine and Blood Grouping Laboratory, Children's Hospital Medical Center; Department of Medicine, Peter Bent Brigham Hospital, Boston, Massachusetts

G. BENDIXEN (321), Medical Department P, Division of Gastroenterology, Rigshospitalet, and Department of Clinical Chemistry Bispebjerg Hospital, Copenhagen NV, Denmark

R. BIANCHI (25), The C.N.R. Laboratory of Clinical Physiology, University of Pisa, Pisa, Italy

GUNNAR BIRKE (415, 427), King Gustaf V Research Institute, Stockholm, Sweden

R. MICHAEL BLAESE (269, 287), Immunophysiology Section of the Metabolism Branch, National Cancer Institute, National Institutes of Health, Bethesda, Maryland

G. BONORRIS (129), Cedars-Sinai Medical Center, Medical Research Institute, Los Angeles, California

E. CARSON (11), Department of Control Engineering, The City University, London EC1, England

A. Y. CHEN (89), Department of Medicine, University of Colorado Medical Center, Denver, Colorado

L. DONATO (25), The C.N.R. Laboratory of Clinical Physiology, University of Pisa, Pisa, Italy

EDWARD C. FRANKLIN (259), Department of Medicine, Rheumatic Diseases Study Group, New York University School of Medicine, New York, New York

TRISTRAM FREEMAN (63), M.R.C. Clinical Research Centre Laboratory, Whittington Hospital, London N19, England

M. CLELIA GANOZA (169), Charles H. Best Institute, Toronto, Ontario, Canada

S. GOLDEN (129), Cedars-Sinai Medical Center, Medical Research Institute, Los Angeles, California

A. H. GORDON (351), National Institute for Medical Research, Mill Hill, London, NW7, England

EVA HEDFORS (427), King Gustaf V Research Institute, Stockholm, Sweden

R. HOFFENBERG (239), M.R.C. Clinical Research Centre, Harrow, Middlesex, England

S. JARNUM (321), Medical Department P, Division of Gastroenterology, Rigshospitalet, and Department of Clinical Chemistry, Bispebjerg Hospital, Copenhagen, Denmark

K. B. JENSEN (321), Medical Department P, Division of Gastroenterology, Rigshospitalet, and Department of Clinical Chemistry, Bispebjerg Hospital, Copenhagen, Denmark

DAVID W. JOHN (207), Department of Radiation Biology and Biophysics, University of Rochester, School of Medicine and Dentistry, Rochester, New York

E. A. JONES (11), Department of Medicine, Royal Free Hospital, London, WC1, England

J. KATZ (129), Cedars-Sinai Medical Center, Medical Research Institute, Los Angeles, California

TERRY LANGER (483), *The Molecular Disease Branch, National Heart and Lung Institute, Bethesda, Maryland

ROBERT I. LEVY (483), The Molecular Disease Branch, National Heart and Lung Institute, Bethesda, Maryland

JERRY LYNN (185), Department of Pathology, University of Pittsburgh School of Medicine, Pittsburgh, Pennsylvania

* Present address: Department of Medicine, Columbia University College of Physicians and Surgeons, Presbyterian Hospital, New York, New York.

A. S. McFARLANE (51), Division of Biophysics, National Institute for Medical Research, Mill Hill, London, NW7, England

P. MANCINI (25), The C.N.R. Laboratory of Clinical Physiology, University of Pisa, Pisa, Italy

CL. MARCHE (337), Pathology Laboratory, Department of Medicine, Bichat University Hospital, Paris, France

G. MARIANI (25), The C.N.R. Laboratory of Clinical Physiology, University of Pisa, Pisa, Italy

LEON L. MILLER (207), Department of Radiation Biology and Biophysics, University of Rochester, School of Medicine and Dentistry, Rochester, New York

HAMISH N. MUNRO (157), Physiological Chemistry Laboratories, Department of Nutrition and Food Science, Massachusetts Institute of Technology, Cambridge, Massachusetts

RENEE NORBERG (427), King Gustaf V Research Institute, Stockholm, Sweden

MURRAY ORATZ (199, 223), Radioisotope Service, Veterans Administration Hospital, and Department of Biochemistry, New York University College of Dentistry, New York, New York

A. PILO (25), The C.N.R. Laboratory of Clinical Physiology, University of Pisa, Pisa, Italy

L.-O. PLANTIN (427), King Gustaf V Research Institute, Stockholm, Sweden

E. B. REEVE (3, 89), Department of Medicine, University of Colorado, School of Medicine, Denver, Colorado

ERWIN REGOECZI (439, 459),* Biophysics Division, National Institute for Medical Research, London, NW7, England

VICTOR M. ROSENOER (11, 111), Children's Cancer Research Foundation, and Harvard Medical School, Boston, Massachusetts

MARCUS A. ROTHSCHILD (111, 199), Radioisotope Service, Veterans Administration Hospital, and Department of Medicine, New York University School of Medicine, New York, New York

* Present address: Department of Pathology, McMaster University, Hamilton, Ontario, Canada.

E. J. SARCIONE (369), Division of Medicine, Roswell Park Memorial Institute, New York State Department of Health, Buffalo, New York

SIDNEY S. SCHREIBER (199), Radioisotope Service, Veterans Administration Hospital, and Department of Medicine, New York University School of Medicine, New York, New York

A. L. SELLERS (129), Cedars-Sinai Medical Center, Medical Research Institute, Los Angeles, California

HERSCHEL SIDRANSKY (185), Department of Pathology, University of Pittsburgh School of Medicine, Pittsburgh, Pennsylvania

J. SOLTOFT (321), Medical Department P, Division of Gastroenterology, Rigshospitalet, and Department of Clinical Chemistry, Bispebjerg Hospital, Copenhagen Denmark

HANS L. SPIEGELBERG (307), Department of Experimental Pathology, Scripps Clinic and Research Foundation, La Jolla, California

KENNETH STERLING (507), Protein Research Laboratory, Bronx Veterans' Administration Hospital, Bronx, New York, and Department of Pathology, Columbia University College of Physicians and Surgeons, New York, New York

WARREN STROBER (269, 287, 483), Immunophysiology Section of the Metabolism Branch, National Cancer Institute, National Institutes of Health, Bethesda, Maryland

Y. TAKEDA (443), Department of Medicine, University of Colorado School of Medicine, Denver, Colorado

P. VESIN (337), Department of Medicine, Argenteuil Hospital, Argenteuil, France

F. VITEK (25),* Institute of Biophysics, Faculty of General Medicine, Charles University, Prague, Czechoslovakia

THOMAS A. WALDMANN (269, 287), Immunophysiology Section of the Metabolism Branch, National Cancer Institute, National Institutes of Health, Bethesda, Maryland

B. WEEKE (321), Medical Department P, Division of Gastroenterology, Rigshospitalet, and Department of Clinical Chemistry, Bispebjerg Hospital, Copenhagen Denmark

* Present address: The C.N.R. Laboratory of Clinical Physiology, University of Pisa, Pisa, Italy.

H. WESTERGAARD (321), Medical Department P, Division of Gastroenterology, Rigshospitalet, and Department of Clinical Chemistry, Bispebjerg Hospital, Copenhagen Denmark

CURTIS A. WILLIAMS (169, 383), The Rockefeller University, New York, New York

PREFACE

In 1839, Professor Magendie devoted a series of lectures to the physiological aspects of blood. He stated "I propose devoting a considerable number of lectures to the study of the blood for I am acquainted with no subject more deserving of attention." These comments are indeed true today with multidisciplinary investigations of the blood, including the plasma proteins, being carried out in centers throughout the world. Over the past few years, one area of many exciting advances has been the study of control mechanisms affecting serum proteins. The object of this volume is to present the current concepts concerning the physiological and pathophysiological factors regulating the distribution, degradation, and synthesis of plasma proteins. The first sections of the book present the assumptions and methodology involved in the various *in vivo* and *in vitro* techniques that have led to our understanding of protein metabolism. Techniques of protein isolation, characterization, labeling, and mathematical analysis of the data are included. In addition, the recently described methods for directly quantitating protein synthetic rates in nonsteady state conditions are considered.

The remainder of the text concerns the factors involved in regulating the serum levels of albumin, acute phase reactants, immunoglobulins, clotting factors, complement and hormone-binding proteins. The controlling factors include such general and specific physiological regulators of protein synthesis and catabolism as levels of specific serum proteins, hormonal regulators, variations of temperature and oncotic pressure, antigenic stimulation, and nutritional factors. It also contains an analysis of the pathophysiological factors including disorders of protein synthesis, distribution, exogenous catabolism, and external loss and stress that produce abnormal levels of serum proteins.

As editors we should like to say how fortunate we have been in securing as authors many of the leading investigators in the field of protein metabolism and to express our appreciation for their participation in writing this work. We also wish to thank the Veterans Administration Department of Medicine and Surgery for their support; Dr. Thomas Chalmers, Assistant Chief Medical Director for Research and Education, who was instrumental in providing this support and who gave us invaluable advice and encouragement;

Dr. Charles Rosenberg, Assistant Director Education Service, Veterans Administration Central Office, who helped throughout the preparation of this text; and Mr. Shamanskin at the Veterans Administration Central Office, Mr. J. V. Sheehan, Hospital Director, and Mr. T. Steele, Fiscal Officer at the Manhattan Veterans Administration Hospital, as well as Miss Judith Rosenblum and Mrs. Sylvia Ser who provided valuable editorial assistance.

Most of all, we wish to thank Bobby. She has helped from the first moment when this text was only an idea. Her encouragement and planning are gratefully and sincerely appreciated.

We wish to thank the following companies for their generous financial support: Abbott Laboratories, North Chicago, Illinois; Buchler Instruments, Fort Lee, New Jersey; Cambridge Nuclear Corporation, Cambridge, Massachusetts; Hoffman-La Roche, Inc., Nutley, New Jersey; International Chemical & Nuclear Corporation, Pasadena, California; Mallinckrodt Pharmaceuticals, St. Louis, Missouri; Mann Research Laboratories, New York, New York; Melpar, Inc., Biological Products Laboratory, Falls Church, Virginia; Neisler Laboratories, Inc., Union Carbide Corporation, Tuxedo, New York; Nuclear Associates, Inc., Westbury, New York; Nuclear Chicago Corporation, DesPlaines, Illinois; Mr. Bernard Palitz, Commercial Alliance Corporation, New York, New York; Ortho Pharmaceutical Corporation, Raritan, New Jersey; Pel Freeze Biologicals, Inc., Rogers, Arkansas; Pentex Biochemicals, Miles Laboratories, Inc., Kankakee, Illinois; Mr. Seymour Schnell, New York, New York.

I

Model Systems and Protein Isolation

1

MEASUREMENTS REQUIRED TO DEFINE BEHAVIOR OF A PLASMA PROTEIN IN VIVO: THE NEED FOR MEASURING PROTEIN SYNTHETIC RATE WITH A CRITICISM OF AVAILABLE METHODS

E. B. Reeve

DEPARTMENT OF MEDICINE,
UNIVERSITY OF COLORADO,
SCHOOL OF MEDICINE,
DENVER, COLORADO

I. Generalized Model of a Plasma Protein System

Figure 1 shows a generalized plasma protein system in which P_p is the total amount of a particular protein in the plasma; P_i the total amount of this protein in the interstitial fluids; S, the rate of protein synthesis; j_1P_p, the rate of protein passage from the plasma to the interstitial fluids (e.g., in g/day); j_3P_p, the rate of protein catabolism; and j_2P_i, the rate of return of

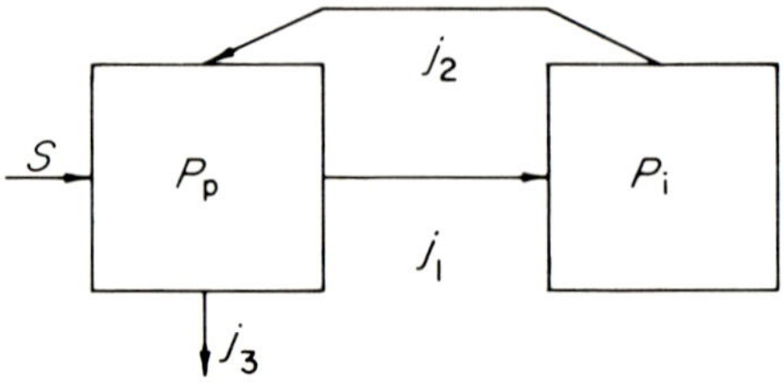

FIG. 1. A generalized plasma protein model. P_p, quantity of plasma protein; P_i, interstitial protein; S, rate of protein synthesis; the j's, fractional rates. See text.

interstitial protein, probably primarily by the lymph. For the moment no assumptions are made about the j's. The equations describing the flow of protein through this system are

$$dP_p/dt = S + j_2 P_i - (j_1 + j_3)P_p \tag{1}$$

$$dP_i/dt = j_1 P_p - j_2 P_i \tag{2}$$

and these equations show that to define this system requires knowledge of the behavior of S, P_p, and P_i and j_1, j_2, and j_3 in various physiological and pathological states.

A. Study of the Generalized System with Proteins Labeled with Radioactive Iodine

Suppose we study this system in a moderately nonsteady state by injecting intravenously an ^{131}I-labeled protein, termed P*. Then the equations describing labeled-protein flow become

$$dP_p^*/dt = j_2 P_i^* - (j_1 + j_3)P_p^* \tag{3}$$

$$dP_i^*/dt = j_1 P_p^* - j_2 P_i^* \tag{4}$$

The tracer studies provide measurements of the time course of P_p^* and thus of the rate of change of P_p^*, dP_p^*/dt. For the first few hours after injection of labeled protein there will be negligible interstitial labeled protein and so in Eq. (3) P_i^* will be zero and $dP_p^*/dt = -(j_1 + j_3)P_p^*$. Thus during this time the rate of change of P_p^* is equal to the sum of the loss rates of labeled protein to interstitial fluids and catabolic sites. If urinary activity is collected or whole-body radioactivity measured, approximate estimates of j_3 may be made by the metabolic clearance method (Berson and Yalow, 1957; McFarlane, 1957; Mosser *et al.*, 1968) later in the experiment. Thus if, as experience suggests, j_3 does not vary greatly, studies with radioactive iodine-labeled proteins allow measurements of P_p^*, dP_p^*/dt, $(j_1 + j_3)$, and j_3 during the early part of a tracer experiment.

Further information is obtained from measurements of native plasma protein concentration and plasma volume, namely P_p, the total amount of the protein in the plasma and dP_p/dt, its rate of change. Thus with the average values of j_1 and j_3 given by the tracer studies, all the protein flow variables of Eq. (1) can be measured during the early part of a tracer experiment, except S and j_2P_i. If either S or j_2P_i could be measured, then the other could be calculated and thus during the first few hours of a tracer experiment the system described by Eqs. (1) and (2) could be completely defined. At present only S, synthetic rate, can be measured and hence the importance of measurements of synthetic rate in nonsteady states.

B. Definition of a Steady State and Its Significance

From Eq. (1) during a steady state $dP_p/dt = 0$ and $S + j_2P_i = (j_1 + j_3)P_p$. Tracer studies show that for many proteins j_3 remains on the average about constant over a few days. Since P_p is constant, we can write

$$S + j_2P_i - j_1P_p = j_3P_p\text{, a constant} \tag{5}$$

Thus without further information we cannot derive any evidence about synthetic rate S from j_3P_p (catabolic rate). Further information has to come from physiological arguments and from simulations of studies with tracer-labeled proteins as shown elsewhere (Reeve and Chen, Chapter 6). A physiological argument is that in a steady state, S must be closely related to j_3P_p, the catabolic rate. If not, total plasma protein would be expected to rise or fall over time, and j_3P_p could not be constant. If $S \cong j_3P_p$, it follows that $j_1P_p \cong j_2P_i$ also. The simulation studies described elsewhere (Reeve and Chen, Chapter 6) provide additional evidence for this. From the relation that $j_1P_p \cong j_2P_i$, it does not follow that j_1P_p or j_2P_i are constant rates. Instead physiological evidence indicates that j_2P_i (lymph protein return) fluctuates during the day so that j_2P_i must fluctuate about some mean value (Reeve and Chen, Chapter 6). Thus j_1P_p (capillary transfer of protein) must also fluctuate about a mean. With this in mind we can rewrite Eqs. (1) and (2) as

$$dP_p/dt = S - j_3P_p + j_2(\omega)P_i - j_1(\omega)P_p \tag{6}$$

$$dP_i/dt = j_1(\omega)P_p - j_2(\omega)P_i \tag{7}$$

where (ω) indicates that j_1 and j_2 are periodic functions of similar if not identical periodicities. Equations (6) and (7) demonstrate what information studies with tracer-labeled proteins can yield in a steady state. The constancy of P_p, and j_3 the fractional catabolic rate of this protein may be taken as evidence of a steady state. Then tracer-labeled proteins allow determination of j_3P_p and thus the synthetic rate of the protein, j_3 (the fractional catabolic rate), $j_1(\omega)$ (an average fractional rate of passage of protein into the interstitial fluids), and $j_1(\omega)P_p$ (an average transcapillary passage of protein), the latter two early in the experiment. During this time, $j_2(\omega)P_i$ (the average lymphatic return rate) must equal $j_1(\omega)P_p$ (the average capillary transfer rate). A second measurement of $j_1(\omega)P_p$ could be obtained by injecting the protein labeled with a second isotope 24 hours or so after the first labeled protein had been injected.

Although steady state studies with tracer-labeled proteins have provided much valuable information, it is well known that control mechanisms are rarely revealed by such studies. To define these, investigations must be made of perturbed systems during transient states.

II. Measurements of Synthetic Rate

Although, as seen above on certain reasonable assumptions in steady states, synthetic rate of a plasma protein can be estimated from measurements with tracer-labeled proteins of catabolic rate, such measurements cannot be made in the transient states required for understanding how a particular protein system is controlled. Direct measurement of synthetic rate is required as well as the information that can be obtained by the other studies. At present two methods of measuring synthetic rate are available: the 6-^{14}C-arginine synthetic rate method (Reeve *et al.*, 1963; McFarlane, 1963; Reeve, 1965); and a mass balance approach using tracer-labeled proteins for plasma proteins showing certain restricting properties (Atencio and Reeve, 1965; Atencio *et al.*, 1969).

A. The 6-^{14}C-Arginine Method

The rate of formation of a plasma protein in a time *T*, providing this rate remains about constant, can be determined from the amount of a ^{14}C-labeled amino acid entering the plasma protein during *T*. If the mean specific activity of the amino-acid at the synthetic site during *T* is known and the amount of newly synthesized protein that has left the plasma during *T* to enter interstitial fluids and catabolic sites can be calculated. The chief difficulty is measuring the mean specific activity at the synthetic site. However, (Reeve *et al.*, 1963; Reeve, 1965) if the labeled amino acid is 6-^{14}C-arginine, both the protein and urea are formed during *T* at a constant rate in the liver, and the 6-^{14}C-arginine is equally available in the liver for protein and urea formation, then the specific activity of the newly formed urea gives a valid estimate of the specific activity of the 6-^{14}C-arginine at the protein synthetic site and, thus, a valid estimate of the specific activity of the 6-^{14}C-arginine entering newly synthesized protein. Thus, in principle, a method was available for measuring the synthetic rate of liver synthesized proteins. Practically, however, this method, which so far has mainly been used for albumin, has presented many difficulties. First, the technically difficult problem must be solved of isolating a fairly pure protein from plasma. This is required so that the specific activity of the protein 6-^{14}C-arginine can be determined and not that of a mixture of the protein and contaminants. This was especially important in an early form of the method in which 6-^{14}C-arginine was injected intravenously (Reeve *et al.*, 1963). However, this form of the method, which had a number of disadvantages (McFarlane, 1963), has now been superseded by a method in which injections of ^{14}C-bicarbonate are given (McFarlane, 1963). These allow the liver cell to synthesize its own 6-^{14}C-arginine. However, large doses of ^{14}C-bicarbonate are required to achieve even a low level of protein labeling.

A second major problem has been to determine the true specific activity of the carbon of *newly synthesized* urea. This might seem simple enough because all that is required is to determine U^*, the total ^{14}C entering urea over the time T, and U, the total weight of urea–carbon formed in the time T. Difficulties lie in the large reservoir of urea in the body fluids, the variable fraction of urea concentrated in the kidneys, and the very active catabolism of urea in the gut by bacterial urease. Thus to determine U^* requires estimates of total ^{14}C-urea in body fluids and urine with that removed by gut breakdown; the determination of U requires estimates of urea carbon excreted in the urine and broken down in the gut and change of urea carbon content of the body (McKinley *et al.*, 1970). Other approaches are based on kinetic principles all of which assume that after an injection of ^{14}C-bicarbonate, ^{14}C-urea is formed and distributed almost instantaneously; while during T, the period of study, urea is synthesized, excreted, and broken down in the gut at constant rates. These assumptions then allow formulation of kinetic models (McFarlane *et al.*, 1965; Wochner *et al.*, 1968; Tavill *et al.*, 1968; Koj and McFarlane, 1968) from which the equivalent of mean specific activity of urea carbon during T can be obtained. However, it is now known that after ^{14}C-bicarbonate injection, ^{14}C-urea is not formed and distributed almost instantaneously (Koj and McFarlane, 1968; Reeve and McKinley, 1970). To allow more precise analysis of the urea system, ^{13}C-urea and ^{14}C-bicarbonate have both been injected and body fluid levels of ^{13}C-urea have been determined with a mass spectrophotometer (Koj and McFarlane, 1968). But addition of technically demanding measurements to an already difficult method provides at best a temporary solution to the problem of measuring the synthetic rate of a plasma protein. A final problem is found in fed animals: Urea synthetic rate may change quite rapidly (McKinley *et al.*, 1970), and a large rapid change will invalidate measurements that use newly synthesized urea to measure the specific activity of 6-^{14}C-arginine at the synthetic site.

It would clearly be valuable if some other substance could be used to measure 6-^{14}C-arginine specific activity at the synthetic site. Recently the mean specific activity of the total CO_2 in the arterial blood over the time T has been used as a substitute (Reeve and McKinley, 1970). This is easy to measure but has the disadvantage that there are not clear and simple kinetic relations between it and the mean specific activity of liver 6-^{14}C-arginine. However, evidence is good that the mean specific activities of total CO_2-carbon and liver 6-^{14}C-arginine are very close to each other (Reeve and McKinley, 1970; Kornberg *et al.*, 1952; Swick, 1958). This method is relatively simple and seems promising.

One limitation of all present forms of the 6-^{14}C-arginine synthetic rate method is that they require protein synthetic rate to remain relatively constant over the 3–4 hr necessary for their application. Thus, if synthetic rate is varying rapidly, as for instance might occur when synthesis of an "acute phase" protein is suddenly stimulated, results may bear little relation to true rates of synthesis. So as yet the 6-^{14}C-arginine method does not allow determination of synthetic rate during sudden transient changes, when measurements would be particularly valuable.

B. A Mass Balance Approach

In our present state of knowledge of the controls of plasma protein systems, valuable information would be yielded by a method that measured repeatedly the synthetic rate of a plasma protein with an error of 10 or even 20%, particularly if the method was relatively simple. Such a method would allow the broad outlines of the controls to be worked out. The finer details could be filled in later when more accurate methods became available. Such a method has been devised for measuring repeatedly the synthetic rate of plasma proteins with properties similar to those of fibrinogen (Atencio and Reeve, 1965; Atencio *et al.*, 1969). These properties are: (1) fractional catabolic rate remains about constant during an experiment; (2) the amount of interstitial protein is much smaller than the amount of plasma protein; and (3) the ratio j_1 (the fractional capillary transfer rate) to j_2 (the fractional lymph return rate) remains nearly constant. It is also necessary that the protein system is well enough described by the linear Eqs. (3) and (4) so that changes in plasma protein concentration allow satisfactory estimates of changes in total plasma and interstitial protein and of catabolic rate. Given these criteria then quite good repeated estimates of the protein synthetic rate over a few hours can be made from studies with radioactive iodine-labeled protein and repeated measurements of protein concentration in the plasma (Atencio *et al.*, 1969).

One proceeds as follows: Initial estimates of total plasma protein $P_p(0)$

and j_1, j_2, and j_3 are obtained by using ^{131}I-labeled protein. Then Eqs. (1) and (2) allow approximate calculation of initial interstitial protein $P_i(0)$, which from our criteria forms only a fraction of $P_p(0)$. Suppose now that synthetic rate is stimulated and repeated measurements are made over intervals of T hours. During the first T hours synthetic rate S must equal the sum of the increase in total plasma protein, ΔP_p over T, which is given by change in concentration multiplied by plasma volume; plus the increase in interstitial protein, ΔP_i over T, which can be calculated from the change in P_p and Eqs. (1) and (2); plus total protein catabolism over T, given by $j_3\bar{P}_p$, where $\bar{P}_p$ is mean total plasma protein during the time T. During the next and subsequent intervals of T hours the same arguments apply so the same measurements and calculations are repeated and yield a series of estimates of synthetic rate. This simple approach allows many estimates of protein synthetic rate to be made with an acceptable amount of effort and is applicable to transient states even when plasma protein concentration is changing rapidly.

Acknowledgments

This research was supported by Grant HE-02262 from the National Institutes of Health, U.S. Public Health Service, Bethesda, Maryland.

References

Atencio, A. C., and Reeve, E. B. (1965). *J. Lab. Clin. Med.* **66**, 20.
Atencio, A. C., Joiner, K., and Reeve, E. B. (1969). *Amer. J. Physiol.* **216**, 764.
Berson, S. A., and Yalow, R. S. (1957). *Fed. Proc. Fed. Amer. Soc. Exp. Biol.* **16**, 13S.
Koj, A., and McFarlane, A. S. (1968). *Biochem. J.* **108**, 137.
Kornberg, H. L., Davies, R. E., and Wood, D. R. (1952). *Biochem. J.* **51**, 351.
McFarlane, A. S. (1957). *Ann. N.Y. Acad. Sci.* **70**, 19.
McFarlane, A. S. (1963). *Biochem. J.* **89**, 277.
McFarlane, A. S., Irons, L., Koj, A., and Regoeczi, E. (1965). *Biochem. J.* **95**, 536.
McKinley, J. E., Gilbert, D. B., and Reeve, E. B. (1970). *Amer. J. Physiol.* **218**, 491.
Mosser, E. L., Lackey, W. W., and Franks, J. J. (1968). *Amer. J. Physiol.* **214**, 179.
Reeve, E. B. (1965). *Tech. Rep. Ser. Int. At. Energy Ag.* **45**, 57.
Reeve, E. B., and McKinley, J. E. (1970). *Amer. J. Physiol.* **218**, 498.
Reeve, E. B., Pearson, J. R., and Martz, D. C. (1963). *Science* **139**, 914.
Swick, R. W. (1958). *J. Biol. Chem.* **231**, 751.
Tavill, A. S., Craigie, A., and Rosenoer, V. M. (1968). *Clin. Sci.* **34**, 1.
Wochner, R. D., Weissman, S. M., Waldmann, T. A., Houston, D., and Berlin, N. I. (1968). *J. Clin Invest.* **47**, 971.

2

AN APPROACH TO THE INVESTIGATION OF THE DYNAMIC STRUCTURE OF A COMPLEX BIOLOGICAL SYSTEM INCORPORATING STATE VARIABLE DIAGRAM ANALYSIS

E. A. Jones, E. Carson, and V. M. Rosenoer

DEPARTMENT OF MEDICINE,
ROYAL FREE HOSPITAL, LONDON, WC1, ENGLAND;
DEPARTMENT OF AUTOMATION ENGINEERING,
THE CITY UNIVERSITY, LONDON, EC1, ENGLAND;
CHILDREN'S CANCER RESEARCH
FOUNDATION AND HARVARD MEDICAL SCHOOL,
BOSTON, MASSACHUSETTS

I. Introduction

The basic principles of the use of models for the investigation of multi-compartmental biological systems are well established (Berman, 1963). To date, however, only fairly simple systems which can be adequately described by fairly simple models have been investigated in this way. In general, experimental data from intact organisms are inadequate to define completely a complex biological system. Nevertheless, the application of the principles of modeling to a complex system may still provide valuable information about its dynamic structure, which cannot readily be obtained by other means. The purpose of this chapter is to illustrate how, starting from basic principles and making a minimum number of assumptions, a very complex system can be modeled in a logical way so that information about the dynamic structure of the system can be derived which is as comprehensive as possible within the limitations of the available experimental data. In particular, the approach described illustrates how the technique of state variable diagram analysis can be applied to a biological system. This is a technique which can be used to facilitate the modeling of any complex dynamic system.

To illustrate the approach, the metabolic pathways taken by ^{14}C in man after its administration as ^{14}C-carbonate by intravenous injection are used as an example of a complex multicompartmental biological system. The intravenous injection of ^{14}C-carbonate can be regarded as a unit impulse input to the system. Following the injection, the plasma specific activity time curves for biosynthesized ^{14}C-urea and ^{14}C-albumin are readily obtained experimentally (Tavill *et al.*, 1968) and can be regarded as impulse responses of the system. These responses are the result of both the input and the dynamic structure of the system. Since the input can be mathematically defined, it is possible to obtain information about the dynamic structure of the system from the experimental curves. To apply this analytical approach, it is first necessary to construct a biological model of the system. This must be based on a description of the metabolic pathways taken by ^{14}C which is as comprehensive as possible based on the best available knowledge of the system. The biological model is described mathematically. The comprehensive mathematical model is then reduced to an extent which enables all the parameter values of the reduced model to be derived from the available experimental data.

II. ^{14}C Metabolic Pathways after Intravenous ^{14}C-Carbonate

After the intravenous injection of ^{14}C-carbonate the ^{14}C label initially becomes incorporated into the extracellular pool of bicarbonate. Most of the ^{14}C label in this form is lost from the lungs as ^{14}C-carbon dioxide in expired air. In contrast to other cells, the liver cell has the capacity to fix ^{14}C selectively in the form of ^{14}C-carbon dioxide (Swick, 1958). Consequently, some of the injected label becomes available for incorporation into intermediary metabolic pathways in the liver cell. For instance, ^{14}C-carbon dioxide becomes incorporated into acetate and thence into the Krebs tricarboxylic acid cycle. Thus, there will be labeling of glucose, fatty acids, glycerol, and amino acids, and hence glycogen, lipids, and proteins. Inevitably some of the ^{14}C label ceases to be metabolically available for periods which vary according to the biological life of these products. Some label, however, may recycle in shorter chain carbon compounds. This phenomenon and persisting extracellular metabolically available ^{14}C-carbon dioxide probably result in some ^{14}C-carbon dioxide being continuously available within the liver cell for reutilization (Jones *et al.*, 1968).

One of the metabolic fates of intracellular ^{14}C-carbon dioxide is condensation with ornithine and ammonia in the Krebs-Hensleit urea cycle, which results in the labeling of hepatic arginine predominantly in the guanidine (6-C) position (Roberts *et al.*, 1957), This labeled carbon atom then has any one of the following metabolic fates: (1) incorporation into the arginine of albumin and other liver-produced plasma proteins; (2) incorporation into the arginine of liver cell proteins, such as hepatic enzymes; and (3) incorporation into urea. The various proteins synthesized in the liver cell have different life spans and after their breakdown 6-^{14}C-arginine is released and is available for reentry into the hepatic pool of arginine. There is a delay of about 30 min in the release of newly synthesized ^{14}C-labeled albumin relative to ^{14}C-urea (Tavill *et al.*, 1968). This protein is then delivered directly into the vascular compartment (Smallwood *et al.*, 1968). These phenomena may apply also to the synthesis and release of other liver-produced plasma proteins. After rapid equilibration in an initial mixing pool, which probably approximates closely to the plasma compartment, each of these proteins is subjected to two processes: (1) distribution to one or more other (extravascular) pools; and (2) catabolism of protein molecules. Catabolism results in the release of 6-^{14}C-arginine which is then available for extrahepatic incorporation into newly synthesized tissue proteins and immunoglobulins and for reentry into the hepatic arginine pool. ^{14}C-arginine filtered at the glomeruli is selectively reabsorbed by the renal tubule (Robinson, 1968).

Newly synthesized ^{14}C-urea is delivered into an initial mixing pool which includes the vascular compartment. Urea in this pool is subjected to three processes: (1) distribution to another pool or pools (Jones *et al.*, 1968); (2) filtration at the glomerulus; and (3) hydrolysis by bacterial urease in the intestine (Jones *et al.*, 1969). A proportion of ^{14}C-urea filtered at the glomerulus is reabsorbed by the renal tubules and enters a complex renal interstitial pool. In this pool there is a concentration (quantity/unit volume) gradient of both labeled and unlabeled urea, with values which are higher in the medulla and lower in the cortex than in the plasma (Regoeczi *et al.*, 1965; Black, 1968). Some ^{14}C-urea in the renal interstitial pool passes back into the initial mixing pool and some passes back into the tubules. ^{14}C-urea which passes from the distal tubules into the collecting tubules is largely excreted into the urine. There is probably reabsorption of a small quantity of ^{14}C-urea from the bladder, which acts as a storage pool (Regoeczi *et al.*, 1965). ^{14}C-carbon dioxide liberated by the catabolism of urea in the intestine reenters the same pool of metabolically available carbon dioxide as ^{14}C introduced as ^{14}C-carbonate by intravenous injection.

III. The Construction of a Biological Model

The construction of a model of the metabolic pathways described requires the representation of each ^{14}C-labeled substance in a particular part of the system in terms of a discrete pool. The pools in this system are then interconnected according to the preceding description of the metabolic pathways (see Fig. 1).

Some of the metabolic pathways are essentially unidirectional. These result in a loss of label from the system, either by fixation of the label in stable compounds or by excretion of the label. Other pathways which involve feedback or recycling of the label tend to conserve ^{14}C in the dynamic parts of the system. The existence of feedback pathways may result in oscillations being superimposed on the impulse response curves.

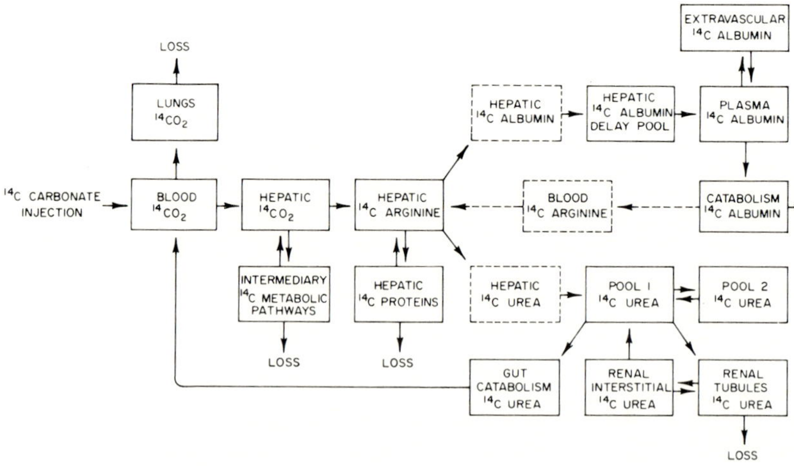

FIG. 1. Simplified biological model of the major pathways taken by ^{14}C in the body after its administration as ^{14}C-carbonate by intravenous injection. A ^{14}C-labeled substance in a particular part of the body is represented as a discrete pool and its metabolic fate is indicated by arrows.

IV. The Formulation of a Mathematical Model

The quantity of ^{14}C-labeled substance present in the system varies with time as a result of the input to and losses from the system. In any individual pool the rate of change of the substance in the pool, both unlabeled (X) and labeled (x), with time may be expressed by a simple equation.

$$\frac{d(X+x)}{dt} = f(\text{inputs}) - f(\text{outputs}) \pm f(\text{disturbances}) \qquad (1)$$

That is, the rate of change of the substance with time is equal to a function of the inputs, minus a function of the outputs, plus or minus a function of the disturbances acting on that pool. Using this approach it is possible to derive general rate equations for all of the pools in the system.

For a complex biological system of the type outlined it may be assumed that, in general, the functions in the rate equations are nonlinear. To facilitate further analysis it is necessary to linearize the equations. An appropriate method is the use of Taylor's series expansion, in which linear excursions about a fixed operating point are considered (Dorf, 1965). This method may only be adopted if it is assumed that (1) the labeled substance behaves in exactly the same manner as the unlabeled substance, and (2) the quantity of labeled substance is small compared with the total quantity of the substance. Linearization results in the exclusion of the expression for the unlabeled substance in the equations and enables the whole system to be described by a set of first-order linear differential equations of the following type.

$$\frac{dx_2}{dt} = K_1 x_1 + K_2 x_2 \qquad (2)$$

That is, the rate of change of the quantity of labeled substance (x_2) in a particular pool with respect to time is given by a function of the input (K_1x_1) and of the output (K_2x_2), where the K values are rate constants. The form of the solution to this type of equation is

$$x_2 = K(1 - e^{-K_2t}) \qquad (3)$$

This solution is consistent with results of many tracer studies of multicompartmental biological systems, the solutions of which can usually be expressed in the form of several exponential functions (Berman, 1963).

The values of the various K terms of the system cannot readily be determined experimentally. It is not possible, therefore, to simulate directly the mathematical model on a computer and hence obtain a solution. However, it is possible by adopting a graphical approach, such as state variable diagram

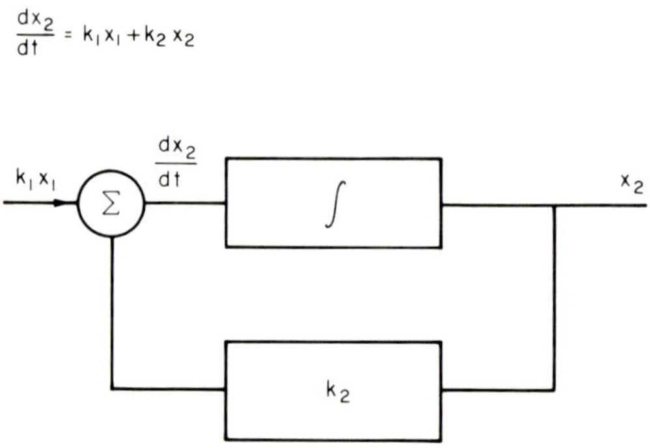

FIG. 2. The ideal analog simulation of a linearized rate equation. The inputs to the summer ($\sum$), $k_1\ x_1$ and $k_2\ x_2$, when summated give dx_2/dt, which when integrated ($\int$) gives the output x_2.

analysis (Dorf, 1965), to reduce the model rationally so that the parameters of the reduced model can be derived from experimental data. State variable diagram analysis involves representing each linear first-order differential equation by its ideal analog simulation diagram (Fig. 2). The diagram represents the analog computer simulation needed to represent the dynamics of each equation. In the example given, the inputs to the summer are K_1x_1 and K_2x_2, which when summated give dx_2/dt. The integration of this function gives the output x_2. By treating each of the linearized equations in this way and interconnecting them according to the biological model the state variable diagram of the whole system is derived.

Reduction and manipulation of this mathematical model is facilitated by converting it into transfer function form, since this results in the replacement of the differential equations by simple algebraic equations. To produce the transfer function of the system, each function in the state variable diagram which is in the time domain is converted into the frequency domain by employing Laplace transformation. The transfer function gives the relationship between input and output in terms of amplification factors in the steady state and factors describing the dynamic relationship between input and output. The derivation of a simple transfer function requires that the initial conditions of the system are zero and the system is linear.

Laplace transformation of the typical first-order differential equation under consideration Eq. (2) gives a result in the frequency domain which is in pure algebraic form.

$$sX_2(s) = K_1X_1(s) - K_2X_2(s) \tag{4}$$

or

$$(s + K_2)X_2(s) = K_1X_1(s) \tag{5}$$

where $s =$ the Laplace operator, $K_1 =$ a constant, $K_2 =$ a constant, $X_1(s) =$ the Laplace transform of the input function, and $X_2(s) =$ the Laplace transform of the output function. In this instance the relationship between $X_2(s)$ and $X_1(s)$ is given by the following expression.

$$\frac{X_2(s)}{X_1(s)} = \frac{K_1}{s + K_2} \tag{6}$$

The expression for $X_2(s)/X_1(s)$ is the transfer function of the particular pool and is defined as the ratio of output to input expressed in Laplace transforms. By converting the whole of the state variable diagram into the frequency domain and expressing the individual functions in the form of block diagrams, the transfer function diagram of the whole system can be derived. Individual functions in the diagram have the following general form.

$$\frac{X_2(s)}{X_1(s)} = \frac{K}{1 + Ts} \tag{7}$$

where $K = K_1/K_2 =$ a gain term which gives the amplification of input to output in the steady state. $T = 1/K_2 =$ a time constant which when combined with the Laplace operator s gives the dynamic relation between input and output. The whole diagram gives the overall transfer function of the system that relates output, the plasma specific activity time curves of ^{14}C-urea and ^{14}C-albumin, to the input, the ^{14}C-carbonate injection.

V. Model Reduction

Complete identification of the comprehensive mathematical model is not possible because of limitations of the experimental data and the restricted methods of dynamic testing of the model, which are applicable to the investigation of a complex biological system in an intact organism. For these reasons it is necessary to reduce (simplify) the model to facilitate further analysis. The first step in this process is to assess the most probable order of magnitude of each time constant governing the dynamic behavior of the ^{14}C-labeled substance in each individual pool. These assessments must in the first instance be based on the best available knowledge of the metabolism of each pool in the system. A time constant is given by the time taken for the output to reach two-thirds of its final value, and these values are expressed in the transfer function diagram in terms of T. For example, in normal subjects the catabolism of ^{14}C-albumin would have a time constant of the order of 10^2 hr; the synthesis of ^{14}C-urea from hepatic ^{14}C-arginine would have a time constant of the order of 10^{-3} hr. Since typical studies of the system involve making experimental observations over a period of about 10 hr (Tavill *et al.*, 1968), it is most unlikely that pools with time constants of the order of 10^{-2} hr or less will show clearly in experimental data. Accordingly, transfer function expressions including these terms were reduced to pure amplification terms (gains), assuming a time constant of zero. Gain terms appear in the transfer function diagram as K values. Furthermore, feedback pathways (pools) in which there are small quantities of ^{14}C-labeled substance (low specific activity) and in which the time constants are of the order of 10^2 hr or more will have a negligible effect on experimental data during the period of study. For instance, it is probably justifiable to neglect the pool involving the transfer to the liver cell of ^{14}C-arginine liberated from the catabolism of all plasma proteins. This ^{14}C-arginine is diluted by the extracellular pool of unlabeled arginine with

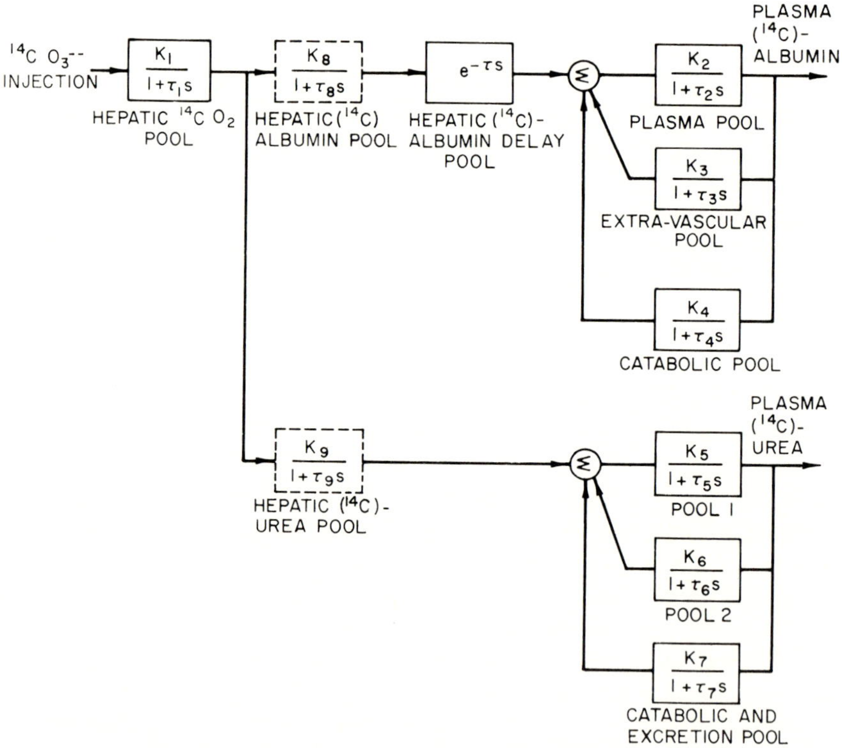

FIG. 3. Transfer function diagram of the reduced model. The K terms are gains, the T terms are time constants, and s is the Laplace operator. The hepatic pools of ^{14}C-albumin and ^{14}C-urea, which represent the dynamics of the synthesis of these two substances from hepatic ^{14}C-arginine, can be reduced to pure gains. The dynamics of hepatic ^{14}C-arginine are incorporated in the gain K_1.

the result that the effects of relabeling of hepatic arginine from this source during a study is likely to be undetectable. Similarly ^{14}C-carbon dioxide liberated in the intestine by the hydrolysis of urea becomes diluted by the large pool of extracellular bicarbonate and consequently relabeling of hepatic carbon dioxide from this source is likely to be negligible.

The results of reducing the comprehensive model utilizing these principles are shown in Fig. 3. The model includes a pure delay pool for the interval between the fixing of the ^{14}C label in hepatic arginine and the appearance of newly synthesized ^{14}C-labeled albumin in the plasma. It incorporates the dynamic structure of the whole system and gives the overall relationship between input to and output from the system.

VI. The Numerical Evaluation of the Reduced Model Parameters

The complete identification of the reduced model requires the numerical evaluation of each of the individual gains and time constants in the transfer function diagram given in Fig. 3. To facilitate this process, it is first necessary to derive equations for the curves of best fit of each set of raw experimental data. The data available comprise not only the plasma ^{14}C-specific activity time curves for urea and albumin which reflect the dynamic behavior of the whole modeled system, but also the disappearance curves from the plasma of intravenously injected ^{13}C-urea and ^{131}I-albumin, which reflect the dynamic behavior of the relevant parts of the system excluding all pools up to and including the synthesis of urea and albumin.

Because the data are the results of tracer studies in a multicompartmental system, it is appropriate to fit the data to the sum of several exponential functions (Berman, 1963). The equation used has the following general form.

$$y = Ae^{-at} + Be^{-bt} + Ce^{-ct} + De^{-dt} \tag{8}$$

where A, B, C, and D are arbitrary constants (intercepts) and a, b, c, and d are the corresponding reciprocals of the time constants (slopes). The number of exponentials chosen should equal the number of pools in the corresponding part of the reduced model.

An appropriate method of obtaining the curve of best fit for each set of experimental points is to use an adaptive technique which minimizes the root mean square error, such as Rosenbrock's hill-climbing technique (Rosenbrock and Storey, 1966) using a digital computer. In practice, it is desirable to make initial approximate estimates of parameter values of the fitted curve by the method of graphical peeling (Matthews, 1957). This enables the field of search for optimal fit to be restricted so that the final fitted parameter values should be consistent with the biological model. Each function in the equation for the fitted curve in the time domain Eq. (8) can be rewritten in the frequency domain. The equation then has the following general form.

$$Y(s) = \frac{A}{s+a} + \frac{B}{s+b} + \frac{C}{s+c} + \frac{D}{s+d} \tag{9}$$

The fitted curve in this form can be equated to the corresponding part of the transfer function diagram. A series of simultaneous equations can then be derived for the model parameters. The number of equations derived should be the same as the number of unknowns, thus enabling values of all the parameters in the transfer function diagram to be calculated. Parameter values obtained in this way have been found to be of the same order of magnitude as those predicted from the currently available data on the metabolic behavior of the system (Carson, 1968).

VII. Discussion

The construction of a model of a complex biological system is a valuable exercise. All the known properties of the system must be assessed in a logical manner and deficiencies in knowledge about the system tend to be more clearly appreciated. The derivation of a mathematical description of the model often requires clarification of the metabolic behavior of parts of the system which may have seemed unimportant or may not have been considered at all.

The biological system described here, like most biological systems, is nonlinear. However, because the injection of ^{14}C-carbonate can reasonably be regarded as a small perturbation to the system, the system can be linearized. This enables the whole system to be expressed in terms of state variable equations. Clearly the rate constants of all these equations cannot be determined experimentally. For instance, it is not possible to make frequent specific activity measurements of compounds within the liver cell. It is, therefore, necessary to reduce the model to facilitate further analysis. Various reduction techniques are more easily applied to a mathematical model when it is expressed in transfer function form as the equations describing the system are then in a pure algebraic form. Reduction is carried out until all of the parameters of the reduced model can be adequately evaluated from the available experimental data. Clearly, the more precise or more varied the experimental data obtained the more precise the evaluation of model parameters will tend to be. The approach affords a means of describing the kinetic behavior of a complex system which is as accurate and comprehensive as possible within the limits of the available experimental data. The reduced model is nonunique and the parameter values calculated, while being consistent, will also be nonunique.

There are various aspects of the experimental data obtained in many tracer studies in biological systems which make accurate analysis difficult. The number of points which can be obtained for any curve is usually limited, particularly in studies on humans, e.g., by the number of blood samples and the volume of blood which can be withdrawn. Ideally for any particular curve it is desirable to obtain more points during periods of time when the effects of pools with short time constants are particularly evident in the experimental data. Experimental points often exhibit a wide scatter, i.e., the data are "noisy." This arises not only from the intrinsic variability of biological data, due to fluctuations in biological functions, but also from technical factors inherent in the measurement of each experimental point. For instance, wide scatter of experimental points is particularly evident in plasma curves of biosynthesized ^{14}C-albumin. This is because the dose of ^{14}C-carbonate permitted for studies or humans is limited to no more than 200 μCi, so that the ^{14}C counts in plasma albumin are usually at the most only up to three times those due to background radioactivity (Tavill *et al.*, 1968). Furthermore, experiments are often not continued until the system returns to a steady state. This would usually be reflected in a clearly definable terminal exponential component of each experimental curve. Thus, the data are often also "short-term."

While it is appropriate to fit the results of tracer studies in biological systems to the sum of several exponential functions (Berman, 1963), the short-term noisy data obtained in many studies limit considerably the techniques which are suitable for curve fitting. One method, already mentioned, which has been found to be applicable to this type of data involves adaptive modeling using a hill-climbing technique based on Rosenbrock's method (Rosenbrock and Storey, 1966). Another technique which is also applicable is the construction of a Bode diagram. This involves the conversion of the experimental data into the frequency domain by Fourier transformation (d'Azzo and Houpis, 1966). The data are then plotted as amplitude in the form of decibels against the logarithm of the frequency and phase against the logarithm of the frequency. The amplitude plot is examined for principle break frequencies, when the amplitude changes at a rate of 20 decibels/decade. This enables the principle time constants of the system to be evaluated. Both plots can be used to assess the main gains of the system. The analysis is limited when high frequency information, which can only be obtained from long-term data, is excluded and also when the frequency of sampling is low. Nevertheless, rough estimates of the principle gains, and time constants of a complex system can often be obtained using this technique. The method has the particular advantage that, in contrast to other conventional methods of analyzing kinetic data from multicompartmental systems, no initial assumptions about the system are required and it affords a means of checking the validity of a reduced model derived by other means. Preliminary results from the use of this method of analysis of ^{14}C metabolic pathways have indicated that the number of dominant time constants and their order of magnitude have been in general agreement with results obtained from time domain analysis (Carson, 1968). The identification of the reduced model adds to knowledge and insight of the system and it allows the response of the system or parts of the system to different types of tracer stimuli to be predicted.

The model itself facilitates the design of further experiments, the results of which would permit definition of the system in greater detail. Ideally the additional experimental data would consist of impulse–response curves for all parts of the system. Of less value are impulse response data reflecting the dynamic behavior of larger subunits of the system, as are given, for instance, by the ^{13}C-urea and ^{131}I-albumin curves. Data obtained after the intravenous injection of ^{14}C-arginine would reflect the behavior of the parts of the system excluding all the pools up to and including the hepatic pool of ^{14}C-carbon dioxide. Unfortunately, this experiment could not be conducted in human subjects on account of the appreciable radiation hazard involved, but it has been carried out in animals (McFarlane, 1963).

VIII. Summary

(1) The metabolic pathways taken by ^{14}C after its administration intravenously as ^{14}C-carbonate have been used as an example of a complex multicompartmental biological system, which is suitable for dynamic analysis.

(2) A comprehensive biological model of the system has been described mathematically in the form of a state variable diagram and subsequently in transfer function form.

(3) A rational method of reducing the mathematical model has been adopted and a method of deriving all of the parameter values of the reduced model from readily obtainable experimental data is outlined.

(4) An adaptive hill-climbing method of curve fitting and the construction of a Bode diagram have been found to be appropriate methods of analyzing typical short-term noisy experimental data.

(5) The methods of analysis described are applicable to any dynamic multivariable system, particularly complex biological systems, and they enable the dynamic structure of complex systems to be evaluated.

Acknowledgments

This work was supported in part by grants from the Medical Research Council, U.K. and the U.S.P.H.S. (Grant C-6516). The authors wish to thank Mr. L. Finkelstein and Mr. C. Barr of the City University, London, for their invaluable advice.

REFERENCES

Berman, M. (1963). *Ann. N.Y. Acad. Sci.* **108**, 182.

Black, D. A. K. (1968). *In* "Renal Disease" (D. A. K. Black, ed.), 2nd Ed., pp. 93–109. Blackwell, Oxford.

Carson, E. (1968). M.Sc. Thesis, City University, London.

d'Azzo, J. J., and Houpis, C. H. (1966). "Feedback Control Systems: Analysis and Synthesis." McGraw-Hill, New York.

Dorf, R. C. (1965). "Time Domain Analysis and Design of Control Systems." Addison-Wesley, Reading, Massachusetts.

Jones, E. A., Craigie, A., Tavill, A. S., Simon, W., and Rosenoer, V. M. (1968). *Clin. Sci.* **35**, 553.

Jones, E. A., Smallwood, R. A., Craigie, A., and Rosenoer, V. M. (1969). *Clin. Sci.* **37**, 825.

McFarlane, A. S. (1963). *Biochem.* J. **89**, 277.

Matthews, C. M. E. (1957). *Phys. Med. Biol.* **2**, 36.

Regoeczi, E., Irons, L., Koj, A., and McFarlane, A. S. (1965). *Biochem. J.* **95**, 521.

Roberts, R. B., Abelson, P. H., Cowie, D. B., Bolton, E. T., and Britten, R. J. (1957). *Carnegie Inst. Wash. Publ.* **607**, 95–112.

Robinson, J. R. (1968). *In* "Renal Disease" (D. A. K. Black, ed.), 2nd Ed., pp. 72–92. Blackwell, Oxford.

Rosenbrock, H. H., and Storey, C. (1966). "Computational Techniques for Chemical Engineers," pp. 48–97. Macmillan (Pergamon), New York.

Smallwood, R. A., Jones, E. A., Craigie, A., Raia, S., and Rosenoer, V. M. (1968). *Clin. Sci.* **35**, 35.

Swick, R. W. (1958). *J. Biol. Chem.* **231**, 751.

Tavill, A. S., Craigie, A., and Rosenoer, V. M. (1968). *Clin. Sci.* **34**, 1.

3

MODELS FOR PROTEIN SYNTHESIS STUDIES

R. Bianchi, L. Donato, P. Mancini, G. Mariani, A. Pilo,

THE C.N.R. LABORATORY OF CLINICAL PHYSIOLOGY, UNIVERSITY OF PISA, PISA, ITALY

*and F. Vitek**

INSTITUTE OF BIOPHYSICS, FACULTY OF GENERAL MEDICINE,
CHARLES UNIVERSITY, PRAGUE, CZECHOSLOVACHIA

I. Introduction

The purpose of this chapter is that of examining the theoretical aspects of plasma protein synthesis studies in man, with special reference to

* Present address: The C.N.R. Laboratory of Clinical Physiology, University of Pisa, Pisa, Italy.

plasma albumin and to the problems of data analysis and experimental design.

A. Rate of Synthesis from Catabolic Studies with Labeled Proteins

In steady state conditions the rate of synthesis of a plasma protein equals its rate of catabolism; hence, the rate at which plasma proteins are synthesized can be estimated from the rate of catabolism of an iodine-labeled protein injected into the circulation. This approach has provided estimates of synthesis rates for plasma albumin, fibrinogen, and γ-globulins.

The main limitations of this approach for the evaluation of synthesis rates of plasma proteins are the need of assuming maintenance of a steady state of synthesis, catabolism, and exchanges between compartments over the several days required to complete the study, and that excretion of iodide derived from catabolism of the labeled proteins mirrors its rate of formation (Donato *et al.*, 1966). These limitations may become particularly important in diseased states and in the presence of renal damage.

Deconvolution techniques (Vitek *et al.*, 1966, 1967a) and use of labeled iodide as a second tracer (Pilo, 1968; Vitek *et al.*, 1967b; Bianchi *et al.*, 1967) partly circumvent the above limitations, but the minimal duration of a study still is of the order of one week.

B. Rate of Synthesis from Incorporation of a Labeled Precursor

If it were possible to obtain the time course of the specific activity of a precursor amino acid at the site of protein synthesis, direct measures of protein synthesis rate would be possible. The first attempt in this direction was that of Weissman *et al* (1961), who derived the specific activity of intracellular ^{14}C-glycine from the specific activity of hippuric acid excreted with the urines.

In 1958, Swick suggested that urea produced by liver cells might be used as an external indicator of the intrahepatic arginine specific activity, since in the Krebs-Henseleit cycle, hepatic arginine is the common precursor of urea and guanido carbon of the arginine which is used for the synthesis of albumin and other plasma proteins.

In 1963 McFarlane and Reeve *et al.* independently proposed the use of ^{14}C-carbonate to label the guanido carbon of intrahepatic arginine pool. With this approach, and under appropriate assumptions, albumin synthesis rate can be estimated in about 10 hr; the method has been more recently applied in man (Tavill *et al.*, 1968), and it has been used in different clinical situations (Rothschild *et al.*, 1969; Craigie *et al.*, 1967).

However, the method has been subjected to two sorts of criticism: on one hand, some of the assumptions on which the calculations were originally based have not been confirmed (Jones *et al.*, 1968); on the other hand, the large amount of blood required for each estimate of the albumin specific activity markedly limits both its applicability in patients and the accuracy with which the experimental functions can be described.

II. Analysis of Models for Albumin Synthesis

A. Method Followed

In order to examine the theoretical aspects of albumin synthesis study, we started out by finding a general solution to the problem. The system was then reduced to a "minimal" model, based on the least number of assumptions and on the simplest experimental design. By evaluating the adequacy of both model and experimental situation to solve the problem of measuring synthesis rate, we successively operated on the assumptions, the model, and the experimental design, to find out the optimal combination of them capable of yielding an acceptable solution. The adequacy of each successively found solution has been judged by comparing the obtained values for fractional albumin synthesis rate with those derived from catabolic studies and by testing their capability of predicting the time course of albumin specific activity.

This approach was made possible by the availability of well-defined experimental curves, thanks to the experimental technique which was developed by Bianchi and Mariani (1970) and which is summarized in the appendix.

All data analysis has been performed on a HP2116B digital computer using the programming language "Basic." Multiexponential fit of the experimental curves was performed with the method of Mancini and Pilo (1970) whenever differentiation, convolution, or deconvolution had to be performed.

B. Assumptions and General Model

There is a sufficient body of evidence to endorse the assumptions and the minimum model schematized in Fig. 1. The assumptions are: (a) Hepatic arginine is the common direct precursor of urea carbon and of the guanido carbon in albumin arginine (Swick, 1958); (b) newly formed urea and albumin molecules are directly released into the circulation, so that their rate of appearance in blood corresponds to their rate of formation (Smallwood *et al.*, 1968).

In fact, it was shown previously, that after injection of ^{14}C-carbonate there is a finite delay before any activity is detectable in plasma albumin. To account for it, and to make possible the comparison of the experimental and predicted curves, the time scale of the experimental curve was shifted

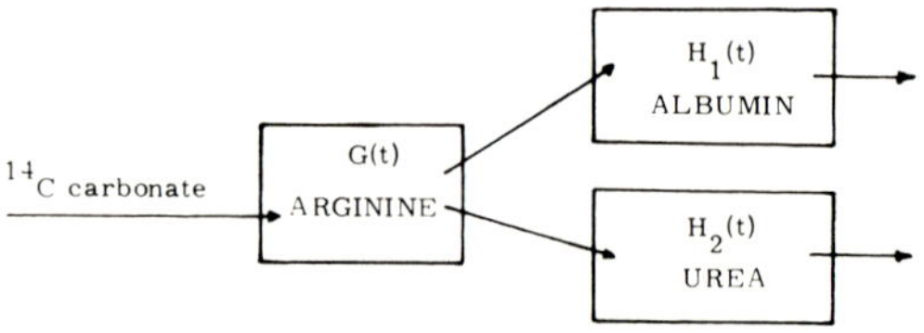

ASSUMPTIONS:

- same precursor for albumin and urea
- steady state conditions
- newly synthesized albumin and urea released directly into plasma

$A(t)$	- specific activity of plasma albumin	$a(s) = \mathcal{L}[A(t)]$
$U(t)$	- specific activity of plasma urea	$u(s) = \mathcal{L}[U(t)]$
S_a	- fraction of albumin pool synthesized per hour	
S_u	- fraction of urea pool synthesized per hour	
$G(t)$	- specific activity of hepatic arginine	$g(s) = \mathcal{L}[G(t)]$
$H_1(t)$	- transit times distribution through albumin pool	$h_1(s) = \mathcal{L}[H_1(t)]$
$H_2(t)$	- transit times distribution through urea pool	$h_2(s) = \mathcal{L}[H_2(t)]$

$$a(s) = S_a g(s) h_1(s)$$

$$u(s) = S_u g(s) h_2(s)$$

$$S_a = S_u \frac{a(s)/h_1(s)}{u(s)/h_2(s)}$$

FIG. 1. Measurement of albumin synthesis from a ^{14}C-labeled precursor.

by taking 0.4 hr after injection as zero time. The relevance of this correction will be discussed in a later section of this chapter. (c) A steady state of albumin and urea kinetics is maintained for the time required to complete the study (from 10 to 20 hr).

If we call $A(t)$ the specific activity of the arginine guanido carbon of albumin [from now on albumin carbon specific activity], $G(t)$ the specific activity of hepatic arginine, $H_1(t)$ the transit time distribution through the plasma albumin pool of newly formed albumin molecules, and S_a the fraction of plasma albumin pool synthesized per day, then

$$A(t) = S_a\,[G(t)]*[H_1(t)] \tag{1}$$

In Laplace notation:

$$a(s) = S_a\, g(s) h_1(s) \tag{2}$$

where $a(s)$, $g(s)$, and $h_1(s)$ are the Laplace transforms of $A(t)$, $G(t)$, and $H_1(t)$, respectively.

Thus, in Laplace notation, the albumin specific activity $a(s)$ equals the product of the two functions $g(s)$ (input) and $h_1(s)$ (output) times the fractional synthesis rate S_a.

If we consider now the possibility of determining S_a on the basis of Eq. (2), we see that: $a(s)$ can be derived from experiment; $h_1(s)$ could also be derived from experiment, since it would correspond to the Laplace transform of the plasma disappearance curve of a bolus of iodine labeled albumin injected intravenously; $g(s)$ cannot be determined because of inaccessibility for sampling of the hepatic arginine pool.

This is a typical case in which the impossibility of sampling from a part of the system makes it impossible to proceed with the estimate of the system parameters.

This difficulty may be overcome by sampling from a different part of the system, namely from the plasma urea pool, which is assumed to be also fed by the hepatic arginine pool.

If we now call $U(t)$ and $H_2(t)$ the plasma urea carbon specific activity after ^{14}C-carbonate injection and the transit time distribution through the plasma urea pool (plasma disappearance curve of a unitary bolus of labeled urea injected into the circulation), respectively, we may write:

$$U(t) = S_u\,[G(t)]^*[H_2(t)] \tag{3}$$

where S_u is the synthesis rate for urea as a fraction of the plasma urea pool. Equation (3) written in the Laplace notation is then

$$u(s) = S_u\, g(s) h_2(s) \tag{4}$$

where $u(s)$ and $h_2(s)$ are the Laplace transforms of $U(t)$ and $H_2(t)$, respectively.

If Eqs. (2) and (4) are taken together, for instance by dividing them one into the other, we see that the inaccessible function $g(s)$ disappears, and it is replaced by two experimentally accessible functions, $u(s)$ and $h_2(s)$. Thus, we have now a general solution to our problem, all in terms of accessible variables:

$$S_a = S_u \frac{a(s)/h_1(s)}{u(s)/h_2(s)} \tag{5}$$

Figure 2 summarizes the situation. After injection of ^{14}C-carbonate and if

General solution:

$$S_a = S_u \frac{a(s)/h_1(s)}{u(s)/h_2(s)}$$

Experimental data:

- $A(t)$—plasma albumin specific activity
- $U(t)$—plasma urea specific activity

Unknowns:

- S_u—fractional synthesis rate for urea
- $h_1(s)$—transfer function for plasma albumin
- $h_2(s)$—transfer function for plasma urea
- S_a—fractional synthesis rate for albumin

FIG. 2. Measurement of albumin synthesis from a ^{14}C-labeled precursor.

we measure both urea and albumin specific activities, we are faced with the problem of the determination of two constants, S_a and S_u, and of the functions $h_1(s)$ and $h_2(s)$, which are related between themselves [S_a with $h_1(s)$ and S_u with $h_2(s)$] in an unknown way.

To proceed, we may choose between two possible routes: *either we complicate the experiment and use additional tracers to determine directly $h_1(s)$ and $h_2(s)$*, whence we could obtain also S_u and solve the system for S_a, *or we put some theoretical constraint on both $h_1(s)$ and $h_2(s)$ or one of them*, so that S_a is univocally determined. Whereas the first route (addition of new tracers) complicates the experiment and poses problems as to their actual biological adequacy, the second one leads us directly into the dangerous and fascinating world of model builders.

C. Minimal Compartmental Model and Single Tracer Approach

This approach was used by the first groups that used the ^{14}C-carbonate method to estimate albumin synthesis (McFarlane *et al.*, 1965).

Figure 3 shows how this was accomplished. The central column lists the data needed to determine S_a on the basis of the general solution. Of these $A(t)$ and $U(t)$ can be obtained experimentally by measuring albumin and urea specific activity.

For urea kinetics, from which $H_2(t)$ and S_u depend, the following assumptions are needed: (1) Body urea behaves as a single mixing compartment (monoexponential decay of a bolus of radioactive urea injected intravenously); (2) shortly after injection, urea specific activity decays as if flash labeling of urea from arginine had occurred; (3) no recycling of label through the urea pool takes place during the study.

The last two assumptions permit one to estimate S_u from the slope of $U(t)$; the first one implies that, within the duration of the study, $U(t)$ will decay as $H_2(t)$.

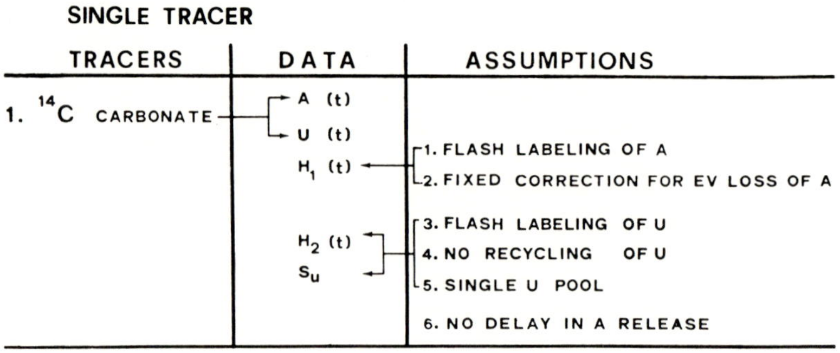

FIG. 3. Minimal compartment model, single tracer: ^{14}C-carbonate. A, albumin; EV, extravascular; U, urea.

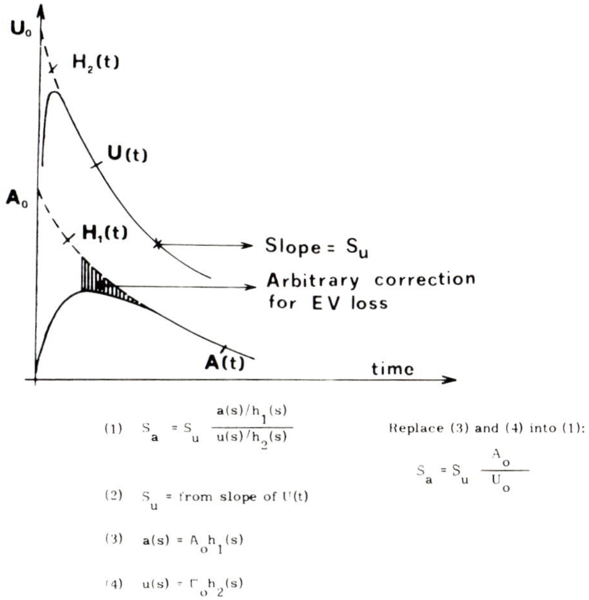

FIG. 4. Graphic representation of minimal compartment model: albumin synthesis from ^{14}C-carbonate (single tracer approach) (see text).

For albumin kinetics, since a single mixing compartment cannot be assumed, a correction was applied to the plasma albumin specific activity values to account for extravascular losses. If it is assumed that with such correction plasma albumin specific activity values decay as if flash labeling had occurred, an estimate for $H_1(t)$ also becomes available.

Figure 4 shows graphically the significance of the above assumptions; $U(t)$ decays as $H_2(t)$; it can be extrapolated to zero time to obtain the ideal zero time specific activity U_0, which would have resulted from flash labeling. S_u is the slope of $U(t)$ itself. $A(t)$, the albumin specific activity curve, is arbitrarily corrected for extravascular losses. The corrected curve extrapolated to zero time gives the ideal zero time specific activity A_0 for flash labeling.

Under these assumptions, the general Eq. (5) reduces to:

$$S_a = S_u \frac{A_0}{U_0} \tag{6}$$

from which the calculation of S_a can be performed with little graphical exercise and even without a slide rule (McFarlane *et al.*, 1965). However, this approach, extremely appealing for its simplicity, is founded on the fragile foundations of an arbitrary compartmental model, based on assumptions which have been shown not to be tenable (Jones *et al.*, 1968).

D. Minimal Compartmental Model and Double Tracer Approach

1. S_u *from Slope of Urea Curve*

At this stage, it is clear that we cannot avoid some complication to cope with our unknowns, which are just too many for the available information. We may, at this level, leave the model unchanged and use an additional

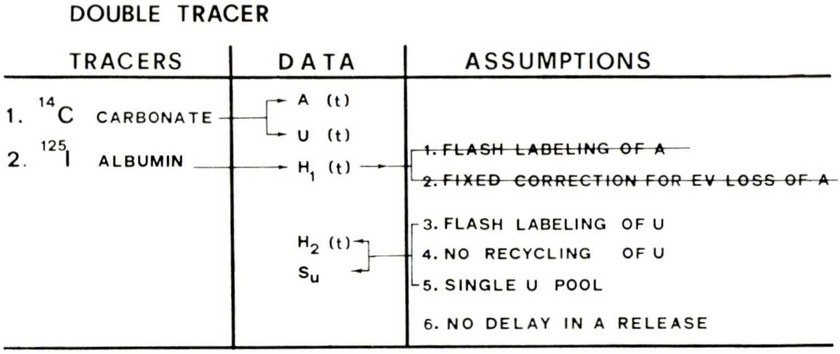

FIG. 5. Minimal compartment model, double tracer: ^{14}C-carbonate and ^{125}I-albumin.

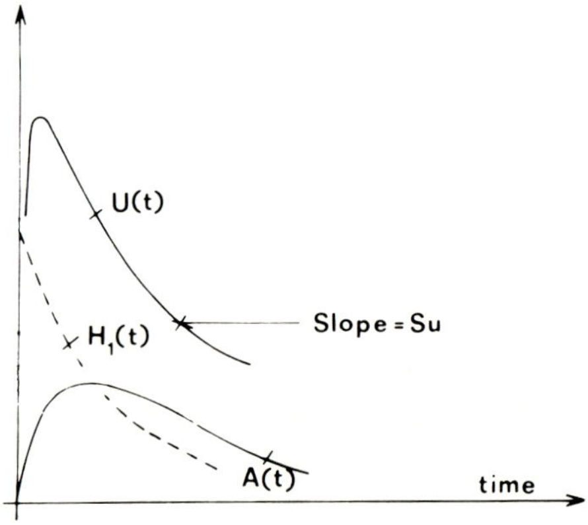

(1) $S_a = S_u \dfrac{a(s)h_2(s)}{u(s)h_1(s)}$

(2) S_u = from slope of $U(t)$

(3) $h_2(s) = 1/(s + S_u)$

(4) $f(s) = u(s)h_1(s)$

Replace (3) and (4) into (1) then antitransform:

$S_a(F'(t) + S_u F(t)) = S_u A(t)$

find S_a by linear least squares analysis for one unknown

FIG. 6. Graphic representation of minimal compartment model: albumin synthesis from ^{14}C-carbonate (double tracer approach). $F(t)$ is the Laplace antitransform of $f(s)$; $F'(t)$ is its first derivative. Both $F(t)$ and $F'(t)$ as well as S_u are considered experimentally determined.

tracer, in order to increase the amount of information derived from experiment and reduce that arising from assumptions.

If the second tracer is radioiodinated albumin injected into the blood stream, then this yields directly the function $H_1(t)$, with the consequence of eliminating the assumptions concerning albumin labeling and distribution kinetics, as it is shown in Fig. 5. We are left now with the assumptions concerning urea kinetics, which still demand some modeling fantasy. The first attempt could be that of leaving unaltered the assumptions of flash labeling, absence of recycling, and single compartment for the urea system. From them it stems that S_u values can be determined from the slope of $U(t)$; $H_2(t)$ is then univocally determined. Figure 6 shows the situation graphically, and the route to obtain S_a. By this method the values obtained for S_a seem reasonable, but when they are introduced into the model they are not capable of yielding a good fit of the experimental curve of albumin specific activity (Fig. 7).

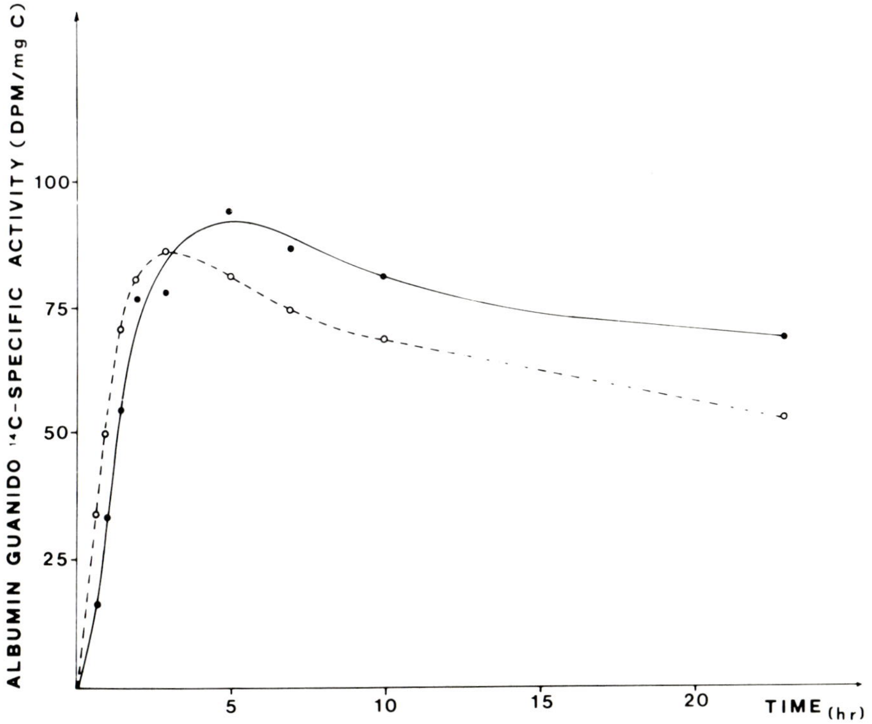

FIG. 7. Comparison of experimental and calculated values for albumin specific activity. The closed circles represent the experimental points and the open circles are the values yielded by the model of Fig. 6.

Thus the model does not work too well. It may be for one of the following reasons: (1) Urea does not behave as a single mixing compartment, hence S_u cannot be estimated from the slope of $U(t)$; (2) urea does behave as a single mixing compartment, but S_u cannot be estimated from the slope of $U(t)$ because the assumptions of flash labeling and/or that of no recycling do not hold.

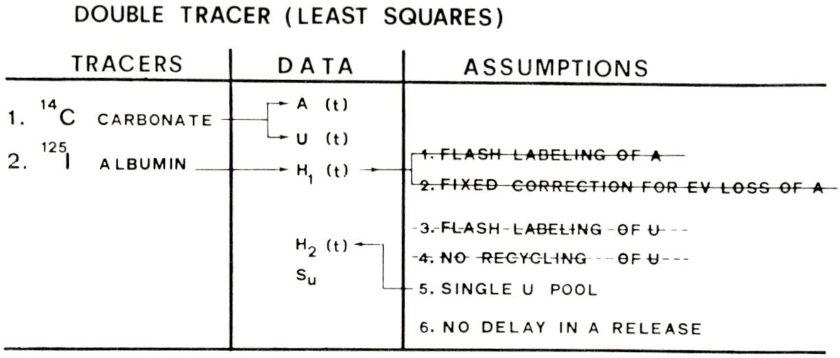

FIG. 8. A second more complicated model may be considered: one in which urea behaves as a single mixing compartment but the assumptions of flash labeling and no recycling of the label are not valid (see text).

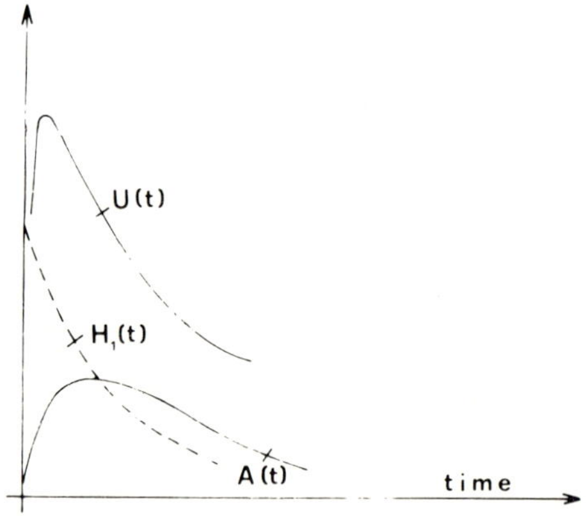

(1) $S_a = S_u \dfrac{a(s)h_2(s)}{u(s)h_1(s)}$

(2) $h_2(s) = 1/(s + S_u)$

(3) $f(s) = u(s)h_1(s)$

Replace (2) and (3) into (1) then antitransform:

$$\frac{S_a}{S_u} F'(t) + S_a F(t) = A(t)$$

find S_a and S_a/S_u simultaneously by linear least squares analysis for two unknowns

FIG. 9. Graphic representation of model in Fig. 8: albumin synthesis from ^{14}C-carbonate (double tracer approach). $F(t)$ is the Laplace antitransform of $f(s)$; $F'(t)$ is its first derivative. Both $F(t)$ and $F'(t)$ can be considered experimentally determined (see text).

2. S_u *as an Unknown*

We shall try first to see if the second hypothesis is tenable. Figure 8 shows the consequent reduction of assumptions: those of flash labeling and no recycling of urea are eliminated, while that of the single urea pool is maintained. As a consequence, S_u is now an unknown, but we have maintained a constraint on $H_2(t)$ in such a way that, if we arrive at determining S_u, $H_2(t)$ will also be univocally determined.

Figure 9 presents the graphical schematization and resulting equations: They are the same as in the preceding case, with the only difference that now both S_a and S_u are unknowns. The problem is solved by simultaneously finding the values of S_a and S_a/S_u for which the $A(t)$ values predicted from the equation give the best fit of the experimental $A(t)$ values (Mancini *et al.*, 1968).

When this approach is used, it is found that the values obtained for S_a and S_u are systematically lower than those for albumin and urea catabolic rates as determined by direct injection of labeled albumin and urea, respectively.

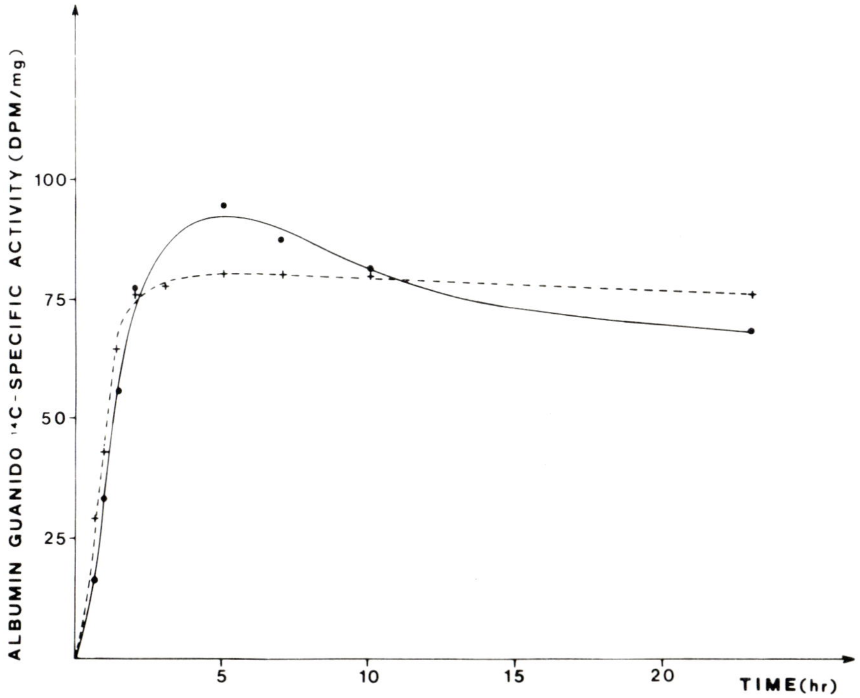

FIG. 10. Comparison of experimental and calculated values for albumin specific activity. The closed circles represent experimental points, while the crosses are the values yielded by the model of Fig. 9.

Figure 10 demonstrates that in this case also the obtained fit of the experimental points of the albumin specific activity curve is not satisfactory.

E. Implications from the Analysis of the "Minimal" Model

The conclusion that the urea pool does not behave as a single mixing compartment is now unescapable. In the attempt to find a satisfactory analytical and experimental system to assess albumin synthesis rate the following approaches can be explored: (1) introduction of a more complex compartmental model for urea kinetics; (2) abandoning the compartmental

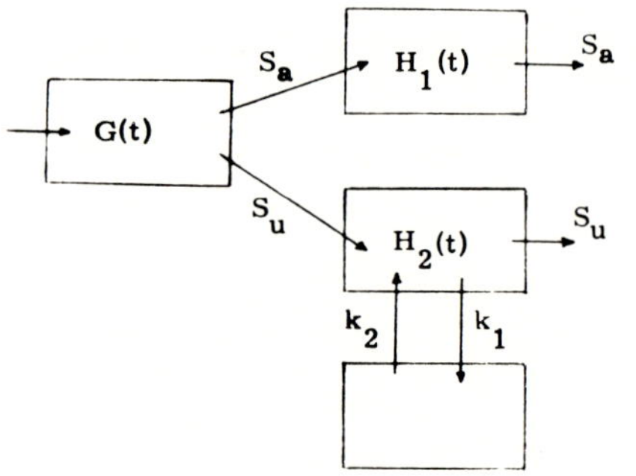

(1) $S_a = S_u \dfrac{a(s)h_2(s)}{u(s)h_1(s)}$

(2) $h_1(s)$: from ^{125}I-HSA

(3) $h_2(s) = \dfrac{s+k_2}{(s+S_u+k_1)(s+k_2)-k_1k_2}$

(4) $f(s) = u(s)h_1(s)$

Replace (3) and (4) in (1) then antitransform:

$$\frac{S_a}{S_u} F'(t) + S_a F(t) + \frac{S_a}{S_u} k_1 \left\{F(t) - k_2 \left[F(t)\right] * \left[e^{-k_2 t}\right]\right\} = A(t)$$

find S_a, S_a/S_u, $k_1 S_a/S_u$ simultaneously by linear least squares analysis once determined k_2 by non linear fitting.

Fig. 11. Two-compartment model for urea kinetics: albumin synthesis (two tracers). $G(t)$, specific activity of hepatic arginine; $H_1(t)$, transit time distribution through the plasma pool of newly made albumin; $H_2(t)$, transit time distribution through the urea plasma pool; S_a, fractional albumin synthetic rate; S_u, fractional urea synthetic rate.

approach and search for a more general integral solution; and (3) using a third tracer to measure directly the urea parameters. Each of the above three solutions has advantages and disadvantages.

F. Two-Compartment Model for Urea

Figure 11 shows a two-compartment model for urea kinetics. It assumes that newly formed urea is released into the plasma compartment which exchanges with an extravascular compartment, and that removal of urea from its pool also takes place from plasma.

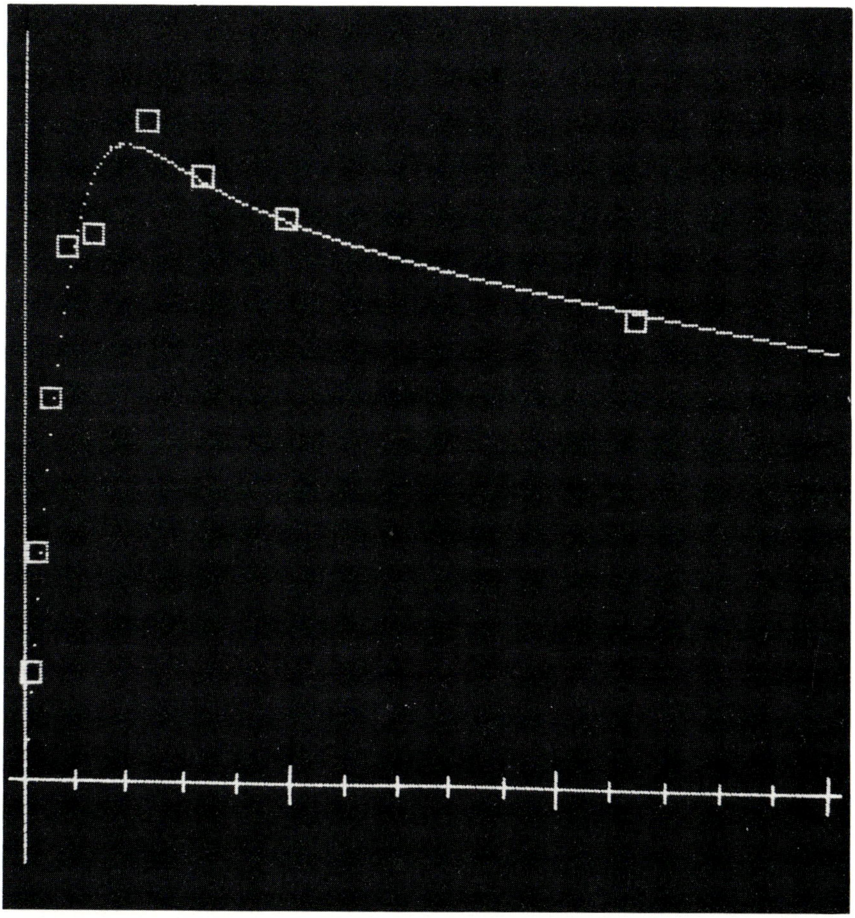

FIG. 12. This figure is a picture of the computer display; the squares represent the experimental points and the dotted line is the best fit yielded by the model of Fig. 11.

Under these constraints, the urea transfer function $H_2(t)$ is univocally defined and it contains three unknown constants (the fractional plasma urea turnover rates and the fractional rate of exchange between the two compartments k_1 and k_2).

When this transfer function is introduced into the general Eq. (5), we obtain an expression (see Fig. 11) with four unknowns (S_a, S_u, k_1, and k_2), which can be simultaneously solved by: (1) fitting with a continuous function the ^{14}C-urea and ^{125}I-albumin specific activity data; (2) convoluting $U(t)$ with $H_1(t)$ which yields $F(t)$; (3) differentiation of $F(t)$ to obtain $F'(t)$; (4) convoluting $F(t)$ into $k_2e^{-k_2t}$; and (5) finding by a least squares method a solution for the four unknowns which minimizes the squares of the differences between predicted and experimental values of $A(t)$. It should be pointed out that while the problem is linear with respect to the three unknowns, S_a, S_a/S_u, $(S_a/S_u)k_1$, the variable k_2 appears in a nonlinear form and must be estimated by iterative procedures (Mancini and Vitek, 1969).

When this analytical sequence is applied to experimental curves, the value obtained for S_a agrees well with that obtained from catabolism. Figure 12 shows that also the fitting of the experimental albumin curve has been adequate. This method also yields values for the exchange constants between the two urea compartments. More parameters of the system are obtained but we must remember that they belong to the model and do not necessarily have biological counterparts.

G. Integral Approach

It has become well known in the protein field, mostly thanks to Nosslin's work (1964) that operating with integrals reduces the number of assumptions required in the analysis of kinetic studies. This approach can also be applied to the present situation.

Figure 13 shows this treatment, schematically: The specific activity curves of hepatic arginine and of plasma urea are plotted and the areas under these curves are outlined. The two areas are equal since hepatic arginine is the only precursor of plasma urea. This identity simply means that the fraction of label that will flow from the hepatic arginine pool into the plasma urea pool is equal to the fraction of total flow of substance out of the arginine pool that will enter the plasma urea pool.*

The area under the plasma urea ^{14}C curve may then be taken as the area

* This is a principle well known in circulation studies, and on the basis of which cardiac output by the indicator dilution method can be measured in a peripheral artery as well as at the aortic root since the areas of the concentration curves at the two sites are identical, provided mixing of indicator and flow has been completed upstream to both sites.

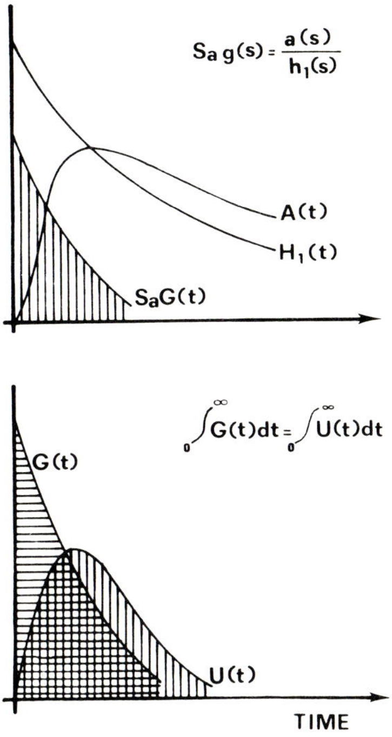

FIG. 13. Albumin synthesis studies, two tracers integral approach. See text.

under the curve of the precursor of serum albumin. On the other hand, from Eq. (1), for albumin specific activity, it is seen that the deconvolution of the ^{14}C-albumin curve $A(t)$ into the iodine-labeled albumin curve $H_1(t)$, the input function to the albumin pool, is obtained; this corresponds to the specific activity curve of hepatic arginine times S_a.

Evaluation in patients have shown that this function drops to 10% of the initial values after 3 hr and to 2% after approximately 10 hr. It may then be easily extrapolated to infinity thus making possible to write:

$$\left[\int_0^\infty S_a G(t)\,dt\right] \Big/ \left[\int_0^\infty U(t)\,dt\right] = S_a \tag{7}$$

from which S_a is readily obtained as the ratio of two areas.

This procedure requires: (1) fitting by continuous functions of the ^{14}C-urea, ^{14}C-albumin, ^{125}I-albumin data; (2) deconvolution of the ^{14}C-albumin specific activity curve into the ^{125}I-albumin curve; (3) calculation of the integral of ^{14}C-urea curve; and (4) calculation of the ratio of the two integrals.

This method is relatively simple and in comparison with the compartmental approach yields less information (S_a only) with a higher reliability, mostly depending on the method used for deconvolution. If a multiexponential fitting is used, it is not necessary to go through steps (2), (3), and (4), since S_a can be directly calculated in terms of the constants of the exponential functions fitting the experimental curves. It is worth pointing out that, thanks to the rapid decay of the specific activity of hepatic arginine, errors in fitting the terminal portions of the experimental curves will have minor influence on the integral of the curve obtained by deconvolution.

H. Three Tracer Approach

The three tracers approach involves the injection of ^{13}C-urea simultaneously with the other two tracers, or, as we have done so far, reinjection of ^{14}C-urea on the third day after the first study together with ^{131}I-labeled albumin to check the maintenance of the steady state conditions for albumin. Using this technique, all assumptions drop since all terms of the general solution are available (Fig. 14). Figure 15 schematizes the experimental curves and the calculations involved.

The following steps have to be carried on to obtain S_a: (1) fitting by continuous functions of the ^{14}C-urea, ^{13}C-urea, and ^{131}I-albumin data; (2) calculation of the fractional urea synthesis rate from the equation

$$S_u = 1/\int_0^\infty H_2(t)dt \tag{8}$$

(3) deconvolution of $U(t)$ into $H_2(t)$ which yields $G(t)$; (4) convolution of $G(t)$ into $H_1(t)$; and (5) solving Eq. (1) for S_a for every experimental value of $A(t)$.

The advantage of this approach is that a value for S_a can be obtained

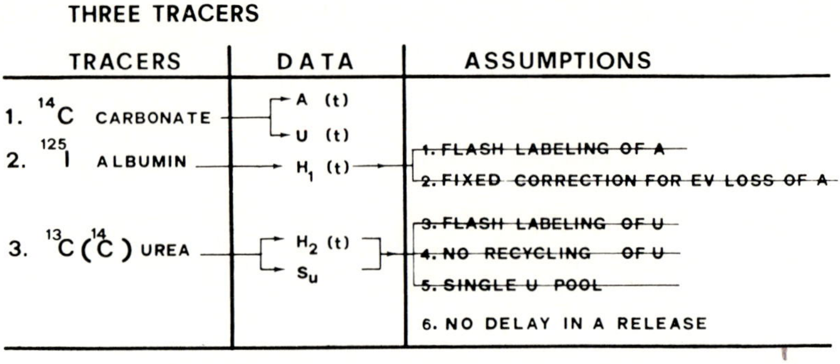

FIG. 14. Three tracers: ^{14}C-carbonate, ^{125}I-albumin, and exogenously-labeled urea.

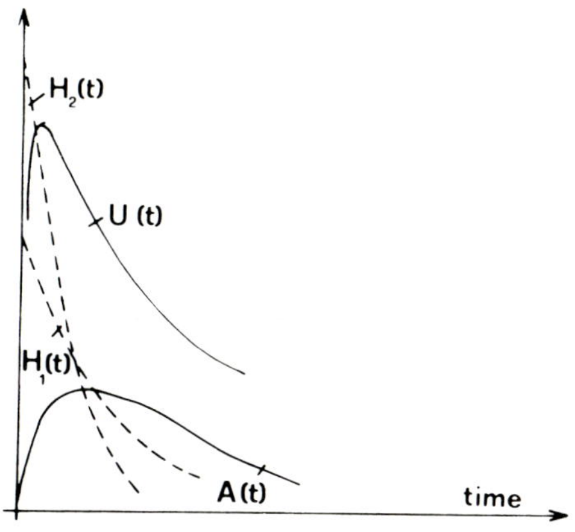

(1) $$S_a = S_u \frac{a(s)/h_1(s)}{u(s)/h_2(s)}$$

(2) $$S_u = 1/\int_0^\infty H_2(t)dt$$

(3) $$g(s) = u(s)/S_u h_2(s)$$

Replace (3) into (1):

$$S_a = \frac{a(s)}{g(s)h_1(s)}$$

where g(s) transfer function for hepatic arginine

FIG. 15. Graphic representation of three tracer approach: albumin synthesis from ^{14}C.

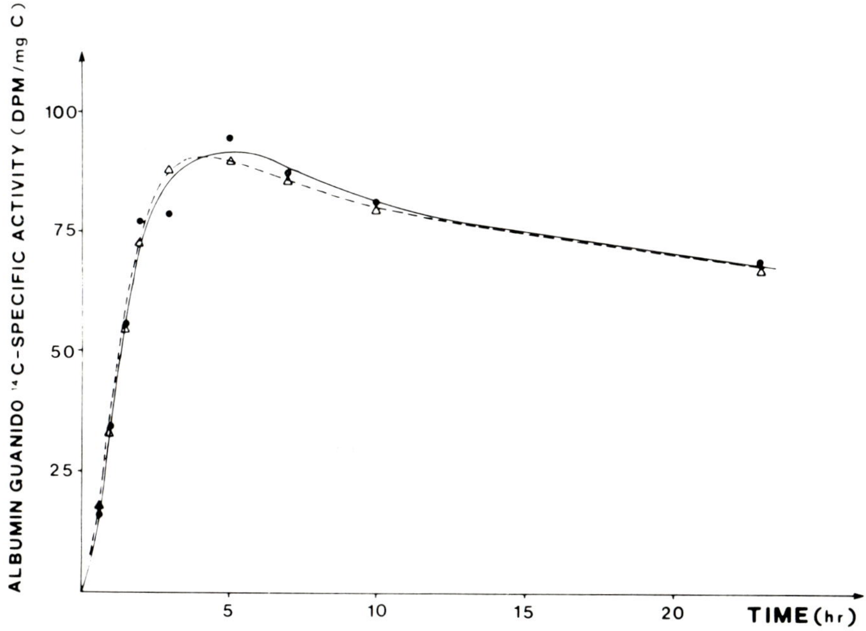

FIG. 16. Comparison of experimental and calculated values for albumin specific activity. The closed circles represent the experimental points and the open triangles are the values yielded by model of Fig. 15.

from each experimental value of $A(t)$, thus making possible to estimate its mean value and standard deviation. The obtained S_a value compares well with the catabolic rate. When the obtained average S_a is introduced into the general equation, the theoretical course of $A(t)$ calculated thereof yields a good fit of the experimental points of albumin specific activity (Fig. 16).

III. Discussion

Figures 17 and 18 show as an example the experimental data obtained in a normal subject in whom intravenous administration of ^{14}C-carbonate and ^{125}I-albumin was followed 2 days later by the injection of ^{14}C-urea and ^{131}I-albumin. Fitting of the data with continuous functions by the multi-exponential method is also shown. Comparison of ^{125}I- and ^{131}I-albumin disappearance curves (Fig. 18) suggests that steady state conditions had been maintained.

In Fig. 19 the values for albumin and (when possible) urea synthesis rate obtained in the same subject with the different methods discussed in the previous sections are compared between themselves and with those obtained with radioiodinated albumin and ^{14}C-urea, respectively.

For albumin synthesis rate the differences in the values obtained by the last three methods are small, and they agree with the catabolic rate. For urea synthesis, the two-compartment model yields a value in good agreement

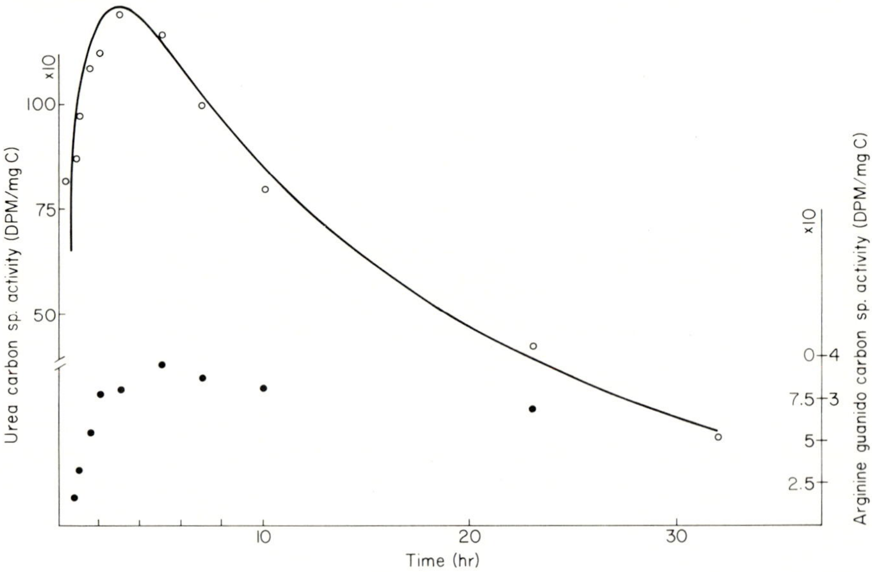

FIG. 17. Urea (open circle) and albumin (solid circle) specific activity curves in a normal subject.

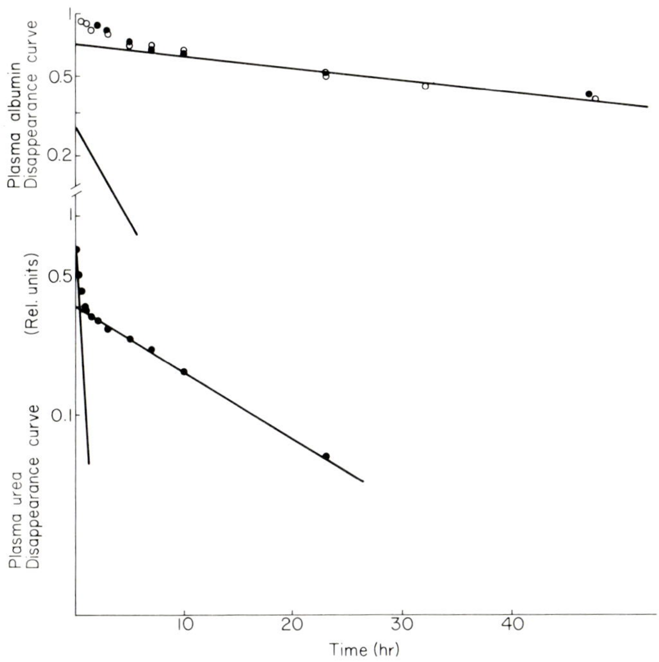

FIG. 18. Plasma ^{125}I-albumin (solid circle) and ^{131}I-albumin (open circle) decay curves in the same patient 48 hr apart. These curves suggest the presence of steady state conditions.

with that measured with ^{14}C-urea. Estimates from the "minimal model" grossly underestimate urea synthesis rate.

Figure 20 shows the values for albumin synthesis rate obtained by the three tracer approach for each experimental point. Mean value and standard deviation of S_a, fractional synthesis rate (FSR) estimates from the first hour on is 13.1 and 0.24%/day, respectively, as compared with 13.9 and 0.43% for catabolic rate.

The possibility of repetitive estimates of S_a by the three tracer method allows some insight into the significance of the delayed release in plasma of newly synthesized albumin. While an arbitrary delay of 0.4 hr was used throughout this study, it was seen that, after the first 2 hr, S_a values were practically identical whether the delay was introduced or not.

Figure 21 summarizes the characteristics, advantages, and disadvantages, of the three methods which have been found to yield satisfactory results.

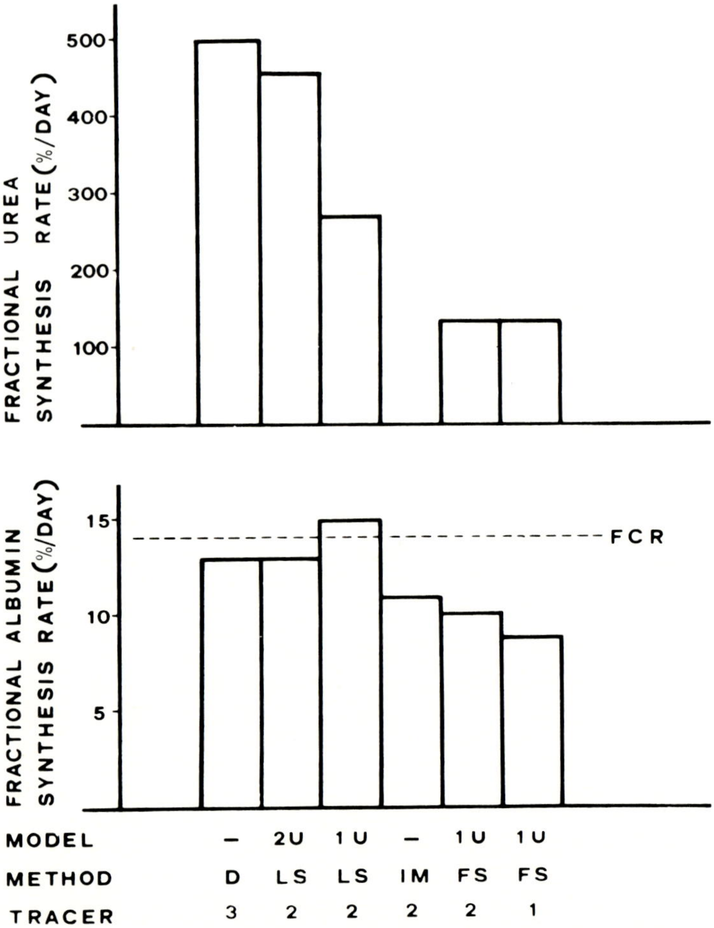

FIG. 19. Graph of fractional urea and albumin synthesis rate. Symbols, –, 1 u, 2 u, refer to the assumption of urea model; they mean, respectively, no assumption, one compartment, two compartment. For the method the following abbreviations were used: D, deconvolution; LS, least squares; IM, integral method; FS, final slope. The number refers to the tracers employed. FCR, fractional catabolic rate (% per day).

The possibility of obtaining an estimate of S_a for every experimental point is a definite advantage of the three tracers method, insofar as the error of the estimate can be determined. In comparison with the integral method, the three tracer approach shares with the two-compartment method the advantage of not requiring the fitting of the ^{14}C-albumin curve by a continuous function. On the other hand, an obvious disadvantage of the three

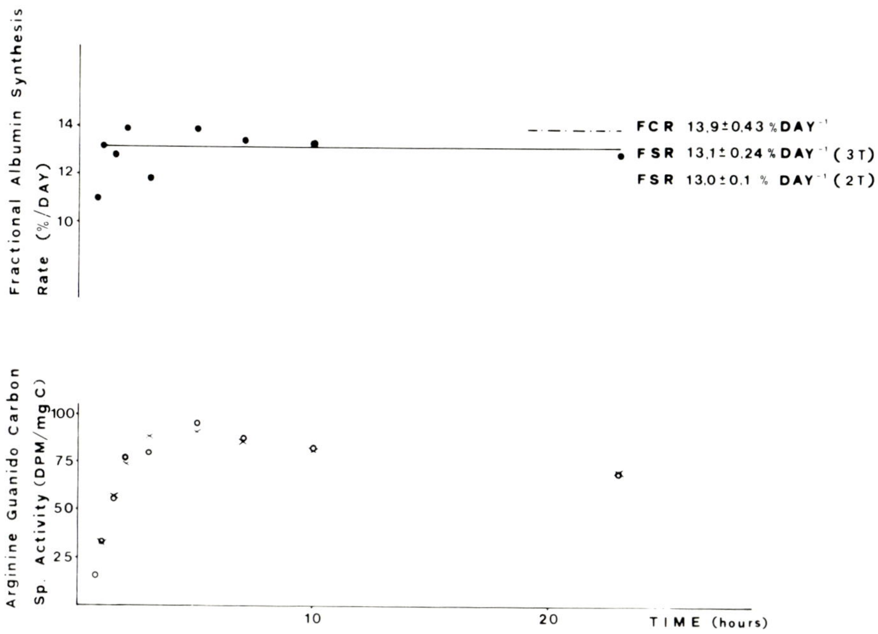

FIG. 20. In the upper part of the figure, the closed circles represent the fractional synthesis rate (FSR), calculated for each experimental point of arginine guanido carbon specific activity by the method of three tracers (3T). The continuous line represents the mean, with the numerical value reported on the right. The fractional catabolic rate (FCR) is also reported. The lower value of FSR is obtained by the method of Fig. 11 employing two tracers (2T).

In the lower part of the figure, the open circles represent the experimental points, while the crosses represent the calculated values according to the method of three tracers (3T).

Characteristics	Method	Disadvantage	Advantage
Two tracers fitting of 2 curves	Two compartments for urea model	Assumption on urea kinetics. Elaboration complicated	No deconvolution
Two tracers fitting of 3 curves	Integral	One deconvolution determination of S_a only	No assumption simple enough
Three tracers fitting of 3 curves	Deconvolution	One deconvolution experiment complicated	No assumption

FIG. 21. Tabular comparison between two tracers and three tracers approach.

tracer method is that of requiring a third tracer and the use of mass spectrometry, or the later injection of ^{14}C-urea and albumin labeled with a different iodine isotope. Both the two-compartment and the integral method avoid this disadvantage, but the cost is the assumption of a definite compartmental model for urea in one case, and the lack of information on urea synthesis rate in the other.

An additional disadvantage that the three tracer method shares with the integral approach is the use of deconvolution, a naturally unstable operation depending on the technique. Its effects are likely to affect less the integral method insofar as it does not operate on the function obtained by deconvolution but on its own integral. On the other hand, the two-compartment method, which does not require deconvolution, demands a definitely heavier mathematical workout of the data, including the use of an iterative procedure. All this considered, however, and provided the data workout is confined to a computer, the two-compartment method seems to be advisable in practice.

Limited computing facilities would recommend the integral approach, while the use of the three tracer method should be confined to the cases in which repeated estimates of S_a during the study are required.

IV. Conclusion

Presently available techniques and knowledge, at chemical analytical and computing level, make possible reliable estimates of albumin synthesis rate to be obtained in man.

The present analysis has shown, as it could be anticipated, that, albumin being a complex system, largely nonaccessible, complex methods are required for its quantitative investigation. Depending on the approach employed the complexities may occur predominantly at the level of the assumptions, of the experiment, or of the data analysis. In any case, for the obtainment of a given amount of information, the cost will be very much the same.

One final point to be stressed is that the use of a digital computer makes all the methods described in this chapter easily applicable.

Appendix

A. Experimental Procedure

As it has been discussed in the present chapter, the use of more refined techniques for data analysis demands that the specific activity curves are defined with acceptable accuracy.

The large amount of blood required for each estimate of the albumin specific activity by the previous methods markedly limits the number of blood samples that can be taken and the accuracy with which the experimental function can be described (Tavill *et al.*, 1968).

In this appendix is reported the protocol that we use for studies in patients, and is described a modification of the original technique, which makes possible to obtain more experimental points with less blood (Bianchi and Mariani, 1970).

B. METHODS

Subjects are kept on a protein-free diet for 12 hr before the study. A dose of 100–200 μCi of sodium ^{14}C-carbonate is injected intravenously, together with 20 μCi of ^{125}I-labeled HSA (electrolytically labeled). Blood samples of 15 ml are taken after 5, 15, 30, 45, and 60 min and then 2, 3, 5, 7, 10, and 23 hr, for the determination of urea carbon and albumin guanido carbon specific activity in plasma and for the study of the plasma disappearance curve of the iodine-labeled albumin.

Since ^{13}C-urea has not been available to us, when the three tracers approach is used, we reinject ^{14}C-urea (5 μCi) on the third day of the experiment, together with 15 μCi of ^{131}I-labeled albumin. Albumin catabolic rate is determined during 8 days by the U/P ratio method (urinary activity excreted in 24 hr/mean plasma activity in the same 24 hr).

In Fig. 22 the method used for determining the carbon specific activity is schematically compared with the previous approach. While the current method resorts to urea splitting with urease and volumetric recovery of the formed CO_2, in the present approach urea is precipitated with xanthydrol to dixanthydryl urea (DXU), which is then directly weighed and counted in gel (4.5% Cab-O-Sil toluene cocktail). The specific activity of urea and/or albumin guanido carbon is then obtained as [^{14}C content of DXU (dpm)/ weight of DXU (mg)] × 35 where 35 is the product of the ratios of DXU/ urea weight and urea/urea carbon weight.

C. COMMENT

The advantages of the method described are the following: (1) Smaller blood samples are required. In our experience, hydrolysis of urea by urease is nearly 100% when the enzyme reacts on urea directly, but it may decrease to only 30% when urease is added after the action of arginase on the albumin hydrolyzate, despite a tenfold increase of the enzyme concentration. On the contrary, using the xanthrydol method, the precipitation of urea formed from the action of arginase on arginine, is always as complete as the precipitation of urea directly from the deproteinized plasma (Engel and Engel, 1947). (2) The method permits the direct measurement in the same sample

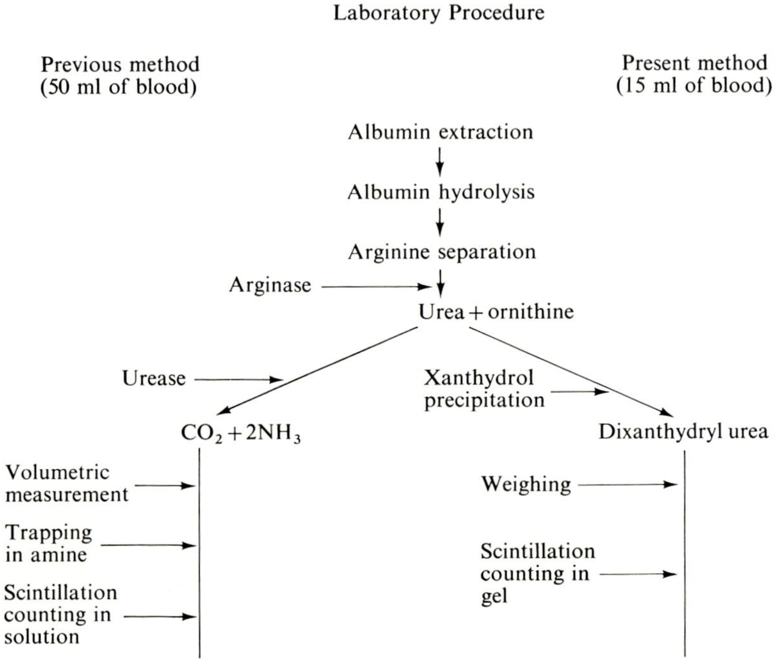

FIG. 22. Comparative scheme between laboratory procedures. See text.

of both ^{14}C content and mass of carbon present as urea. This increases the accuracy with which the specific activity is obtained, since the determination of the mass by weighing is certainly performed with greater simplicity and precision than the manometric estimation of the CO_2 liberated by the reaction of urease on urea. The sevenfold increase of dixanthydryl urea weight in respect to that of urea increases the reliability of the estimation of the urea carbon mass. (3) The suspension counting technique results in considerable higher detection efficiency than counting $^{14}CO_2$ in homogenous solution, due to the absence of quenching effect and β-ray absorption (Helf *et al.*, 1960). (4) While a 100% trapping of $^{14}CO_2$ released by urease is essential for the correct measurement of specific activity in the traditional method, an incomplete precipitation in the xanthydrol method has the only adverse effect of reducing the counting rate, but it does not affect the estimation of urea and albumin guanido carbon specific activities, since the same dixanthydryl urea precipitate is weighed and then counted.

In conclusion, the present chemical technique makes possible a better definition of the experimental curves, both permitting a larger number of blood samples to be taken, and increasing the accuracy of each estimate. This makes possible the use of more refined methods for data analysis.

ACKNOWLEDGMENTS

The authors would like to thank L. Fusani, R. Pasqualini, and F. Bonaguidi for their invaluable technical assistance.

REFERENCES

Bianchi, R., and Mariani, G. (1969). *J. Nucl. Biol. Med.* **4**, 135.

Bianchi, R., Federighi, G., Giagnoni, P., Giordani, R., Navalesi, R., and Donato, L. (1967). *In* "Physiology and Pathophysiology of Plasma Protein Metabolism," 4th Int. Symp. (G. Birke, R. Norberg, and L. O. Plantin, eds.), Vol. 12, p. 195. Macmillan (Pergamon), New York.

Craigie, A., Jones, E. A., Rosenoer, V. M., Smallwood, R. A., and Tavill, A. S. (1967). *In* "Physiology and Pathophysiology of Plasma Protein Metabolism," 4th Int. Symp. (G. Birke, R. Norberg, and L. O. Plantin, eds.), Vol. 12, p. 61. Macmillan (Pergamon), New York.

Donato, L., Matthews, C. M. E., Nosslin, B., Segre, G., and Vitek, F. (1966). *J. Nucl. Biol. Med.* **10**, 2.

Engel, M. G., and Engel, F. L. (1947). *J. Biol. Chem.* **167**, 535.

Helf, S., White, C. G., and Shelley, R. N. (1960). *Anal. Chem.* **32**, 238.

Jones, E. A., Craigie, A., Tavill, A. S., Simon, W., and Rosenoer, V. M. (1968). *Clin. Sci.* **35**, 553.

McFarlane, A. S. (1963). *Biochem. J.* **89**, 277.

McFarlane, A. S., Irons, L., Koj, A., and Regoeczi, E. (1965). *Biochem. J.* **95**, 536.

Mancini, P., and Pilo, A. (1970). *Comput. Biomed. Res.* **3**, 8.

Mancini, P., and Vitek, F. (1969). Internal Report No. 21. Dept. of Biophysic and Biomathematic of C.N.R., Clinical Physiology Laboratory, Pisa.

Mancini, P., Pilo, A., and Vitek, F. (1968). Internal Report No. 6. Biophysic and Biomathematic Group, Clinica Medica Generale, Univ. of Pisa.

Nosslin, B. (1964). *In* "Metabolism of Human Gamma-Globulins (Gamma-ss-Globulins)" (S. B. Andersen, ed.), p. 34. Blackwell, Oxford.

Pilo, A. (1968). Internal Report No. 5. Biophysic and Biomathematic Group, Clinica Medica Generale, University of Pisa.

Reeve, E. B., Pearson, J. R., and Martz, D. C. (1963). *Science* **139**, 914.

Rothschild, M. A., Oratz, M., Zimmon, D., Schreiber, S. S., Weiner, I., and VanCaneghem, A. (1969). *J. Clin. Invest.* **48**, 344.

Smallwood, R. A., Jones, E. A., Craigie, A., Raia, S., and Rosenoer, V. M. (1968). *Clin. Sci.* **35**, 35.

Swick, R. W. (1958). *J. Biol. Chem.* **231**, 751.

Tavill, A. S., Craigie, A., and Rosenoer, V. M. (1968). *Clin. Sci.* **34**, 1.

Vitek, F., Bianchi, R., and Donato, L. (1966). *J. Nucl. Biol. Med.* **10**, 121.

Vitek, F., Bianchi, R., Mancini, P., and Donato, L. (1967a). "Physiology and Pathophysiology of Plasma Protein Metabolism," 4th Int. Symp. (G. Birke, R. Norberg, and L. O. Plantin, eds.), Vol. 12, p. 29. Macmillan (Pergamon), New York.

Vitek, F., Bianchi, R., and Federighi, G. (1967b). *J. Nucl. Biol. Med.* **11**, 139.

Weissman, S., Tschudy, O. P., Bacchus, H., and Eubantis, M. (1961). *J. Lab. Clin. Med.* **57**, 136.

4

SYNTHESIS AND DEGRADATION OF PLASMA PROTEINS

A. S. McFarlane

DIVISION OF BIOPHYSICS,
NATIONAL INSTITUTE FOR MEDICAL RESEARCH,
MILL HILL, LONDON, N.W.7, ENGLAND

I. Introduction

The expectation that it would be possible to measure rates of synthesis and degradation of plasma proteins by mainly or wholly independent procedures appears now to have been realized making the time opportune to review briefly the situation with special reference to the validity of both methods. The importance of these procedures being essentially independent lies, of course, in the prospect of being able to determine unequivocally whether one or both rates have changed in experimentally induced unbalanced states.

II. Measurement of Rates of Synthesis of Liver-Produced Plasma Proteins

The ^{14}C-carbonate method (McFarlane, 1963b) of doing this depends on the combination of $^{14}CO_2$ inside the parenchymal cells of the liver with ammonia and ornithine to give 6-^{14}C-arginine which then becomes available both for incorporation into proteins and for the production of urea according to the Krebs-Henseleit cycle.

The following relationship holds for any specified interval of time. The mean specific activity of the precursor carbon integrated over the interval in question is equal to two fractions, namely,

$$\frac{\text{total } ^{14}\text{C activity appearing in urea carbon}}{\text{mass of urea carbon synthesized in same time}}$$

and

$$\frac{\text{total } ^{14}\text{C activity appearing in guanidine carbon of protein arginine}}{\text{mass of corresponding protein carbon synthesized in same time}}$$

The fractions being equal, it is only necessary to measure total ^{14}C activities appearing in urea and protein arginine and the absolute synthesis rate of urea to be able to calculate the synthesis rate of the protein. The interval chosen is usually 4–5 hr after injection of ^{14}C-carbonate, although most of the labeled products are formed in the first few minutes. If shorter intervals are used, a difficulty arises which has its origin in the fact that, whereas ^{14}C-urea is liberated immediately from the liver into the body water, the arrival of labeled proteins in the plasma is delayed by up to half an hour. With intervals of 4–5 hr or longer, errors due to this are negligible; a minimum of 4–5 hr is also required for measuring the absolute synthesis rate of urea.

The main advantage of injecting ^{14}C-carbonate rather than 6-^{14}C-arginine as originally proposed by Reeve *et al.* (1963) and McFarlane (1963a) is that liver-produced proteins in the plasma are approximately an order of magnitude "hotter" than carcass proteins, including plasma γ-globulins, at the same time due to the fact that endogenous liver produced 6-^{14}C-arginine is first diluted with an excess of "cold" dietary arginine before it is used to produce proteins of extra hepatic origin. This experimental fact (McFarlane, 1963a) has important consequences concerning relative degrees of reutilization of the isotopic carbon and hence both the accuracy of the method and the proportion of injected isotope persisting in the body, the latter being liable to constitute a radiation hazard in clinical investigation. In spite of the low efficiency of utilization of the isotope for protein synthesis when given as carbonate, the administration in this form is both cheaper and safer, and accuracies achieved in the measurement of synthesis rates are higher.

The ^{14}C-carbonate method of measuring synthesis rates of liver-produced proteins is currently being used in several laboratories in relation to patients and to experimental animals. ^{14}C-Carbonate should not be given to children because of the radiation hazard, but it has been given to at least 100 adult humans at dose levels of 100–200 μCi, various groups in and around London having made more than 50 such measurements. For reasons which will be discussed accuracies are probably not better than $\pm 10\%$. Fractional catabolic rates measured with iodine-labeled proteins in animals in equilibrium are more accurate but represent mean rates measured over several days, whereas a valuable feature of synthesis estimates is that they refer to intervals of a few hours following the injection of the carbonate.

A. Rate of Synthesis of Urea

Paradoxically, accuracy of protein synthesis rate measurements appears to be limited mainly by uncertainties surrounding the parameters of urea metabolism about which it was thought naively that nearly everything was known. Rates of urea excretion are not acceptable as indices of urea synthesis because of endogenous catabolism as first pointed out by Walser and Bodenloss (1959). To overcome this difficulty McFarlane *et al.* (1965) recommended an injection of ^{13}C-urea along with the ^{14}C-carbonate, estimating the fractional synthesis rate of urea from the slope of the plasma ^{13}C-urea exponential assuming the patient or animal to be in nitrogen equilibrium during the few hours of the measurement. In practice satisfactory single ^{13}C-urea exponentials are almost invariably obtained during this interval. Wochner *et al.* (1968), on the other hand, argued that since endogenous catabolism cannot involve discrimination between labeled and unlabeled urea, it is permissible to use the specific activity of urinary rather than plasma urea in the formula for calculating protein fractional synthesis rates. Since there is a difficulty in relating the properties of urea in the urine with urea synthesized at a specified time, it is necessary, using their method, to integrate the time course of urinary urea specific activities to infinite time, which in practice means an interval of 24–48 hr following the ^{14}C-carbonate injection. Nadkarni *et al.* (1968) compared these alternative procedures using a variety of patients and obtained essentially the same result in each patient which is in broad agreement with our own experience as illustrated in Figs. 1 and 2.

B. Total ^{14}C Radioactivities Incorporated into Urea and into the Guanidine Carbon of Protein Arginine

Specific activities of the guanidine carbon of arginine in the plasma protein as measured at approximately 300 min should be increased by some 20% to allow for new ^{14}C-albumin and ^{14}C-fibrinogen that have been lost from

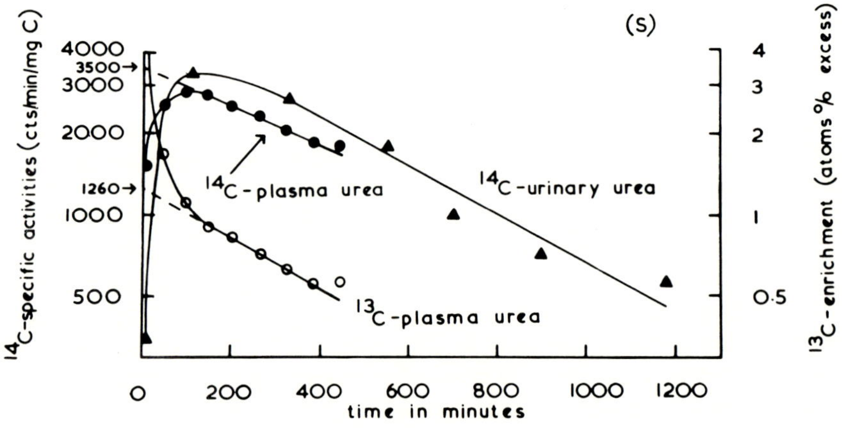

FIG. 1. Urea activities in the plasma and urine at different times after injecting ^{14}C-carbonate and ^{13}C-urea into subject S who had received a transplanted liver 5 months earlier (McFarlane *et al.*, unpublished observations).

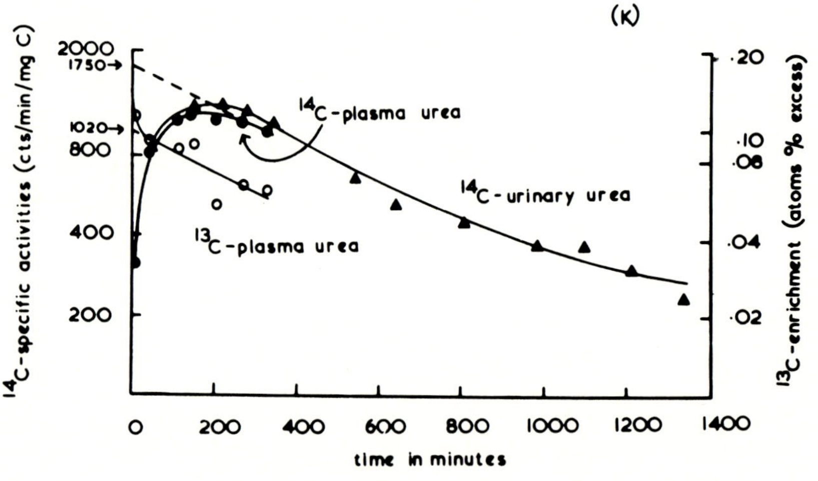

FIG. 2. Urea activities in the plasma and urine at different times after injecting ^{14}C-carbonate and ^{13}C-urea into subject K who had received a transplanted liver 5 months earlier (McFarlane *et al.*, unpublished observations).

the plasma compartment in this interval. Since a good deal of work is involved in measuring these losses accurately, *ad hoc* increases of this order were used until Koj and McFarlane (1968) proposed a procedure for making individual assessments based on losses of the corresponding iodine-labeled proteins injected along with the ^{14}C-carbonate. These authors provided equations summarizing the results of a correlation of iodine with ^{14}C-protein losses in a population of rabbits and more recently similar data shown in Fig. 3 have been obtained in collaboration with E. A. Jones for 22 patients at the Royal Free Hospital, London.

The mathematical problem may be formulated in general terms as follows: If the quantity of a substance (e.g., ^{14}C-albumin) in a central compartment (such as the plasma) is known at successive intervals of time and also the concentration–time curve of a unit pulse of the same substance (e.g., ^{131}I-albumin), then it is possible to calculate the rate of input of the substance by a procedure which includes *deconvolution analysis*. This applies both to ^{14}C-urea arriving in the body water and to ^{14}C proteins arriving in the plasma following an injection of ^{14}C-carbonate; a first step is to resolve the experimental concentration curves by graphical analysis or by computer fitting. Whereas graphical analysis was used by Koj and McFarlane, the new human data was computer analyzed by an adaptive (hill-climbing) technique which minimizes the root mean square error (Rosenbrock and Storey, 1966). ^{13}C-Urea curves were fitted to three exponentials as described by Jones *et al.* (1968), an interesting and unexplained feature of their results being the finding that the two urea pools were of almost equal size. Iodine-labeled albumin and fibrinogen curves were fitted to three exponentials and

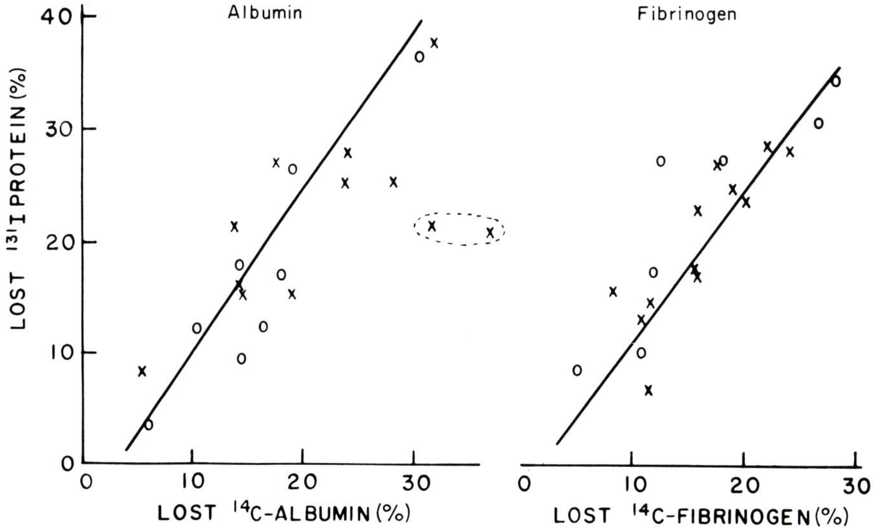

FIG. 3. Calculated cumulative losses of ^{14}C-labeled proteins from the plasma 5 hr after injecting ^{14}C-carbonate into patients who also received the corresponding iodine-labeled protein. For details of the analytical methods used and equations representing the regression of ^{14}C-protein losses in terms of ^{131}I-protein losses see text. The percent ^{131}I protein lost from the plasma is plotted against the percent of ^{14}C-labeled protein lost from the plasma.

all ^{14}C curves to four exponentials. Each pair of curves was then subjected to numerical inversion of the convolution integral relating them and integration of the derived function to give the curve of presumptive labeled-protein (or labeled-urea) concentration, assuming that no losses had occurred from the plasma or the urea initial mixing compartments, respectively. Essentially the same procedures have been recommended by Donato *et al.* (1967) and Wraight (1969) for analyzing catabolic data obtained after injecting iodine-labeled proteins.

Protein loss in 300 min, defined as the proportion of new labeled protein which has left the plasma compartment in this interval expressed as a percentage of the labeled protein found in this compartment at the end of the interval, is given by the expression: ^{14}C-albumin loss $= 3.172 + 0.696 \times$ ^{131}I-albumin loss, and for ^{14}C-fibrinogen $= 1.878 + 0.736 \times$ ^{131}I-fibrinogen loss where the ^{131}I-protein losses are expressed as percentages of the concentration in the plasma a few minutes after injecting the ^{131}I protein. These equations are valid for normals and for patients suffering from a wide variety of conditions including liver disease. Using these equations, measured ^{14}C-albumin and ^{14}C-fibrinogen specific activities at 300 min had to be increased by 22.4 and 19.0%, respectively, in the experiment illustrated in Fig. 1 and by 23.8 and 19.2%, respectively, in the experiment shown in Fig. 2.

C. Complicating Effects of Isotopic Reutilization

Protein synthesis rates in our two studies, which incidentally were carried out on patients 5 months after they had received transplanted livers, were not significantly different whether based on plasma ^{13}C-urea slopes or integrated urinary ^{14}C-specific activities; this is in agreement with Nadkarni *et al.*'s finding, although the slopes they used were those of endogenously produced ^{14}C-urea. Two potential sources of error can be detected in the results, both probably arising from reutilization of the label and more clearly discernible in Fig. 2. Plasma ^{14}C-urea specific activities in this figure show no maximum in the first 2–3 hr such as is visible in the more normal situation in Fig. 1, although a well-defined single exponential slope is visible at an early stage in the curve of ^{13}C-urea enrichment. A marked "tailing" is also evident by 24 hr on the semilog plot of urinary urea specific activities which complicates the evaluation of the integral of this function to infinity as required by Wochner *et al.*'s method of analysis.

Jones *et al.* (1968) measured declining plasma $^{14}CO_2$ activities in man after injecting ^{14}C-carbonate, and considered their results to show that whereas plasma $^{14}CO_2$ continues to be used in the production of ^{14}C-urea for some hours after the initial pulse of so-called "flash-labeling," later

$^{14}CO_2$ levels are too low to account satisfactorily for the persistently shorter half-life that was observed of injected ^{13}C-urea compared with endogenous ^{14}C-urea. It seems highly likely that this difference in half-life is closely linked with both phenomena in Fig. 2 to which attention has been directed.

When synthesis of labeled urea happens to be unduly prolonged, the reason why this occurs from time to time not being fully understood, the plasma urea method, requiring as it does extrapolation of the ^{14}C-urea exponential to zero time to obtain the total ^{14}C-urea activity, can only be applied by relying on the ^{13}C-urea slope. This could be regarded as an ideal situation for applying the urinary urea procedure. In doing so, however, a fundamental difficulty in measuring protein synthesis rates at a particular time is merely being circumvented by substituting mean values measured over 24 hr or more. Such a substitution would not be permissible in many circumstances, especially where an acute unbalance is being studied. It is precisely in dynamically changing situations of this kind that synthesis rates derived indirectly from catabolic measurements with iodine-labeled proteins can be seriously misleading.

D. Advantages of Using ^{13}C-Urea

In view of the above, the use of ^{13}C-urea appears to be essential when short-term measurements of synthesis rate are required and is particularly valuable in situations in which allowance has to be made for the effects of unusually prolonged synthesis (especially of urea) or of marked reutilization of the label. Unfortunately, in planning experiments of this kind it is not always possible to foresee when such effects will arise.

III. Measurement of Rates of Catabolism of Plasma Proteins

These rates are now almost exclusively based on the use of proteins labeled with iodine by methods which continue to be improved upon (Rosa *et al.*, 1965; Katz and Bonorris, 1968). Catabolic rates obtained in this way represent mean rates measured over several days; the proposal by Campbell *et al.* (1956) for measuring an almost instantaneous rate at the so-called "equilibrium time," has not found any wide application. This limitation constitutes a particular difficulty when attempts are made, as they have been in several laboratories, to demonstrate close agreement between synthetic and catabolic rates in adult subjects or animals in nutritional balance. The order of agreement observed (within $\pm 10\%$) is probably as good as can reasonably be expected in view of the different intervals to which the two rates refer.

A. Renal Clearance of Iodide

In collaboration with Dr. A. Koj, the author has recently investigated methods of shortening the interval required for catabolic studies with a view to making it approach that required for synthetic ones. In measuring catabolic rates, the requirement that plasma and urinary activities must be available over several days to calculate fractional catabolic rates arises from the erratic renal clearance of iodide from the body water. The difficulty can be circumvented if fluctuations in the pool of iodide activities in the body water are also measured. To assess the latter requires the injection of labeled iodide along with the injected protein using a different isotope of iodine. This is essentially as proposed by Donato *et al.* (1967) and Vitek *et al.* (1967) for catabolic measurements in patients suffering from renal insufficiency.

The technique proved not to be particularly complicated, and numerous measurements have been carried out in patients using as little as 30–40 μCi ^{131}I protein and 10–15 μCi ^{125}I-iodide. The result, both in patients and experimental animals, was quite unexpected: Namely, the total iodide activity released in the first few hours of catabolism is much less than would be anticipated from the prevailing level of the iodoprotein in the plasma and the assumption that a constant fraction of the plasma protein is catabolized in unit time (Berson and Yalow, 1954).

B. Intravascular Site of Catabolism

This result contrasts with an earlier finding in rabbits (McFarlane, 1962) that a constant fraction of the protein-bound iodine in the plasma appeared daily in the form of iodide from the first day, provided the injected labeled albumin had been biologically screened in another rabbit. The new experiments are more exact, since alternatively labeled iodide was injected to measure parameters of iodide distribution and excretion, but a major part of the discrepancy is probably due to the fact that biological screening for 1 or 2 days was inadequate to remove all traces of a more rapidly catabolized component. Labeling techniques now available are such that labeled proteins are being produced that contain sensibly no denatured material, whereas earlier preparations have been shown to contain significant amounts even after screening. The new technique of testing labeled proteins is described in Section III, C.

Iodine-labeled protein solutions to be used for measuring catabolic rates within a few hours of injection must contain very low proportions of non-protein-bound isotope. These should be below 0.2% and lower values have been regularly achieved with freshly iodinated albumins and fibrinogens.

The proportion increases during storage, even of sterile solutions in the cold (McFarlane, 1967), and the original low value of the stored protein must be reestablished by exchange resin treatment immediately before injection.

Many explanations have been considered for the gradual build-up to a constant fractional catabolic rate in the 24–36 hr after injection of an undenatured protein but none that has so far wholly convinced Dr. Koj or myself. Most of them involve a departure from the now almost universal assumption that plasma proteins are catabolized at an intravascular site. This has usually been qualified with the words "catabolized in, or in close proximity to, the circulating plasma." Reeve and Roberts (1959), in particular, have illustrated the situation diagrammatically with a catabolic compartment which is in close dynamic exchange with the circulating plasma. As more short-term data on catabolism materializes it should be possible to give this compartment some badly-needed definition.

C. Detection of Small Proportions of Polymerized Proteins

The slow rates of release of iodide in the first few hours from catabolism of undenatured proteins makes the situation favorable for detecting small proportions of more rapidly catabolized contaminants. This is illustrated in Fig. 4, in which the accumulation of iodides in the body water is shown in an experiment using a mixture of iodine-labeled human fibrinogen and albumin. A marked "bulge" is evident on the curve of iodide from the fibrinogen (observation by McFarlane and Regoeczi, as reported by McFarlane, 1969). The effect of screening fibrinogen is to remove this bulge; a simpler method is to precipitate the "low-solubility" or "cryo"-polymer

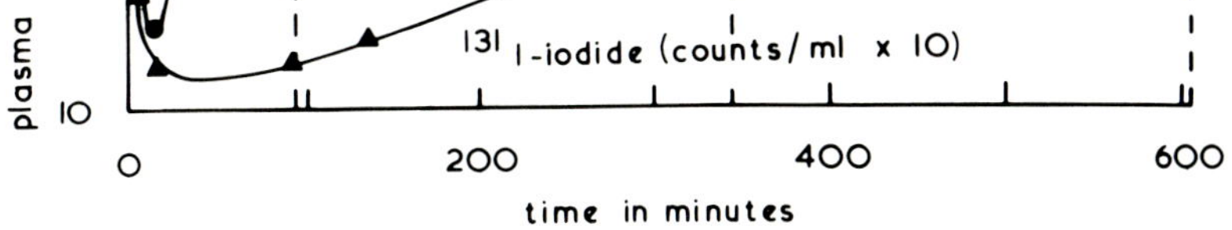

FIG. 4. Iodide activities per ml of plasma at different times after injecting ^{131}I-albumin (▲—▲) and ^{125}I-fibrinogen (●—●) into a patient (McFarlane, 1969).

preferably beforehand with ammonium sulphate at 18% saturation. Human low-viscosity fibrinogen which is precipitated at 18–25% saturation shows no similar bulge after labeling and injection. Unfractionated rabbit fibrinogens generally contain less of the objectionable polymer but still require to be fractionated or screened. The rabbit polymer has been separated and shown after labeling to be catabolized much more rapidly than the monomer.

Not all forms of denatured protein are necessarily dealt with by the body in such a fashion, but the above findings show the presence of even 1% of a more rapidly catabolized component can be recognized in a labeled protein if attention is paid to iodide activities in the body water in the first few hours. In effect, the recipient "screens" the protein, thus providing two measurements of catabolic rate: one before and one after removal of the contaminant. To measure the proportion of the contaminant accurately requires that all the iodide liberated by its catabolism should have been delivered into the body water before the measurement of the second catabolic rate begins and this is not always possible. Indeed, the cryofibrinogen illustration is a particularly favorable one in this respect presumably because of the large particle size of the contaminant.

D. Short-Term Measurements of Catabolic Rate

In the first few hours of labeled protein catabolism, most of the liberated iodide accumulates in the body water; hence accurate estimates of total iodide activities in the plasma or body water are mandatory in relation to catabolic rate measurements. Later, as the body water pool of iodide reaches saturation, it becomes important to be able to rely on constancy of iodide slopes or excretion rates. These appear to fluctuate especially with changes in the pool size of cold iodide in the body water. For this reason fractional catabolic rates prevailing over a short interval are probably more accurately measured in the first few hours than say at 24 hr in spite of giving results which in terms of current thinking are very difficult to interpret. This suggests that a fuller understanding of what is happening both to the protein and to the iodine isotope liberated by its catabolism in the first few hours after injecting the protein is urgently required and could indeed fundamentally alter our whole outlook on plasma protein catabolism. The fact that organic intermediates should be considered in relation to the liberation of iodide from iodotyrosines has not been overlooked. It appears, however, that the widespread distribution of tissue deiodinases effectively precludes the possibility that the iodide which is missing in the first few hours escaped detection by being stored in organic form.

REFERENCES

Berson, S. A., and Yalow, R. S. (1954). *J. Clin. Invest.* **33**, 377.

Campbell, R. M., Cuthbertson, D. P., Matthews, C. M. E., and McFarlane, A. S. (1956). *Int. J. Appl. Radiat. Isotop.* **1**, 66.

Donato, L., Vitek, F., Bianchi, R., and Federighi, G. (1967). *J. Nucl. Biol. Med.* **1**, 11.

Jones, E. A., Craigie, A., Tavill, A. S., Simon, W., and Rosender, V. M. (1968). *Clin. Sci.* **35**, 553.

Katz, J., and Bonorris, G. (1968). *J. Lab. Clin. Med.* **72**, 966.

Koj, A., and McFarlane, A. S. (1968). *Biochem. J.* **108**, 137.

McFarlane, A. S. (1962). *Lancet* **i**, 131.

McFarlane, A. S. (1963a). *Biochem. J.* **87**, 22P.

McFarlane, A. S. (1963b). *Biochem. J.* **89**, 277.

McFarlane, A. S. (1967). *In* "Physiology and Pathophysiology of Plasma Protein Metabolism," 4th Int. Symp. (G. Birke, R. Norberg, and L. O. Plantin, eds.), p. 87. Macmillan (Pergamon), New York.

McFarlane, A. S. (1969). *Proc. Roy. Soc. Med.* **62**, 1127.

McFarlane, A. S., Irons, L., Koj, A., and Regoeczi, E. (1965). *Biochem. J.* **95**, 536.

Nadkarni, G. D., Singh, B., and Jeejeebhoy, K. N. (1968). *Int. J. Appl. Radiat. Isotop.* **19**, 685.

Reeve, E., and Roberts, J. E. (1959). *J. Gen. Physiol.* **43**, 445.

Reeve, E. B., Pearson, J. R., and Martz, O. C. (1963). *Science* **139**, 914.

Rosa, U., Scassellati, G., Pennisi, F., Ambrosini, C., Liberatori, J., Federighi, G., Donato, L., and Bianchi, R. (1965). *In* "Radioaktive Isotope in Klinik und Forschung," Vol. VI, p. 17. Urban & Schwarzenberg, Munich.

Rosenbrock, H. H., and Storey, C. (1966). *In* "Computational Techniques for Chemical Engineers," Ch. 4, pp. 48–97. Pergamon Press, Oxford, England.

Vitek, F., Bianchi, R., and Federighi, G. (1967). *J. Nucl. Biol. Med.* **11**, 139.

Walser, M., and Bodenlos, L. J. (1959). *J. Clin. Invest.* **38**, 1617.

Wochner, R. D., Weissman, S. M., Waldmann, T. A., Houston, D., and Berlin, N. I. (1968). *J. Clin. Invest.* **47**, 971.

Wraight, E. P. (1969). *Phys. Med. Biol.* **14**, 463.

5

TECHNIQUES FOR PROTEIN SEPARATION

Tristram Freeman

M.R.C. CLINICAL RESEARCH CENTRE LABORATORY,
WHITTINGTON HOSPITAL,
LONDON N.19, ENGLAND

I. Introduction

It has been realized for over 50 years that there is a multiplicity of serum proteins, and from the outset it has been appreciated that each protein has its own rate of metabolism and controlling mechanisms. The measurement of concentration of discrete plasma proteins and their isolation has therefore been an integral part of all studies on metabolism, since any conclusions about metabolism are dependent on these two; and since in my opinion they constitute the largest source of error in such studies, it seems worth considering them in a little more detail.

The errors introduced into metabolic studies by quantitation are usually slight, and the development of immunological techniques permits a reasonably accurate measurement of many proteins in small quantities of serum. Whatever method is used for measurement the problems of expressing results in absolute terms are very real. Because completely pure proteins are difficult to obtain, the standards used often contain contaminents or lack the antigenic determinants present in native protein. Thus the values obtained may be erroneously high or low: This is probably of little significance within a laboratory, but may lead to discrepancies when comparing results with other workers. To overcome this problem, a scheme has been initiated by the International Federation of Clinical Chemistry to produce reference preparations for use in laboratories throughout the world; thus there is some hope that in the not too distant future inconsistencies of this nature will be reduced.

In relation to metabolic studies using radioisotopes there are two separate requirements for fractionation, depending on the type of study. In experiments where a labeled precursor is supplied and many proteins become labeled, the requirements are different from those where a single protein is labeled *in vitro*. In both cases a pure protein is required, but for *in vivo* experiments it is necessary to be able to handle multiple samples and to obtain a good yield; it is not necessary to preserve the integrity of the protein. When labeling *in vitro*, a small quantity of pure protein can be spread over many experiments so that the yield is less important but it is vital that the resultant protein is essentially the same as native protein.

II. Requirements

The isolation of undenatured protein has some special problems. These lie in the word "undenatured": It is difficult to define what is meant in physical terms by a denatured protein, and since there are no single tests available for the absence of denaturation, it is difficult to be certain that any protein which is prepared is free of some alteration. The methods available for detecting alterations in plasma proteins will be discussed later.

Unfortunately, it is inevitable that our knowledge of the techniques suitable for preparation of undenatured proteins will be highly colored by the few proteins that have been studied. In my experience there are four methods that, in general, do not seem to damage proteins, these are: electrophoresis; ammonium sulphate precipitation; gel filtration; and DEAE chromatography. In my view preparation methods employing organic solvents, whether these be alcohol, acetone, or ether, tend to cause fairly gross alterations in the protein molecule. I have on one occasion worked with an ether precipitated albumin which appeared to be undenatured, but I have not yet handled a protein prepared by Cohn fractionation which did not show a considerable degree of alteration. Other precipitation techniques using specific ion interactions with, for instance, zinc, aluminum, or polyphosphates have, in my experience, not yielded satisfactory results. Similarly the Rivanol precipitation system which is now fairly extensively used, can be used satisfactorily for some proteins, but not for others as will be described.

Each of the four techniques that appear not to cause denaturation (electrophoresis, salt precipitation, gel filtration, DEAE chromatography) have advantages and disadvantages. Except for a few special instances it is not possible to produce a pure protein by any one of these techniques alone. It is necessary to use a combination of two or more in order to obtain pure material. Nor is it necessarily possible to get each serum protein isolated using these techniques alone; for some proteins it may be necessary to use other methods.

The preparation of each protein is a problem in itself, and has to be considered as such. The choice of which technique to use first very often depends on the quantity of serum that can be handled by each method, for example, the resolution obtained by electrophoresis, and in particular electrophoresis combined with molecular sieving, is very much greater than can be obtained by salt precipitation; however, it is difficult to apply more than 1 g of protein to an electrophoretic system, but salt precipitation methods can easily handle liters of whole serum. For many of the well characterized proteins several techniques exist in the literature for isolation, and it is obviously preferable to use one of these recognized techniques or a modification of them, than to start from scratch designing an isolation method. [The reader is referred to Schultz and Heremans (1966) for an excellent account of protein fractionation methods in general.] However, there are many serum proteins, perhaps over 40, that have not yet been isolated with a view to metabolic studies. With these particular unknown quantities in mind the description of fractionation techniques have been written.

III. Quantitative Analytical Technique

Immunoelectrophoretic Technique

The first essential in isolating any plasma protein is to be able to measure the quantity of that protein in each of the steps in the fractionation. Furthermore it is desirable to be able to measure the quantity of contaminating proteins in each successive fractionation step. The best technique for doing this is the quantitative immunoelectrophoretic technique originally devised by Laurell (1965) and modified by Clarke and Freeman (1966, 1968). In brief, this method consists first of a simple electrophoresis in agarose. When this is complete a second electrophoresis is performed at right angles to the first, so that the separated antigens are moved into a bed of antiwhole human serum. The human proteins migrate until optimum proportions are reached, thus it is important that this second dimension electrophoresis be performed for an excess time (24 hr at 2.5 V/cm). When performed in this manner the area under each peak is directly proportional to the concentration of human antigens and inversely proportional to the concentration of antibody to that protein in the antiwhole human serum used. Thus, to achieve quantitative results, the same antiserum must be used for a series of experiments, and some standard must be used. When performing fractionation of human serum it has been found convenient to use the original whole serum as standard, so that the areas found in fractions are proportional to the yield of each protein in each fraction. This technique has been described in detail (Clarke and Freeman, 1968), and suitable apparatus is available commercially (tanks and power packs, M.B.I., Wotton Road, Kingsnorth Industrial Estate, Ashford, Kent; antiwhole human serum, Burroughs Wellcome & Co., The Wellcome Building, 183 Euston Road, London N.W.1; electronic planimeter, Chemical Electronics Co., C.W.S. Hall, Durham Road, Birtley, Co. Durham).

An example of the use of this analytical technique is shown in Figs. 1–3; the initial serum is shown in Fig. 1, and the results of ammonium sulphate precipitation is shown in Fig. 2. Fraction A shown in Fig. 2 was taken and submitted to a preparative electrophoresis in polyacrylamide using the technique described by Brownstone (1969). The resultant material is shown in Fig. 3; two principal peaks were obtained from the polyacrylamide, the first and largest one is shown in Fig. 3A; the region between the peaks is shown (B) and the second peak is shown in Fig. 3C. From these five analytical runs a very considerable amount of information can be obtained about the isolation procedure. The analytical technique has a particular advantage in that the sensitivity of an immunological reaction is used, but does not require a multiplicity of specific antibodies. In other words, it is easy to make or buy antiserum against whole serum; it is not possible to buy, and it is extremely difficult to make antisera against all the proteins found in serum.

IV. Salting-out Techniques

The history of protein fractionation by addition of neutral salts is now nearly 100 years old, and during that time a large number of different neutral salts have been tried and an equally large number of techniques and modifica-

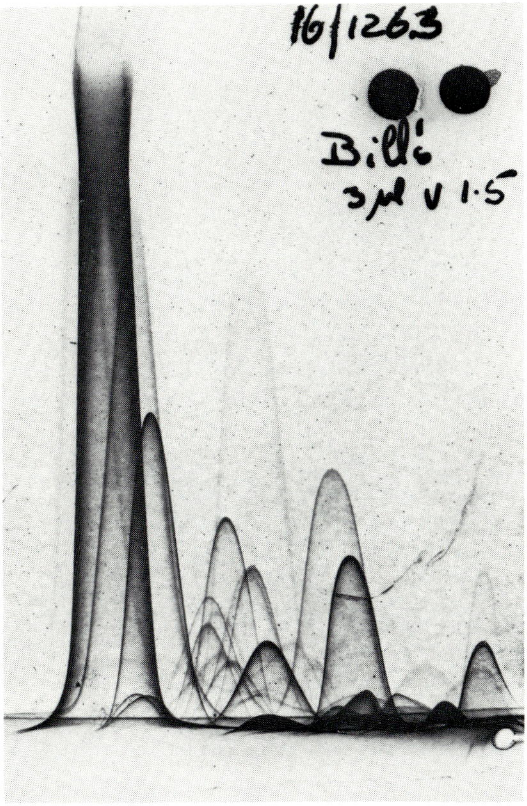

FIG. 1. The whole serum used in a preparation of human albumin; separated by the Laurell quantitative immunological technique.

tions have been described. It is now clear that no matter what modifications are introduced the technique has a poor resolution compared to the other techniques that will be described. Furthermore, the precipitation characteristics of any one protein can be altered by the presence of other protein or proteins. Nevertheless this technique is inexpensive and is applicable to very large quantities of serum, thus it forms a useful initial step by which some of the contaminating proteins can be removed. It is, perhaps, somewhat surprising that a protein that has been rendered insoluble by the addition of neutral salts should remain undenatured, but no evidence has, to my knowledge, been found of denaturation directly attributable to salt precipitation, with the exception of the alteration caused to the lipoproteins. A large number of papers are available on this technique, and good recent results are

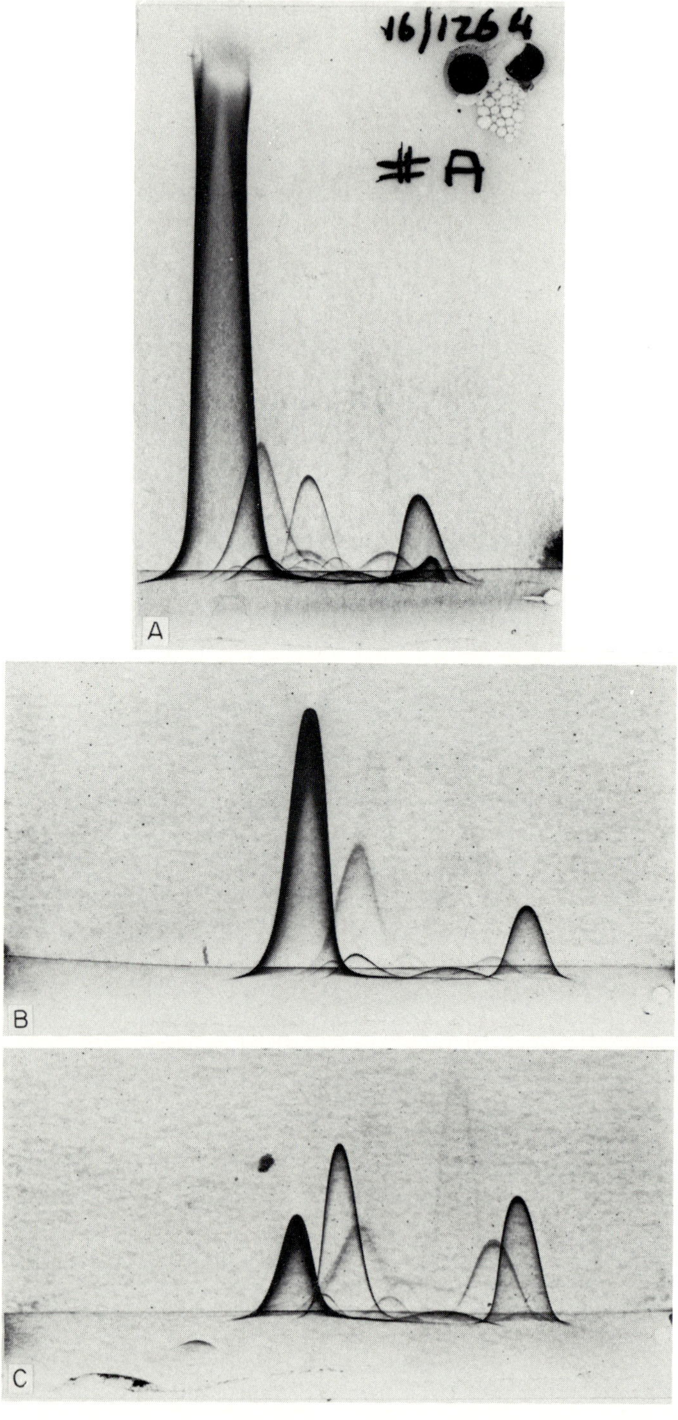

FIG. 2. Three fractions of the serum shown in Fig. 1 obtained by ammonium sulphate precipitation.

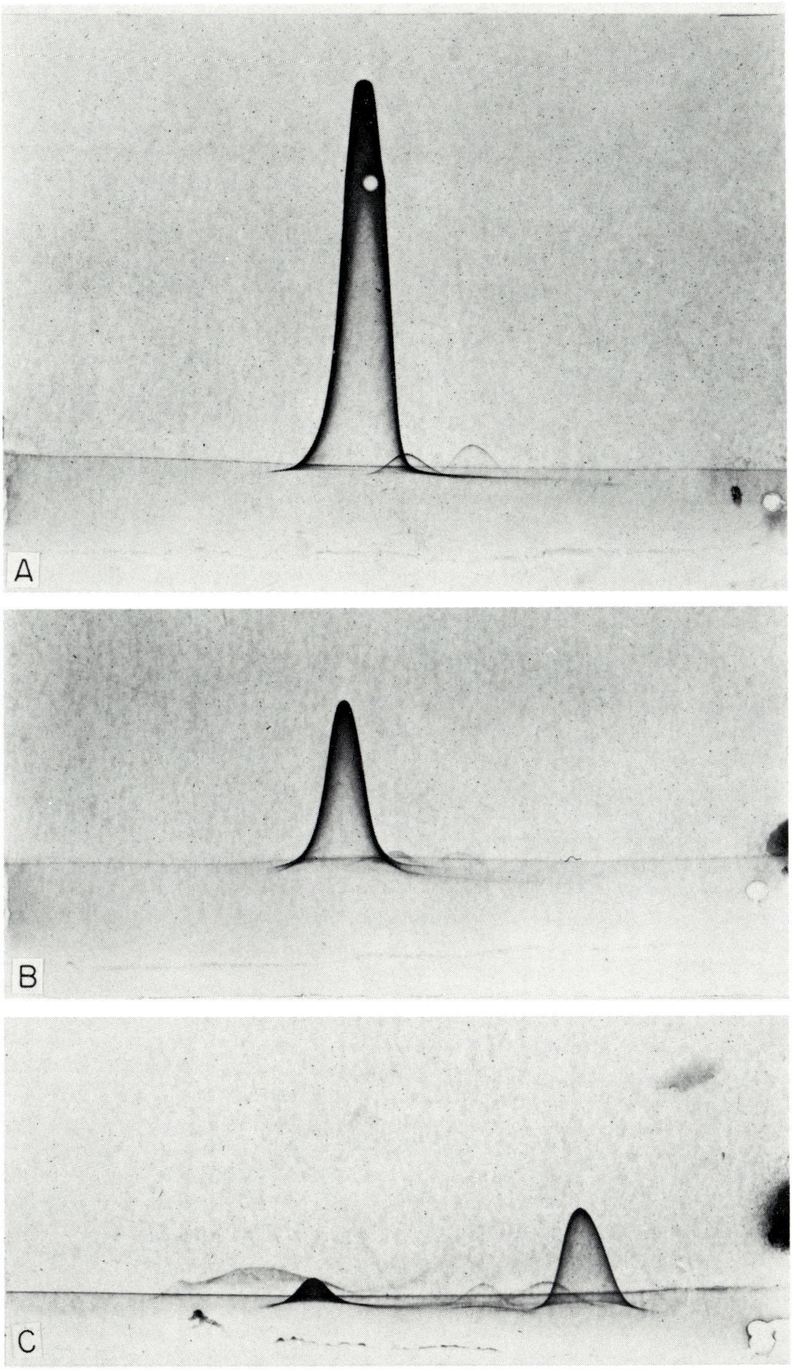

FIG. 3. Three fractions obtained from the whole serum using a preparative polyacrylamide technique (Brownstone, 1969).

given by Fuller (1953), Keller and Block (1960), Pennell (1960), Pedersen (1945), and Mosse (1959).

The most commonly used salt for this procedure is ammonium sulphate, because it has a high solubility at room temperature and its solubility is only slightly changed by alterations in temperature. Nevertheless the old habit of expressing concentrations of ammonium sulphate in terms of percent saturation is probably less good than defining the temperature and molar concentration. It is also important to define the pH at which precipitation is performed; in general, this should be below neutral because proteins are less soluble near their isoelectric point and because above pH 7.0, free ammonia can be given off which certainly causes protein denaturation. The resolution of this technique can be improved (1) by multiple precipitations; the precipitate at a given concentration is taken into solution by the addition of water, and reprecipitated by the addition of further ammonium sulphate. This has been found to be more effective than washing the precipitate in the appropriate ammonium sulphate solution. (2) Greater resolution can be obtained by altering the ammonium sulphate concentration slowly. This is best done by dialysis against the appropriate solution. Although this does result in a more effective fractionation, the increase in trouble and time taken is such that it is now seldom used. Table I gives the quantity of solid ammonium sulphate to be *added* to 1 litre to give the required concentration.

TABLE I
WEIGHT OF AMMONIUM SULPHATE (MW 132.15) TO *ADD* TO 1 LITER[a]

Required molarity	1.0	1.2	1.4	1.6	1.8	2.0	2.2	2.4	2.6	2.8	3.0	3.2	3.6	4.0
Required % saturation (20°C)	25	30	35	40	45	50	55	60	65	70	75	80	90	100
g Ammonium sulphate to *add* to 1 liter	144	176	208	243	277	312	351	390	430	474	516	560	657	760

[a] For % saturation: $x = 533\,(S_2 - S_1)/100 - 0.3 \times S_2$; for molarity: $x = \dfrac{533\,(S_2 - S_1)}{4 - 0.3 \times S_2}$, where S_2 = desired concentration, S_1 = present concentration, and x = weight to add to 1 liter.

V. Fractionation by Gel Filtration

The separation of proteins according to their molecular size is now a well established technique, and has the advantage that it can be scaled-up to handle something in the order of 10–20 g of protein. It is also useful as a method of removing the salt or small molecules from protein solution. This technique has been extensively reviewed by Porath (1960), Tiselius, Porath, and Albertsson (1963), and Laurent and Killander (1964), and is also well described by the manufacturers of the supporting medium.

The term "gel filtration," although commonly used, is really a misnomer; the supporting media, whether they be crossed-linked dextran, agarose, or polyacrylamide beads, are so constructed that space within the bead is available to the smaller molecules but not the larger ones; thus the total space available to a small molecule in a given packed column is much larger than the space available to a large molecule. Thus when a solution containing large and small molecules is passed through the column the larger molecules will pass through before the small molecules because of the decreased available space. Thus the term "exclusion chromatography" is probably a rather better one. It will be seen that although the fractionation is largely by molecular size, it is influenced by molecular shape.

The most commonly used commercially available gels are those produced by Pharmacia (Great Britain) Ltd., Sinclair House, The Avenue, London W.13, under the name of Sephadex. Gels are available ranging from G200, which has an exclusion limit of a molecular weight round about 200,000 and provides good separation of proteins from this molecular size downward, to G25 which is used principally for changing the suspending medium, since all protein molecules will be excluded from the matrix of this size gel, but the salt and water can exchange freely. Thus by passage of a protein solution through a G25 column one can effectively change the solution in which the protein is suspended.

Whatever the nature of the supporting medium, the techniques for using them remain very much the same. There are two main problems in this type of separation. The first is to get a reasonable rate of flow through the column, and second is to get the proteins to pass through the column as a uniform band; in other words, when a colored macromolecule solution is passed through the column, this should be seen to pass through as a uniform horizontal band and not to run down one side of the column.

Good flow rates are dependent on several factors. (1) Beads of uniform size should be used to form the column; this can be achieved by passing the dry beads through a standard sieve of varying sizes, and selecting those of uniform size. (2) It is essential that prior to packing a column the beads should be fully swollen. This can be done by either leaving the beads in buffer solution for about 5 days prior to packing the column, or allowing them to stand in 5 *M* urea for at least 24 hr prior to packing. Of these two, it is my own opinion that the 5-day treatment in buffer solution is the most reliable. (3) Careful packing of the column is well worthwhile, since with care a single column will perform many separations. In our experience, the most satisfactory technique is to have at the bottom of each column a stainless steel mesh, on top of which is placed a layer about 5 mm thick of Ballantini beads. These are allowed to settle through buffer onto the stainless steel mesh, when done in this way the larger glass beads settle first followed by a finer layer on top. Above this layer of glass beads a similar 5 mm layer of G25 is placed, this is also allowed to settle through buffer. Finally a slurry of the appropriate fully swollen Sephadex is poured into the column which is one-third full of buffer solution. This is allowed to settle by gravity until a definite and firm top layer can be seen. The fluid above the top layer of the Sephadex is then removed and a further quantity of slurry poured onto the column. When this has also settled and formed a top layer, the flow through the column can be started, preferably using some form of pump. The flow rates should never be more than 1 ml/cm^2/hr; this should be continued and further Sephadex poured onto the top of the column until packing is complete. It is advantageous, at this stage, to run through a colored high molecular weight material, such as albumin marked with bromphenol blue and/or a blue high molecular weight dextran (supplied by Pharmacia Ltd., Uppsala, Sweden). This allows one to see that the column is running true and not down one side.

In order to obtain protein running through as a band and not faster along one side of the column than the other, it is vitally important that the column should be absolutely vertical from the moment of packing the column—this is best done by putting a plumbline alongside the column and leaving it there until the column is firmly established. If the column should be moved at any time, it is again essential to make certain that it is absolutely vertical. In general, it is advisable to pack the column at the temperature at which it will normally run, either at room temperature or in the cold room. It is sometimes possible to move a column successfully from room temperature to cold room temperature, but it is seldom possible to do the reverse because the flow rate decreases rapidly. Finally the column is connected to a buffer reservoir which should not be more than about 10 cm above the top of the column. Although various authors (Flodin and Killander, 1962) have successfully used a higher hydrostatic head, it is our experience that this merely causes over-packing of the column and reduced flow rates.

Once a column is well established and known to be functioning properly, it can be maintained in that condition for a number of months, doing repeated fractionations on the same column. It is, however, necessary under these circumstances to prevent bacterial growth. If the column is to be left for any appreciable time, either at room temperature or in the cold room, it should be emptied of all protein solution, and a bacterious static or bactercidal solution should be run onto the column. This can successfully be eluted from the column when it is required for use again. Sodium azide has been used with success in this laboratory for a number of years.

Protein solutions are applied at the top of the column without disturbing the surface of the Sephadex. This is most easily done by placing a small filter paper disc, slightly smaller than the inner diameter of the column on top of the Sephadex; the protein solution is then applied with a Pasteur pipette or by a pump just above the filter paper disc. To minimize mixing, salt or sugar can be added (10%) to increase the specific gravity of the solution to be applied. Columns of this nature can be run with a variety of buffer solutions; in general, it is useful to keep the salt concentration high to limit the extent of protein–protein interaction. We have used for many years a 1 *M* salt sodium chloride solution with 0.1 *M* phosphate buffer pH 7, and sodium azide 1 part in a 1000 to prevent bacterial growth. This concentration of azide should be increased if the column is not used for a period.

There are a variety of modifications of this technique that sometimes can be usefully applied. The two principal ones are, first, reverse flow gel filtration; in other words, the buffer solution is pumped from the bottom of the column, the protein solution applied to the bottom, and the fractions selected from the top of the column. The advantage of this technique is that high flow rates can be maintained for a very long period. In our experience good flow rates can be maintained with the normal downward flow, provided that at no time is the column allowed to flow at more than 1 ml/cm^2/hr. The second principal modification is recycling chromatography (Porath and Bennich, 1962); the resolution obtainable on these columns can be improved either by increasing the length of the column, or by recycling that portion of the column in which one is interested. Because columns over 2 meters in length are difficult to operate and difficult to maintain at good flow rates, the recycling procedure is advisable; this can be done with either the upward flow system or the easier downward flow system.

VI. Fractionation by Absorption Chromatography

The absorption of proteins onto solid material is a technique that has long been recognized and used in the fractionation of proteins. However, until the introduction of true iron exchange chromatography by Sober *et al.* (1956), the absorption methods employed gave a poor yield and frequently resulted in gross denaturation of protein. A variety of substituted celluloses are available for use, of which only diethylaminoethyl cellulose (DEAE) and carboxymethyl cellulose (CM) have been used extensively in the production of metabolic quality proteins. It is thought that the proteins are bound to the substituted cellulose by electrostatic bonds; thus, the elutions can be obtained either by altering the pH, hence altering the number of charge groups on the protein molecule or by increasing the salt concentration to provide competition for the available charges. Details of this technique and the materials available can be found in the works of Sober and Peterson (1962) and Keller and Block (1960). It is also available from the commercial manufacturers of these celluloses.

Despite the low degree of substitution of the cellulose the weight of protein retained is good; some losses occur and are rather variable from protein to protein, but in our experience the material that is eventually eluted shows no evidence of denaturation. This, however, must be qualified in that each protein is a separate problem, and evidence of denaturation may be found with other proteins using this technique. Chromatography columns can be run using either a continual gradient of pH or salt concentration or a series of different eluting fluids either by changing the pH or increasing salt concentration. The usual procedure is to run a small gradient column to establish the condition under which a particular protein is eluted, and then to scale this up using a batch elution system. Using the latter system it is possible to handle quite large quantities of human serum. Nevertheless it is not possible to obtain any pure protein by this technique alone, save the immunoglobulin IgG.

Preparation of Column

The success of this technique is largely dependent upon the care taken in the preparation of column and sample. Most commercially available adsorbents, whether they be in powder, flock, or microgranular form, require the removal of fine particles before use. This may best be done by repeatedly removing the supernatant fluid after the large particles have settled by gravity in a weak suspension. Most cellulose adsorbents need to be cleaned prior to use or reuse; the washing instructions supplied by the manufacturer should be followed. Before use, both the adsorbent and the protein solution should be equilibrated with the starting buffer. While many buffer systems have been described, the most commonly used is a phosphate buffer starting with 0.05 *M* phosphate pH 8 and the final buffer being 0.3 *M* phosphate, also pH 8. Washing should continue until the pH of the supernatant fluid remains at pH 8 (but not less than three changes of wash solution). Similarly the protein solution must be equilibrated with this buffer. This can be achieved either by extensive dialysis or by passing the protein solution through a Sephadex G25 column that has previously been equilibrated with the appropriate buffer.

Packing of the column is performed in much the same way as a gel filtration column; briefly, a thin slurry of the cellulose is poured on top of a column and a third filled with the appropriate buffer solution. This is allowed to settle under gravity and further slurry is added to complete the packing. Some workers recommend the use of a positive pressure of up to 10 psi to promote efficient flow through the column. It is our experience that by careful packing and exclusion of the finer particles, good flow rates can be maintained without applying a positive pressure. The quantity of cellulose required for each ml of protein is somewhat variable depending on the nature of the protein to be applied and adsorbent used; in general, 1 g (dry weight) of DEAE cellulose is sufficient for 50 mg of protein.

Elution can be carried out either using stepwise increases of molarity of decrease of pH, or a gradient system can be used. A variety of devices are available commercially for producing a gradient. A simple system can be made in the laboratory by connecting two containers with a narrow tube: The column is connected to the first (sealed) container which holds the starting buffer; as this flows onto the column, the fluid is replaced from the second container holding the final buffer solution. By varying the size and shape of these containers, a variety of gradients can be produced quite simply.

VII. Preparative Electrophoresis

The clear separations observed in analytical electrophoresis have led many research workers to try to scale-up the column of serum applied to produce a preparative technique. Unfortunately many additional difficulties are encountered in this process.

A. Electrophoresis in Carrier-Free Systems

The original free boundary electrophoresis introduced by Tiselius is not readily adaptable as a preparative procedure; because there is no complete separation of proteins, the principal methods developed have been zone electrophoresis, the problem being to stabilize the separated protein zones without the addition of carrier agents. Three solutions to this have been found: (1) the use of density gradient columns, usually generated in sucrose [this has been well described by Svensson (1960)]. (2) The second principle, suggested by Tiselius and worked on by Hjerten, involves the use of a horizontal slowly rotating tube; this device effectively prevents the appearance of convection currents due to differences in density. It is, however, a somewhat troublesome piece of apparatus to use routinely. (3) The system is a free flow electrophoresis in a thin film of buffer. A curtain is allowed to pass between two cooled plates with a charge at either side and the protein fed in at one point at the top. Ideally the proteins should migrate in the field in their passage down the plate in a fanlike fashion, giving a quantity of fractionated proteins at the bottom. In practice considerable difficulty has been found with this technique in that it is vitally important to maintain absolutely constant temperature flow rates and current throughout the experiment, or the system becomes unstable. This technique was originally described by Barollier *et al.* (1958) and again by Hannig (1964). There is a commercially available apparatus based on Hannig's design (Elphor Va P1: Bender & Hobein, Munich, W. Germany).

B. Zone Electrophoresis in Supporting Media

Zone electrophoresis has been extensively used and a variety of different media have been tried; these include, glass powder, starch granules, agar or agarose, cellulose, and pevicon, either in slabs or columns. Detailed description of this procedure has been given by Heide *et al.* (1959) and by Hannig (1967). Finally a variety of techniques have been described using the gels which have some molecular sieving effect, such as starch gel and polyacrylamide. The principal difficulty in this latter technique is, despite the increased resolution, that it is more difficult to obtain a protein from the gel; this can be overcome by electrophoresing the proteins from the gel into either liquid or to a more convenient medium such as pevicon. A preparative polyacrylamide technique has recently been described (Brownstone, 1969) which is capable of separating quite large quantities of protein with a good resolution. This technique was used to produce the separations shown in Fig. 3. In general, the preparative electrophoresis is a useful step in preparing pure proteins; however, it alone is not sufficient in that any fraction obtained by this technique contains mixtures of several proteins. Further, the quantity of protein applied is quite limited; therefore, one of its principal functions is as a secondary step to an initial fractionation by some other technique.

VIII. Protein Concentration

Whatever method is used for fractionation, it is usually necessary to concentrate a protein solution during the course of fractionation. A variety of methods have been described; these include freeze-drying dialysis against a high molecular weight polymer such as carbowax or dextran, the addition of a dry Sephadex G25 powder, and ultrafiltration. It is our impression that the least damaging to proteins is ultrafiltration, provided that a system is used as is shown in Fig. 4. This consists of the usual dialysis tubing into the bottom of which is tied a glass container of known volume. A negative pressure not exceeding 500 mm of mercury is arranged outside the dialysis tubing. The purpose of the glass container is twofold: (1) The concentration will

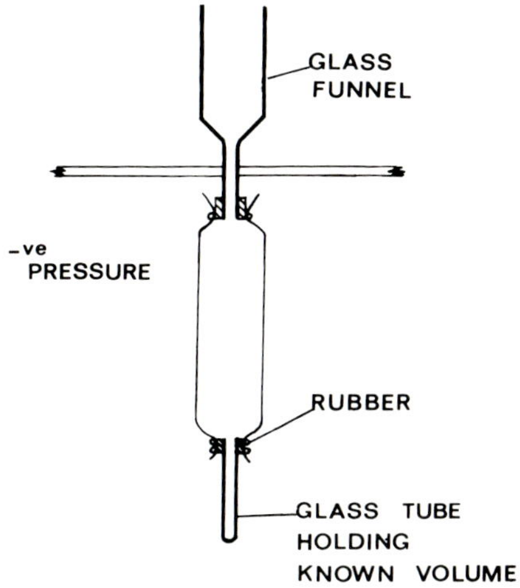

FIG. 4. A diagram of a simple concentrator; designed to avoid over high protein concentrations.

continue until a predetermined volume is reached; this permits ultrafiltration to be carried out overnight. (2) It prevents the very high protein concentration that can usually be observed at the bottom of a simply tied-off dialysis sac. There is some evidence to show that this very high concentration can be deleterious to some plasma proteins. Detailed quantitative analysis of diluted whole serum shows no significant change in relative concentration after this type of procedure. However, the use of radioiodinated proteins showed that there are some losses on the dialysis membrane; these are roughly proportional to the area of membrane. Thus, if only small quantities of protein are to be concentrated, it is important to maintain the area of membrane low.

It is important to reemphasize that each protein is an individual problem, and the methods that have proved satisfactory for one protein may not necessarily give an undenatured product with a different protein. For example, IgG globulin can be prepared in a number of different ways, the most common being the use of diethylaminoethyl (DEAE) cellulose. It can also, however, be prepared using Rivanol precipitation. Since the chromatographically produced protein is a selection of the IgG molecules, experiments were performed using other methods of preparation to see if the "whole immunoglobulin" behaved in the same way as the DEAE fraction normally used. To do this, serum was taken and precipitated with Rivanol so that all proteins other than immunoglobulin and transferrin were rendered insoluble. These were spun off, the Rivanol removed using activated charcoal and the transferrin and gamma separated using the block electrophoresis. Fig. 5

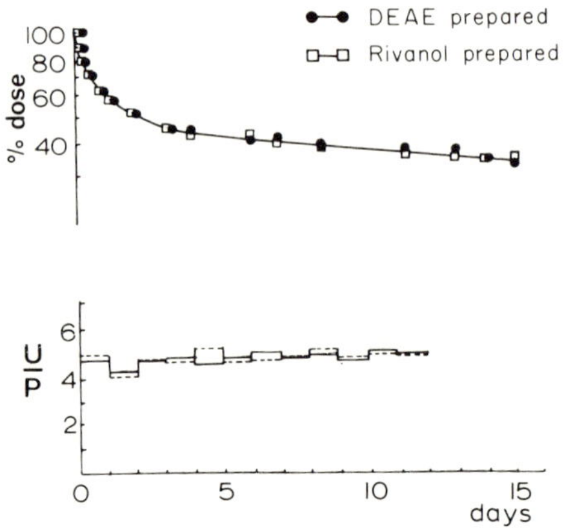

FIG. 5. Plasma decay curves of human IgG prepared using DEAE (^{131}I) and Rivanol (^{125}I) studied in the same subject.

shows the results obtained from such an experiment and you will see there is no significant difference between the results obtained from DEAE gamma and from Rivanol gamma. This, and subsequent experiments of a similar nature, made use of two different iodine isotopes for protein labeling in one subject. However, if the transferrin produced as a byproduct from the above experiment is compared with transferrin produced on DEAE cellulose, a rather different result emerges (Fig. 6). It will be seen that the Rivanol-produced transferrin is degraded at a significantly faster rate than the chromatographically produced transferrin so that, although Rivanol appears

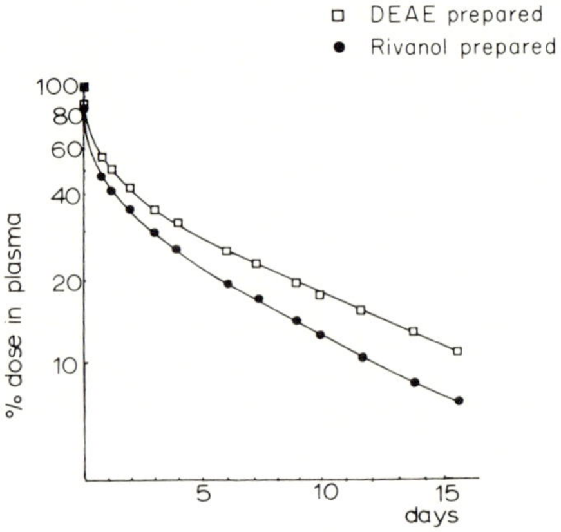

FIG. 6. Plasma decay curves of human transferrin prepared using DEAE (^{131}I) and Rivanol (^{125}I). Indicating that the Rivanol transferrin was denatured.

a satisfactory technique to use for immunoglobulin, it causes significant damage to transferrin molecules. In both experiments daily fractional breakdown rates based on urinary excretion of iodine were calculated and the results were as one would expect; both IgG molecules were catabolized at the same rate, whereas the Rivanol prepared transferrin showed a considerably increased daily breakdown.

IX. Tests for Purity

Having produced what one believes to be a pure protein, there are two problems. First, is it pure? Second, is it denatured? By far the commonest evidence of purity is that of immunoelectrophoresis; Fig. 7 shows a typical immunoelectrophoretic run. Most people will have seen such a preparation

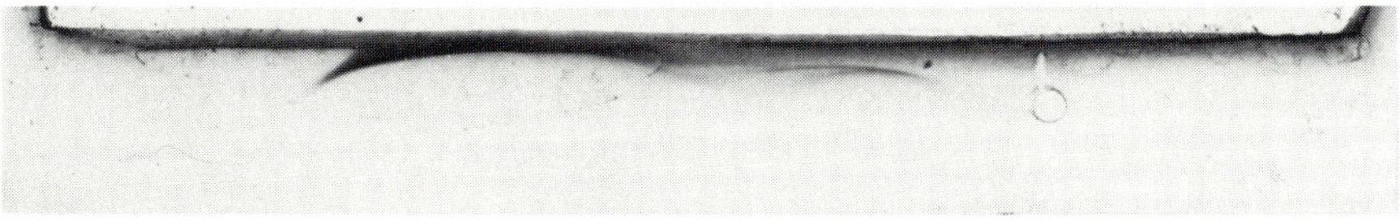

FIG. 7. Conventional immunoelectrophoresis of a protein preparation, showing one main band and some minor contaminants.

with precipitation arc given as evidence of purity, but one essential feature is missing: No adequate control is run alongside it. A second run (Fig. 8) with the protein preparation on one side but using whole serum on the other side with the same antiserum, reveals that the immunoprecipitation with whole serum is no different from that which appeared to be a good protein preparation; in fact this particular protein preparation is whole serum, totally unfractionated, thus, Fig. 7 was particularly poor evidence of purity. Better evidence of purity is demonstrated in Fig. 9 where the serum is shown to react well with whole serum giving a multiplicity of precipitation lines, but a single line only with the pure protein. Such results are more convincing when the Laurell type of quantitative immunoelectrophoresis is used (Fig. 10) (Laurell, 1965; Clarke and Freeman, 1966, 1968). This gives some indication of the proportion of contaminent in the pure material.

X. Evidence of Denaturation

Evidence for the absence of denaturation is much more difficult to obtain. Except in exceptional circumstances the pure protein should be uniform by all criteria. For example, it should be electrophoretically homogeneous, it should be homogeneous in size, and homogeneous on column chromatography.

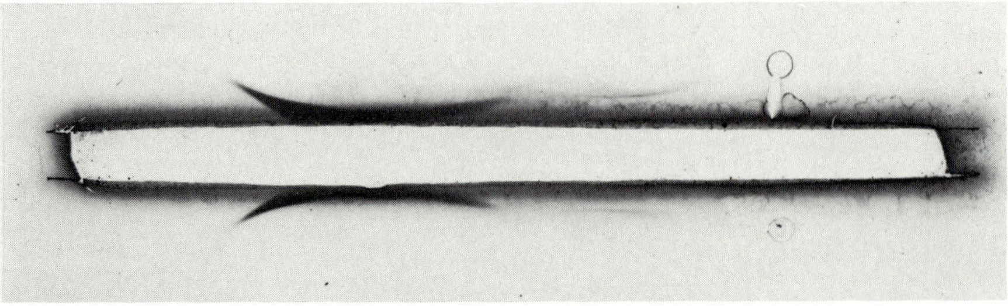

FIG. 8. The same protein as in Fig. 7 against antiwhole serum but in this instance a control of normal serum was put in right slot. This indicates that the protein preparation in Fig. 7 was no more pure than whole serum.

Any inhomogeneity is usually evidence of denaturation. However, occasions arise when it is not possible to detect the presence of denaturation by existing physical and chemical means. Fig. 11 shows such an example: This was a comparison between two similar albumins, one of which had been irradiated with 56 K-rads. These two protein preparations were identical by immunoelectrophoresis, gel chromatography, starch gel electrophoresis, and on DEAE. Nevertheless it is quite clear from using the two proteins labeled with different isotopes in the same subject that one is degraded faster than the other; this is borne out by the daily breakdown rate which was calculated from the urinary radioactivity, showing that the irradiated protein was catabolized faster than the untreated albumin.

From this it is apparent that some *in vivo* test is required; in the experiment just described the catabolism of the irradiated proteins calculated from the urinary radioactivity started at a high value and gradually fell to a value approaching that of native protein. This falling fractional catabolic rate is evidence of denaturation. However, the converse is not necessarily true; a protein that is uniformly denatured may show a constant fractional rate,

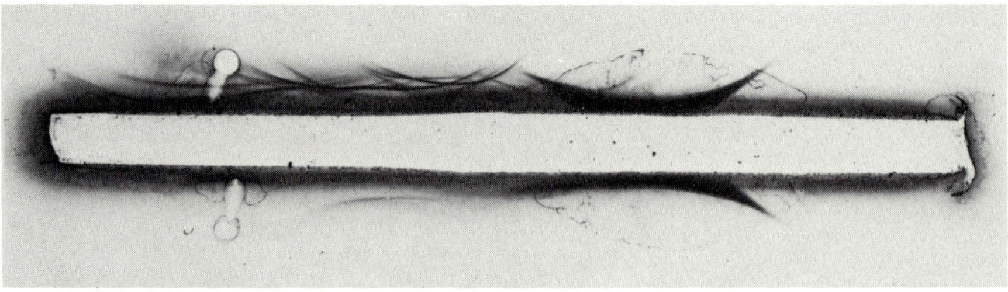

FIG. 9. An albumin preparation and whole serum against anti-whole human serum. The multiplicity of precipitation lines against whole serum shows that the antiserum was polyvalent and that the albumin preparation probably has only "minor contaminants."

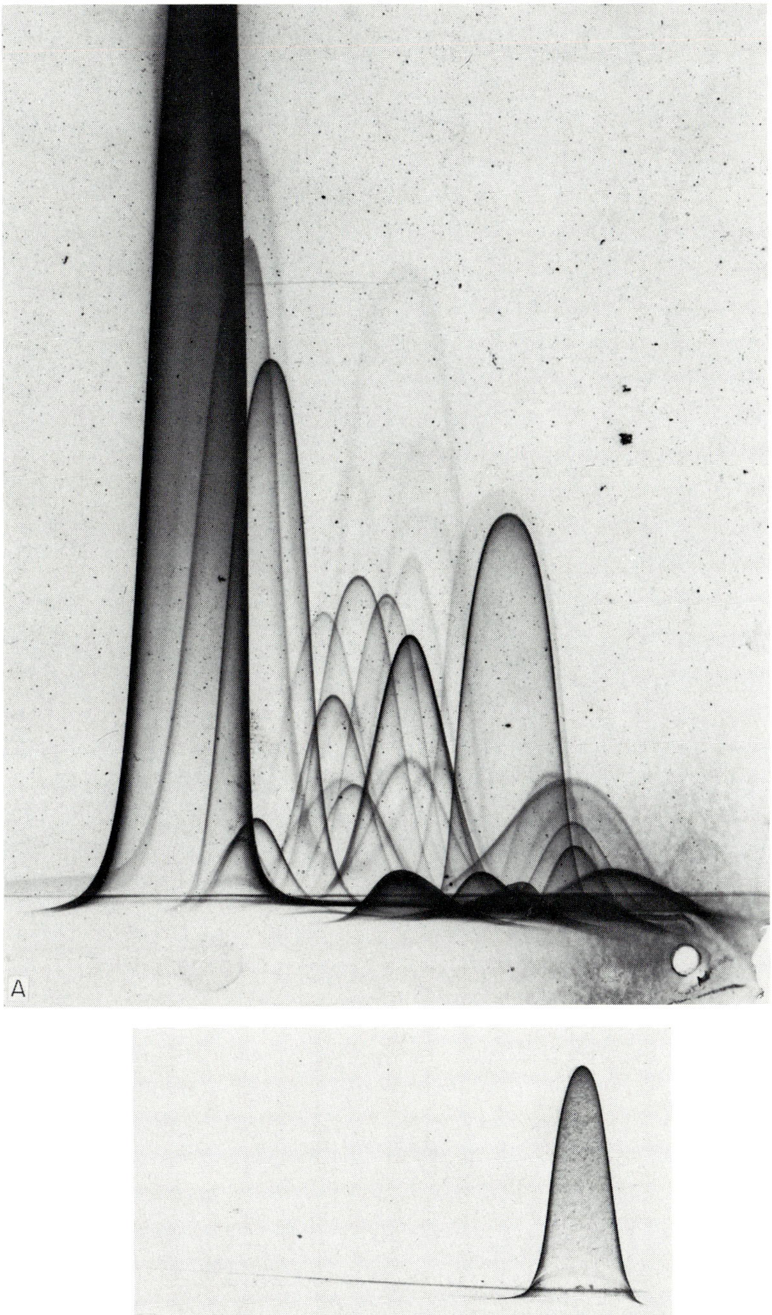

FIG. 10. A preparation of B_{1A} compared to whole serum using Laurell quantitative immunoelectrophoresis.

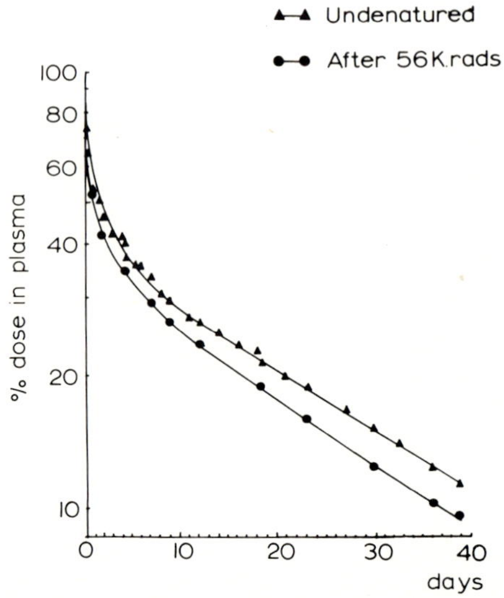

FIG. 11. Plasma curves of two human albumins in the same subject; one albumin (^{125}I) had been irradiated (56 K-rads). This *in vivo* difference could not be demonstrated by any *in vitro* technique.

even though it is catabolized faster than native protein. Where good quality proteins are available for comparison this can easily be observed, but when a new protein is being examined it may be difficult to be certain that the material under study is not all equally altered. A different type of experiment is shown in Fig. 12. A good quality rabbit albumin was labeled with ^{125}I and whole serum with ^{131}I, the two being injected together. Albumin was

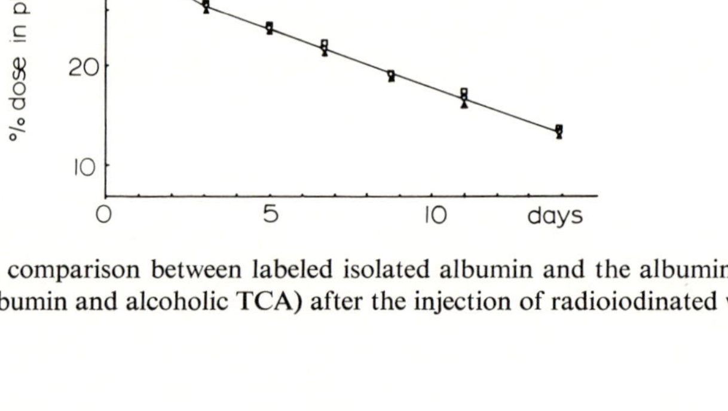

FIG. 12. A comparison between labeled isolated albumin and the albumin isolated (by specific antialbumin and alcoholic TCA) after the injection of radioiodinated whole serum.

isolated from subsequent samples in two ways: first by alcoholic trichloroacetic acid (TCA) precipitation of proteins other than albumin, and second by precipitating the albumin using a specific antialbumin. All these methods give similar results. In this study the daily fractional catabolism could not be calculated because the radioactivity was derived from the catabolism of all the serum proteins; thus, comparisons depend on the plasma curve only. This experiment indicates that the purified albumin was no more degraded than the albumin in whole serum which had not been fractionated in any way. It does not, however, indicate that the radioactive labeled albumin behaved as native albumin. The final proof can only be obtained by the comparison of iodine labeled protein with ^{14}C proteins labeled biosynthetically in the same animal; this type of experiment was done by McFarlane and his coworkers in 1956 (Campbell *et al.*, 1956; Cohen *et al.*, 1956). It should be stressed that when iodine is introduced into a protein molecule this molecule is essentially altered, so it is necessary when working with new proteins to consider the possibility that the metabolic behavior of the protein molecule may have been altered.

It is clear that proper analysis of fractions obtained using any system is important. Further systematic separations with proper analysis of the resultant fractions, should make easier the problem of isolating new proteins. For example, a pool of normal serum was fractionated on Sephadex G150, subdividing into 15 fragments each of which was analyzed by the quantitative immunoelectrophoresis of Laurell (Smith and Freeman, 1970). When all the information is put together, a three-dimensional model can be constructed in which molecular size forms one axis, and electrophoretic mobility forms the second axis. Such a model is shown in Fig. 13. Considerable time was spent checking whether the proteins which appear in one fraction were the same or different from those appearing in the subsequent fractions. Two methods were used: (1) The mixing of adjacent Sephadex fractions to obtain either an immunological reaction of identity or a reaction of nonidentity. These were absolutely clear-cut, and there were no instances of partial identity. (2) A single Sephadex fraction was labeled with ^{125}I, and mixed with other fractions; the presence of radioactivity was demonstrated by subsequent autoradiography. By this technique, 63 immunologically distinct plasma proteins were demonstrated, excluding the immunoglobulins.

In conclusion, it is hoped that this analysis of molecular sieving as a method of fractionation, taken together with other fractionation methods studied in a similar way, will provide in the future enough information to be able to plan in advance the simplest procedure for isolating any plasma protein. The protein fractionation is an essential part of any metabolic study, and I would suggest that, no matter how erudite one's mathematical approach or how cunning one's experimental design, if the protein isolation is poor, the subsequent data will be equally poor.

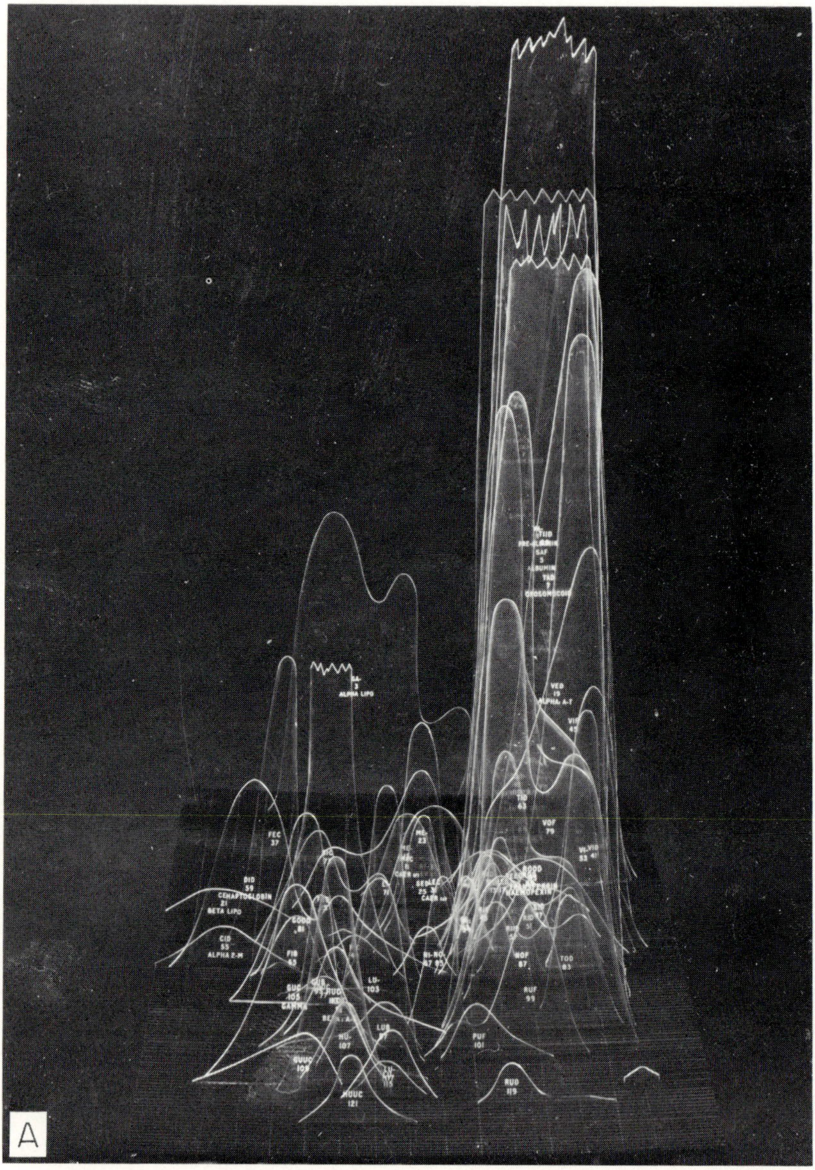

FIG. 13. (A and B) Photographs of a model constructed from the results obtained performing Laurell quantitative immunoelectrophoresis on 15 fractions obtained by gel filtration on Sephadex G150.

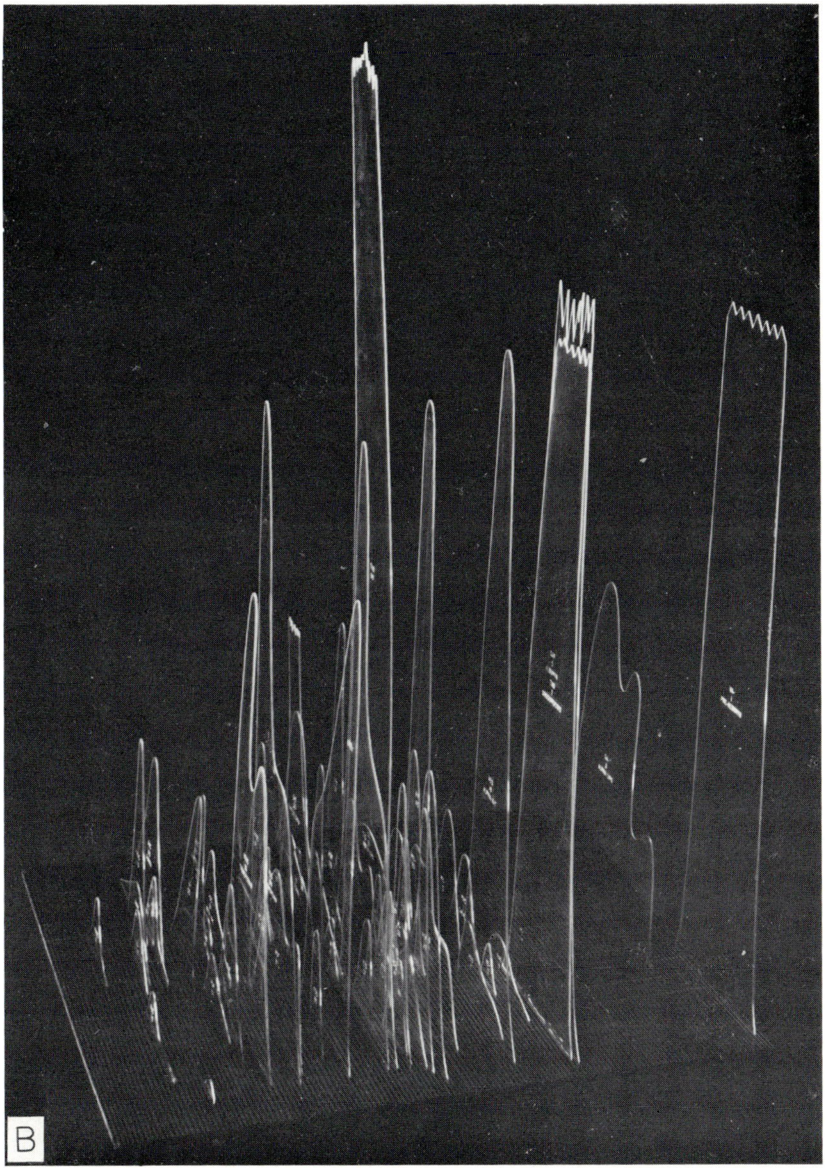
B

References

Barollier, J., Watske, E., and Gibbian, H. (1958). *Z. Naturforsch. B.* **13**, 754.

Brownstone, A. D. (1969). *Anal. Biochem.* **27**, 25.

Campbell, R. M., Cuthbertson, D. P., Matthews, C., and McFarlane, A. S. (1956). *Int. J. Appl. Radiat. Isotop.* **1**, 66.

Clarke, H. G. M., and Freeman, T. (1966). *Protides Biol. Fluids Proc. Colloq.* **14**, 503.

Clarke, H. G. M., and Freeman, T. (1968). *Clin. Sci.* **35**, 403.

Cohen, S., Holloway, R. C., Matthews, C., and McFarlane, A. S. (1956). *Biochem. J.* **62**, 143.

Flodin, P., and Killander, J. (1962). *Biochim. Biophys. Acta* **63**, 403.

Fuller, J. F. (1953). *In* "The Proteins" (H. Neurath and K. Bailey, eds.), Vol. 1, p. 85. Academic Press, New York.

Hannig, K. (1964). *Z. Physiol. Chem.* **338**, 211.

Hannig, K. (1967). "Electrophoresis: Theory, Methods, and Applications" (M. Bier, ed.), Vol. 2. Academic Press, New York.

Heide, K., Haupt. H., and Schmidtberger, R. (1959). *Behringwerk-Mitt.* **37**, 29.

Keller, S., and Block, R. J. (1960). "A Laboratory Manual of Analytical Methods in Protein Chemistry" (P. Alexander and R. F. Block, eds.), Vol. 1, p. 65. Macmillan (Pergamon), New York.

Laurell, C. B. (1965). *Anal. Biochem.* **10**, 358.

Laurent, T. C., and Killander, J. (1964). *J. Chromatogr.* **14**, 317.

McFarlane, A. S. (1958). *Nature (London)* **182**, 53.

Mosse, J. (1959). *C. R. Acad. Sci.* **249**, 2638.

Pedersen, K. O. (1945). "Ultracentrifugal Studies of Serum and Serum Fractions." Almquist & Wiksell, Uppsala.

Pennell, R. B. (1960). *Plasma Proteins* **1**, 9.

Peterson, E. A., and Sober, H. A. (1960). *Plasma Proteins* **1**, 105.

Petersen, E. A., and Sober, H. A. (1962). *In* "Methods in Enzymology" (S. P. Colowick and N. O. Kaplan, eds.), Vol. 5, p. 3. Academic Press, New York.

Porath, J. (1960). *Biochim. Biophys. Acta* **39**, 163.

Porath, J., and Bennich, H. (1962). *Arch. Biochem. Biophys. Suppl.* **1**, 152.

Schultze, H. E., and Heremans, J. F. (1966). "Molecular Biology of Human Proteins," Vol. 1. Elsevier, Amsterdam.

Smith, J., and Freeman, T. (1970). *Biochem. J.*

Sober, H. A., Gutter, F. J., Wyckoff, M. M., and Peterson, E. A. (1956). *J. Am. Chem. Soc.* **78**, 756.

Svensson, H. (1960). "A Laboratory Manual of Analytical Methods in Protein Chemistry" (P. Alexander and R. J. Block, eds.), Vol. 1, p. 195. Macmillan (Pergamon), New York.

Tiselius, A., Porath, J., and Albertsson, P. A. (1963). *Science* **141**, 13.

II

Protein Distribution

6

REGULATION OF INTERSTITIAL ALBUMIN

E. B. Reeve and A. Y. Chen

DEPARTMENT OF MEDICINE, UNIVERSITY OF COLORADO MEDICAL CENTER,
DENVER, COLORADO

I. Introduction

Interstitial plasma protein plays an important part in the distribution of body water and the regulation of plasma volume and forms a store of plasma proteins and substances bound to them. Though we often estimate the total quantity of several interstitial proteins we have little understanding of how these quantities are regulated. Below some factors controlling interstitial plasma protein are examined with emphasis on interstitial albumin and fibrinogen.

II. Constancy of Body Albumin

A. Constancy of Plasma Albumin and Total Protein in Health

Repeated measurements of total albumin in plasma in a healthy subject over 3 weeks are summarized in Table I. They were made by McIntosh *et al.* (unpublished studies) during a study of albumin metabolism with ^{131}I-albumin. The remarkable constancy of total albumin in plasma is shown. Chien *et al.* (1966) repeated measurements of plasma volume, blood volume, and total serum proteins on 41 healthy men after an interval of 16 years. The average age at the first measurement was 24 years. Both blood and plasma volumes were remarkably constant in spite of weight increase. The mean total protein (excluding fibrinogen) was 226 (SD 28 g) at the first measurement and 226 (SD 25 g) 16 years later.

TABLE I

VARIATION IN SOME DAILY MEASUREMENTS ON HEALTHY MALE, D.M.[a]

	Weight	Plasma volume	Hematocrit	Plasma albumin	Total plasma albumin
N[b]	22	18	18	18	18
Mean	71.6 kg	3455 ml	45.02%	4.36 g %	150.3 g
CV[c]	0.7%	4.25%	3.02%	3.21%	2.58%

[a] Measurements of plasma volume were made with ^{131}I-albumin at the beginning and end of these studies and initial and final red cell volumes were then calculated from hematocrit. The latter agreed closely. Intervening plasma volumes over the 22-day period of observations were calculated from hematocrit measurements and the mean of the initial and final red cell volumes.

[b] *N* = number of observations.

[c] CV = coefficients of variation.

B. Constancy of Interstitial Albumin

A mathematical argument shows that constancy of plasma albumin implies constancy of interstitial albumin, provided albumin synthetic rate equals albumin catabolic rate.

Writing A_P for total albumin in plasma, A_i for total interstitial albumin then quite generally

$$dA_P/dt = S + j_2(t)A_i - [j_1(t) + j_3(t)]A_P \tag{1}$$

$$dA_i/dt = j_1(t)A_P - j_2(t)A_i \tag{2}$$

Here the j's, the fractional rates of transfer or breakdown of albumin (see Fig. 1A), may be constant or time dependent, while S is the liver secretion rate of albumin. Since A_P is constant $dA_P/dt = 0$ and thus in health

$$S - j_3(t)A_P + j_2(t)A_i - j_1(t)A_P = 0 \tag{3}$$

The most reasonable physiological assumption is that albumin synthesis and catabolism are equal or $S = j_3(t)A_P$ whence $j_2(t)A_i - j_1(t)A_P = 0$. Thus from Eq. (2) $dA_i/dt = 0$ and A_i is constant. Hence

$$j_1(t)/j_2(t) = A_i/A_P = \text{constant} \tag{4}$$

Equation (4) states that for both interstitial and plasma albumin to remain constant, j_1, the fractional transcapillary transfer rate of albumin, and j_2, the fractional return rate, may fluctuate but they must fluctuate proportionately and in phase. In fact we know from measurements of thoracic duct lymph flow and protein content that lymph albumin return, $j_2 A_i$, does fluctuate (Yoffey and Courtice, 1956a); therefore, j_2 must be some periodic function of time rising and falling about a mean value. A periodic function that is easy to define and is sufficiently close to reality to allow study of the effects

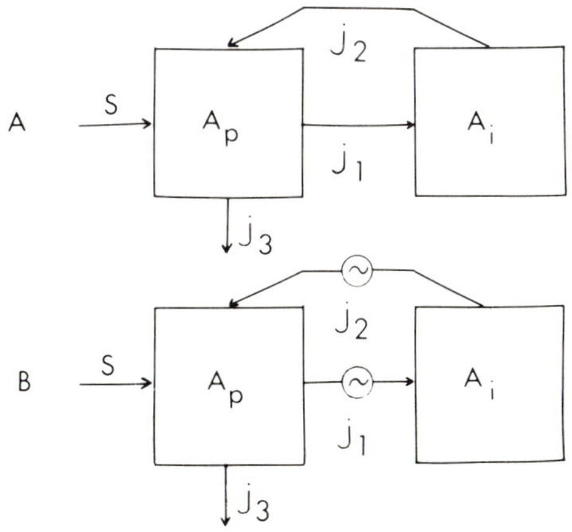

FIG. 1. (A) Generalized model of the albumin system. A_p is total plasma albumin, A_i is total interstitial albumin, S is synthetic rate, j_1, j_2, and j_3 are fractional rates that may be constant or time dependent. (B) The above model with j_1 and j_2 periodic functions of time (see text).

of fluctuation is $b + d \sin \omega t$. We have therefore studied solutions of Eqs. (1) and (2) with periodic coefficients of this kind (Fig. 1B).

It turns out that Eqs. (1) and (2) with periodic coefficients are very difficult to solve analytically (Moulton, 1958) but can be solved on an analog computer. We have therefore made a number of simulations of the behavior of ^{131}I-albumin and ^{131}I-fibrinogen in man and animals when capillary exchange and lymph return fluctuate. Figure 2 shows simulation of the simpler ^{131}I-fibrinogen system in a typical rabbit (Atencio and Reeve, 1965) by solving Eqs. (1) (omitting S) and (2): first when the j's are held constant (the continuous lines), and second when j_1 and j_2 are periodic (the dashed lines). In each graph the top continuous line represents plasma ^{131}I-fibrinogen when $j_1 = 0.55$/day, $j_2 = 1.84$/day, $j_3 = 0.41$/day. The equation describing plasma ^{131}I-fibrinogen is $0.7e^{-0.3t} + 0.3e^{-2.5t}$ with t in days. The dashed lines show the effects of keeping j_3 constant at 0.41/day but making j_1 and j_2 periodic;

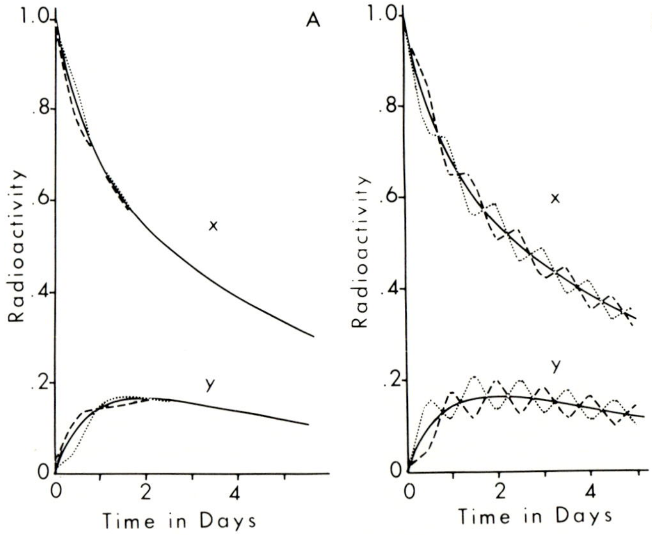

FIG. 2. Simulation of plasma, x, and interstitial, y, ^{131}I-fibrinogen curves in a typical healthy rabbit. All the continuous lines present results for an animal in which j_1 (fractional transcapillary transfer rate), j_2 (fractional lymphatic return rate), and j_3 (fractional catabolic rate) are constant at 0.55/day, 1.84/day, and 0.41/day, respectively. The dotted and dashed lines give results for animals in which j_1 and j_2 are periodic, but j_3 remains constant at 0.41/day. (A) $j_1 = b + d \sin \omega t$, with $b = 0.55$/day, $d = 0.4$/day, and $\omega = 12.56$, while $j_2 = Kj_1$, with $K = 3.345$ obtained by dividing 1.84 by 0.55. The value of ω gives a frequency of two complete sine waves/day. (B) The same values are taken for b, d, ω, and K but now $j_1 = b - d \sin \omega t$, $j_2 = K(b + d \sin \omega t)$. The dashed curves show findings when $\sin \omega t$ in the expression for j_1 starts from zero and increases, the dotted curves when it starts from zero and decreases.

j_1 becomes $b + d \sin \omega t$ and j_2 becomes Kj_1. The values of b, d, ω, and K are given in the figure legend but are so chosen that both j_1 and j_2 fluctuate with a period of half a day about the same mean value as used for the constant coefficients and the fluctuations are quite large, about $\pm 70\%$ of the mean value. Two dashed tracings are shown with each continuous line. These, as noted in the figure legend, correspond to the sine wave starting from zero on its up-swing or starting from zero on its down-swing.

On the left are shown the effects of making $j_1(t)/j_2(t)$ constant with both showing parallel fluctuations. Notice how little effect this has on the form of the plasma and interstitial ^{131}I-protein curves. The only differences from the constant coefficient curves are small changes in slope during the first day and this is also true of simulations of ^{131}I-albumin behavior. Many of us may have studied curves like these variable coefficient curves without knowing it. But now note the effect if $j_1(t)/j_2(t)$ is not a constant. We show the effect on the right of making $j_1(t) = b - d \sin \omega t$ when $j_2(t) = K(b + d \sin \omega t)$ so that $j_1(t)$ is directly out of phase with $j_2(t)$. The effects on the ^{131}I-fibrinogen curve are spectacular and similar spectacular effects are seen in ^{131}I-albumin simulations. Labeled protein curves like those in Fig. 2B are never seen in studies in animals or men. These and many other simulations strongly suggest that ordinarily significant fluctuations in interstitial protein do not occur. Such fluctuations must, of course, be separated from persisting increases or decreases in interstitial protein that follow persisting increases or decreases in plasma protein, as for instance might result from changes in synthetic rate (Atencio *et al.*, 1969).

III. Functional Anatomy of Interstitial Protein

The walls of a typical capillary consist of an inner layer of endothelium surrounded by an outer layer of basement membrane (Majno, 1965; Bruns and Palade, 1968a). The endothelial tube is formed of flattened cells joined to each other at intercellular junctions at which the two cells overlap each other often in complicated fashion. A major part of the overlap by electron microscopy seems to enclose a channel between endothelial cells of perhaps 100 Å in cross section (Landis and Pappenheimer, 1961) but except in inflammation (Cotran, 1967) in the capillaries of many tissues these potential channels between the inside and outside of the capillary seem to be closed or nearly closed by thickened parts of adjacent cell membranes, the zonulae occludentes, coming into contact. The endothelium contains varying numbers of vesicles 500–1000 Å in diameter (Bruns and Palade, 1968a) some fused with its inner and others with its outer surface which seem by pinocytosis to transport fluid globules between the two cell surfaces (Bruns and Palade,

1968b; Shea and Karnovsky, 1969). We may distinguish three types of capillaries on the bases of porosity and density of basement membranes (Bennett *et al.*, 1959): (1) those with interendothelial gaps of perhaps 1000 Å in diameter not closed by zonulae occludentes and with an incomplete basement membrane as seen chiefly in the liver sinusoids; (2) those with intercellular junctions in close apposition and a fine but complete basement membrane but with fenestrae or pores, lined only by a very thin membrane, passing almost completely through the endothelium. These are typically seen in the intestine (Clementi and Palade, 1969); and (3) those with intercellular junctions that seem to be tightly closed, with no apparent pores through endothelial cells but with very large numbers of vesicles in their endothelial walls and with a fairly dense basement membrane. This type of capillary is seen particularly in muscle and skin (Bruns and Palade, 1968a).

It is difficult to get a clear idea of the ultramicro and molecular structure of the interstitial supporting structures which lie between capillaries and cells. These vary from tissue to tissue and contain several kinds of fibers and "ground substance" of uncertain composition. In health there seems to be no free interstitial fluid, but interstitial water is held in the form of a gel (Rusznyak *et al.*, 1967c). Probably all interstitial spaces contain varying amounts of collagen, elastin, and other less well defined fibers (Schubert, 1966). For instance, the electron micrographs of Bruns and Palade (1968a) show a particularly rich network of collagen and other fibers lying in the interstitium between capillaries and muscle fibers. It seems likely that hyaluronic acid, an amino-hexose polymer of very high molecular weight (MW $\sim 10^7$), forms part of the ground substance lying between fibers (Laurent, 1966a). This is chemically combined with about 2% by weight of protein and at physiological pH is negatively charged and binds diffusible cations and perhaps protein cations (Catchpole *et al.*, 1966). In concentrations as low as 0.1 to 1.0 g/liter the long coiled molecules of hyaluronic acid become intertwined and form a gel which resists the passage of water and solutes and excludes proteins from a large fraction of the gel volume (Schubert, 1966). Thus a 1 g/liter solution of hyaluronic acid excludes protein from 102 ml of its volume (Schubert, 1966). According to Laurent (1966b) transport in connective tissue is then analogous to transport through a column of polysaccharide gel grains (Laurent and Killander, 1964). Large molecules will be excluded from a large fraction of the total volume and will pass along the main channels of water flow, smaller molecules will be able to pass into the inter- or intramolecular interstices (Polson and Katz, 1969).

Penetrating the interstitial spaces are the lymphatic capillaries (Casley-Smith, 1967). These may be 2–5 times the diameter of blood capillaries and often have a poorly developed basement membrane. Proteins readily enter them carried by fluid that enters the easily opened interendothelial cell

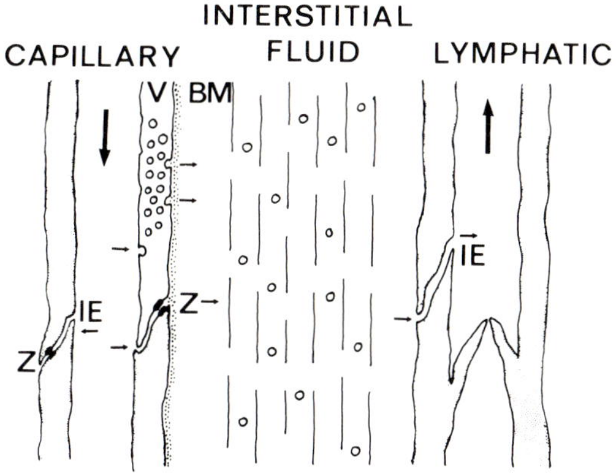

FIG. 3. Diagrammatic presentation of the functional anatomy of interstitial protein. Shown is a capillary, the intervening connective tissue containing fibers, and a lymphatic. V, vesicles in capillary endothelium, some opening on both sides of the endothelium; BM, basement membrane; IE, interendothelial junctions, mainly closed in the capillary by zonulae occludentes, Z. The small arrows show possible paths of protein passage. In the liver the interendothelial junctions are much wider. In the intestine fenestrae are present perhaps made from fusion of vesicles. These are not shown in the diagram.

junctions. The endothelial cells are connected by fibrils to the tissue cells so movement or swelling of tissues pulls interendothelial cell junctions open while muscle contraction causes endothelial overlaps and closes them. Thus protein-containing fluid readily enters the smallest lymphatic while muscle movements and the lymphatic valves force it on into the lymphatic system. The large lymphatics have thicker walls and though easily permeable to water and crystalloids are almost impermeable to proteins the size of albumin (Mayerson *et al.*, 1962).

Figure 3 summarizes the functional anatomy of interstitial protein as described in this section.

IV. Measurement of Interstitial Proteins in Tissues

Only a few estimates of the quantities of interstitial proteins in different tissues have been reported. In one approach (Dewey, 1959) following injection of ^{131}I-albumin or ^{131}I-γ-globulin into a group of rats, the animals were killed at intervals and their tissues were examined for total ^{131}I-albumin and circulating ^{131}I-albumin, the difference giving interstitial albumin. By assuming that distribution is described by linear differential equations the quantity of albumin in the different tissues was calculated. However, the rat catabolizes albumin quite rapidly (Campbell *et al.*, 1956) and when this occurs we find that the linear differential equation compartmental model approach (Matthews, 1957) may give considerable underestimates.

Figure 4 summarizes results of Dewey (1959) using ^{131}I-albumin and ^{131}I-γ-globulin and this approach. The estimated quantity of protein in each tissue is represented by a horizontal bar, the width of the bar on the ordinate representing the average transcapillary inflow rate (g plasma/hr), the length of the bar on the abscissa representing the mean passage time in hours of albumin molecules through the tissue. Dewey (1959) found essentially no difference between the distribution of albumin and γ-globulin which is about 2.5 times the molecular weight of albumin. Figure 4 shows that liver, kidney, and lung content of albumin is small, but flow of albumin through the tissue is rapid while skin and muscle content of albumin is high but tissue flow is slow. Estimated total interstitial albumin was 1.48 times that in plasma. Using a qualitatively similar approach Rothschild and co-workers (1955) earlier found that the major part of interstitial albumin in man was distributed in skin and muscle and only a minor part in heart, lung, liver, kidney, and spleen.

Sellers and co-workers (1966) used a second approach in which albumin was extracted from the tissues of young rats with the aid of specific antibody and measured directly. They also found the major distribution of albumin was in skin and muscle but the ratio of extravascular to vascular albumin was 2.9:1, roughly twice that found by Dewey (1959). It is to be noted that Sellers and co-workers findings would include intracellular albumin which fluorescent-labeled antibody studies have shown occurs in a number of cells

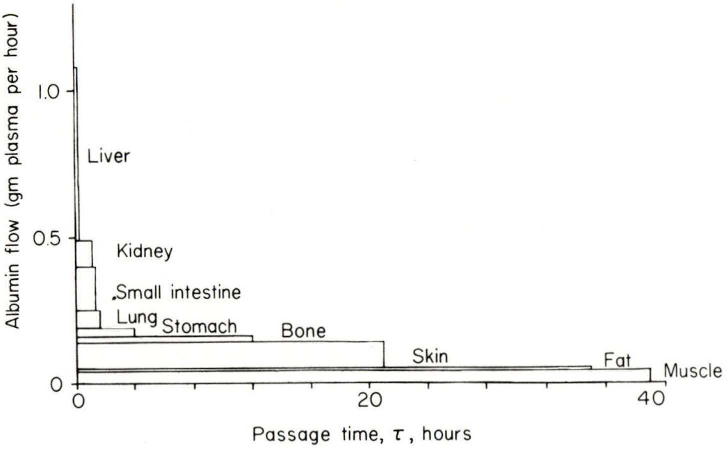

FIG. 4. Calculations of rat interstitial albumin distribution made from Dewey (1959). For description see text.

(Gitlin, 1957), and young animals may have more interstitial protein (Atencio and Reeve, 1965) than older animals.

A. Location of Interstitial Albumin in Tissues

The volume of the larger lymphatics is small (Mayerson, 1963) and they may contain only 2 or 3% of the total interstitial albumin. In the tissues through which albumin flows rapidly the volume of fluid available to albumin in the small lymphatics and tissue fluids is quite small. Thus, Morris (1956) has calculated 3 ml of liver lymph and interstitial fluid/100 g liver suggesting that liver interstitial albumin is primarily confined to the spaces of Disse and the septal lymphatics. We calculate from data (Morris, 1956) that in a 2.5 kg cat, the abundant lymphatics of the intestine with their intravillal interstitial space maximally distended by large saline infusions, perhaps have a volume of 40 ml. Ordinarily the volume must be much smaller. The tissues of slow interstitial albumin flow present a different picture. Thus in muscle lymphatic capillaries are relatively sparse and are inserted in the periphery of muscle fiber bundles (Rusznyak *et al.*, 1967a) which suggests that the chief part of muscle interstitial albumin is in the connective tissue between muscle fibers. The same may be true of skin (Rusznyak *et al.*, 1967a) which, though it is penetrated by a richer lymphatic network, contains much collagen and hyaluronic acid. If situated primarily in the interstitial fluids of muscle and skin, albumin would be well placed for osmotic regulation of plasma volume.

B. Measurement of Total Interstitial Albumin and Fibrinogen

In living men and animals it is not possible to analyze tissues for their protein content. Fortunately a well-known approach dependent on injecting indicator and applying input–output principles can be applied (Reeve and Bailey, 1962). This approach may be called the "convolution method" and is also used for measuring cardiac output and pulmonary blood volume (Zierler, 1962). If indicator is injected at a known rate into tubes through which a protein solution is flowing and the time course of the amount of indicator in the tubes is determined, then the frequency function of passage of protein through the tubes and the total protein can be determined. Limitations are that flow rate through the tubes should be constant or fluctuate about a mean and the geometry of the tubes should remain constant. In our case the indicator is ^{131}I-protein injected into the plasma, the tubes of flow are the innumerable streams of protein molecules carried through the tissue fluids and lymph and total interstitial indicator is total interstitial ^{131}I-protein. When, as often happens, the time course of the plasma ^{131}I-protein curve is described by a sum of exponentials $\sum C_i e^{-a_i t}$, appropriate compartmental analysis (Matthews, 1957; Reeve and Bailey, 1962) can be applied and gives the same estimates of total interstitial protein as the convolution method of which it is a special case. But it should be noted that assigning values of the C_i's and a_i's to exponential curves is difficult even with computer aid and the compartments obtained for interstitial protein have no physical meaning but are arbitrary divisions of the frequency distribution function. The convolution approach, however, is valid even when the input is not described by a sum of exponentials.

Figure 5 shows the distribution of albumin (Takeda and Reeve, 1963) and fibrinogen (Takeda, 1966) passage times in healthy men determined with ^{131}I-albumin and ^{131}I-fibrinogen. The area under these curves multiplied by mean transcapillary protein flux gives total interstitial albumin or fibrinogen. Compare human interstitial albumin passage time distribution with the rat albumin passage time distribution of Fig. 4 calculated from Dewey's results. The general similarity is striking and it seems probable that the long paths with slow flow encompassing a major portion of interstitial albumin in man represent albumin primarily of skin and muscle. Similarly the short paths with rapid flow containing very little albumin must represent the rapid passage of a minor portion of the albumin through the interstitium and small lymphatics of liver, kidneys, lungs, and intestines. Without studies of the type made by Dewey (1959) it would not be possible to say this, for the method only gives a frequency diagram of passage times of interstitial albumin molecules. It says nothing about their anatomical location.

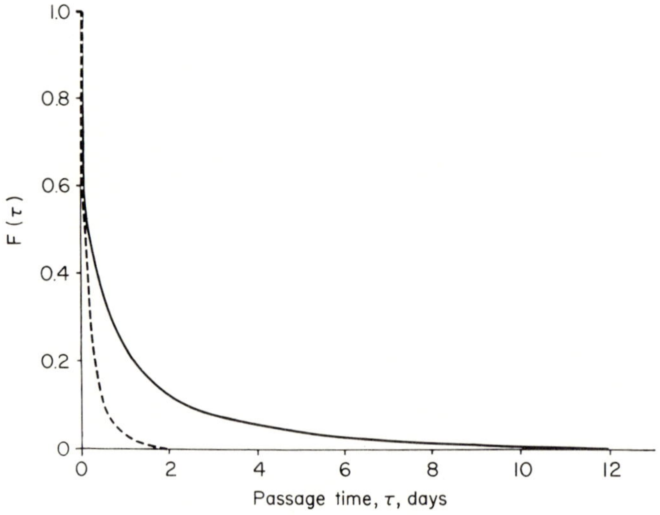

FIG. 5. Distribution of albumin (———) and fibrogen (– – – –) passage times through the interstitial fluids of man.

From Fig. 5 in man total interstitial albumin is about 1.38 times total albumin in plasma (Takeda and Reeve, 1963) and total interstitial fibrinogen is about 0.19 times total plasma fibrinogen (Takeda, 1966). In rabbits these ratios are somewhat different (Reeve and Roberts, 1959; Atencio and Reeve, 1965). Later we consider reasons for the difference between interstitial albumin and fibrinogen.

V. Control of Interstitial Protein—Mathematical Evaluation

So far we have concluded that amounts of interstitial albumin and probably a number of other interstitial proteins remain about constant. It also appears that the main part of these proteins is in the interstitial spaces and the smallest lymphatics and only a minor part is in the larger lymphatics. Now, passage of proteins through the interstitial fluids is dependent on interstitial fluid flow and this fluid flow is in turn dependent on capillary filtration rate, fluid reabsorption through the capillary, tissue movements leading to translocation of interstitial fluid and lymph flow. Landis and Pappenheimer (1961) review the mass of evidence showing that even quite small changes in mean hydrostatic pressure or colloid osmotic pressure across the capillary membrane can markedly alter capillary filtration rate. Much evidence indicates that lymph flow from different tissues may increase five- or tenfold or even more above resting levels (Yoffey and Courtice, 1956a; Morris, 1956; Drinker and Yoffey, 1941). Thus our problem may be formulated as: How is interstitial protein (for instance albumin) maintained about constant in the face of widely different flows of interstitial water through the tissues and changing volumes of interstitial water? The major part of interstitial water is in muscle and skin and here water is thought to enter the interstitial space primarily by "small" pores, perhaps of average radius 45 Å, and to only a small extent by larger pores of 200 Å radius or greater (Landis and Pappenheimer, 1961). The capillary equivalent of the small pores is perhaps the intercellular junctions (Landis and Pappenheimer, 1961; Cotran, 1967; Bruns and Palade, 1968b; Shea and Karnovsky, 1969). In other tissues such as liver and intestine the capillary small pore system may be much less important for water distribution and the chief sites of entry of interstitial water may be the larger pores mentioned earlier (Bennett *et al.*, 1959; Clementi and Palade, 1969).

To think about the interactions of interstitial water and protein requires two models, the first describing water flow and the second describing albumin flow in a segment of tissue. These are shown in Fig. 6.

Before proceeding further with the analysis of these two models we summarize the general approach used below. First are presented the known steady state relations governing net segmental water inflow, segmental water volume,

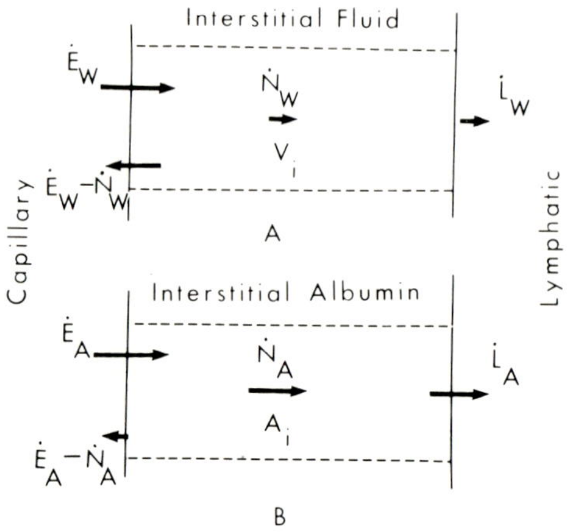

FIG. 6. Models of the control of water (A) and albumin (B) in a segment of connective tissue lying between capillaries and lymphatics. The upper model shows the rates of water entry, $\dot{E}_W$, water return to the capillaries, $\dot{E}_W - \dot{N}_W$, and lymph exit, $\dot{L}_W$, with the net rate of passage of water into the segment, $\dot{N}_W$, and the volume of water in the segment, V_i. The lower model shows in the same segment the rates of albumin entry, $\dot{E}_A$, of albumin return to the capillaries, $\dot{E}_A - \dot{N}_A$, and of lymph exit of albumin $\dot{L}_A$, with the net rate of albumin entry into the segment, $\dot{N}_A$, and the total albumin in the segment, A_i. For further discussion see text.

and segmental lymph outflow. Both segmental water volume and segmental lymph outflow depend on segmental fluid pressures P_i. We next examine the relations between net segmental albumin inflow, net albumin in the segment, and net lymph albumin outflow and show that the water and protein systems interact with each other in lymph outflow. This allows us to write the steady state $A_i = \dot{N}_A(V_i/\dot{L}_W)$, where A_i is the total interstitial albumin in the segment, $\dot{N}_A$ is the net inflow rate of albumin, V_i is the water volume in which the segmental albumin is dissolved, and $\dot{L}_W$ is the lymph outflow rate from the segment. We can thus qualitatively examine the effects on interstitial albumin, A_i, of changes of interstitial volume and lymph output, both functions of P_i, and of net albumin inflow, $\dot{N}_A$. We now formulate and examine the models.

Conceptually isolate in a tissue a stream of water molecules which enter the tissue fluids at a rate $\dot{E}_W$ and leave at a rate $\dot{L}_W$, the rate of lymph flow. As shown in Fig. 6A, a proportion, ordinarily large, of the entering molecules is short-circuited back across the capillary membrane at a rate $\dot{E}_W - \dot{N}_W$, so $\dot{N}_W$ is the net entry rate of water into the tissue segment. Though diffusion is responsible for water molecule exchange across pores, net flow of water from capillary to lymphatic is driven by pressure and osmotic differences, the latter dependent on large colloids (Landis and Pappenheimer, 1961). We can thus write an approximate equation describing the net rate of water inflow into the stream as

$$\dot{N}_W = K_1(P_c - P_i - \pi_c + \pi_i) \tag{5}$$

where P_c is mean capillary pressure, P_i is mean interstitial pressure, π_c is capillary osmotic pressure, π_i is mean interstitial fluid osmotic pressure, and K_1 is the conductance relating fluid flow across the capillary membrane to the pressure differences. K_1 will depend on pore size and shape, fluid viscosity, friction in the pores, etc.

The rate of lymph efflux from the stream is

$$\dot{L}_W = K_2(P_i - P_l) \tag{6}$$

where P_l is pressure at the point where the lymphatics enter the veins and K_2 is a conductance relating lymph flow to pressure difference. Note that both P_c and P_i, but not P_l, will be affected by local tissue pressure changes as from muscle movements.

The mass balance equation describing V_i, the segmental volume of interstitial water available for dissolving protein, is

$$dV_i/dt = \dot{N}_W - \dot{L}_W \tag{7}$$

In a steady state when interstitial volume is constant $\dot{N}_W = \dot{L}_W$ or net water entry rate equals net lymph exit rate. Using this relation and rearranging Eqs. (5) and (6) we get a steady state solution for interstitial fluid pressure, P_i

$$P_i = [K_1(P_c - \pi_c + \pi_i) + K_2 P_l]/(K_1 + K_2) \tag{8}$$

This states that interstitial pressure is a complicated function depending on the conductances K_1 and K_2, the pressures P_c and P_l, and the osmotic pressures π_c and π_i. Guyton (1965) has shown that interstitial fluid volume increases with interstitial pressure first slowly and then more rapidly. His complex relation may be written for later use

$$V_i = f_1(P_i) \tag{9}$$

We also require a model of interstitial protein which is shown in Fig. 6B. Conceptually isolate in the stream of water molecules pictured in Fig. 6A a stream of albumin molecules which enter the tissue fluids at a rate $\dot{E}_A$ and leave in the lymph at a rate $\dot{L}_A$. The net entry rate of albumin into the stream, $\dot{N}_A$, is $\dot{E}_A$ minus the capillary reabsorption rate. If A_i is the total albumin in the stream we can write a mass balance equation defining it, thus

$$dA_i/dt = \dot{N}_A - \dot{L}_A \tag{10}$$

This equation is exactly analogous to Eq. (2) presented earlier except that it describes only one small segment of total interstitial albumin and uses different notation. It may be written in the earlier notation

$$dA_i/dt = j_1 A_p^- j_2 A_i \tag{10a}$$

provided that j_1 is now regarded as the fractional rate applying only to the entry of albumin into the segment and j_2 is the fractional loss rate of albumin from the segment. However, the models of Fig. 6 and Eq. (10) now allow us to interpret $j_1 A_p$ as $\dot{N}_A$, the net influx of albumin. Also, j_2, the fractional removal rate of A_i, is clearly $\dot{L}_A/A_i$ or the rate of lymph removal of albumin divided by total albumin in the segment. Further lymph removal may be pictured as a "clearance," a certain volume L_W of the interstitial fluid volume being cleared in unit time. Then $j_2 = \dot{L}_W/V_i$ where V_i is the interstitial fluid volume of the segment in which albumin is dissolved at the same concentration as in lymph. Thus, Eq. (10a) becomes

$$dA_i/dt = \dot{N}_A - (\dot{L}_W/V_i)A_i \tag{11}$$

Equation (11) provides the conditions for A_i to remain constant. They are

$$A_i = \dot{N}_A \cdot V_i \dot{L}_W = \text{a constant} \tag{12}$$

Equation (11) points to the variables important in regulating A_i, interstitial albumin. These are (1) the factors controlling the net rate of entry of albumin $\dot{N}_A$ into the interstitial fluid segment, and (2) the factors controlling the rate of lymph flow $\dot{L}_W$ and the volume of the interstitial stream V_i. Though Eq. (12) only applies to a minute segment of the interstitial fluids, it is easy conceptually to imagine a mean tube which when multiplied by a suitable factor represents the average behavior of all the tubes in one or several tissues.

Also note that the albumin concentration in lymph is given by $\dot{L}_A/\dot{L}_W$ where $\dot{L}_A$ is the rate of albumin and $\dot{L}_W$ the rate of water outflows in lymph. From Eqs. (10) and (11) $\dot{L}_W = (\dot{L}_W/V_i)A_i$ whence in the steady state lymph albumin concentration $= A_i/V_i$. Thus from Eq. (11)

$$A_i/V_i = \dot{N}_A/\dot{L}_W = \text{steady state lymph albumin concentration} \tag{13}$$

A. Control of Net Protein Inflow Rate, $\dot{N}_A$, into a Segment of Tissue

Two extreme possibilities may be visualized. In the first protein inflow is independent of water inflow. This would occur if the protein entered by vesicular transport or by diffusion through pores through which little water flowed, the main water inflow occurring through "small" pores impervious to protein. In either case net protein inflow is described by $F(c_p - c_i)/t$, where c_p is plasma protein and c_i is average interstitial protein concentration.

The factor F turns concentration difference into a flux. For vesicular transport F is $n\bar{v}$, where n is the number of vesicles passing from lumen to periphery through the capillary endothelium in the time t, and $\bar{v}$ is the mean vesicular volume, while for diffusion F is $DA/\Delta x$ of the Fick diffusion equation (Landis and Pappenheimer, 1961). Vesicular transport is believed to be mainly responsible for protein entry into the interstitial fluids of muscle and skin by Bruns and Palade (1968b) and others (Shea and Karnovsky, 1969).

In the second possibility protein transport through the capillary membrane is entirely dependent on carriage by water streams through large pores and these form the sole interstitial water inflow. This seems to be almost true of the liver lymph which is a low pressure filtrate of plasma with albumin concentrations four-fifths or more of those in the plasma. In this case net protein inflow $\dot{N}_A$ must increase directly with net water inflow $\dot{N}_W$. Mixtures of both extreme possibilities may occur. Thus, fenestrae occur in intestinal capillary endothelia but the albumin concentration in the lymph of about half that in plasma suggests in addition either molecular sieving (Landis and Pappenheimer, 1961) or additional water passage through "small" pores.

B. Steady State Control of Interstitial Albumin in Some Tissues

Equation (12), $A_i = \dot{N}_A(V_i/\dot{L}_W)$, throws light on this. Consider first the two terms in parentheses $(V_i/\dot{L}_W)$. V_i is the volume of fluid in the interstitial water stream available to protein. We have seen that V_i increases in a complicated fashion with interstitial fluid pressure P_i and have written $V_i = f_1(P_i)$ [Eq. (9)]. $\dot{L}_W$ is the lymph water flow and this also is highly dependent on P_i [Eq. (6)] and may be written $\dot{L}_W = f_2(P_i)$. Since both V_i and $\dot{L}_W$ increase with P_i, we do not expect great changes in the ratio $V_i/\dot{L}_W$, given by $f_i(P_i)/f_2(P_i)$, even with considerable changes in P_i. Thus the most important factor controlling A_i, the total interstitial albumin in a segment, will be the control of the net inflow rate of albumin $\dot{N}_A$. If $\dot{N}_A$ is little affected by water inflow, as in vesicular transport, then $\dot{N}_A$ should remain about constant and therefore from the above discussion so should total interstitial albumin. If, on the other hand, $\dot{N}_A$ increases with interstitial water inflow, then A_i must increase. The latter occurs as we have seen when liver and intestinal lymph flow increase, but Fig. 4 has shown that the fraction of the total interstitial albumin in liver and intestines is small and even a two- or threefold increase would not be of much significance. On the other hand, the large quantities of interstitial albumin in skin and muscle demand a mechanism insensitive to large changes in interstitial water inflow which makes the vesicular transport proposed by Bruns and Palade (1968b) very attractive.

Equation (13), which defines steady state lymph albumin concentration as $\dot{N}_A/\dot{L}_W$, where $\dot{N}_A$ is net albumin inflow and $\dot{L}_W$ is lymph water flow, allows some test of the above arguments. If net albumin inflow remains constant Eq. (13) predicts lymph albumin concentration should vary inversely with the rate of lymph flow. In agreement lymph flow from a foot increases with exercise while its protein content falls (Drinker and Yoffey, 1941). On the other hand, if net albumin inflow increases with lymph flow, lymph albumin concentration might actually increase, remain unchanged, or fall slightly depending on the relation between $\dot{N}_A$ and $\dot{L}_W$. In the burned paw of the dog, lymph flow temporarily increases and so does its protein concentration (Yoffey and Courtice, 1956b; Roberts and Courtice, 1969). The burn would be expected to widen the capillary interendothelial cell spaces and permit the free filtration of protein (Majno, 1965; Cotran, 1967). When lymph flow from the liver is greatly increased by hepatic vein congestion, liver lymph albumin concentration remains high (Morris, 1956). These two examples show that albumin inflow rate increases with capillary filtration rate when the filtrate can carry significant quantities of protein and provide further evidence in favor of the separation of transcapillary water and protein transport in healthy muscle and skin.

C. Control of Interstitial Protein in Transient States

Figure 7 presents a very simple diagram of interstitial protein distributed in tubes of flow which illustrates the difficulty of maintaining interstitial protein in the face of sudden large changes of interstitial water and lymph flow. Suppose lymph flow suddenly increases tenfold. The simple model of

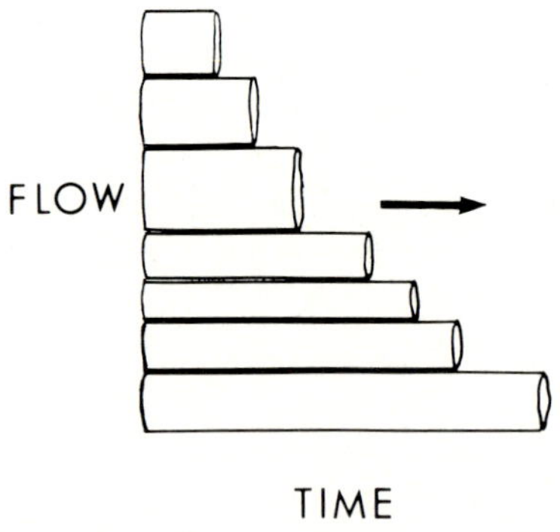

FIG. 7. Interstitial protein is pictured as distributed in pipes of different passage times and of different diameters representing different flow rates. The origin (left end) of all the pipes is at the capillary endothelium, the termination (right end) is at the point of interstitial protein return into the circulation. For the amount of protein to stay constant in a pipe, protein inflow must equal protein outflow. A sudden inflow of a dilute protein solution from the capillaries will wash much of the protein contained in the pipe into the circulation.

Fig. 7 predicts a washout of interstitial protein into the blood stream unless the washout is balanced by an equal inflow from the circulation. The latter would not be expected to occur with muscle and skin albumin, if transfer is by vesicular transport. How then can large and dangerous transfers of skin and muscle albumin into the circulation be prevented when large increases in passages of interstitial water occur? For instance, these might occur in the skin during continued warming or in the muscle during prolonged exercise. A possible answer is the interstitial network of fibers and of hyaluronic acid illustrated in Fig. 8. This must entangle many protein molecules and allow only slow transfer of these from the interstitial fluids to the blood stream. In this way only slow changes in the interstitial colloid osmotic pressure would be permitted.

VI. Factors Responsible for the Different Distributions of Different Plasma Proteins in the Interstitial Fluids

It is customary to express the total amount of a given interstitial protein as a fraction of the amount in the plasma. It has been seen that this ratio is much smaller for fibrinogen (MW 360,000) (Takeda, 1966) than for albumin (MW 68,000) (Takeda and Reeve, 1963) which suggests that molecular size is important; this is confirmed by Norberg's finding (personal communication) that α_2-macroglobulin (MW $\sim$ 800,000) is almost excluded from the interstitial fluids. The entrance of protein to the interstitial fluids will be controlled by the sizes of pores or vesicles responsible for their transport and by the permeability of the basement membrane (Cotran, 1967). Our

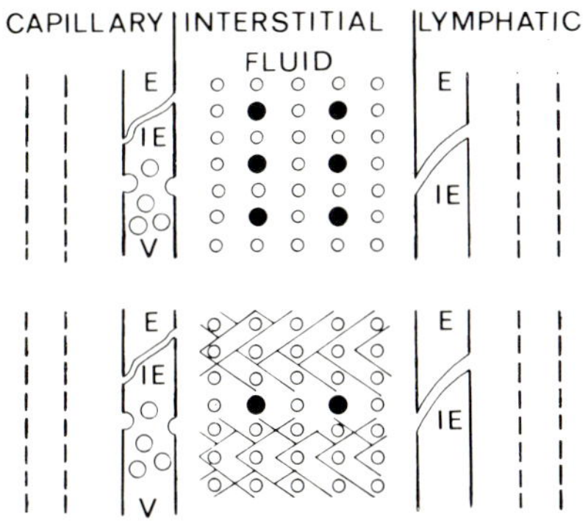

FIG. 8. Diagram of effects of interstitial space network on distribution and movement of interstitial proteins. Above: random distribution of a large (●) and small protein (○) in the interstitial spaces. Below: redistribution of protein molecules on the basis of size with restraint of passage through interstitial spaces network. E, endothelium; IE, interendothelial spaces; V, vesicles. Compare with Fig. 3.

unpublished studies with labeled albumin and fibrinogen in the intact animal show that albumin penetrates the capillaries at about twice the fractional rate of fibrinogen indicating a twofold difference in permeability. Once having entered the interstitial fluids a second factor controls distribution. This, the ultramicrostructure of the interstitial spaces, if comparable to polysaccharide gel beads interspersed with fibers (Laurent, 1966b), will impose smaller volumes of distribution on the larger molecules. So far this factor has not been studied. Differential adsorption of proteins by charged interstitial macromolecules if it occurs would also greatly alter the interstitial distribution of different proteins.

VII. Discussion

Above we have outlined present knowledge of some controls of interstitial protein in health and presented a steady state relation between the interstitial protein in a segment, its net rate of entry, the volume of water in the segment, and the rate of lymph outflow from the segment. This relation has allowed us to make some useful qualitative statements. Our goal, however, is to understand the *regulation* of interstitial protein. This implies that we can write a set of equations which quantitatively describe in time the factors controlling entry of an interstitial protein, its distribution in interstitial fluids, and its return to the circulation. A valuable beginning to this approach has been made by Guyton (Guyton and Coleman, 1967) but it is clear that much further experimental work is required to define the physiological mechanisms sufficiently well so that mathematical formulation becomes possible. The present availability of well-tested methods for separating and labeling pure undenatured proteins (Freeman and Hunter, 1967) provides a powerful tool for these studies.

We do not review here present knowledge of exudates and transudates and of the estimates of the relative proportions of interstitial and plasma proteins in different diseases, though the steady state relation given above does throw some light on these. Instead as a spur to further investigation we mention a few common disturbances of distribution that demand definition and understanding. Thus, after a large hemorrhage, interstitial protein shifts quite rapidly into the plasma (see Fig. 9) though thoracic duct lymph flow decreases (Furneaux and Tracy, 1969). Present knowledge throws no light on this. When animals or men are fed a low protein diet proportionately more interstitial than plasma albumin is thought to be lost (Hoffenberg, 1965). The explanation of this is unknown. When plasma albumin concentration is low, lymph flow increases (Rusznyak *et al.*, 1967b). The effect of this increased flow in albumin distribution and the time course of any redistribution are unknown. Alterations in tissue hyaluronic acid may occur,

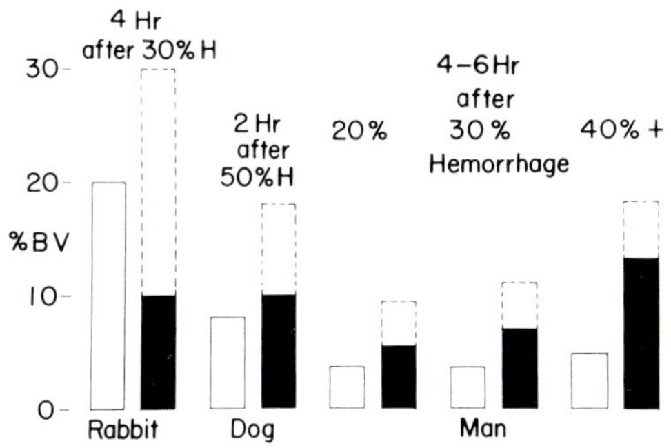

FIG. 9. Restoration of blood volume (BV) 2–6 hr after hemorrhage (H) in rabbit (Reeve, unpublished observations), dog (Reeve *et al.*, 1953), and man (Grant and Reeve, 1951). Hemorrhages were fairly rapid and as shown varied between 20–50% of blood volume. Restoration of blood volume is pictured as restoration of a volume of saline, the open rectangles, and a volume of protein containing plasma, the solid rectangles. In the animals the values were obtained from measurements of red cell volume, plasma volume, and plasma total protein concentration before and after hemorrhage and in the shed blood. The approach used for man is described in Grant and Reeve (1951). It is seen that animals and man add protein equivalent to that in a volume of plasma equal to 5–15% of their *blood volume* within a few hours after severe hemorrhage. Since after severe hemorrhage liver protein synthesis is depressed, this protein must have been transferred from interstitial fluids.

as perhaps in thyroid disease. The effect of altered distribution of hyaluronic acid or other connective tissue constituents on interstitial protein distribution is not known. More knowledge of the above problems would throw new light on the pathophysiology of common and dangerous diseases.

VIII. Summary

Evidence is given of the constancy of albumin in the plasma in health. Physiological and mathematical arguments and analog computer simulations show that this implies constancy of interstitial albumin. Plasma proteins pass through capillaries to enter the interstitial fluids and are returned by lymphatics to the blood stream. Current knowledge of the structure of capillary walls, interstitial spaces, and lymphatic walls is reviewed insofar as it bears on distribution and flow of interstitial protein. Measurements of interstitial albumin in particular tissues and in the whole body are examined and interstitial protein is shown to be distributed among areas of rapid flow but small interstitial fluid volume, as in the viscera, and slow flow but large interstitial fluid volume, as in the skin and muscles. Simple steady state models of interstitial water and interstitial protein in tissue segments are described. These predict that for interstitial protein in a segment to remain constant, protein inflow must be independent of water inflow. Vesicular transport of protein through the capillary endothelia of skin and muscle, suggested by the studies of Palade and collaborators, provides such independence with a mechanism for maintaining the large mass of interstitial protein in skin and muscle about constant. In the interstitial segments of liver and intestines protein inflow varies with water inflow but these sites contain only a small fraction of the total interstitial protein and so these variations are inconsequential. In transient states a mechanism is required to prevent sudden increases of interstitial water inflow transferring much interstitial protein into the plasma. The network of connective tissue fibers probably delays such a transfer and provides this mechanism. Factors controlling the distribution of proteins of different sizes in the interstitial fluids are briefly noted, as well as some common disturbances of interstitial protein distribution, which cannot be explained on present knowledge.

Acknowledgments

This research was supported by Grant HE-02262 from the National Institutes of Health U. S. Public Health Service, Bethesda, Maryland.

Several discussion with Dr. Arthur C. Guyton have been helpful in developing some of the ideas expressed here.

References

Atencio, A. C., and Reeve, E. B. (1965). *J. Lab. Clin. Med.* **66**, 20.

Atencio, A. C., Chao, P.-Y., Chen, A. Y., and Reeve, E. B. (1969). *Amer. J. Physiol.* **216**, 773.

Bennett, H. S., Luft, J. H., and Hampton, J. C. (1959). *Amer. J. Physiol.* **196**, 381.

Bruns, R. R., and Palade, G. E. (1968a). *J. Cell Biol.* **37**, 244.

Bruns, R. R., and Palade, G. E. (1968b). *J. Cell Biol.* **37**, 277.

Casley-Smith, J. R. (1967). *In* "New Trends in Basic Lymphology" (J. M. Collette, ed.), pp. 19–39. Birkhauser, Basel.

Campbell, R. M., Cuthbertson, D. P., Matthews, C. M., and McFarlane, A. S. (1956). *Int. J. Appl. Radiat. Isotop.* **1**, 66.

Catchpole, H. R., Joseph, N. R., and Engel, M. B. (1966). *Fed. Proc. Fed. Amer. Soc. Exp. Biol.* **25**, 1124.

Chien, S., Usami, S., Simmons, R. L., McAllister, F. F., and Gregersen, M. I. (1966). *J. Appl. Physiol.* **21**, 583.

Clementi, F., and Palade, G. E. (1969). *J. Cell Biol.* **41**, 33.

Cotran, R. S. (1967). *In* "Physical Bases of Circulatory Transport: Regulation and Exchange" (E. B. Reeve and A. C. Guyton, eds.), pp. 249–275. Saunders, Philadelphia, Pennsylvania.

Dewey, W. C. (1959). *Amer. J. Physiol.* **197**, 423.

Drinker, C. K., and Yoffey, J. M. (1941). "Lymphatics, Lymph and Lymphoid Tissue," Ch. 4. Harvard Univ. Press, Cambridge, Massachusetts.

Freeman, T., and Hunter, W. M. (1967). *In* "Handbook of Experimental Immunology" (D. M. Weir, ed.), Chs. 17 and 18. Davis, Philadelphia, Pennsylvania.

Furneaux, R. W., and Tracy, G. D. (1969). *Aust. J. Exp. Biol. Med. Sci.* **47**, 447.
Gitlin, D. (1957). *Ann. N.Y. Acad. Sci.* **70**, 122.
Grant, R. T., and Reeve, E. B. (1951). *Med. Res. Counc.* (*Gt. Brit.*) *Spec. Rep. Ser.* **277**.
Guyton, A. C. (1965). *Circ. Res.* **16**, 452.
Guyton, A. C., and Coleman, T. G. (1967). *In* "Physical Bases of Circulatory Transport: Regulation and Exchange" (E. B. Reeve and A. C. Guyton, eds.), Ch. 11. Saunders, Philadelphia, Pennsylvania.
Hoffenberg, R. (1965). *Tech. Rep. Ser. Int. At. Energy Ag.* **45**, 217.
Landis, E. M., and Pappenheimer, J. R. (1961). *In* "Handbook of Physiology, Section 2: Circulation," (W. F. Hamilton, ed.), Vol. 2, Ch. 29. Am. Physiol. Soc., Washington, D.C.
Laurent, T. C. (1966a). *Fed. Proc. Fed. Amer. Soc. Exp. Biol.* **25**, 1037.
Laurent, T. C. (1966b). *Fed. Proc. Fed. Amer. Soc. Exp. Biol.* **25**, 1128.
Laurent, T. C., and Killander, J. (1964). *J. Chromatogr.* **14**, 317.
McIntosh, D., Mosser, E. L., Lackey, W., and Franks, J. J. Unpublished observations.
Majno, G. (1965). *In* "Handbook of Physiology, Section 2: Circulation," (W. F. Hamilton, ed.), Vol. 3, Ch. 64. Am. Physiol. Soc., Washington, D.C.
Matthews, C. M. E. (1957). *Phys. Med. Biol.* **2**, 36.
Mayerson, H. S. (1963). *In* "Handbook of Physiology, Section 2: Circulation," (W. F. Hamilton, ed.), Vol. 2, Ch. 30. Am. Physiol. Soc., Washington, D.C.
Mayerson, H. S., Patterson, R. S., McKee, A., LeBrie, S. J., and Mayerson, P. (1962). *Amer. J. Physiol.* **203**, 98.
Morris, B. (1956). *Quart. J. Exp. Physiol.* **41**, 326.
Moulton, F. R. (1958). "Differential Equations," Ch. 17. Dover, New York.
Norberg, R. (1969). Unpublished observations.
Polson, A., and Katz, W. (1969). *Biochem. J.* **112**, 387.
Reeve, E. B., and Bailey, H. R. (1962). *J. Lab. Clin. Med.* **60**, 923.
Reeve, E. B., and Roberts, J. E. (1959). *J. Gen. Physiol.* **43**, 415.
Reeve, E. B., Gregersen, M. I., Allen, T. H., Sear, H., and Walcott, W. W. (1953). *Amer. J. Physiol.* **175**, 211.
Roberts, J. C., and Courtice, F. C. (1969). *Aust. J. Exp. Biol. Med. Sci.* **47**, 421.
Rothschild, M. A., Bauman, A., Yalow, R. S., and Berson, S. A. (1955). *J. Clin. Invest.* **34**, 1354.
Rusznyak, I., Foldi, M., and Szabo, G. (1967a). "Lymphatics and Lymph Circulation," 2nd ed., Ch. 5. Macmillan (Pergamon), New York.
Rusznyak, I., Foldi, M., and Szabo, G. (1967b). "Lymphatics and Lymph Circulation," 2nd ed., Ch. 6. Macmillan (Pergamon), New York.
Rusznyak, I., Foldi, M., and Szabo, G. (1967c). "Lymphatics and Lymph Circulation," 2nd ed., pp. 370–374. Macmillan (Pergamon), New York.
Schubert, M. (1966). *Fed. Proc. Fed. Amer. Soc. Exp. Biol.* **25**, 1047.
Sellers, A. L., Katz, J., Bonorris, G., and Okuyama, S. (1966). *J. Lab. Clin. Med.* **68**, 177.
Shea, S. M., and Karnovsky, M. J. (1969). *J. Theoret. Biol.* **24**, 30.
Takeda, Y. (1966). *J. Clin. Invest.* **45**, 103.
Takeda, Y., and Reeve, E. B. (1963). *J. Lab. Clin. Med.* **61**, 183.
Yoffey, J. M., and Courtice, F. C. (1965a). "Lymphatics, Lymph and Lymphoid Tissue," Ch. 3. Harvard Univ. Press, Cambridge, Massachusetts.
Yoffey, J. M., and Courtice, F. C. (1956b). "Lymphatics, Lymph and Lymphoid Tissue," Ch. 8. Harvard Univ. Press, Cambridge, Massachusetts.
Zierler, K. L. (1962). *In* "Handbook of Physiology, Section 2: Circulation," (W. F. Hamilton, ed.), Vol. 1, Ch. 18. Am. Physiol. Soc., Washington, D.C.

7

THE EXTRAVASCULAR TRANSPORT OF ALBUMIN

Victor M. Rosenoer and Marcus A. Rothschild

CHILDREN'S CANCER RESEARCH FOUNDATION,
BOSTON, MASSACHUSETTS; VETERANS ADMINISTRATION HOSPITAL,
NEW YORK UNIVERSITY, SCHOOL OF MEDICINE, NEW YORK,
NEW YORK

I. Introduction

Serum albumin is produced in the liver; about 30 min is required for its passage from the site of assembly on the membrane-bound polysomes in the parenchymal cells to its appearance in the plasma. Thereafter, it is distributed to various vascular pools at rates reflecting the rate of blood flow to the area or organ under study. Thus, after injecting labeled albumin intravenously, there is rapid equilibration over the head and liver, reflecting blood flow of 750 and 1400 ml/min to the organ. Equilibration within the muscle blood pool is much slower requiring 8–10 min: This organ mass, which contains 2–4% of the vascular volume, has an average blood flow of only 1–7 ml/min/100 g tissue (Bauman and Rothschild, 1956).

Despite the relative rapidity with which substances are distributed throughout the intravascular fluid, the extravascular fluid must be considered as the predominant factor in the maintenance of cell growth and function—constituting the "milieu interieur" of Claude Bernard in which the body cells live and with which they exchange nutrients and waste products. Starling (1896) pointed out that since the capillary endothelial membranes were relatively impervious to proteins, but almost freely permeable to water and solutes of small molecular weight, a colloid osmotic gradient would develop between the protein-rich extravascular fluids, a gradient which would counterbalance the hydrostatic pressure gradient between the intra- and extravascular spaces. The balance of forces would be in favor of fluid loss to the extravascular spaces at the level of the arterial ends of the capillaries and, at the venous ends, the decreased hydrostatic pressure and increased colloidal osmotic pressure would ensure that the direction of fluid flow was reversed. The role attributed to the lymphatic vessels was that of hydrodynamically inert drains slowly drawing off the residual excess extravascular fluid and returning it by way of the major lymph vessels to the venous circulation.

In Starling's hypothesis the capillary walls were regarded as being practically impermeable to plasma proteins although, as Drinker and Field pointed out (1931), the protein content of the extravascular spaces and of the lymph which drains them was known to be far from negligible. Drinker and his colleagues challenged the general conviction that the walls of typical blood capillaries throughout the body did not permit the escape of more than traces of plasma protein. Impressed by the considerable concentration of protein in the lymph from cannulated lymphatic channels in the dog's neck, limbs, and kidneys, and in thoracic duct lymph, they proposed a hypothesis complementary to that outlined by Starling. According to Drinker's views, the filtrate from the blood capillaries in practically all parts of the body contains all the proteins of blood albeit in fairly high dilution. This filtrate is subject to concentration and to dilution depending upon capillary blood pressure and other varying contingencies. Water and salts are reabsorbed by blood vessels: Protein enters the lymphatics together with water and salts in the concentration existing in the tissue fluid at the moment of lymphatic entrance and is returned to the blood vessels by way of the lymphatic vessels.

II. Proteins of the Extravascular Tissue Spaces

Specific immunoprecipitation, immunoelectrophoretic, and immunofluorescent techniques have amply confirmed the presence in the extravascular tissue spaces of nearly all the different plasma proteins. Thus, Gitlin *et al.* (1955) using a specific anti-albumin serum has proved that "myoalbumin" is identical with serum albumin while Szabo *et al.* (1963) using immunoelectrophoresis demonstrated that the patterns of lymph from various sites display only minor differences and collectively resemble dilute plasma.

Studies by Lewis (1921) and by Field and Drinker (1931) with foreign proteins which, after intravenous injection, were rapidly recovered in the lymph, had earlier suggested that the predominant part of the plasma proteins found in extravascular sites was derived from the blood rather than from local synthesis. Similar behavior was observed in the case of serum albumin using tracer dyes which are firmly bound to the albumin molecules both *in vivo* and *in vitro*. The extravascular escape of circulating albumin tagged with any of these dyes, e.g., T-1824 (Evans blue) is reflected in the rapid decline of the plasma-concentration curve and is accompanied by the appearance of labeled albumin in lymph draining various body sites, the bluish discoloration of the dermal tissues and microscopically observable extravasation of the dye into the pericapillary areas. Inumerable repetitions of these studies using isotopically labeled plasma proteins, following the pioneer studies of Abdou and Tarver (1951) and Sterling (1951), have amply confirmed that plasma proteins begin to leave the circulation immediately after intravenous injection. Mancini and his colleagues (1962) have localized histologically the extravasated protein in rat dermal connective tissue after the intravenous injection of homologous serum whose proteins were labeled with Rhodamine B. Sixty minutes after the injection, labeled serum proteins could be seen outside the capillary spreading along the fibers and through the connective tissue spaces of the surrounding tissues.

The quantitative importance of the extravascular circulation of the plasma proteins is considerable—the mass of extravascular protein in equilibrium with the plasma being as great or greater than that present in the plasma (Rothschild *et al.*, 1955; Beeken *et al.*, 1962). The velocity of circulation of extravascular protein, however, is slow compared with that in the intravascular compartment. As calculated from plasma protein kinetics, the extravascular pool of protein is recycled only about once daily instead of once per minute for the intravascular pool. The major portion of the extravascular circulation is eventually channeled through the thoracic duct. Thus, Forker *et al.* (1952) showed that a thoracic duct fistula produced a loss of about two-thirds of the plasma protein from the blood stream per day in unanesthetized rats and a loss of about half of the plasma protein per day in anesthetized dogs. In this study, the rats lived for about 11 days, losing a total of 10.7 g of plasma proteins in this period (an amount equivalent to 7.3 times their estimated intravascular protein mass).

Observations by Clark and Clark (1933) on the movements of tracer particles injected into the tissue spaces of tadpole's tail and by McMaster and Parsons (1939a,b) on particles injected into the mouse ear tip lymphatics have demonstrated that the interstitial fluid under normal conditions does not flow freely through a system of intercommunicating lacunae, but is restrained by submicroscopic channels. Gersh and Catchpole (1960) have put forward a two-phase concept of interstitial material supported by electron microscopic evidence. The ground substance remains in osmotic equilibrium with the plasma, despite great fluctuations in the state of hydration, by varying the distribution of water between the two phases—a semisolid gel phase in which water is tightly bound to the matrix—the proteins and acid polysaccharides of the ground substance, and a fluid sol phase contained in inclusion vacuoles embedded within the matrix. It appears that the movement of particles and large molecules is restricted to the sol phase. Ogston (1962) and Ogston and Phelps (1961) have shown that for steric reasons large protein molecules such as albumin are not permitted to occupy the layer bound to the highly hydrophilic long-chain molecule of hyaluronic acid. The exclusion principle operating in the interstitial spaces is exactly the same as that obtaining in gel filtration where advantage is taken of the property to achieve a molecular sieve.

III. Role of the Lymphatics in the Return of Proteins to the Circulation

Experiments in which labeled or foreign proteins injected intravenously are identified in the contents of the draining lymphatic vessels within a few minutes, while providing excellent evidence of capillary permeability to these proteins, do not provide evidence that the lymphatic route is the sole means of return to the venous system of extravasated plasma proteins. Lewis (1921) was able to detect horse serum injected subcutaneously into dogs in thoracic duct-lymph within 40 min but in the bloodstream only after 3 1/2 hr. Conklin (1930) failed to find any sign of immediate absorption into the blood of foreign proteins injected into the lymph spaces of the frog's back, while similar findings were reported by Field and Drinker (1931) following the injection of horse serum into dogs. Barnes and Trueta (1941), in a more dramatic demonstration, reported that severing of all the lymphatics draining the site of snake venom injection into the rabbit's leg prior to the injection, protected the animals against the toxic effects of the injected mixture of enzyme proteins—an effect unlikely if there had been a significant direct venous pathway of return. Courtice and Simmonds (1949) and Courtice and Steinbeck (1950, 1951) found that plasma albumin, labeled with Evans blue (T-1824), injected into the pleural or peritoneal cavities, was almost quantitatively absorbed by the lymphatics. McMaster (1937) has given a morphological demonstration of the role of the lymphatics in the reabsorption of protein from interstitial tissues. Intradermally injected vital dyes with high affinities for serum albumin, e.g., Evans blue, Trypan blue, patent blue V, are removed from the injection site, together with bound albumin, along a network of tortuous channels extending not along the venous circulation but toward the nearest lymph nodes.

IV. Delivery of Newly Synthesized Albumin to the Plasma

Although the evidence available strongly suggests the predominant role of the lymphatic pathway in the return of albumin from the extravascular space to the vascular system in peripheral tissues, this is by no means true of all tissues. The delivery of newly synthesized albumin from its synthetic site in the liver provides a model which should be considered.

In a study of albumin synthesis using McFarlane's (1963) ^{14}C-carbonate technique, Tavill *et al.* (1968) presented evidence which suggested that newly synthesized albumin might also reach the plasma by an indirect route. Intravenously injected ^{14}C as sodium ^{14}C-carbonate is incorporated by way of the Krebs-Henseleit cycle into the guanidine carbon of arginine in the liver, where it is available for the synthesis of albumin and urea. Despite their synthesis from a common pool of labeled arginine, 6-^{14}C-arginine-labeled albumin appeared in the plasma 30–60 min after ^{14}C-urea. This delay could be explained if the urea and albumin entered the intravascular compartment by different routes. Urea (MW 60) would be more likely to pass directly into the sinusoids than albumin (MW 66,000), the ingress of which might be impeded by the sinusoidal wall. If this were so, albumin might then have to pass from the space between the liver cell and the sinusoidal wall (the space of Disse) to the hepatic lymphatics and thoracic duct before being delivered into the intravascular compartment.

Clearly the permeability characteristics of the sinusoidal endothelium are a major factor determining the route taken by a protein molecule after its release from the liver cell. Electron microscopy has shown the presence of large gaps between the endothelial cells, some as large as several thousand angstroms (Fawcett, 1955; Hampton, 1958; Bennett *et al.*, 1959; Suzuki *et al.*, 1961), far larger than the maximum diameter of the albumin molecule (approximately 160 Å, Fruton and Simmonds, 1958) at present under consideration. It is well established that the macromolecular constituents of plasma leak more readily from the hepatic sinusoids than from any other capillary circulation in the body (Yoffey and Courtice, 1956; Grotte *et al.*, 1951; Szabo *et al.*, 1963). However, the maximum size of the molecule to which the sinusoid is freely permeable is undetermined (Grotte, 1956; Mayerson *et al.*, 1960), although Grotte *et al.* (1960) have demonstrated the rapid appearance in liver lymph of intravenously injected particles of methyl methacrylate with a radius of about 500 Å (300–700 Å).

Smallwood *et al.* (1968) evaluated the role of the lymphatic system in the delivery of newly synthesized albumin to the plasma. Eight adult dogs, lightly anesthetized with intermittent intravenous sodium pentobarbitone, were studied. Blood samples were taken from a cannula in the right femoral artery and thoracid duct lymph was collected from a cannula inserted into the duct at the level of the seventh rib, any subsidiary mediastinal lymphatics being ligated to ensure complete hepatic lymph collections. Preliminary experiments using dye injections into the liver revealed no major route of hepatic lymph drainage other than the thoracic duct. At time zero, 400 μCi of sodium ^{14}C-carbonate and an accurately weighed quantity of ^{131}I-labeled human albumin (approximately 5 μCi) were injected intravenously. At 10 min blood was taken for estimation of the plasma volume from the plasma ^{131}I activity. After 5 hr, a further weighed quantity of ^{131}I-albumin was given to remeasure the plasma volume and additional blood was taken to measure the plasma albumin concentration and plasma albumin-^{14}C-arginine specific activity. The albumin ^{14}C-arginine specific activity in blood and lymph was determined by the methods described by Tavill *et al.* (1968) and the total ^{14}C-arginine activity of albumin in the plasma at 5 hr was calculated from the product of the measured plasma volume, albumin concentration, and specific activity: that of the lymph from the total volume of lymph collected; its albumin concentration; and its specific activity. In order to calculate the total ^{14}C-arginine albumin delivered into the intravascular compartment, a correction was applied to the activity present at 300 min to allow for losses from the plasma by distribution to the extravascular pools and by catabolism. This correction factor was determined in each dog from the plasma disappearance curves of ^{131}I-albumin injected intravenously at zero time. The error involved in utilizing the data obtained using ^{131}I-human albumin in this study could in no way modify the interpretation of the data.

TABLE I

New Albumin Transfer to Thoracic Duct Lymph

	^{131}I-Albumin in lymph (% injected dose)	^{14}C-Albumin in lymph (% corrected iv ^{14}C-albumin)
Controls		
Dog 1	8.7	8.8
Dog 2	4.8	3.3
Dog 3	5.7	3.3
Dog 4	3.1	2.5
Dog 5	3.8	2.3
Hepatic venous obstruction		
Dog 6	17.2	13.3
Dog 7	10.2	7.2
Dog 8	11.3	11.4

The results of this study are given in Table I. In control dogs 1–5 only a small proportion (4.04 ± 1.19%, mean ± SE) of the newly synthesized 6-^{14}C-arginine-labeled albumin appeared in the lymph (Table I). The proportion of the intravenously injected ^{131}I-albumin appearing in the lymph in the same experimental period was also small, the correlation between the ^{14}C- and ^{131}I-albumin recoveries in each dog being statistically significant ($r = 0.94$; $0.05 > P > 0.01$).

In dogs 6–8 with partial hepatic venous obstruction the proportion of labeled albumin recovered in the lymph was significantly higher than in the control animals ($P < 0.01$). Again there was close similarity between the ^{14}C- and ^{131}I-albumin recoveries in each animal, so that the overall correlation (dogs 1–8) was highly significant ($r = 0.95$; $P < 0.001$).

If delivery of newly synthesized plasma protein into the intravascular compartment were predominantly by way of the hepatic lymphatics and thoracic duct, relatively little 6-^{14}C-arginine-labeled protein would be found in the plasma in an experiment in which the thoracic duct was drained. The very low recoveries of ^{14}C-albumin in the thoracic duct lymph and the high recoveries in the plasma in the present study indicated clearly that the lymphatic system was not the principal route by which this protein reached the intravascular compartment. Further, the close agreement between the recoveries of newly synthesized ^{14}C-albumin and intravenously injected ^{131}I-albumin from the lymph in each animal suggested that the newly synthesized albumin reaching the thoracic duct could be adequately explained by leakage from the intravascular compartment during the experiment. The slightly higher ^{131}I-albumin recoveries in most of the dogs may reflect the fact that all the ^{131}I-protein was administered as a single intravenous injection whereas the ^{14}C-albumin was released more slowly into the intravascular compartment (Tavill *et al.*, 1968). However, a systematic overestimate of the correction factor used in calculating the total ^{14}C-albumin appearing in the plasma, attributable to the use of human rather than canine albumin, would lead to an underestimate of the proportion of ^{14}C-albumin recovered in the lymph.

It is difficult to envisage protein delivery from the liver cell to the plasma by a route which bypasses the lymphatics, unless it be by immediate entry into the vascular compartment through the hepatic sinusoidal wall. Further, the close agreement between the proportion of biosynthesized ^{14}C-albumin and intravenously injected ^{131}I-albumin recovered in the lymph in each dog must indicate that the hepatic sinusoidal wall is freely permeable to this protein. The reported differences with respect to plasma proteins between the composition of plasma and of hepatic lymph [the hepatic lymph protein concentration being slightly lower and the proportion of albumin in the lymph protein slightly higher than in plasma (Brauer, 1963)] probably arise from admixture of sinusoidal (space of Disse) lymph with that from the peribiliary plexus, rather than from the sinusoidal endothelium presenting a perceptible barrier to plasma protein movements. Smallwood *et al.* proposed a model in which the hepatic sinusoidal wall is freely permeable to albumin and fibrinogen, the sinusoid and space of Disse forming a single mixing pool, the distribution of these proteins between the hepatic venous and lymphatic outflow tracts being governed by the relative rates of flow in these tracts.

V. Models of Distribution of Albumin in the Interstitial Tissues

This diversion into a highly specialized situation has been made to stress the differences between the hepatic model and that assumed to occur in other tissues. Thus, Areskog and his co-workers (1964) in their study on the kinetics of distribution of ^{131}I-albumin in the dog heart–lung preparation adopted a model in which the passage of albumin from blood to the extravascular space was due to filtration alone and any protein so filtered could only leave the extravascular space by way of the lymph. This assumption is unlikely to be valid as it implies that material which has entered the extravascular space cannot diffuse back across the capillary wall. Duncan *et al.* (1958) have developed a more realistic model which takes into account this objection. The sites of entrance of albumin into the tissue are at the arterial and venous ends of the capillaries: Exit occurs at the sites of entrance and also via the lymphatics. Water filters from serum to tissue at the arterial ends of capillaries and from tissue to plasma at the venous ends, at each site transporting some albumin. Gross diffusion of albumin between plasma and tissue occurs in each direction at both arterial and venous capillary sites. In the development of this model, it is assumed that interstitial fluid passes into the lymphatics without change in composition and is removed from the tissue as lymph. Further, it is assumed that mixing within the tissue is instantaneous and that the transfer processes are of the first order with respect to the plasma and tissue concentrations. From this model and current concepts of filtration and diffusion through porous membranes (Pappenheimer, 1953) Duncan and his co-workers have derived a quantitative formulation of the steady-state ratio of the concentration of albumin in the interstitial water of the tissue to its concentration in plasma (see Appendix). Schultze and Heremans (1966) have considered a simplified version of this model, regarding the hydrostatic return flow of protein via the veins as virtually inoperative merely "complicating the already cumbersome expression," while Johnson (1966) has discussed a further simplification of the model.

From these models, it is apparent that in the absence of lymph flow, even very low capillary permeability would allow equalization of the concentrations of albumin on either side of the capillary wall and, conversely, an increase in the rate of lymph flow in relation to the rate of transfer of albumin from the capillary to the interstitial water would be accompanied by a decrease in the ratio of the interstitial to serum albumin concentration. An increase in the permeability of the capillary walls for larger molecules—without an accompanying increase in the flow rate of fluid—would increase the ratio of interstitial to serum albumin concentration and at the limit when no restriction to the passage of albumin remained, the protein would attain the same concentration in the tissue fluid as in the plasma. The hepatic mixing pool discussed by Smallwood *et al.* (1968) clearly represents one aspect of the model proposed by Duncan *et al.* (1958).

In the light of this discussion, it becomes obvious that the ratio of interstitial to serum albumin concentration can at best give an estimate of the ratio between the rate constants at which albumin leaves and returns to the vessels. The concentration ratio has been used as first approximation of the permeability coefficient for a high molecular weight substance in tissues under study. Thus, Grotte (1956) studied the lymph:plasma concentration ratios for dextrans of various molecular dimensions in an attempt to quantitate capillary permeability in various tissues. However, Mayerson and his coworkers (1960) recognizing the importance of fluid flow rates in these tissues, chose radioactive iodinated albumin as a common denominator and designated the "relative permeability coefficient" as the ratio between the dextran to albumin lymph:plasma concentration ratios. Variations in lymph flow, filtration, and reabsorption will affect the concentration ratios both of dextran and of albumin, the absolute value of the "relative permeability coefficient" being essentially unchanged. In sum, the tissue:plasma concentration ratio can be used with profit only if one wishes to compare the permeability coefficients of several substances studied simultaneously, so as to cancel the effects of variations in the movement of water.

The influence of flow rate is convincingly demonstrated in an experiment by Grotte (1956) who measured the passage rates of dextran molecules of different molecular weights from the dog's plasma into the leg lymph in terms of the lymph:plasma equilibrium concentration ratios. When the capillary filtration rate was doubled by venous congestion, the concentration of dextran in the lymph dropped appreciably while the molecular sieving became more selective with a significant shift of the distribution in lymph to lower molecular weights. It was also clear that despite the decreased lymph concentration, for dextrans of a given molecular size, the total amount removed by the lymphatic route increased—a situation in the limb analogous to that discussed by Smallwood *et al.* in relation to venous congestion of the liver.

It is now quite clear that plasma proteins normally leave the vascular compartment in all capillary beds and return to the vascular compartment either directly by diffusion and to a lesser degree by hydrostatic return flow or indirectly by way of the lymphatics. The relative role of the two return pathways will depend predominantly upon the relative barriers presented to plasma protein movements at the capillary and lymphatic sites and the relative flow rates in the capillaries and lymphatics concerned.

The problem of the mechanisms and morphological aspects of the passage of plasma proteins through capillary walls has been extensively reviewed—most recently by Schultze and Heremans (1966) who conclude that protein molecules of small size, comprising those of albumin, and most of the plasma globulins, are sifted out by restricted diffusion through the action of the basement membrane. Larger molecules may escape in limited quantities by bulk flow through leaks consisting of patent intercellular junctions which may normally exist in larger numbers and with wider dimensions in the sinusoidal vessel type such as those in the liver or in ordinary capillaries in inflammatory states.

The lymphatic vessels appear to behave as essentially open vessels, presenting little resistance to the inflow of extravascular fluid with its contained protein. McMaster (1947) has shown by direct measurements on the cutaneous lymphatics of the mouse's ear that there is a small gradient of pressure (1.9 against 1.2 cm water) from the interstitial spaces to the lumen of the lymphatic capillaries. Electron microscopic studies suggest that the intercellular junctions of the lymphatic capillary endothelium yield easily to the pressure of the surrounding tissue fluid, so as to allow the passage of fluid, macromolecules and even particles as large as red blood cells (Mayerson, 1963).

VI. Transport of Albumin into Ascitic Fluid

The transport of albumin into abnormal collections of fluid has long been a source of interest, for the accumulation and disappearance of these effusions may occur rapidly and not always in relationship to understandable disease processes. The only abnormal fluid collection which has been studied in detail in terms of protein is ascites. Electrophoretic and immunoelectrophoretic analysis of ascitic fluid have revealed repeatedly that there is a strong qualitative resemblance in protein patterns between this fluid and plasma although there is a tendency for the albumin and α_1-globulin fractions to be present in slightly higher concentrations and for the α_2-, β-, and γ-globulins to be present in slightly lower concentrations (*vide* Schultze and Heremans, 1966). However, when the total protein content of ascitic fluids of different origin is compared with that of edema fluid in the same disease states, it can be seen that marked discrepancies may occur. In the nephrotic syndrome and in the hypoproteinemic edema of malnutrition, both types of effusion have a similar low protein content: about 5% of that in the plasma. In the ascites of hepatic cirrhosis or of congestive heart failure or of hepatic venous congestion there is 3–10 times as much protein as in the corresponding edema fluid.

The origin of the albumin of ascites is, therefore, a matter of some interest. If labeled albumin is injected intravenously, it appears in the ascitic fluid during the first 24 hr but reaches equilibrium only slowly (Berson and Yalow, 1954). On the other hand, radioiodinated albumin injected into the ascitic fluid may require as long as 30 days for distribution equilibrium with plasma and other tissues. Depending upon the extent of the effusion and its etiology, the amount of albumin exchanged daily between ascitic fluid and plasma may vary from 0.3 to 3.0 g—a quantity reflecting the high permeability of the intestinal capillaries, the peritoneal lymphatics, and the hepatic sinusoids. However, these observations do not permit the elucidation of the pathway by which albumin enters ascitic fluid—whether directly by way of the very permeable liver sinusoids reaching the peritoneal cavity by way of the liver capsule or indirectly by way of the plasma—or by way of a space rapidly in equilibrium with the plasma.

Zimmon and his co-workers (1969) have studied this problem by comparing the amount of newly biosynthesized ^{14}C-albumin, following the intravenous injection of sodium-^{14}C-carbonate, which appeared in the ascitic fluid with that which appeared in the plasma. Intravenously injected ^{131}I-albumin served as a marker for the transfer of plasma albumin into the ascites and permitted a simple calculation of the amount of plasma derived ^{14}C-albumin in the ascitic fluid. Eight patients with alcoholic cirrhosis and ascitic (ranging in volume from 1250 ml to 19 liter) were studied. In two of the patients the thoracic duct lymph was collected. Plasma and ascites were sampled simultaneously 2 hr after the intravenous injection of 200 μCi of sodium ^{14}C-carbonate and 10 μCi of ^{131}I-albumin and the samples analyzed for

TABLE II

NEW ALBUMIN TRANSFER TO ASCITES IN 2 HR

	Ascites volume (liters)	Total ^{131}I-albumin plasma/ascites (%)	Total ^{14}C-albumin plasma/ascites (%)
Mean values 8 patients	1.25–19.7	0.9	6.2

^{14}C-arginine of albumin and ^{131}I-albumin by standard procedures, the results being expressed in terms of the plasma pool.

While 0.4–2.2% of the plasma ^{131}I-albumin appeared in the ascitic fluid during the 2-hr observation period, between 4.0 and 11.7% of the newly biosynthesized plasma ^{14}C-albumin appeared in the ascitic pool in the same period. Transfer of ^{14}C-albumin from the plasma could account for only 6–25% of the ascitic ^{14}C-albumin. In the two patients with thoracic duct lymph drainage, there was no reduction in the proportion of ^{14}C-albumin recovered in the ascitic fluid. However, it was possible to demonstrate that more newly synthesized albumin was transferred into the lymph than

TABLE III

NEW ALBUMIN TRANSFER TO LYMPH IN 2 HR

	Total ^{131}I-albumin plasma/lymph (%)	Total ^{14}C-albumin plasma/lymph (%)
2 Studies	2.8–3.8	6.1–13.5

could be accounted for by plasma to lymph transfer (2.8–3.8% of the plasma ^{131}I-albumin appearing in the lymph while 6.1–13.5% of the newly synthesized ^{14}C-albumin was recovered in the lymph over the same time interval).

The results of this study indicated clearly that all of the albumin moving into ascitic fluid in patients with ascites could not have been derived from the systemic circulation. A more direct extravascular route either by direct lymphatic drainage or by direct loss through the hepatic capsule must have been involved. In animals with experimental suprahepatic venous obstruction, most of the ascitic fluid produced appears to reach the peritoneal cavity by way of the liver capsule (McKee *et al.*, 1948; Volwiler *et al.*, 1950; Hyatt and Smith, 1954). More recently Tanikawa (1968) has provided direct evidence for this transcapsular route by infusing colloidal gold into the portal vein of rats. Electron microscopic studies showed the particles extending from the liver surface.

It appears, then, that a proportion of the albumin synthesized in patients with cirrhosis of the liver may be transported directly into the peritoneal cavity by a route which bypasses the systemic circulation. This route may be of great importance in the regulation of ascites accumulation and certainly must be considered in the kinetic analysis of albumin distribution in synthetic and catabolic studies in patients with ascites.

VII. Delay in the Delivery of Albumin from Its Synthetic Site in the Liver

It was of interest that in the study by Zimmon *et al.* (1969) no ^{14}C-albumin appeared in the lymph in the first 30 min—an interval which corresponds with the delay discussed by Smallwood et *al.* (1968) in the appearance of newly synthesized ^{14}C-albumin in the plasma. This delay was probably dictated by events which occur before the release of albumin from the liver cell. Studies on the synthesis and secretion of α-chymotrypsinogen by the pancreatic exocrine cell have shown that 1 min or less is required to synthesize this particular protein whereas approximately 45 min are required for transport to its storage site in the zymogen granules (Siekevitz and Palade, 1960; Palade, 1966). Peters (1962), from a study of the incorporation of 1-^{14}C-leucine into serum albumin from various fractions of rat liver cells, found that newly formed immunologically identifiable albumin appeared in the microsomes of the "rough" endoplasmic reticulum within 1.5 min and appeared to pass to the microsomes of the "smooth-surface" endoplasmic reticulum 3–4 min later, remaining bound to the smooth endoplasmic reticulum for about 15 min. Little soluble ^{14}C-albumin was detected intracellularly. Thus, the delay in the appearance of newly synthesized labeled albumin appears to be due to the delayed release of the fully formed molecule, rather than the time taken to synthesize it. However, immunofluorescent techniques have either failed to demonstrate appreciable amounts of albumin in liver cells (Gitlin *et al.*, 1953) or have shown albumin in only a small percentage of cells (Hamashima *et al.*, 1964). This may reflect a low rate of albumin production per cell, or simply the technical problems of producing a satisfactory visualization of intracellular albumin.

Appendix

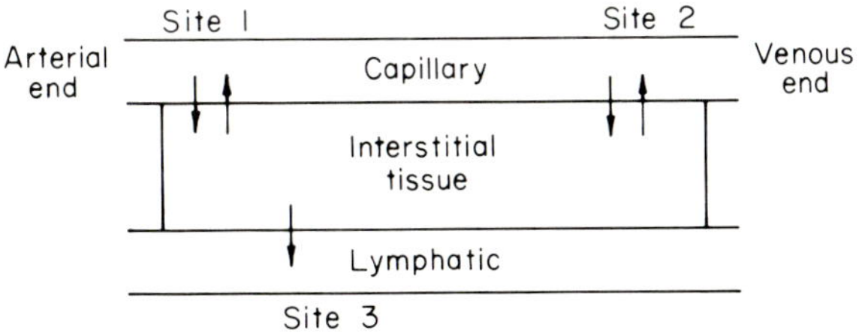

FIG. 1. Sites of entry and exit of albumin in the interstitial tissues.

In the model (Fig. 1) to be analyzed, the following assumptions have been made:

(1) Albumin enters the interstitial tissues at the arterial and venous ends

of the capillaries (sites 1 and 2) and leaves at the sites of entry and also by way of the lymphatics (site 3).

(2) Mixing within the interstitial tissue is instantaneous.

(3) Albumin enters and leaves the interstitial tissues by filtration and by diffusion and these transfer processes are of the first order with respect to the plasma albumin and tissue albumin concentrations.

(4) Water filters from the plasma to tissue at site 1 and from the tissue to the plasma at site 2, at each site transporting some albumin.

(5) Diffusion of albumin between plasma and tissue occurs in both directions at sites 1 and 2.

(6) Interstitial fluid passes into the lymphatics without changes in composition and is removed from the tissue as lymph.

From these assumptions we may write

$$\frac{dy}{dt} = k_1 p - k_2 y \tag{1}$$

where y = concentration of albumin in the tissue under consideration,
p = concentration of albumin in the plasma,
k_1 = amount of albumin transferred into 1 g tissue per day expressed as a fraction of the amount in 1 ml plasma,
k_2 = amount of albumin transferred from 1 g tissue per day expressed as a fraction of the amount in 1 g tissue,
t = time in days.

Now, let T represent albumin transport and let superscripts i and o indicate whether transport is into or out of the tissue and let subscripts d and f indicate whether transport is by diffusion or by filtration, respectively.

Then

$$\frac{dy}{dt} = {}^{\mathrm{i}}T_{\mathrm{d}} + {}^{\mathrm{i}}T_{\mathrm{f}} - {}^{\mathrm{o}}T_{\mathrm{d}} - {}^{\mathrm{o}}T_{\mathrm{f}} \tag{2}$$

Let A = area at a given site for the filtration of water in 1 g tissue, A' = restricted pore area at a given site for the passage of albumin in 1 g tissue, $\Delta\chi$ = length of path of passage through the capillary wall, F = rate of flow of water through exchange surface at a given site in 1 g tissue, D = diffusion coefficient for albumin, and use the site numbers as subscripts to identify the location of each A, A', $\Delta\chi$, and F.

Then, based upon the current concepts of diffusion and filtration through porous membranes,

$${}^{\mathrm{i}}T_{\mathrm{d}} = D\frac{A_1'}{\Delta\chi_1}p + D\frac{A_2'}{\Delta\chi_2}p \tag{3}$$

$$^{\mathrm{i}}T_{\mathrm{f}} = \frac{A_1'}{A_1} F_1 p \tag{4}$$

$$^{\mathrm{o}}T_{\mathrm{d}} = D \frac{A_1'}{\Delta\chi_1} y + D \frac{A_2'}{\Delta\chi_2} y \tag{5}$$

$$^{\mathrm{o}}T_{\mathrm{f}} = \frac{A_2'}{A_2} F_2 y + F_3 y \tag{6}$$

Substituting in Eq. (2) we obtain

$$\frac{dy}{dt} = \left[D \frac{A_1'}{\Delta\chi_1} + D \frac{A_2'}{\Delta\chi_2} + \frac{A_1'}{A_1} F_1 \right] p - \left[D \frac{A_1'}{\Delta\chi_1} + D \frac{A_2'}{\Delta\chi_2} + \frac{A_2'}{A_2} F_2 + F_3 \right] y \tag{7}$$

Comparison of Eqs. (7) and (1) reveals that

$$k_1 = D \frac{A_1'}{\Delta\chi_1} + D \frac{A_2'}{\Delta\chi_2} + \frac{A_1'}{A_1} F_1 \tag{8}$$

and

$$k_2 = D \frac{A_1'}{\Delta\chi_1} + D \frac{A_2'}{\Delta\chi_2} + \frac{A_2'}{A_2} F_2 + F_3 \tag{9}$$

whence

$$\frac{k_1}{k_2} = \frac{D \dfrac{A_1'}{\Delta\chi_1} + D \dfrac{D_2'}{\Delta\chi_2} + \dfrac{A_1'}{A_1} F_1}{D \dfrac{A_1'}{\Delta\chi_1} + D \dfrac{A_2'}{\Delta\chi_2} + \dfrac{A_2'}{A_2} F_2 + F_3} \tag{10}$$

Now, in the steady state the concentration of albumin in the plasma and tissue fluid is not changing so that $dy/dt = 0$ and $k_1 p = k_2 y$ thereby giving

$$\frac{k_1}{k_2} = \frac{y}{p} \tag{11}$$

Thus, Eq. (10) represents the steady state ratio between the concentration of interstitial tissue albumin to its concentration in the plasma.

ACKNOWLEDGMENTS

Supported in part by U.S.P.H.S. grant C-6516.

REFERENCES

Abdou, I. A., and Tarver, H. (1951). *J. Biol. Chem.* **190**, 781.

Areskog, N. H., Arturson, G., and Grotte, G. (1964). *Acta Physiol. Scand.* **62,** 209.

Barnes, J. M., and Trueta, J. (1941). *Lancet* **1**, 623.

Bauman, A., and Rothschild, M. A. (1956). *J. Lab. Clin. Med.* **48**, 20.

Beeken, W. L., Volwiler, W., Goldsworthy, P. D., Garby, L. E., Reynolds, W. E., Stogsdill, R., and Stemler, R. S. (1962). *J. Clin. Invest.* **41**, 1312.

Bennett, H. S., Luft, J. H., and Hampton, J. C. (1959). *Amer. J. Physiol.* **196**, 381.

Berson, S. A., and Yalow, R. S. (1954). *J. Clin. Invest.* **33**, 377.

Brauer, R. W. (1963). *Physiol. Rev.* **43,** 115.

Clark, E. R., and Clark, E. L. (1933). *Amer. J. Anat.* **52,** 273.

Conklin, R. E. (1930). *Amer. J. Physiol.* **95**, 98.

Courtice, F. C., and Simmonds, W. J. (1949). *J. Physiol.* (*London*) **109**, 117.

Courtice, F. C., and Steinbeck, A. W. (1950). *Aust. J. Exp. Biol. Med. Sci.* **28**, 161.

Courtice, F. C., and Steinbeck, A. W. (1951). *Aust. J. Exp. Biol. Med. Sci.* **29**, 451.

Drinker, C. K., and Field, M. E. (1931). *Amer. J. Physiol.* **97**, 32.

Duncan, L. E., Cornfield, J., and Buck, K. (1958). *Circulation Res.* **6**, 244.

Fawcett, D. W. (1955). *J. Nat. Cancer. Inst.* **15**, 1475.

Field, M. E., and Drinker, C. K. (1931). *Amer. J. Physiol.* **97**, 40.

Forker, L. L., Chaikoff, I. L., and Rheinhardt, W. O. (1952). *J. Biol. Chem.* **197**, 625.

Fruton, J. S., and Simmonds, S. (1958). "General Biochemistry," 2nd Ed., p. 150. Wiley, New York.

Gersh, I., and Catchpole, H. R. (1960). *Perspect. Biol. Med.* **3**, 282.

Gitlin, D., Landing, B. H., and Whipple, A. (1953). *J. Exp. Med.* **97**, 163.

Gitlin, D., Nakasato, D., and Richardson, W. R. (1955). *J. Clin. Invest.* **34**, 935.

Grotte, G. (1956). *Acta Chir. Scand. Suppl.* **211**, 1.

Grotte, G., Knutson, R. C., and Bollman, J. L. (1951). *J. Lab. Clin. Med.* **38**, 577.

Grotte, G., Juhlin, L., and Sandberg, N. (1960). *Acta Physiol. Scand.* **50**, 287.

Hamashima, Y., Harter, J. G., and Coons, A. H. (1964). *J. Cell. Biol.* **20**, 271.

Hampton, J. C. (1958). *Acta Anat.* **32,** 262.

Hyatt, R. E., and Smith, J. R. (1954). *Amer. J. Med.* **16**, 434.

Johnson, J. A. (1966). *Amer. J. Physiol.* **211**, 1261.

Lewis, J. H. (1921). *J. Amer. Med. Ass.* **76**, 1342.

McFarlane, A. S. (1963). *Biochem. J.* **89**, 277.

McKee, F. W., Schloerb, P. R., Schilling, J. A., Tishkoff, G. H., and Whipple, G. H. (1948). *J. Exp. Med.* **87**, 457.

McMaster, P. D. (1937). *J. Exp. Med.* **65**, 347.

McMaster, P. D. (1947). *J. Exp. Med.* **86**, 293.

McMaster, P. D., and Parsons, R. J. (1939a). *J. Exp. Med.* **69**, 247.

McMaster, P. D., and Parsons, R. J. (1939b). *J. Exp. Med.* **69**, 265.

Mancini, R. E., Vilar, O., Dellacha, J. M., Davidson, O. W., Gomez, C. J., and Alvarez, B. (1962). *J. Histochem. Cytochem.* **10**, 194.

Mayerson, H. S. (1963). *Sci. Amer.* **208**, (6), 80.

Mayerson, H. S., Wolfram, C. G., Shirley, H. H., and Wasserman, K. (1960). *Amer. J. Physiol.* **198**, 155.
Ogston, A. G. (1962). *Arch. Biochem. Biophys. Suppl.* **1**, 39.
Ogston, A. G., and Phelps, C. F. (1961) *Biochem. J.* **78**, 827.
Palade, G. E. (1966). *J. Amer. Med. Ass.* **198**, 815.
Pappenheimer, J. R. (1953). *Physiol. Rev.* **33**, 387.
Peters, T. (1962). *J. Biol. Chem.* **237**, 1186.
Rothschild, M. A., Bauman, A., Yalow, R. S., and Berson, S. A. (1955). *J. Clin. Invest.* **34**, 1354.
Schultze, H. E., and Heremans, J. F. (1966). "Molecular Biology of Human Proteins." Elsevier, New York.
Siekevitz, P., and Palade, G. E. (1960). *J. Biophys. Biochem. Cytol.* **7**, 619.
Smallwood, R. A., Jones, E. A., Craigie, A., Raia, S., and Rosenoer, V. M. (1968). *Clin. Sci.* **35**, 35.
Starling, E. H. (1896). *J. Physiol.* (*London*) **19**, 312.
Sterling, K. (1951). *J. Clin. Invest.* **30**, 1228.
Suzuki, T., Yasuda, H., Sekiguchi, H., Taguchi, S., Yamaguchi, H., and Takaki, F. (1961). *Tokyo Jikeikai Med. J.* **8**, 32.
Szabo, G., Gergely, J., and Magyar, Z. (1963). *Experientia* **19**, 98.
Tanikawa, K. (1968). "Ultrastructural Aspects of the Liver and its Disorders," pp. 120–127. Springer, Berlin and Igaku Shoin Ltd., Tokyo.
Tavill, A. S., Craigie, A., and Rosenoer, V. M., (1968). *Clin. Sci.* **34**, 1.
Volwiler, W., Grindlay, J. H., and Bollman, J. L. (1950). *Gastroenterology* **14**, 40.
Yoffey, J. M., and Courtice, F. C. (1956). "Lymphatics, Lymph and Lymphoid Tissue," 2nd Ed., pp. 125, 238. Arnold, London.
Zimmon, D. S., Oratz, M., Kessler, R., Schreiber, S. S., and Rothschild, M. A. (1969). *J. Clin. Invest.* **48**, 2074.

8

STUDIES ON THE EXTRAVASCULAR ALBUMIN OF RATS

J. Katz, A. L. Sellers, G. Bonorris, and S. Golden

CEDARS-SINAI MEDICAL CENTER, MEDICAL RESEARCH INSTITUTE,
LOS ANGELES, CALIFORNIA

I. Introduction

Plasma proteins are present in the circulating blood and in the extracellular fluid. While the nature and distribution of extravascular proteins are of considerable physiological interest, our knowledge about them is limited and indirect. Most of the extravascular proteins are in interstitial water, with smaller amounts in other body fluids such as lymph, cerebrospinal fluid, and aqueous humor. Since obtaining interstitial fluid is difficult, most studies of extravascular proteins have dealt with lymph, edema, ascites, and other accessible body fluids. Due to sampling difficulties, the mass of extravascular proteins and the rate of their exchange with circulating plasma proteins have been estimated indirectly, using isotopic methods. The most common procedure is to calculate the size of the extravascular pool from the decay curve of plasma protein activity following the injection of labeled protein. These isotopoc methods are based on several assumptions which will be discussed in Section II, C.

The most widely studied plasma protein has been albumin. It is generally held that in humans, about 40% of the total albumin is in blood, with the remainder in extravascular fluids. In rats, estimations of extravascular albumin differ, depending on the method used. From the plasma decay curve, the calculated extravascular albumin mass is less than, or at most, equal to, that in the circulation (Katz *et al.*, 1963); by another procedure (see below) the mass of extravascular albumin is at most one and a half times that in the circulation (Katz *et al.*, 1963). However, it is well known that to deplete circulating albumin, extensive blood loss or heavy proteinuria must occur. The stability of plasma albumin concentration in rats in spite of heavy albuminuria has been shown by Katz *et al.* (1963) and is illustrated in Fig. 1. In these experiments, transient proteinuria was induced by renin injections, and the animals excreted in 24 hr an amount of albumin equal to 70% of that in blood. The plasma concentration was only slightly decreased and soon returned to normal levels. There was little change in the specific activity of iodoalbumin (Fig. 1), indicating that albumin repletion in the circulation was not due to increased albumin synthesis. This experiment suggested that extravascular albumin stores are larger than previously estimated.

We decided therefore to determine extravascular albumin mass by extracting it quantitatively from rat tissues and determining its mass by immunochemical assay. Our previously published studies (Sellers *et al.*, 1966, 1968) show extravascular albumin mass to be about three times that in blood, an amount at least twice that estimated by indirect isotopic methods. In view of this discrepancy, a reexamination of our analytical methods was important. Results presented below show that our methods actually underestimated the content of extravascular albumin, and in the rat only 20–25% of body albumin is in the circulation.

Extraction of extravascular albumin from tissues of rats injected with radioactive iodoalbumin permits one to follow its specific activity in various tissues. According to current concepts of albumin metabolism (see below), the extravascular albumin of normal rats is considered to be a single

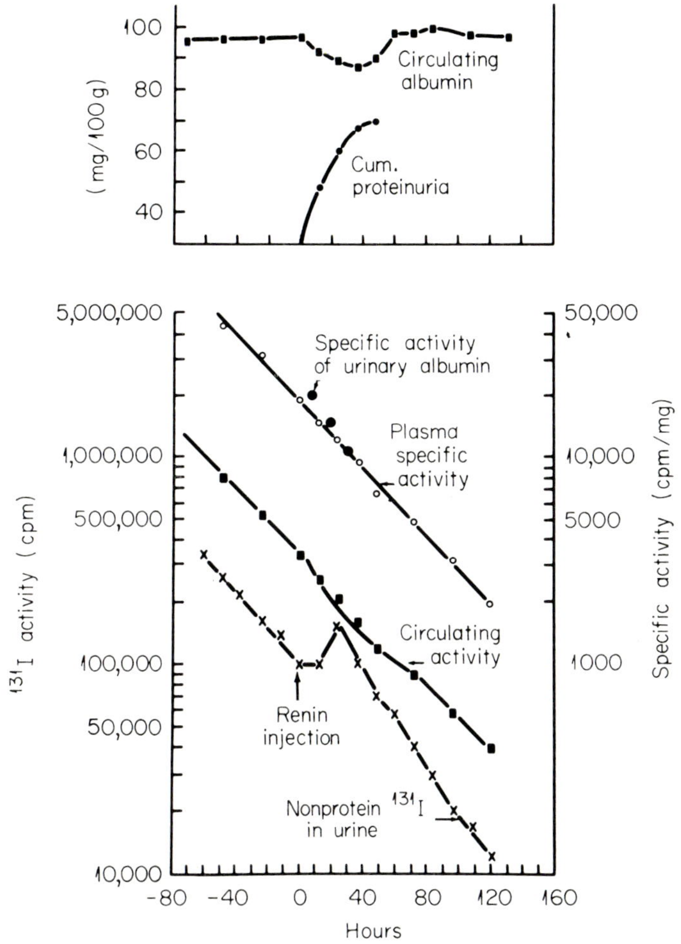

FIG. 1. Effect of renin proteinuria on circulating albumin. The rat was injected with ^{131}I-albumin 4 days prior to the injection of renin (from Katz *et al.*, 1963).

homogenous pool, and the specific activity of the albumin in this pool, to be higher than that in the circulation. The difference in specific activity should, in theory, be more pronounced in nephrotic rats with proteinuria. Our findings were that in both normal rats (Sellers *et al.*, 1966) and nephrotic rats (Sellers *et al.*, 1968), the specific activities of vascular and extravascular albumin became equal. More recently, we have also examined the kinetics of labeling extravascular albumin in several tissues and find it to differ markedly from tissue to tissue. Thus our findings are not consistent with the usually accepted model of plasma albumin metabolism.

In this chapter we (1) review briefly the commonly used methods for the estimation of extravascular albumin, (2) examine methods for the direct determination of extravascular albumin, (3) report the albumin content of tissues and the albumin concentration in extracellular water, and (4) present the kinetics of labeling of extravascular albumin in carcass, muscle, skin, and gut.

II. The Extravascular Protein Pool

The methods for determination of the extravascular pool were derived from mathematical analysis of the plasma protein specific activity decay curve following the injection of labeled protein. A rigorous and lucid exposition of this analysis is provided in the recent text by Schultze and Heremans (1966). However, the principles of the analysis can be readily comprehended in an intuitive way. The basic assumptions on which the calculations are based and the changes in specific activity of the extravascular pool following injection of tracer into the blood are outlined below.

A. The Compartmental Model

The exponential decay curve of radioactive plasma protein is explained in the simplest way by assuming a model such as depicted by Fig. 2a. Protein is present in the vascular (blood) compartment and in one or two homogenous extravascular compartments which exchange with it. The number of extravascular compartments is given by the number of exponential curves into which the plasma curve can be resolved. It is usually one in the rat and two in the human.

The model was first proposed and analyzed by Berson and his co-workers (Berson *et al.*, 1953; Rothschild *et al.*, 1955). It was subsequently elaborated by Matthews (1957), Lewallen *et al.* (1959), Reeve and Roberts (1959), and, more recently, by Nosslin (1965). It is assumed that (1) the system is in a steady state, (2) newly formed protein enters the vascular compartment directly, (3) catabolism occurs within or in "close proximity" to the vascular pool, and (4) mixing within the pools is instantaneous.

After injection of labeled protein into the blood, the specific activity in the plasma pool will decrease, and that in the extravascular pool will increase. At some instant the specific activities will be equal and henceforth will decline in both pools. The specific activity in the pool into which newly formed unlabeled albumin is delivered will be less than that in the other. Thus the specific activity in the extravascular pool will exceed that in plasma, but after a certain time the ratio of the specific activities will become constant (distribution equilibrium). The kinetics of the changes in specific activity will

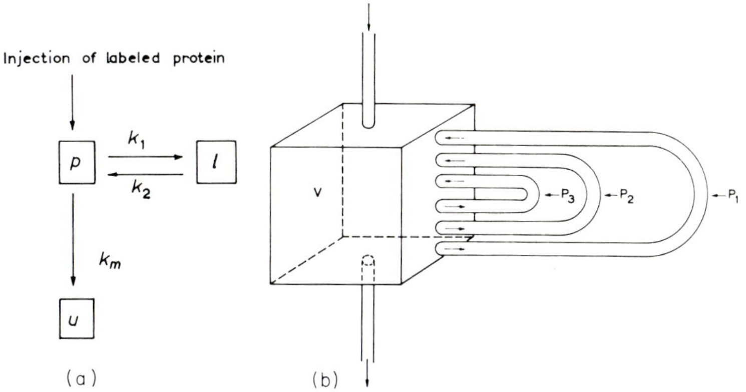

FIG. 2. (a) Multicompartmental model of protein metabolism consisting of one vascular and one extravascular pool. *P*'s, vascular compartment; u, urine; l, extravascular compartment; *k*'s exchange constants from pools indicated. (b) Pipe model of Reeve and Bailey.

depend on the relative rates of exchange and replacement (or loss). If loss (or synthesis) is high relative to exchange rates, the difference in specific activities will be large. If exchange is much more rapid than replacement, the specific activities will be nearly equal. It is also apparent that since the masses in the pools are constant the protein masses in the two pools will be inversely proportional to the fractional rates of exchange, k_1 and k_2 in Fig. 2a.

This system is described by a set of linear differential equations. Solution of these equations provides expression for all the rates of Fig. 2a, in terms of the intercepts and slopes of the resolved exponential plasma curve. Thus the relative masses in the pools and their specific activities can be calculated from the data obtained by sampling only blood or alternately by sampling blood and collecting the radioactivity excreted in the urine.

B. THE PIPE MODEL

Reeve and Bailey (1962) proposed an alternate model of extravascular protein, illustrated in Fig. 2b. In this model protein escapes from the capillaries of various tissues into the interstitial space and is returned to the blood via the lymph. The extravascular pool is envisaged as consisting of a series of pipes which are traversed by protein, with different transit times. The model has the appeal of representing the anatomy of the extravascular system in a more realistic way than does the compartmental model. Mathematical analysis of this system is much more complex than that of the model of Fig. 2a, and the reader is referred to Reeve and Bailey's paper (1962) and the text of Schultze and Heremans (1966), where the calculations are illustrated in detail. Provided the number of "pipes" is sufficiently large, this model yields identical curves and prediction of average specific activity as does the compartmental model, and calculations of the mass of extravascular protein yield identical values.

C. Methods to Calculate Extravascular Protein

Two general methods, based on the model of Fig. 2a, have been derived. In addition, two special methods applicable only under certain limiting conditions have also been used.

1. *From Slopes and Intercepts of the Plasma Decay Curve*

The solution of the differential equations and calculation have been presented in the literature in several forms. The notation of Nosslin (1965) appears the most compact and is in Eq. (1)

$$\frac{\text{Plasma pool}}{\text{Total pool}} = \frac{\sum(c_i/b_i)^2}{\sum(c_i/b_i{}^2)} \tag{1}$$

where the c's and b's represent the intercepts and slopes of the plasma curve. For a model with one extravascular pool, the expression equals

$$\frac{(c_1/b_1 + c_2/b_2)^2}{c_1/b_1{}^2 + c_2/b_2{}^2}$$

Since the protein mass in plasma is known, that in the extravascular pool can be calculated.

Relative specific activities of the protein in the two pools after distribution equilibrium is presented according to Matthews' (1957) notation by Eq. (2):

$$\frac{\text{S.A. of plasma protein}}{\text{S.A. of E.V. protein}} = \frac{k_2}{k_2 - b_1} = \frac{c_1 b_2 + c_2 b_1}{c_1 b_2 + c_2 b_1 - b_1} \tag{2}$$

k_2 is the fractional rate of exchange of extravascular protein (Fig. 2a) and b_1 the terminal slope of the plasma curve. The equation was also derived by Berson and Yalow (1957).

2. *From Plasma and Extravascular Activity*

The method proposed by Campbell *et al.* (1956) is based on the consideration (see above) that when total activity in extravascular protein is maximal, the average specific activities in all pools are equal. It requires, in addition to the plasma curve, the curve for extravascular protein. This may be obtained by total body counting or, if urinary excretion of isotope is determined, by difference: Extravascular activity equals injected dose minus activity in plasma minus activity excreted. The curve usually crosses over the plasma curve, is fairly flat at the region of the maximum, and finally becomes parallel to the plasma curve. When total activity in the extravascular pool is maximal, the specific activities of the two protein pools are equal, and hence protein masses in both pools are proportional to their total activities and to the ordinates of the two curves. The method is of limited accuracy due to the flatness of the curve at maximum and the difficulty of estimating this point.

3. *Special Methods for Determination of Extravascular Albumin*

Berson and Yalow (1957) pointed out that in humans the rate of exchange between pools is much higher than the replacement rate, and hence the specific activities of albumin at distribution equilibrium are nearly equal throughout the body. Thus once equilibrium distribution is attained, the ratio of the dose retained to that in plasma equals the ratio of protein masses.

Another method was used by Sterling (1951). It was assumed that at distribution equilibrium, the specific activity of albumin was uniform and also that catabolism proceeds uniformly at all sites. The fraction of total albumin which is in plasma is then proportional to the intercept on the ordinate of the terminal slope of the plasma curve if the sum of the intercepts is set to unity. It is generally held, however, that catabolism does not occur uniformly in all pools but predominantly in close proximity to the plasma pool. Under conditions of large loss of plasma protein directly from the circulation, as in proteinuria, the method of Sterling is not valid, even if specific activity throughout the body were the same.

III. Experimental Procedure

A. Albumin and Antibody

Albumin was prepared from rat serum by precipitation with 5% trichloroacetic acid followed by extraction with 85% ethyl alcohol. The albumin was precipitated by adding 2 volumes of ethyl ether to the alcohol extract. The precipitate was dissolved in water, dialyzed, and lyophilized.

This albumin, when examined by acrylamide gel electrophoresis, showed in addition to the major band a band between the β- and α-globulin area. As determined by scanning, it contained from 3 to 7% of the protein. Attempts to remove this minor component by ammonium sulfate fractionation, chromatography on carboxymethyl cellulose or preparative acrylamide gel electrophoresis were unsuccessful. This suggests that the compound is a dimer of albumin, possibly formed during electrophoresis, but this is not established. This compound also appeared on immunoelectrophoresis on agarose gels, showing a second arc with our antibody preparation.

Iodination with ^{131}I or ^{125}I of albumin was performed by the ICl technique of McFarlane (1956) and, in later work, by electrolytic iodination (Katz and Bonorris, 1968). The iodoprotein labeled with ICl was screened by injection into rats which were bled after 2–3 days. Iodoalbumin prepared electrolytically was used without screening.

Antibody to rat albumin was obtained from rabbits. The antibody was precipitated from rabbit serum by 25% saturated ammonium sulfate and the precipitate dissolved in 0.1 *N* sodium chloride.

B. Extraction of Extravascular Protein

To correct for residual blood in tissues, rats were injected (in the tail vein) with 0.1–0.2 μCi of rat or (commercial) human labeled iodoalbumin and used 3–6 min later. The rats (under ether anesthesia) were shaved, the abdomen opened, and as much blood as possible aspirated from the aorta. The body was divided into skin, carcass, gastrointestinal tract, and other viscera. The carcass and skin were ground in a meat grinder and 25 g extracted in a high-speed blender with 250 ml of ice cold 0.1% sodium deoxycholate solution, pH 7.0. Gut and viscera were extracted directly in the blender without grinding. The extracts were immediately centrifuged in a refrigerated centrifuge and the turbid supernatants frozen in a dry ice–acetone bath. For analysis, the frozen extracts were thawed and immediately centrifuged at high speed in a refrigerated Servall centrifuge. The clear supernatants were analyzed promptly. Prompt analysis is essential with gut extracts to avoid breakdown.

Samples of the extract containing from 40 to 80 μg of albumin were incubated overnight with excess antibody at 37°C. The precipitate was washed and extracted with 5 ml of 5% trichloroacetic acid in 90% ethanol (very little albumin or activity were removed by a second extraction). Samples of the TCA extract were used for protein assay and, if specific activity was measured, assayed for ^{131}I or ^{125}I.

The content of blood albumin was determined by radioassay of samples of the extracts and of plasma and corrections for albumin due to blood contamination applied. Protein in the extracts was determined according to Lowry *et al.* (1951).

C. Extracellular Water

Rats were anesthetized with ether, bilaterally nephrectomized, and injected via the tail vein with 0.5 ml of 0.9% NaCl containing (without carrier) about 10μCi $Na_2{}^{35}SO^4$ and 50 μCi mannitol-l-T (New England Nuclear). From 1/2 to 3/4 hr later, the rats were injected intravenously with 0.2 μCi of human albumin ^{131}I, and a few minutes later the rats were bled and tissues separated, ground, and extracted as described above. Samples of the extracts were treated with 1/4 volume of 10% $AgNO_3$ in 50% trichloroacetic acid and the supernatants clarified by high-speed centrifugation. This treatment removed all the ^{131}I. Plasma and blood were similarly treated. Residual blood in the tissues was determined from the ^{131}I count in extracts and in blood. ^{35}S and tritium content of the $AgNO_3$-treated extracts were assayed in a liquid scintillation counter. The sulfate space was calculated as follows:

$$\text{Sulfate space} = \frac{^{35}\text{S cpm/g tissue} - \text{ml residual blood} \times {}^{35}\text{S cpm/ml blood}}{^{35}\text{S cpm/ml plasma}}$$

Plasma was taken as 100% water, neglecting the partial volume of the proteins. Mannitol space was calculated in exactly the same manner from the tritium counts. The sulfate or mannitol spaces were expressed as ml/100 g of blood-free tissue.

D. Other Procedures

Disc electrophoresis was performed on acrylamide gel and the discs scanned with a Canalco Scanner. Immunoelectrophoresis was performed on agarose gels.

The albumin content was also determined by an immunodiffusion procedure according to Mancini *et al.* (1965). Agarose gel containing antibody was used. Diffusion was for 20 hr. Areas of the precipitate were determined by projecting with a photographic enlarger and a measuring ring diameter.

IV. Results

A. Extraction and Determination of Albumin in Tissue

1. *Extraction*

The radioactivity from tissues of animals injected with radioactive albumin was all recovered in the deoxycholic acid extracts. Total protein content was quite variable and ranged from 30 to 80 mg/g for all tissues. The electrophoretic patterns on acrylamide gel were also variable. Typical patterns are shown in Fig. 3. The fractions represent both plasma and soluble tissue proteins. The major band corresponded in position to that of albumin. On immunoelectrophoresis in agarose gel against antibody to plasma albumin, one prominent arc due to plasma albumin was seen (Fig. 4).

2. *Determination*

Four methods described in Section II were examined for the estimation of tissue albumin: (1) scanning of acrylamide electrophoretic gels; (2) alcohol–trichloroacetic acid (TCA) extraction—total protein is precipitated with

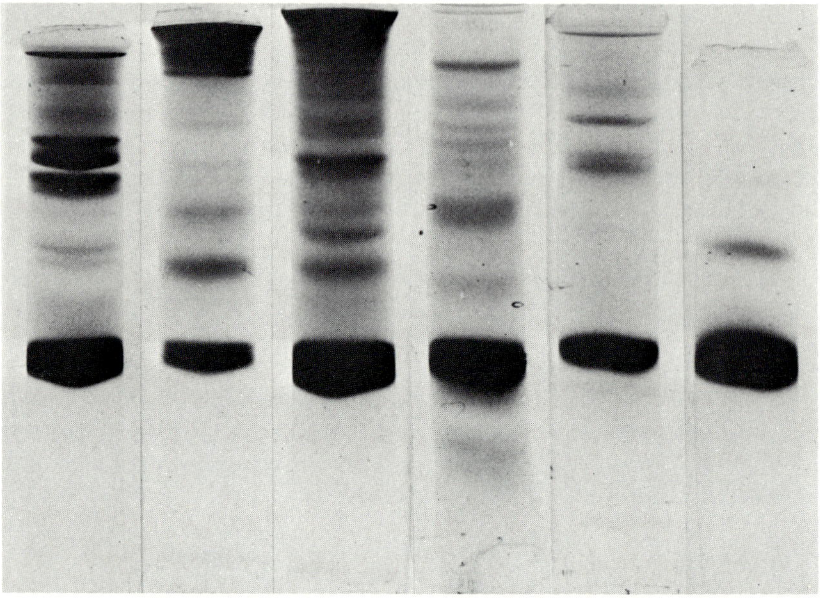

FIG. 3. Acrylamide gel disc electrophoresis of tissue extracts. From left to right: plasma, carcass, skin, gut, ascitic fluid, and rat albumin. Discs overloaded to bring out minor components.

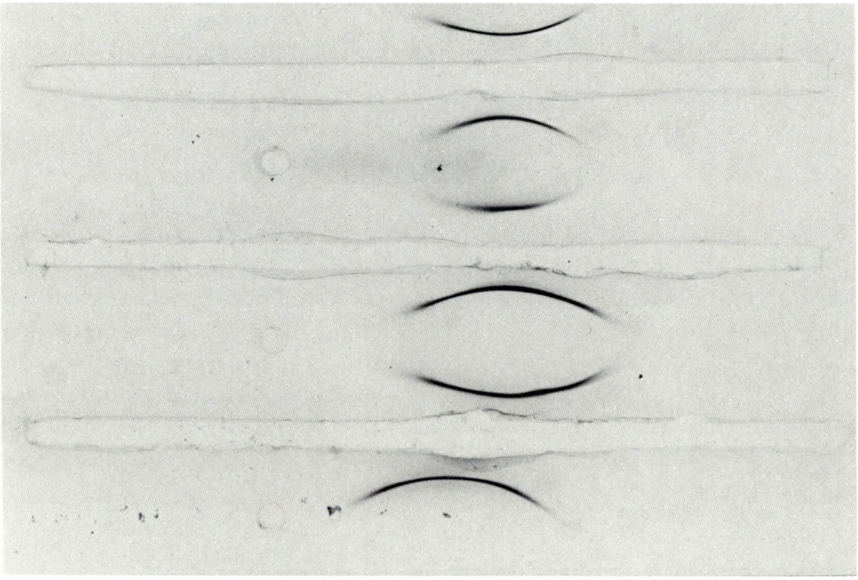

FIG. 4. Immunoelectrophoresis on agarose against antialbumin antibody. From top to bottom: plasma, gut, carcass, and skin extracts.

TABLE I

COMPARISON OF METHODS FOR ESTIMATION OF TISSUE ALBUMIN[a]

Method No.		1	2	3	4	
Tissue	No. of rats	Disc electro-phoresis scan (mg/g)	TCA–alcohol (mg/g)	Antibody–TCA–alcohol (mg/g)	Immuno-diffusion (mg/g)	Ratio 3/4
Plasma	8	31.6[b] (30.0–33.2)	32.2 (29.6–33.8)	29.3 (27.3–32.1)	32.8 (31.5–34.0)	0.89
Carcass	8	8.5[b] (7.6–9.5)	5.4 (4.5–5.9)	2.2 (2.0–2.9)	3.8 (3.2–4.8)	0.58
Muscle	4	9.6[c]	5.6 (6.0–6.1)	2.4 (2.2–2.6)	4.4 (3.8–5.0)	0.55
Skin	8	14[b] (12–18)	8.1 (7.5–9.7)	5.1 (4.3–5.7)	8.3 (7.8–9.3)	0.62
Stomach and small intestine	2	26[c]	9.2	6.9	8.7	—
Small intestine	6	22[c]	10.7 (10–11.7)	7.0 (6.2–8.1)	7.9 (7.0–9.2)	0.89

[a] Values not corrected for blood in tissues. Albumin in residual blood is 0.2–0.4 mg. Parentheses indicate averages and ranges.

[b] Four experiments.

[c] Two experiments.

TCA, and albumin extracted with acid–alcohol from the precipitate; (3) antibody–TCA–alcohol method—albumin is precipitated with antibody and the precipitate extracted with alcohol–TCA; (4) radioimmunodiffusion—albumin was determined from the area of precipitate in agarose gel containing antibody. In previous work (Sellers *et al.*, 1966, 1968) method 3 (antibody–TCA–alcohol) was used, and it was important to evaluate that procedure. In Table I, the four methods are compared.

With plasma, all four methods gave similar values for albumin. Three of the methods (1, 2, and 4) gave virtually identical values (32.2 ± 0.6 mg/ml). The value by the antibody–TCA–alcohol method was somewhat less (29.3 mg). With a large number of determinations by methods 2 and 3, we obtained consistently lower extimates by 5–10% with the antibody–TCA–alcohol method (method 3).

When the four methods were applied to tissue extracts, markedly different results were obtained (Table I). The values by scanning of the acrylamide gel were by far the highest, and seemed unreasonably high especially for the small intestine. The reason for the apparent high "albumin" content of tissues by this method is of interest but was not pursued. Obviously scanning of electrophoretic patterns for tissue extracts is not suitable for the estimation of plasma albumin.

Comparison of the three remaining methods (Table I) shows that in all tissues the antibody–TCA–alcohol method gave the lowest results. The alcohol–TCA method gave the highest values.

There are some tissue proteins soluble in acid alcohol (Munro and Downie, 1964). These are present in very small amounts in normal plasma. The TCA–alcohol procedure is excellent for albumin under normal conditions but is not generally valid. Hemoglobin is also partially soluble in TCA–alcohol, and with hemolyzed blood we have obtained spuriously high values for albumin. Tissue extracts contain acid–alcohol–soluble proteins, and the method overestimates the true albumin content.

Of special interest is the comparison between methods 3 and 4, both employing antibody. In method 3 recovery of albumin depends on both precipitation with antibody and extraction with TCA–alcohol from the antibody–albumin precipitate, whereas in method 4 the assay of albumin is based only on precipitate formation. Use of a radioimmunoassay, based on competition with antibody of labeled and unlabeled albumin (to be described elsewhere) gave essentially the same results as the immunodiffusion assay. It appears then that method 4 gives a true estimate of immunologically potent albumin in tissues. On the other hand, the method used by us previously, using simultaneously two criteria for albumin—immunological precipitation and solubility in TCA–alcohol—gave low levels for extravascular albumin. It appeared likely that this was due to the fact that part of the albumin in the

TABLE II

EXTRACTION OF RADIOACTIVE ALBUMIN WITH ALCOHOL–TCA

Days after injection[a]	Tissue	Total activity (cpm/g)	Distribution of activity (%)				
			TCA–alcohol		Antibody–TCA–alcohol		
						Antibody precipitate	
			TCA soluble	TCA-alcohol[b] soluble	Antibody soluble	TCA–alcohol soluble	Residue
5	Plasma	272,000	5.5	94	11	78	11
5	Carcass	36,000	16.0	86	25	47	36
5	Skin	65,000	7.6	88	19	68	12
5	Small intestine	61,000	11.0	91	14	77	10

[a] Average of two rats.
[b] There was negligible activity in the residue.

antibody precipitate was insoluble in TCA–alcohol. This is demonstrated in Table II.

Tissue extracts from rats previously injected with iodoalbumin ^{125}I were prepared, and the distribution of radioactivity between fractions soluble and insoluble in TCA–alcohol determined. With plasma, 94% of total radioactivity was precipitated with trichloroacetic acid, and practically all of this activity was extracted with alcohol. Precipitation with antibody was somewhat less than with trichloroacetic acid (89%), but after precipitation only 78% could be extracted by repeated acid alcohol extraction, and 11% remained insoluble.

In tissues, as expected, the fraction of TCA-soluble radioactivity due to nonprotein-bound iodine is higher than in plasma. It ranged from 7% in skin to 16% in carcass. Practically all the TCA precipitate was soluble in alcohol. On the other hand, in skin and carcass extracts, 20–25% of the protein-bound activity was not precipitated by antibody, and an appreciable fraction of the activity in the precipitate could not be extracted with alcohol. In carcass, about 40% of the antibody-precipitated activity was rendered insoluble in TCA–alcohol. The finding that most of the albumin in tissue extracts is soluble in TCA–alcohol but becomes partially insoluble after precipitation with antibody is surprising. Why tissue albumin is modified in this respect from that in plasma remains to be established.

3. *Specific Activity*

In previous work we have used the antibody–TCA–alcohol method to follow the specific activity of tissue albumin. Comparison of this method with that based on immunodiffusion is presented in Table III. The specific activity was determined from the radioactivity precipitated with TCA and the albumin content of the tissue extracts. Specific activities of plasma albumin were essentially the same whether obtained by the TCA–alcohol method, antibody–TCA–alcohol, or immunodiffusion. In tissues, specific activity by the two methods using antibody was virtually identical. As expected, the TCA–alcohol method gave values 70% of those of antibody.

To conclude, the method used by us in previous work underestimates tissue albumin, but it provides a valid and reliable estimate of the specific activity of this albumin.

B. Distribution of Extravascular Albumin

1. *Albumin Mass*

In Table IV, the results of previous determinations (Sellers *et al.*, 1966, 1968) and more recent unpublished experiments are summarized. According to these estimates, between 25 and 30% of the body albumin of normal and

TABLE III

DETERMINATION OF SPECIFIC ACTIVITY OF TISSUE ALBUMIN

Method No.		2		3		4	
		TCA–alcohol		Antibody–TCA–alcohol		Immuno-diffusion	
Days after iodoalbumin injection[a]	Tissue	(cpm/mg)	(%)	(cpm/mg)	(%)	(cpm/mg)	(%)
5	Plasma	7900	100	7660	100	8070	100
	Carcass		75		101		104
	Skin		76		100		102
	Small intestine		70		90		93
4	Plasma	12,000	100	10,600	100	11,400	100
	Carcass		65		107		106
	Skin		66		98		90
	Small intestine		50		99		87

[a] Average of two rats.

nephrotic rats was in the circulation. The value rose to 35% in protein-depleted rats. These determinations were made by the antibody–TCA–alcohol method which greatly underestimates tissue albumin. From the data of Table I, it appears that albumin in tissue is underestimated by one-third and in blood, by one-tenth. Using these correction factors, the distribution of albumin between blood and tissues is recalculated. In normal and nephrotic rats, about 20% of total albumin is in the blood. By immunodiffusion from the experiments of Table I a value of 22% is obtained. The mass of extravascular albumin in normal rats ranges from 430 to over 500 mg/100 g body weight.

Comparison of these values with the extravascular albumin mass calculated by the two isotopic methods of Section I, C is of interest. Most of the animals of experiment 2 of Table IV were injected with radioactive iodoalbumin and the decay of the radioactivity of plasma albumin followed. The intercepts of the plasma curve (presented below; Fig. 6) are 0.28 and 0.72 and the slopes, 0.013 and 0.10 (fraction of dose per hour). From Eq. (1),

$$\frac{\text{Plasma pool}}{\text{Total pool}} = \frac{\left[\dfrac{0.72}{0.10} + \dfrac{0.28}{0.013}\right]^2}{\dfrac{0.72}{0.10^2} + \dfrac{0.28}{0.013^2}} = 0.48$$

TABLE IV

VASCULAR AND EXTRAVASCULAR ALBUMIN AS DETERMINED BY THE ANTIBODY–TCA–ALCOHOL METHOD

Expt. no.	1		2		3		4	
Reference	Sellers *et al.* (1966)		New observations		Sellers *et al.* (1966)		Sellers *et al.* (1968)	
Condition	Normal		Normal		Protein-depleted		Nephrotic	
No. of rats	40		36		7		34	
Body weight (g)	122		151		118		194	
Albumin	(mg/100 g)	(%)	(mg/100g)	(%)	(mg/100 g)	(%)	(mg/100 g)	(%)
(1) By the antibody–alcohol–TCA method								
Vascular	117	26	131	31	82	35	56	27
Extravascular	338	74	291	69	148	65	149	73
(2) Corrected estimates[a]								
Vascular	130	20	145	25	90	29	60	21
Extravascular	510	80	435	75	220	71	225	79

[a] Multiplied by 1.1 for vascular and by 1.5 for extravascular albumin. Values rounded out to 5 mg.

This ratio compares with 0.25 from Table IV. The extravascular albumin mass from Eq. (1) would be about 160 mg as compared to 430 mg, determined directly. In previous experiments (Katz *et al.*, 1963) the ratio by Eq. (1) was about 0.6, and values of between 0.5 and 0.6 follow from data published by Matthews (1957).

Urinary excretion was not determined in these experiments, and thus the method by Eq. (2) cannot be used in this experiment. Previous estimates (Katz *et al.*, 1963) based on a large number of animals, for the ratio (^{131}I in extravascular pool)/(^{131}I in plasma) when the total ^{131}I in the extravascular pool was maximal, were between 1.2 and 1.5. Accordingly, the extravascular albumin mass would be at most 220 mg, one-half that obtained by direct determination.

2. *Concentration of Albumin in Tissues*

To determine the concentration of albumin in the fluid of tissues, extracellular water content must be known. To estimate the extracellular water, the mannitol and sulfate space of the tissues was determined. The results are summarized in Table V. The mannitol and sulfate space in total body and the various tissues, with the exception of gut, agreed closely and appear to measure the extracellular fluid. The extracellular water represented 26% of the total blood-free body. Total body water as measured by the dilution of tritiated water was about 75% of total body mass (including blood). The extracellular water of muscle was 20%, somewhat less than in the total body. The extracellular water of skin was 40% of the blood-free organ, a value much higher than that found in other tissues studied. Total water content of skin as determined by dessication was 73–78%. Thus extracellular water constitutes over half of total skin water.

The apparent sulfate space of gut was rather variable and frequently exceeded the total volume of the organ. It was much greater than the mannitol space. Rinsing out of the intestines removed most of the sulfate and one-fourth to one-third of the mannitol (Table V). It appears that sulfate is excreted into the lumen, but this does not account altogether for the large apparent space. It may be that, in the gut, sulfate penetrates the cells or exchanges with sulfate of mucopolysaccharides. The apparent mannitol space of rinsed gut was also quite variable (16–27%). The determination of the extracellular water of stomach and intestine is thus not valid with sulfate and not certain with mannitol.

In Table VI, the distribution of albumin in tissues and the concentration of albumin in the extracellular water of tissues are presented. The values were calculated from the data of Tables I and V. Carcass (mainly muscle) contains nearly half of the extravascular albumin, and skin, one-quarter. The concentration of albumin in extravascular water of muscle and skin is

TABLE V

SULFATE AND MANNITOL SPACE

		% Blood-free mass[a]	
		Sulfate	Mannitol
Total body		27	26
		(26–29)	(23–28)
Carcass		24	23
		(22–26)	(19–26)
Muscle		20	20
		(17–22)	(17–22)
Skin		39	40
		(37–42)	(34–42)
Viscera[b]		26	29
		(22–28)	(25–30)
Stomach	(Unrinsed)	125	30
and		(69–160)	(25–41)
small		44	22
Intestine	(Rinsed)	(40–48)	(16–27)

[a] Parentheses indicate averages and ranges from eight rats.

[b] Includes liver, heart, lung, spleen, and genitalia.

about 20 mg/ml, but since the extracellular water content of skin is much higher than muscle, the albumin content of skin is high. The small intestine, including stomach, constitutes 4–5% of total body weight but contains about 10% of the extravascular albumin. About 20% of the albumin was rinsed out and may be in the lumen. The concentration of albumin in (rinsed) gut cannot be estimated with precision, since the extracellular water content is uncertain. However, a conservative estimate is 25 mg/ml, and it may even be 30 mg/ml, a value nearly equal to its concentration in plasma. Thus the concentration of albumin in extracellular body water, with the exception of the intestine, is about half that in plasma. In the small intestine, it is at least 75% that in plasma. It is reported (Schultze and Heremans, 1966) that the concentration of albumin in thoracic lymph is 40–50% that in plasma, and of intestinal lymph, 70–80%. It is likely that the composition of lymph represents the albumin concentration of the tissues it drains. It is of interest that the tissue concentrations shown in Table VI agree well with those reported for lymph. Taking the average concentration of albumin in extracellular water in the body as half that in plasma and the volume of the extracellular space as 26% (Table VI), we obtain an estimate of extravascular albumin of over three times that in plasma, in reasonable agreement with

TABLE VI

DISTRIBUTION AND CONCENTRATION OF EXTRAVASCULAR ALBUMIN IN TISSUES

Tissue	Body weight (%)	Extracellular water[a] (ml/g tissue)	Albumin			Conc. in extracellular water (mg/ml)
			mg/g tissue[b]	mg/100 g body	% of total	
Blood	7	—	—	—	—	—
Plasma	—	—	33	130	22	33
Carcass	60	0.23	36	220	38	16
Muscle	—	0.20	4.1	—	—	20
Skin	16	0.40	8.0	130	22	20
Small intestine and stomach	5	0.2–0.3	8.4	50	8	25–35
Other viscera	12	—	5.0[c]	60	10	—
Blood-free body	—	0.26	4.9	450	—	18
Total	100			590		

[a] From Table I, corrected for residual blood.
[b] From Table V.
[c] Estimated.

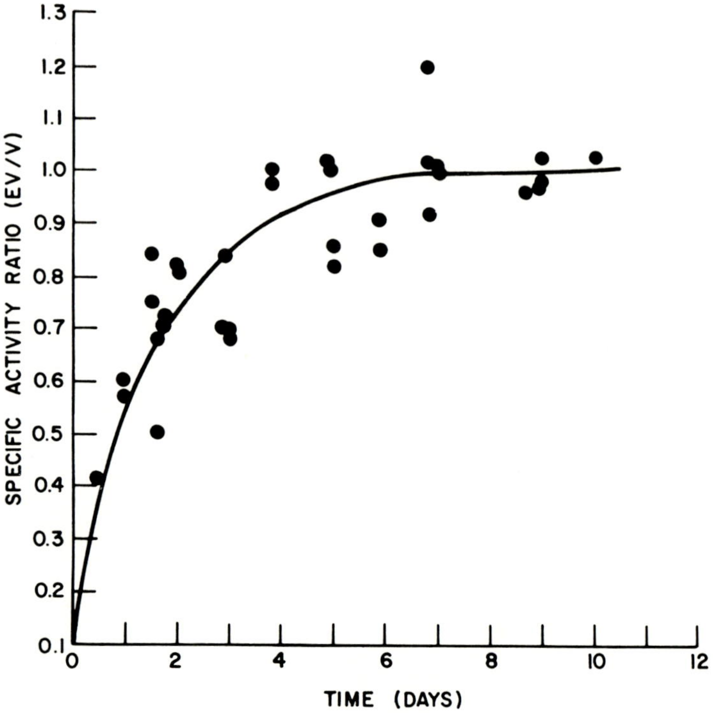

FIG. 5. Ratio of specific activities of plasma and extravascular albumin (from Sellers *et al.*, 1966).

direct determination. The high concentration of albumin in the intestines is of interest and warrants further studies.

C. The Kinetics of Labeling of Extravascular Albumin

We have previously shown (Sellers *et al.*, 1966) that the specific activity of estravascular albumin became equal to that of plasma 4–6 days after injection of radioactive iodoalbumin (Fig. V). It appeared of interest to study the course of labeling of the individual tissues, and the results are shown in Fig. 6. The curves are strikingly different. The specific activity of carcass (mainly muscle) albumin increased very rapidly, equaling that of circulating albumin after about 12 hr. For the next few hours, activity somewhat exceeded that of plasma, but after 2 days specific activities in muscle and circulating blood became equal and remained equal thereafter.

The specific activity of skin albumin increased for about 24 hr, attaining a value about one-third that in plasma, and remained fairly constant for the next 2 days. Due to the rapid fall in specific activity of plasma albumin, the specific activities became equal on the fifth day. Labeling of gut albumin

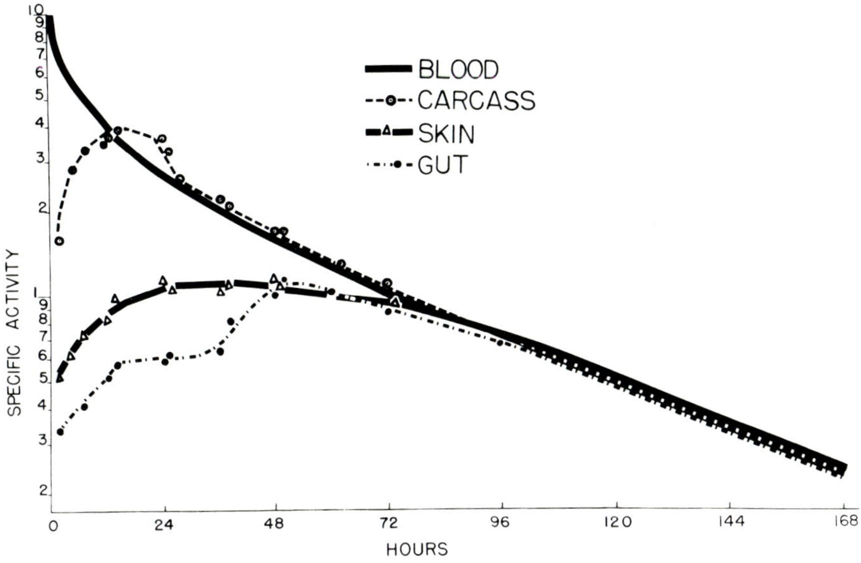

FIG. 6. Specific activities of albumin in plasma and tissues. Rats injected with screened ^{131}I-albumin. Each point average of two rats.

was the slowest, and the specific activity curve indicates very complex kinetics.

The average specific activity of the total extravascular albumin pool exhibits much simpler apparent kinetics than those of its individual constituent pools. Maximum specific activity of total extravascular albumin was attained about 39 hr after injection, and the specific activities of vascular and average extravascular pools became equal 5–6 days after injection. Average specific activity differs greatly from the actual activities in the individual tissues. These results show the assumption of a homogenous extravascular pool in the rat to be questionable and the exchange rates between vascular and extravascular pool to have little meaning.

Comparison of the specific activity calculated by Eq. (2) with these results is of interest. From the slopes and intercepts of Fig. 6, a specific activity ratio of (extravascular albumin)/(vascular albumin) of 1.53 is obtained by Eq. (2). In a previous experiment (Katz *et al.*, 1963) with 15 normal rats, an even higher average specific activity ratio of 2.1 (range 1.8–3.0) was obtained. The difference between theoretical and observed ratios is high significant.

D. EXTRAVASCULAR ALBUMIN IN NEPHROTIC RATS

Rats with severe aminonucleoside nephrosis excrete 10–15 mg/hr/100 g of albumin in the urine, and their plasma albumin may decrease below 10 mg/ml. They are edematous and frequently accumulate large volumes of ascitic fluid. The albumin concentration of ascitic fluid is from 0.2 to 0.4 mg/ml, and while the amount varies greatly, it rarely exceeds 10% of total body albumin. Total body albumin of these rats is markedly depressed, but distribution between vascular and extravascular space is not greatly different from that in normal rats (Table IV).

Since the replacement rate of albumin in these animals is greatly increased over normal, a large difference between specific activities of vascular and extravascular specific activities would be expected according to Eq. (2). From previous determination of slopes and intercepts of the plasma curve (Katz *et al.*, 1963), the calculated specific activity ratios calculated by Eq. (2) ranged from 3 to 4 for rats with mild proteinuria to as high as 8, when proteinuria was severe. However, we found (Sellers *et al.*, 1968) that 3 days after iodo-albumin injection, the specific activity of extravascular albumin, with the exception of that in ascitic fluid, became equal to that in plasma. The specific activity of ascites albumin attained values up to eight times that in plasma.

Results with another group of rats in which proteinuria was very severe and the terminal half-life of plasma albumin was 14 hr (as compared to 50–60 hr in normal rats and about 25 hr in the group of nephrotic animals previously studied) is shown in Fig. 7. Relative specific activities of albumin in carcass and skin and ascitic fluid are shown. Specific activities of tissue albumin become nearly equal to that in plasma within 2–3 days, but that in ascitic fluid was 10–18 times that in plasma. The skin of these animals was

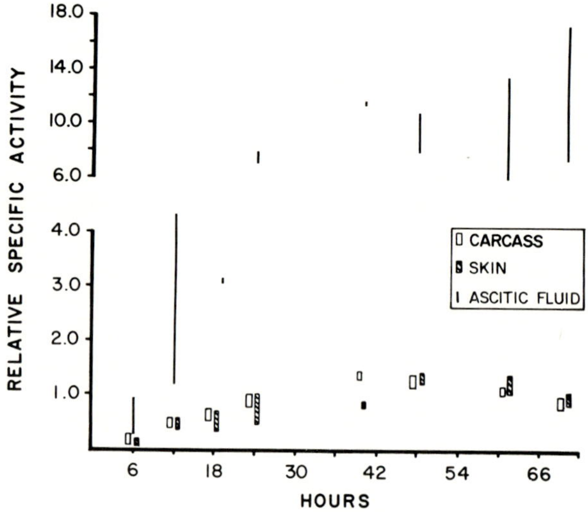

FIG. 7. Relative specific activities of ^{125}I-albumin of aminonucleoside nephrotic rats. The lines indicate the values obtained with two animals.

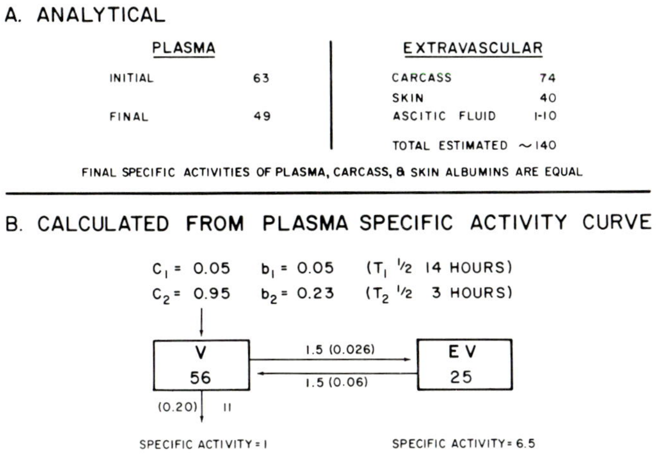

FIG. 8. Albumin metabolism in nephrotic rats (experiment of Fig. 7); distribution of albumin (mg/100 g body weight) and specific activity, observed and calculated from Eqs. (1) and (2).

very edematous, oozing water, but the specific activity of its albumin differed markedly from that in ascitic fluid.

The albumin content of skin and carcass was determined and total extravascular albumin mass estimated. The results are summarized in Fig. 8 and compared to those calculated by Eq. (1). Albumin loss was about 20% of the plasma pool per hour. There was a decrease in plasma albumin by about 20% in 3 days, or about 0.3%/hr. Thus loss exceeded replacement by 1–2%, and calculation based on the assumption of steady state must be used with caution. Using conservative estimates of slopes and intercepts, it is calculated that the albumin in the extravascular pool should be only half that in plasma, and the specific activity, 6.5 times that in plasma. Even if these values are only rough approximations, the discrepancy between calculated and observed values is striking.

V. Results by Other Workers

Our estimates on the size and exchange characteristics of extravascular albumin are at variance with generally held conclusions, and hence a comparison with results obtained by others is important. Unfortunately there are hardly any reports in the literature in which albumin was extracted from tissue and quantitatively analyzed. The few available studies are indirect estimates based on the distribution of radioactivity, and results of different investigators seem to conflict.

Oeff and Koenig (1956) determined the activity remaining in the rat body by total body counting after exhaustive perfusion to remove the blood. Two days after injection of ^{131}I-iodoalbumin, 11% of the dose was in blood, 34% was in the cadaver, and 50% in urine. Assuming equal specific activities in vascular and extravascular albumin, it follows that 25% of body albumin is in blood, a figure in agreement with our findings. However, Matthews *et al.* (1961), using essentially the same method, reported approximately equal amounts of radioactivity in blood and in carcass and skin.

Rothschild *et al.* (1955) estimated from the ^{131}I content of biopsy samples that skin contained 8–10 mg/g of albumin, but muscle only 1–1.2 mg. The value for skin is similar to that observed by us, but that in muscle is much less than found in rats (Table I). These workers estimated that plasma contained 40% of total albumin, skin 18%, and muscle 15%.

The only work in which interstitial fluid was directly sampled by micropuncture is that of Creese *et al.* (1962). They inserted fine capillaries between muscle fibers in the guinea pig and collected a few microliters of interstitial fluid. Fifteen minutes after injection of labeled iodoalbumin, the radioactivity of the fluid was negligible, indicating that it was essentially blood-free. One day after isotope injection, the radioactivity of the fluid was 44% that of plasma. Electrophoresis showed the albumin content to be 42% that in plasma, while the composition of the globulin fraction was highly variable. These results indicate that the concentration of albumin in muscle water of guinea pig is quite similar to the rat.

There are numerous studies of the kinetics of albumin exchange between ascitic fluid and plasma. Gitlin (1957) sampled edema and ascitic fluid of nephrotic children and found specific activities to be some 30 times that in plasma. He apparently believed that these specific activities prevail throughout the extravascular pool. Most other investigators, such as Berson and Yalow (1954), Schoenberger *et al.* (1956), and Kushner *et al.* (1957), found the albumin of ascitic fluid of humans to attain specific activities equal to or only slightly higher than that in plasma.

VI. Discussion

Our results indicate that one-quarter of body albumin is in blood, a value markedly different from that estimated by others. Whether this distribution pattern occurs only in rats or applies to other species remains to be determined. In man, it seems to be firmly established that blood albumin constitutes about 40% of the body pool. The total body albumin content of rats, 500–600 mg/100 g, is not so different from that found in other species (400 mg in man), and the difference in distribution per se is of limited physiological interest. The most interesting and puzzling aspect of our work is the discrepancy between the direct observations and theoretical calculated values, based on conventional isotopic models of plasma protein metabolism. If our observations were to be confirmed, the multicompartmental model becomes untenable for rats and possibly also for other animal species.

Discrepancies between observed and calculated albumin masses and specific activities are of such magnitude that they cannot be reasonably accounted for by errors or imperfections of our methods. This is especially the case with nephrotic rats, where observed and calculated values differ by a factor of from three to six. It is difficult to conceive of a type of error that consistently, in several tissues and over a hundred observations, yields a ratio of unity.

It is difficult to explain our findings, and we are not able to suggest an alternate model of albumin metabolism that will accommodate all known observations. Various types of models were explored with the aid of a computer. A very close fit to the specific activities curves of Fig. 3 (as well as urinary excretion) was obtained by a model consisting of three extravascular pools (for carcass, skin, and gut), provided inflow of newly synthesized albumin occurs directly into all pools, roughly in proportion to their masses. It is, however, well established (Woolley and Courtice, 1962; Smallwood *et al.*, 1968) that nearly all newly synthesized albumin enters the circulation via the hepatic vein, and such a model is therefore untenable.*

Our results indicate that the rates and kinetics of exchange between circulating albumin and that in muscle, skin and gut differ greatly. The conclusion from Fig. 6 is that the concept of a uniform homogenous extravascular pool with a single exchange rate has no physiological reality. Of special interest is the labeling of albumin in ascitic fluid. It appears that albumin, once it enters the peritoneal cavity, exchanges very slowly with the rest of the body albumin and that kinetics of labeling of this fraction is radically different from that in tissue.

Our results indicate the importance of studying extravascular albumin by direct methods, both by extraction and micropuncture. Recent advances in quantitative immunochemistry permit precise determination of plasma proteins in small biopsy samples.

* Very recently, Zimmon *et al.* (1969) reported that in patients with cirrhosis of the liver and ascites, newly synthesized albumin labeled with $^{14}CO_2$ appeared in ascitic fluid 4–17 times faster than albumin from the circulation (labeled with ^{131}I). Also, newly synthesized albumin appeared 3 times faster in thoracic duct lymph than albumin from the circulation. These experiments suggest some transfer of albumin formed in liver directly into the extravascular pool, but it is obscure what pathway could be involved. It is of great interest to determine whether the phenomenon occurs under normal conditions.

ACKNOWLEDGMENTS

Supported by USPHS Grant AM-07633-06.

REFERENCES

Berson, S. A., and Yalow, R. S. (1954). *J. Clin. Invest.* **33**, 377.
Berson, S. A., and Yalow, R. S. (1957). *Fed. Proc. Fed. Amer. Soc. Exp. Biol.* **16**, 135.
Berson, S. A., Yalow, R. S., Schreiber, S. S., and Post, J. (1953). *J. Clin. Invest.* **32**, 746.
Campbell, R. M., Cuthbertson, D. P., Matthews, C., and McFarlane, A. S. (1956). *Int. J. Appl. Radiat. Isotop.* **1**, 66.
Creese, R., D'Silva, J. L., and Shaw, D. M. (1962). *J. Physiol.* (*London*) **162**, 44
Gitlin, D. (1957). *Ann. NY. Acad. Sci.* **70**, 122.
Katz, J., and Bonorris, G. (1968). *J. Lab. Clin. Med.* **72**, 966.
Katz, J., Bonorris, G., and Sellers, A. L. (1963). *J. Lab. Clin. Med.* **62**, 910.
Kushner, D. S., Bronsky, D., Dubin, A., Maduros, P., and Armstrong, S. H. (1957). *J. Clin. Med.* **49**, 440.
Lewallen, C. G., Berman, M., and Rall, J. E. (1959). *J. Clin. Invest.* **38**, 66.
Lowry, O. H., Rosenbrough, N. J., Farr, A. L., and Randall, R. J. (1951). *J. Biol. Chem.* **193**, 265.
McFarlane, A. S. (1956). *Biochem. J.* **62**, 135.
Mancini, G., Carbonara, A. O., and Heremans, J. F. (1965). *Immunochemistry* **2**, 235.
Matthews, C. M. E. (1957). *Phys. Med. Biol.* **2**, 36.
Matthews, C. M. E., Bruce-Robertson, A., and Humphrey, J. H. (1961). *In* "Plasma Proteins and Gastrointestinal Tract in Health and Disease" (M. Schwartz and P. Vesin, eds.). Munksgaard, Copenhagen.
Munro, H. N., and Downie, E. D. (1964). *Arch. Biochem. Biophys.* **106**, 516.
Nosslin, B. (1965). *In* "Metabolism of Human Gamma Globulin" (S. B. Andersen, ed.), p. 103. Blackwell, Oxford.
Oeff, K., and Koenig, A. (1956). *Z. Gesamte Exp. Med.* **127**, 484.
Reeve, E. B., and Bailey, H. R. (1962). *J. Lab. Clin. Med.* **60**, 923.
Reeve, E. B., and Roberts, J. E. (1959). *J. Gen. Physiol.* **43**, 415.
Rothschild, M. A., Bauman, A., Yalow, R. S., and Berson, S. A. (1955). *J. Clin. Invest.* **34**, 1354.
Schoenberger, J. A., Kroll, G., Eckert, E. L., and Mark, R. M. (1956). *J. Lab. Clin. Med.* **47**, 227.
Schultze, H. E., and Heremans, J. F. (1966). "Molecular Biology of Human Proteins," Vol. 1. Elsevier, Amsterdam.
Sellers, A. L., Katz, J., Bonorris, G., and Okuyama, S. (1966). *J. Lab. Clin. Med.* **68**, 177.
Sellers, A. L., Katz, J., and Bonorris, G. (1968). *J. Lab. Clin. Med.* **71**, 511.
Smallwood, R. A., Jones, E. A., Craigie, A., Raia, S., and Rosenaer, V. M. (1968). *Clin. Sci.* **35**, 35.
Sterling, K. (1951). *J. Clin. Invest.* **30**, 1228.
Woolley, G., and Courtice, F. C. (1962). *Aust. J. Exp. Biol. Med. Sci.* **40**, 121.
Zimmon, D. S., Oratz, M., Kessler, R., Schreiber, S. S., and Rothschild, M. A. (1969). *J. Clin. Invest.* **48**, 2074.

III

Subcellular Synthesis Site and Amino Acid Regulation

9

FACTORS IN REGULATION OF LIVER PROTEIN SYNTHESIS

Hamish N. Munro

PHYSIOLOGICAL CHEMISTRY LABORATORIES,
DEPARTMENT OF NUTRITION AND FOOD SCIENCE,
MASSACHUSETTS INSTITUTE OF TECHNOLOGY,
CAMBRIDGE, MASSACHUSETTS

I. Introduction

The liver continues to be the tissue in the mammals whose metabolic reactions are more frequently examined than those of any other organ of the body. Furthermore the liver, because of its strategic position astride the portal vein, is subject to larger variations in nutrient supply than any other part of the body except the intestinal mucosa. The objective of this chapter is to relate what is known about regulation of protein synthesis in the liver to the case of one group of proteins secreted by the liver, the plasma proteins, and notably to plasma albumin formation. Four aspects will be presented: First, the relationship between rates of albumin synthesis in different animals will be related to the composition of their livers. Second, the site of albumin synthesis in the liver cell will be considered. Third, the regulation of liver protein synthesis in relation to changes in amino acid supply will be described. Finally, diurnal rhythms in liver protein metabolism will be examined.

II. Species Differences in Rates of Albumin Synthesis and in Liver Composition

There is a wide variation in the adult size of different mammals, and various parameters of metabolism have been shown to change in a systematic manner in relation to the body weight of the animal. The available data on this subject have recently been reviewed (Munro, 1969). First, there is often a striking relationship between the mature body size of an animal and the dimensions of its organs. In the case of the liver, for example, the percentage of body weight accounted for by this organ decreases from 6% in the case of the shrew (body weight 7 g) to about 1% in the case of the elephant (body weight 6000 kg). It has also been recognized for many years that the intensity of metabolism decreases progressively with increasing body weight; indeed, the energy output per kg of body weight of an adult man is five times less than the energy output per kg of an adult rat (Brody, 1945). A similar relationship holds for many aspects of protein metabolism, including the turnover of those plasma proteins that have been examined in different species (Munro, 1969). Since the plasma proteins are mostly made in the liver, it is appropriate to ask whether the protein-synthesizing machinery in the liver cell also undergoes systematic differences in abundance related to the body size of the animal.

TABLE I

PLASMA ALBUMIN TURNOVER, LIVER CELL COMPOSITION PER MG DNA AND COMPOSITION OF LIVER MICROSOMES OF DIFFERENT MAMMALS ARRANGED IN ASCENDING ORDER OF WEIGHT[a]

Species	Half-life of plasma albumin (days)	Whole liver composition (mg/mg DNA)			Microsome fraction (% of total dry matter)	
		RNA	Phospholipid	Protein	RNA	Phospholipid
Mouse	1.2	4.45	14.2	86	9.6	24.2
Rat	2.5	3.05	9.8	60	7.3	22.9
Rabbit	5.7	2.44	11.6	77	6.2	29.2
Dog	8.2	1.69	9.4	70	—	—
Cow	20.7	1.29	8.7	47	5.9	31.3

[a] From Munro and Downie (1964).

With this in mind, we compared the effect of species size on the turnover of plasma albumin and on the RNA content of the livers of the same species (Table I). With increasing weight of species, it is evident that albumin turnover becomes progressively slower. Furthermore, this is paralleled by a reduction in the RNA content of the average liver cell, so that the RNA: DNA ratio falls from 4.45 in mouse liver to 1.29 in cow liver. In order to quantitate the relationship between albumin turnover and liver RNA content, it is necessary also to take account of the relatively smaller livers of the larger animals with their fewer cells. When this is done, it is found that the amount of RNA in the liver per kg of the weight of the animal diminishes by a factor of four from the rat to man, while the rate of albumin turnover slows down by a factor of five. Thus, the reduction in protein synthesis rate parallels fairly closely the abundance of cellular RNA, an integral component of the protein-synthesizing apparatus of the cell. This diminution in liver RNA is primarily a reflection of the reduction in cell ribosome population. Table I shows that the microsomes, to which the ribosomes secreting albumin are attached, show a progressively smaller RNA content as one passes from the liver of the mouse to that of the cow. This type of analytical study has been extended by us to other tissues of the body. For example, the thyroid glands of different species show similar systematic changes in RNA content, indicative of reduced rate of thyroglobulin synthesis in the larger species.

III. Sites of Formation of Secreted and Retained Proteins

It is generally accepted that specific proteins are made in the cell cytoplasm by using the amino acid sequences encoded in the different messenger RNA species secreted from the nucleus. Thus serum albumin is made by ribosomes attached in polysome formation to one messenger, whereas ferritin is made by polysomes containing another messenger species. These two proteins have different fates. The albumin is secreted into the plasma, whereas the ferritin is retained within the liver cell. For this reason, it has been suspected for more than a decade that proteins secreted from the cell are made on the polysomes seen by electron microscopy to be attached to the membranes of the endoplasmic reticulum, whereas retained proteins are made on polysomes seen to be lying free in the cytoplasm. Several investigators have recently sought to verify this hypothesis by identifying the proteins made by these two types of polysomes. Redman (1968) first showed that membrane-attached polysomes are the main or exclusive site of albumin biosynthesis in the liver cell. Using an immunological precipitation method to identify the albumin, he showed that the membrane-attached polysomes carried many nascent chains of albumin, whereas the free polysomes were nearly devoid of these. This conclusion has been confirmed by Takagi and Ogata (1968).

We have recently made a comparison of the efficiency of these polysomes in different locations to synthesize ferritin and albumin (Hicks *et al.*, 1969). The free polysomes and the total (free and membrane-bound) polysomes were prepared from rat liver and incubated in a cell-free protein-synthesizing system with ^{14}C-leucine. At the end of incubation, albumin and ferritin were isolated by immunological procedures and the amount of radioactivity in each specific protein was compared with the total incorporation into protein by the cell-free system. Table II shows the amounts of albumin and of

TABLE II

RELATIVE INCORPORATION OF ^{14}C-LEUCINE BY A CELL-FREE PROTEIN-SYNTHESIZING SYTEM INTO FERRITIN AND SERUM ALBUMIN USING FREE POLYSOMES AND TOTAL POLYSOMES ISOLATED FROM RAT LIVER[a]

:Protein	No. of expts.	Uptake into protein as % of total peptide incorporation		Difference (%)
		Total polysomes	Free polysomes	
Ferritin	5	0.08	0.13	+65
Albumin	2	0.35	0.16	−56

[a] Summarized from data of Hicks *et al.* (1969).

ferritin made by the total liver polysome population and by the free polysome population incubated under comparable conditions. The total polysome population is seen to be more efficient for the synthesis of albumin, whereas the free polysomes are more effective in the incorporation of precursor into ferritin. Redman (1969) has also found that ferritin is preferentially synthesized on the free polysomes of the liver. Ganoza and Williams (1969) have confirmed that the division into membrane-attached and free polysomes is functionally distinct in the case of a number of secreted plasma proteins and retained liver proteins.

We can accept these data to mean that the messenger RNA for secreted proteins is present only in polysomes attached to the membranes of the endoplasmic reticulum, and the messenger for retained proteins is located in the free polysomes of the cell. However, this does not explain how the two classes of messenger RNA find their way to the appropriate locations in the cytoplasm.

IV. Amino Acid Supply and Liver Protein Synthesis

After consuming a meal containing protein, the liver is subjected to an extensive increase in amino acid supply through the portal vein (Denton and Elvehjem, 1954). Elwyn (1970) has quantitatively evaluated the role of the liver in determining how much of this incoming load of amino acids is destroyed or sequestered in the liver, or transformed into plasma proteins. Using dogs with cannulas in the portal vein, the splenic artery (representing arterial blood going to the liver) and the hepatic vein leaving the liver, he was able to monitor continuously the exchange of amino acids across the liver during the entire 12-hr period after a large meal of meat. He found that 57% of the absorbed amino N is converted to urea, some 6% leaves the liver as plasma proteins, and only 23% enters the general circulation as free amino acids; the remaining 14% not accounted for is presumably temporarily retained in the liver as hepatic protein. These data indicate that, when large amounts of protein are fed, the systemic circulation is protected against excessive changes in free amino acid concentration by immediate adaptive responses within the liver. Presumably, a meal containing less protein would result in less urea formation and a larger proportion of the incoming amino acids would be allowed to pass into the general circulation, but this has not been experimentally tested as yet.

These adaptive reactions of the liver demonstrate that this organ plays an important role in the utilization and fate of dietary protein. The mechanisms involved in the adaptation of the liver to such influxes of amino acids have been examined by many investigators. They involve a transient accumulation of liver protein during absorption that is later dissipated during the post-absorptive period. We have explored the way in which an increase in flow of amino acids to the liver stimulates protein synthesis. Fasting rats were fed by stomach tube with nutritionally complete or tryptophan-deficient mixtures of amino acids and an hour later they were killed (Fleck *et al.*, 1965). Microsomes isolated from the livers of animals fed the complete mixture showed a greater capacity for *in vitro* protein synthesis than did microsomes prepared from the animals fed with the tryptophan-deficient mixture (Table III). Since the microsome fraction owes its amino acid incorporating capacity to its polysome content, the state of aggregation of the polysomes in the liver was examined after feeding each of these amino acid mixtures. The total population of polysomes, that is, both free and membrane-attached, was separated into different classes on a sucrose gradient, and it was observed that animals fed with the amino acid mixture lacking tryptophan had fewer heavy polysomes and more monosomes and disomes than had the livers of rats given the nutritionally complete amino acid mixture. This finding indicates that omission of tryptophan from the administered amino acid mixture

TABLE III

In Vitro UPTAKE OF ^{14}C-LEUCINE BY MICROSOMES FROM RATS FED 1 HR BEFORE DEATH WITH EITHER A COMPLETE OR TRYPTOPHAN-DEFICIENT AMINO ACID MIXTURE[a,b]

Actino-mycin	Incubation time (min)	Incomplete amino acids	Complete amino acids	Difference
		(counts/min/mg microsome protein)		
None	10	3690	4630	+940
	40	4600	5580	+980
Pretreated	10	2020	2540	+520
	40	2520	3550	+1030

[a] Reproduced from Fleck *et al.* (1965), with permission.
[b] Same animals were pretreated with actinomycin D.

leads to disaggregation of the liver polysomes and to consequent accumulation of free ribosomes; this loss of polysomes provides an adequate reason for the reduced capacity of microsomes isolated from the livers of the tryptophan-deficient rats to carry out *in vitro* protein synthesis. In subsequent experiments (Wunner *et al.*, 1966), it was shown that the disaggregation caused by feeding the tryptophan-deficient mixture could be rapidly reversed by administering the missing tryptophan at a later time. These experiments demonstrate that protein biosynthesis in the liver is sensitive not only to a change in amino acid supply coming to the organ, but also to the quality of the amino acid mixture.

The intracellular site of the response to amino acids has also been explored. Animals were pretreated with actinomycin-D before feeding the amino acid mixtures described above. The capacity of the liver microsomes to incorporate ^{14}C-leucine *in vitro* and the polysome patterns obtained from the liver were then examined. The response of both of these parameters to the complete and to the tryptophan-deficient amino acid mixtures was still evident after actinomycin administration. Table III illustrates the results obtained in the case of the liver microsomes. These findings indicate that the response of the liver protein-synthesizing mechanism is independent of increased secretion of messenger RNA and in consequence is likely to be dependent only on cytoplasmic regulation. It should, therefore, be possible to construct a cell-free cytoplasmic system that mimics the changes in protein synthesis and in polysome aggregation when amino acid supply is varied. This in fact has been done (Baliga *et al.*, 1968) and it has been shown that the predicted changes can be produced in response to addition of a complete mixture of amino acids to the system, but that deletion of any one of the amino acids results in a failure of polysome aggregation and of amino acid incorporation. These experiments with the reconstructed cytoplasmic system for protein synthesis confirm that changes in polysome profile can be brought about by alterations in amino acid supply and that the participation of the nucleus is not essential. There is, however, one difference between the cell-free system and the whole animal in their responses to amino acid mixture. Further study of polysome aggregation in intact rats has shown that feeding amino acid mixtures deficient in any other essential amino acid except tryptophan fails to cause disaggregation of polysomes in the way that deficiency of tryptophan does (Pronczuk *et al.*, 1968). It is suggested (Munro, 1968) that the special sensitivity to tryptophan of the polysome population in the liver of the intact animal occurs because tryptophan is normally the

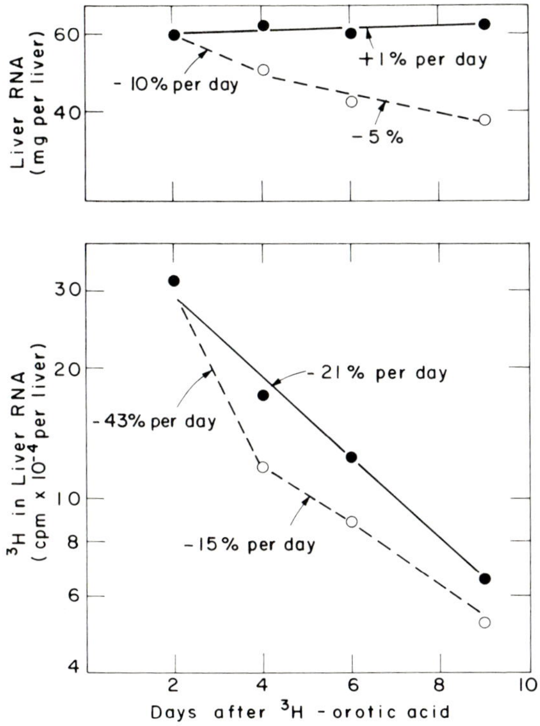

FIG. 1. Total amount of RNA per liver and rate of loss of ^{3}H-RNA from the livers of rats receiving a diet adequate in protein (●——●) and of rats immediately after changing to a protein-free diet (○ - - - ○). All animals were fed the diet containing adequate amounts of protein until 2 days after orotic acid-^{3}H administration, when one group was then given a similar amount of a protein-free diet. The amounts of total RNA and of ^{3}H-RNA per liver are plotted on a logarithmic scale. Each point represents results from the pooled livers of three animals (Enwonwu and Munro, 1970).

least abundant amino acid in the free amino acid pool, and consequently becomes the rate-limiting factor in protein synthesis. In the case of the cell-free system, it seems that the concentrations of individual amino acids can be lowered to trace amounts and thus each amino acid can be made limiting under *in vitro* conditions.

Coincident with these changes in liver polysome aggregation, amino acid supply appears to be responsible for alterations in RNA turnover in the cytoplasm of the liver cell. The rats fed the tryptophan-deficient amino acid mixture show not only more monosomes in the liver cells, but also accumulation of ribosomal subunits (Wunner *et al.*, 1966). Dissociation of ribosomes into subunits is associated with activation of their latent ribonuclease, from which it can be predicted that administration of the tryptophan-deficient amino acid mixture should lead to increased breakdown of liver RNA. This has received confirmation from early studies of liver RNA metabolism in rats receiving different intakes of protein or amino acids (Clark *et al.*, 1957; Munro and Clark, 1959) and also by recent studies on the effect of protein intake on the rate of breakdown of liver RNA (Enwonwu and Munro, 1970). In these latter experiments, the RNA of the liver was prelabeled by injecting orotic acid into rats, and then the rate of loss of the label was used as a measure of RNA breakdown. If protein was suddenly withdrawn from the diet of such animals, there was a temporary acceleration in breakdown rate (Fig. 1) which was found to coincide with a transient accumulation of subunits in the liver cell sap (Munro *et al.*, 1964).

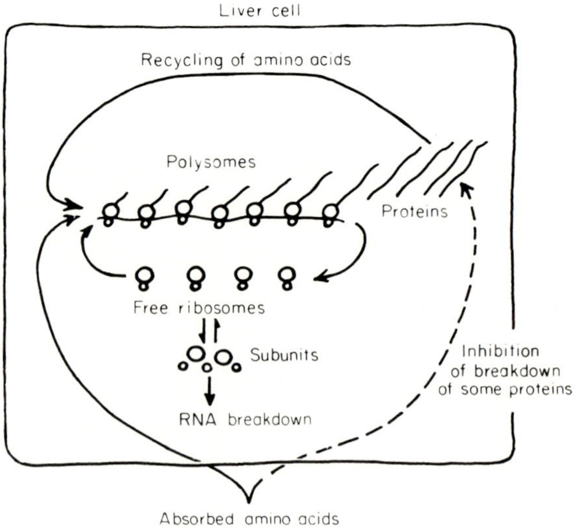

FIG. 2. Scheme showing interrelationship of amino acid supply to protein synthesis and turnover and RNA turnover in the liver cell (Munro, 1970).

From these various pieces of evidence, we can construct a picture of the events occurring in the protein and RNA metabolism of the liver whenever a meal containing protein is consumed (Fig. 2). The increased amounts of amino acids passing up the portal vein accelerate liver protein synthesis; protein accumulates in that organ not only because of this effect but also because breakdown of some liver enzymes involved in amino acid catabolism is retarded through the increased levels of their substrates. The increased rate of liver protein synthesis causes polysome aggregation, with a resultant reduction in monosomes and in the population of ribosome subunits in the liver, which in turn reduces liver RNA breakdown. Since RNA synthesis continues unrestricted, there is an accumulation of RNA in parallel with the increase in liver protein. When absorption of amino acid ceases, the polysomes disaggregate and the liver loses both protein and RNA at an accelerated rate.

V. Diurnal Rhythms in Liver Protein Metabolism

The picture described in the preceding section suggests that liver protein metabolism should show rhythmic changes throughout the day in response to successive meals containing protein. This prediction has been verified experimentally by Fishman *et al.* (1969), who correlated the diurnal cycles in liver polysome aggregation and in the activity of the liver enzyme tyrosine aminotransferase in the case of rats subjected to a 12-hr alternating cycle of light and dark. Such rats adopt a pattern of feeding that begins toward the end of the light period, reaches a maximum intake during the early hours of darkness, and becomes minimal shortly before the light comes on again. In relation to this feeding pattern, the liver polysome pattern shows disaggregation in the middle of the light period, starts to aggregate shortly after feeding commences, attains a maximum in the middle of the dark period, and begins to disaggregate again by the start of the next light period (Fig. 3). Tyrosine aminotransferase, an enzyme that causes degradation of tyrosine, shows an almost parallel diurnal pattern of increase and decrease in activity in the liver. These rhythms in protein metabolism are discussed in detail by Wurtman (1970) who summarizes the evidence that both the polysome changes and the changes in tyrosine aminotransferase activity are dependent on intermittent intake of tryptophan in the diet.

It is legitimate to ask whether all liver proteins share in these rhythmic fluctuations in protein synthesis. This has not so far been tested. Wilson *et al.* (1967) have concluded that starvation has less effect on synthesis of plasma albumin by rat liver than on the synthesis of other liver proteins, and attribute this to more stable messenger RNA for albumin. On the other hand, Kirsch *et al.* (1968) have shown a rapid response of albumin synthesis when

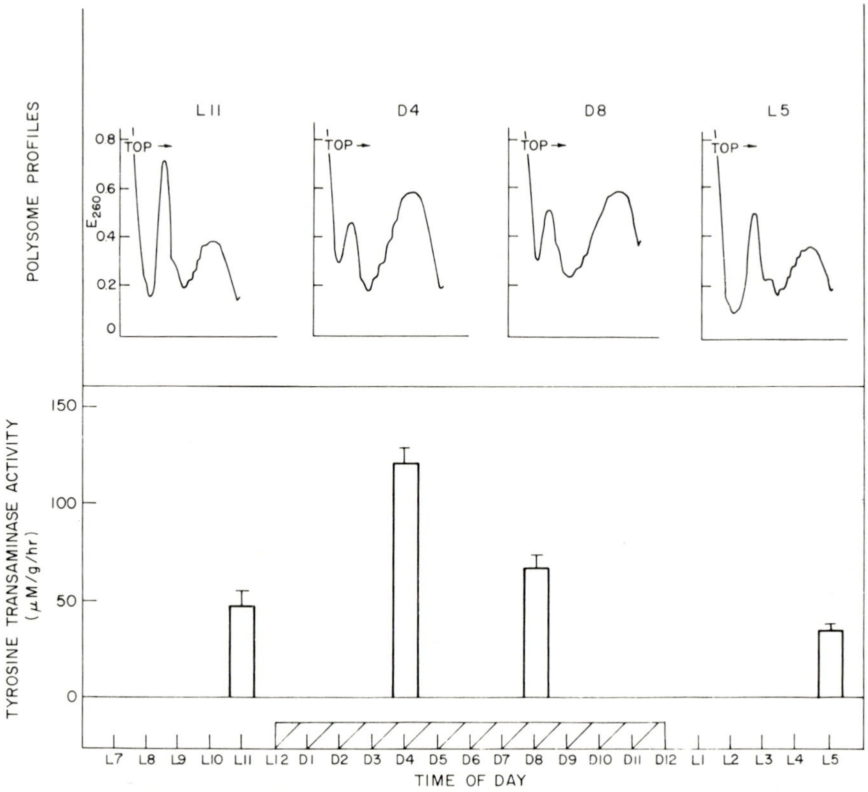

FIG. 3. Hepatic polysome profiles and tyrosine transaminase activity among groups of six rats fed Purina Chow and killed at each time indicated. Vertical bars indicated standard errors of the mean, and hatching along the abscissa indicates the daily dark period (Fishman *et al.*, 1969).

protein-depleted rats are refed protein, and Rothschild *et al.* (1969) have obtained evidence from liver perfusion that rate of albumin formation is rapidly stimulated by addition of amino acids, particularly tryptophan, to the fluid used to perfuse the livers of starved rabbits. It would thus be interesting to determine whether the synthesis of albumin and other plasma proteins is subject to diurnal rhythms coincident with feeding patterns.

REFERENCES

Baliga, B. S., Pronczuk, A. W., and Munro, H. N. (1968). *J. Mol. Biol.* **34**, 199.
Brody, S. (1945). "Bioenergetics and Growth." Reinhold, New York.
Clark, C. M., Naismith, D. J., and Munro, H. N. (1957). *Biochim. Biophys. Acta* **23**, 581.

Denton, A. E., and Elvehjem, C. A. (1954). *J. Biol. Chem.* **206**, 449.
Elwyn, D. (1970). *In* "Mammalian Protein Metabolism" (H. N. Munro, ed.), Vol. 4, p. 523. Academic Press, New York.
Enwonwu, C. O., and Munro, H. N. (1970). *Arch. Biochem. Biophys.* (in press).
Fishman, B., Wurtman, R. J., and Munro, H. N. (1969). *Proc. Nat. Acad. Sci. US.* **64**, 677.
Fleck, A., Shepherd, J., and Munro, H. N. (1965). *Science* **150**, 628.
Ganoza, M. C., and Williams, C. A. (1969). *Proc. Natl. Acad. Sci., U.S.*, **63**, 1370.
Hicks, S. J., Drysdale, J. W., and Munro, H. N. (1969). *Science* **164**, 584.
Kirsch, R., Frith, L., Black, E., and Hoffenberg, R. (1968). *Nature* **217**, 579.
Munro, H. N. (1968). *Fed. Proc. Fed. Amer. Soc. Exp. Biol.* **27**, 1231.
Munro, H. N. (1969). *In* "Mammalian Protein Metabolism" (H. N. Munro, ed.), Vol. 3, p. 133. Academic Press, New York.
Munro, H. N. (1970). *In* "Mammalian Protein Metabolism" (H. N. Munro, ed.), Vol. 4, p. 299. Academic Press, New York.
Munro, H. N., and Clark, C. M. (1959). *Biochim. Biophys. Acta* **33**, 551.
Munro, H. N., and Downie, E. D. (1964). *Nature*, **203**, 603.
Munro, H. N., McLean, E. J. T., and Hird, H. J. (1964). *J. Nutr.* **83**, 458.
Pronczuk, A. W., Baliga, B. S., Triant, J. W., and Munro, H. N. (1968). *Biochim. Biophys. Acta* **157**, 204.
Redman, C. M. (1968). *Biochem. Biophys. Res. Commun.* **31**, 845.
Redman, C. M. (1969). *J. Biol. Chem.* **244**, 4308.
Rothschild, M. A., Oratz, M., Mongelli, J., Fishman, L., and Schreiber, S. S. (1969). *J. Nutr.* **98**, 395.
Takagi, M., and Ogata, K. (1968). *Biochem. Biophys. Res. Commun.* **33**, 55.
Wilson, S. H., Hill, H. Z., and Hoagland, M. B. (1967). *Biochem. J.* **103**, 567.
Wunner, W. H., Bell, J., and Munro, H. N. (1966). *Biochem. J.* **101**, 417.
Wurtman, R. J. (1970). *In* "Mammalian Protein Metabolism" (H. N. Munro, ed.), Vol. 4, p. 445. Academic Press, New York.

10

IDENTIFICATION OF PROTEINS MADE ON MICROSOMES AND FREE RIBOSOMES OF RAT LIVER

Curtis A. Williams and M. Clelia Ganoza

THE ROCKEFELLER UNIVERSITY, NEW YORK, NEW YORK;
CHARLES H. BEST INSTITUTE, TORONTO, ONTARIO, CANADA

I. Introduction

Research on the mechanism and the regulation of protein synthesis in mammalian and other eukaryotic cells has not progressed as rapidly as it has in lower organisms, notably in certain bacteria. Not only has dissection of the biochemical mechanism proven more refractory in eukaryotes (Lengyel and Söll, 1969), but it has not yet been possible to subject these specialized cells and tissues to the rapid and refined genetic analysis possible with microbial systems (Epstein and Beckwith, 1968). Moreover, since many of the primary regulatory events in higher organisms occur in tissues or organ systems remote from the cell type under consideration, the relevance of *in vitro* experiments on regulation is always open to question. These disadvantages have not prevented significant progress, however, when it has been possible to identify and study the synthesis of specific proteins.

We have devised a cell-free system from hepatic tissues in which the synthesis of a number of specific proteins can be detected by immunochemical methods. These studies have shown a partitioning of rat liver ribosomes according to function (Williams and Ganoza, 1969; Ganoza and Williams, 1969). In this cell-free system, ribosomes which are isolated attached to endoplasmic reticulum appeared to synthesize exclusively exportable proteins. Ribosomes isolated free of membranes synthesize only intracellular proteins.

This observation has profound implications with respect to messenger RNA-ribosome recognition processes at the molecular level. Selective recognition phenomena of this type in protein synthesis are not commonly considered. From a structural point of view, these and other studies reported here give added support to the notion that exportable protein at time of their synthesis are transferred into endoplasmic microtubular structures (Siekevitz and Palade, 1960).

II. Methods

The general scheme followed in all of these experiments is outlined in the diagram below.

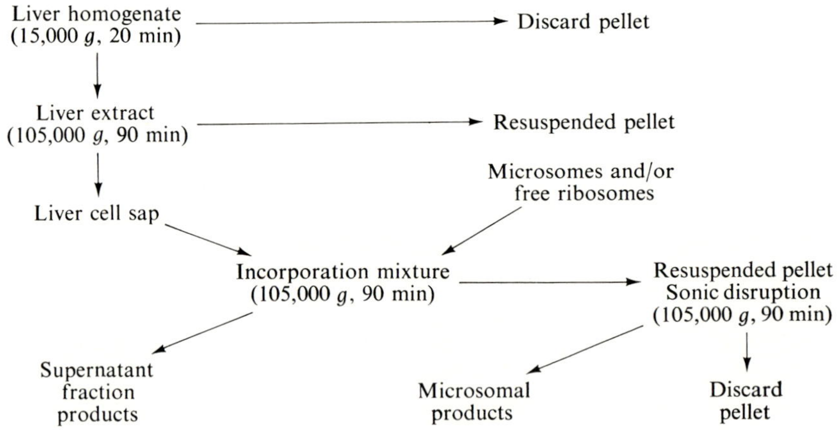

Homogenization of liver tissue and preparation of ribosomal fractions are the critical steps for the kinds of *in vitro* incorporation studies we discuss here. Reproducible quantitative results depend on the use of a loose fitting (0.8–0.9 mm clearance), smooth Teflon pestle used with careful attention to time, temperature, rpm, and volume (Ganoza *et al.*, 1965). The preparation of ribosomal fractions on a discontinuous sucrose gradient, as described by Ganoza and Williams (1969), is a method modified from Wettstein *et al.* (1963).

The composition of the incorporation mixture and the conditions of incubation has varied somewhat to accommodate other experimental requirements or conditions. The most frequently employed incorporation mixture was adapted from Campbell *et al.* (1960) and contained 27 m*M* Tris (HCl) buffer, pH 7.8, 0.5 m*M* GTP, 8 m*M* GSH, 2m*M* ATP, 8 m*M* PEP, 6 m*M* $MgCl_2$, 50 m*M* KCl, and 0.06 m*M* each of the nonradioactive amino acids. In addition, each milliliter of the incubation mixture contained 40 μg pyruvate kinase, 2.5 mg RNA equivalents of ribosomes (either free or bound), 0.01 mg liver tRNA, and 2.5–6.0 mg protein of dialyzed 105,000 *g* supernatant fraction of the liver homogenate. Incubation was at 37°C for 50 min with 2.5–4.0 μCi ^{14}C-amino acids per ml.

When the supernatant fraction products and the microsomal products are recovered, they are assayed for incorporated radioactivity; the total incorporation mixture is assayed before separation of the fractions; in some experiments the discard pellet containing the voided microsomes is resuspended and assayed. Radioactivity is most conveniently determined by collecting hot trichloroacetic acid precipitates of the samples on millipore filters. These are transferred to planchets, dried, and counted in a thin-window, gas-flow counter.

Concentration of products is required for analysis by diffusion in gels. Products are concentrated at least 20 times by lyophilization after dialysis or by precipitation of the protein by ammonium sulfate at 75% saturation followed by extensive dialysis.

Antisera against rat plasma proteins and liver cell sap antigens were produced in rabbits by intramuscular and subcutaneous injection of antigen in complete Freund's adjuvant. Immunization was continued several weeks by alternate-week series of three intravenous injections until suitable antisera were obtained. The objective was to obtain a highly complex, multispecific antiserum which would give a large number of reaction lines by IEA. Anticell sap antiserum is treated to remove any antibody to plasma proteins. Stepwise additions of small aliquots of rat plasma, incubation overnight at 4°C, and centrifugation is continued until no reaction with plasma proteins can be detected. The quantity and quality of antisera required for extensive work of this nature makes most commercial products too expensive or otherwise unsatisfactory.

IEA is performed in barbital buffered ($\mu = 0.04$, pH 8.2) agar (1.5%) on microscope slides. Electrophoresis is continued for 90 min at 6 V/cm (Williams, 1970). Diffusion and precipitation is continued for 24–48 hr depending on the density and number of reaction lines. It is frequently desirable to fill the antigen well up to four times (1–2 μliter each time) in order to assure sufficient radioactive antigen representing minor components. Antiserum is sometimes concentrated for use in developing IEA patterns with concentrated antigens. This was always necessary with antisera to liver sap antigens. Nonradioactive carrier antigens were used to adjust antigen concentration to antigen–antibody proportions optimal for sharp and dense reaction lines, although this is not possible for all reactions in a given analysis.

Autoradiograms are prepared by placing a 1 × 3 inch NoScreen plate (Kodak) in contact with the washed, dried IEA pattern on the microscope slide. Details of this technique are given by Thorbecke *et al.* (1970).

III. Experimental Results

The intensity of autoradiographic traces depends on the amount of radioactivity and its concentration into discrete zones. Figure 1 demonstrates the effect of different amounts of carrier plasma antigens (P/10, diluted; and P, undiluted) on the autoradiography of IEA patterns. Identical amounts of *in vitro* labeled microsomal products (M*) were used and the plate was developed with rabbit antiserum to rat plasma proteins (aP). With carrier diluted 10 times (left side) only the albumin and transferrin reaction are near antigen–antibody equivalence and only those reactions give good autoradiograms. With undiluted carrier the albumin reaction is in great antigen excess, the isotope is dispersed and the autoradiogram is almost lost. The transferrin autoradiogram is also broader and fainter in this pattern. On the other hand, five or six other reactions are brought sufficiently close to immunological equivalence by the addition of undiluted carrier to give good autoradiograms. Thus, we know that many plasma proteins are labeled *in vitro* even though a single IEA pattern may not reveal them. The dense radioactive "fog" on the left side of the autoradiogram in Fig. 1 corresponds to the hazy precipitate on the agar slide. Although there is usually some denatured material bound to the agar toward the anode from the starting well, much of the haze is thought to represent several specific immunological reactions in extreme antibody excess. The addition of more antigen, in the form of carrier, not only sharpens up some reactions but partially clears the background "fog." In the light of this finding we content ourselves with the demonstration of labeled albumin and/or transferrin using dilute carrier or no carrier. When these are found labeled we assume that a large group of plasma proteins are synthesized *in vitro*.

Early in the course of this work it became apparent that the labeled plasma proteins we detected could be positively identified only in the supernatant fraction of a sonicated suspension of the microsomal pellet. Only a small trace

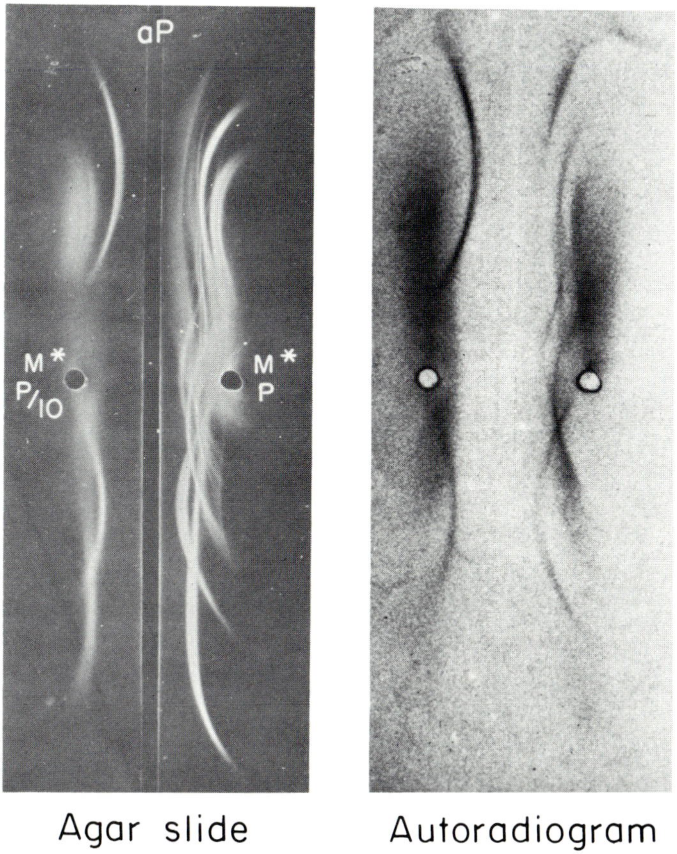

FIG. 1. IEA patterns (left) and autoradiograms of radioactive rat plasma proteins released from microsomal pellet (M*) by sonication after cell-free incorporation of ^{14}C-amino acids. After M* had been added to the starting wells and absorbed into the agar, plasma proteins were added as carrier (P/10, ten times diluted; P, undiluted plasma). IEA pattern was developed by addition of rabbit antiserum to rat plasma proteins (aP) to the central trough. The anode is at the top of the figure, thus the albumin reaction is large arc nearest the top. The transferrin reaction is the dense arc below the starting well. The dense arc nearest the cathode is the γ-globulin precipitate. The diameter of wells and width of trough are 1 mm.

of labeled albumin was detected in the soluble fraction of the total incubation mixture even though a large portion of the incorporated amino acids was in the protein of this fraction.

We reasoned, however, that the label incorporated into the incubation supernatant fraction also represented synthesis of intact native protein, and that it could be characterized with a suitable antiserum. An antiserum prepared against liver cell sap antigens (aL) detected numerous soluble antigens and gave a complex but clear IEA pattern and autoradiogram (Fig. 2). *In vitro* labeling of these antigens was not so rapid as with serum proteins, and although several were detected on the autoradiogram using the customary amounts of antigen and antiserum, excessively long exposure times (2–3 months) were required for convincing demonstrations of specific incorporation. The experiment shown in Fig. 2 was performed with oversize (2 mm) wells and lateral troughs so that four times the usual amount of reactants could be employed. This is an expensive procedure, however, and once the demonstration was made we returned to smaller scale experiments. No additional carrier is required for the analysis of liver antigens in the supernatant fraction since nonradioactive cell sap proteins were added to the incubation mixture and were concentrated along with the newly synthesized products. It is clear that many more reactions are in immunochemical equivalence than appear radioactive. This can be explained by two arguments: (1) the normal turnover and therefore the *in vitro* rate of incorporation into different components probably varies greatly among these sap proteins; (2) some newly labeled products are probably unassembled subunits of native proteins and they either are not recognized by the antibodies to intact antigens or they are migrating differently in the electric field, placing their reactions in extreme antibody excess (and therefore not visible).

Up to this point in our experiments only the total microsomal fraction pellet had been used in incorporation mixtures. It was becoming apparent that there were either separate compartments of protein synthesis associated with different categories of protein, or there was significant nonspecific binding of labeled byproducts and debris to the antigens precipitated by antiliver antibodies.

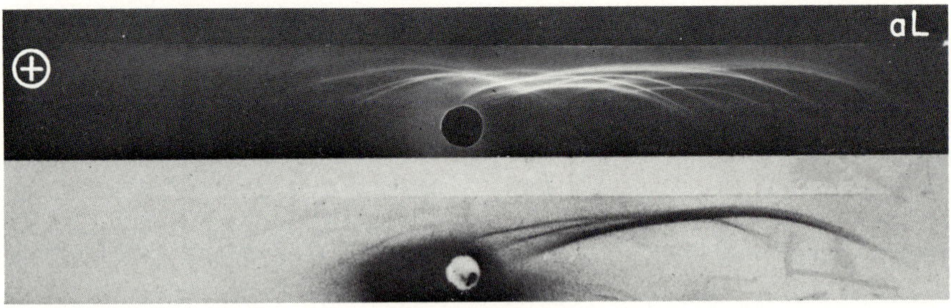

FIG. 2. IEA pattern (top) of rat liver cell sap antigens in the supernatant fraction of a cell-free incorporation mixture (S*). The autoradiogram of the IEA plate (bottom) indicates those proteins into which a significant amount of ^{14}C-amino acids was incorporated during the 50-min incubation. The IEA pattern was developed with rabbit antiserum specific for soluble rat liver antigens (aL). The anode (+) is to the left. The diameter of the well and the width of the trough are 2 mm.

An experiment was performed to test the possibility that the autoradiograms represented nonspecific binding or coprecipitation either of labeled debris or of proteins of another category. Two identical incubation mixtures were prepared, one with thirteen ^{14}C-amino acids (reconstituted protein hydrolyzate from Schwartz Bioresearch) and the other with ^{12}C-amino acids. Products were prepared as in the general diagram above. The cold supernantant products (S) were mixed with labeled microsomal products (M*); the labeled supernatant products (S*) were mixed with the cold microsomal

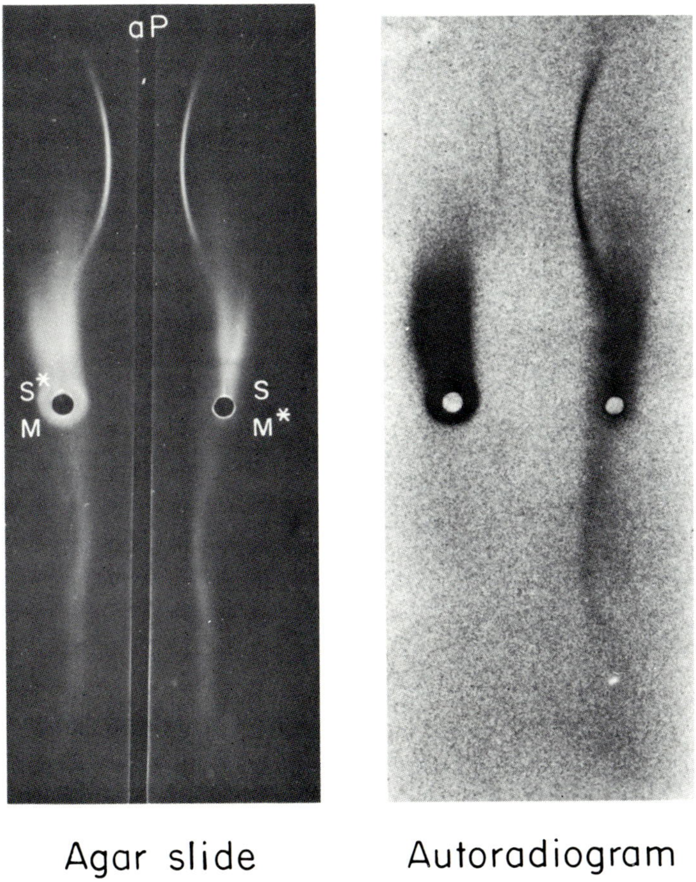

FIG. 3. IEA patterns (left) and autoradiogram of ^{14}C supernatant fraction products (S*) mixed with ^{12}C microsomal products (M), left on plate, and ^{12}C supernatant fraction (S) with ^{14}C microsomal products (M*), right on plate. No carrier plasma proteins were added to the mixtures. The patterns were developed with antiplasma proteins (aP), thus the autoradiogram demonstrates that significant amounts of plasma proteins are obtained only from microsomes after cell-free incubation. Only faint label is contributed to the albumin reaction by S*.

products (M). These biochemically identical mixtures were incubated for 50 min at 37°C and were then concentrated for analysis. No additional carrier was employed. Figure 3 shows the IEA pattern and autoradiograms of the product concentrates developes with antiserum to plasma proteins (aP). Only the mixture containing M* has significant radioactivity associated with the serum proteins. There is a faint trace of labeled albumin in the S*/M mixture and the possibility that this is nonspecific coprecipitation is not ruled out. Figure 4 shows IEA patterns and autoradiograms of the product

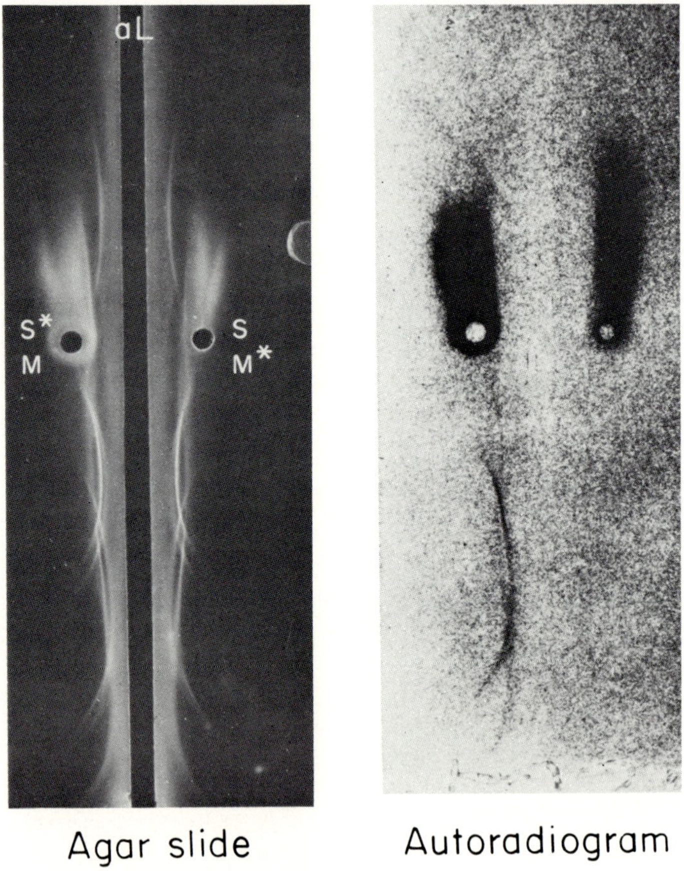

FIG. 4. IEA patterns (left) and autoradiogram of ^{14}C supernatant fraction products (S*) mixed ^{12}C microsomal products (M), left on plate, and ^{12}C supernatant fraction (S) with ^{14}C microsomal (M*), right on plate. No carrier was added to the mixtures. The patterns were developed with antiliver sap (aL), thus the autoradiogram demonstrates that the soluble liver antigens detected by this antiserum are found only in the supernatant fraction of the cell-free incubation mixture.

concentrates developed with antiserum to liver antigens (aL). Only the S* contains identifiable labeled liver antigens. The S/M* mixture, which gives an identical IEA pattern, shows no detectable radioactivity associated with the reactions in the pattern. Since no significant nonspecific coprecipitation of radioactive products or debris was detected, we conclude that the soluble liver sap proteins are recovered only in the supernatant fraction of the incubation mixture and that the serum or secretory proteins are recovered almost totally from the sonically ruptured microsomes.

This latter observation is consistent with the reports of others (Peters, 1959; Campbell *et al.*, 1960) that plasma proteins synthesized *in vitro* remain associated with the microsomes. It is also consistent with the model for the production and export of proteins by secretory cells proposed by Siekevitz and Palade (1960), and the unidirectional extrusion across the membrane of protein made by ribosomes bound to the endoplasmic reticulum (Redman and Sabatini, 1966; Jamieson and Palade, 1968).

If proteins made by membrane-bound ribosomes remain associated with microsomes, then proteins released to the supernatant fraction must be made on unbound or free ribosomes. To test this hypothesis ribosomes were freed from the microsomal membranes by treatment of the liver extract with 0.6% deoxycholate (DOC) prior to the 105,000 *g* centrifugation, a method modified from Wettstein *et al.* (1963). During incubation these ribosomes should release labeled plasma proteins directly to the supernatant. Figure 5 shows the IEA patterns and autoradiograms of the supernatant proteins from the DOC ribosome incubation and from the untreated pellet or microsome incubation. The patterns were developed with antiserum to rat serum proteins (aP). It is seen that the labeled serum albumin and a trace of transferrin are released to the supernatant fraction of the DOC ribosome incorporation mixtures but only the usual trace of albumin is released by the microsome incubation.

We assumed, therefore, that the labeled liver proteins in the fraction of the total ribosome incubations were made on unbound, normally free ribosomes in the liver cell. Accordingly, free ribosomes were prepared (Ganoza and Williams, 1969). Electron micrographic examination of free ribosomes revealed a small contamination of free ribosomes by fragments of rough membranes. Phospholipid was not detected in the preparation, however, indicating less than 10% contamination by bound ribosomes.

Incorporation mixtures were made with the free ribosomes and the total ribosomal pellet from the same liver homogenate. Products were isolated and concentrated as shown in the general diagram in Methods. IEA patterns and autoradiograms are shown in Figs. 6 and 7. For IEA of sonicated pellet products (SON-P) of free ribosomes (F) by antiserum to plasma proteins (aP), carrier plasma diluted 30 times was added to the well. Since only traces

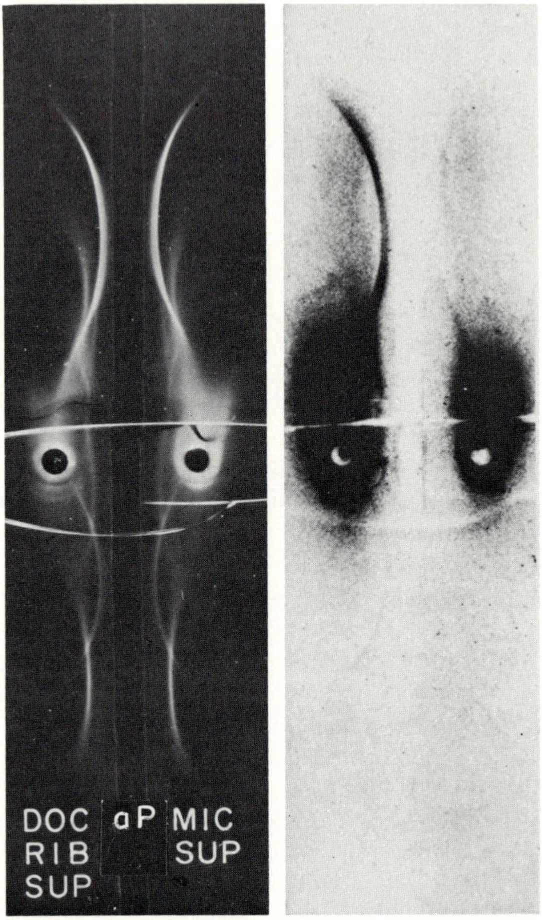

FIG. 5. IEA patterns (left) and autoradiograms of supernatant fractions (SUP) from incubations with deoxycholate-treated microsomes (DOC/RIB) and the normal microsomal pellet (MIC). The patterns were developed with antiplasma proteins (aP). The autoradiogram demonstrates that serum albumin and a trace of transferrin, normally obtained only from the microsomes after sonication, are released to the supernatant fraction if the microsomal ribosomes are freed from the membranes with DOC.

of plasma protein would be expected, this was done to assure detection of any labeled albumin produced by the small contamination of microsomes in the free ribosome preparation. No carrier was necessary for the IEA pattern of the sonicated pellet products of the total ribosome incubation (T/SON-P) developed by antiplasma antibodies (aP). Autoradiograms of the plasma proteins in the T/SON-P are very strong even for reactions barely visible in the IEA patterns.

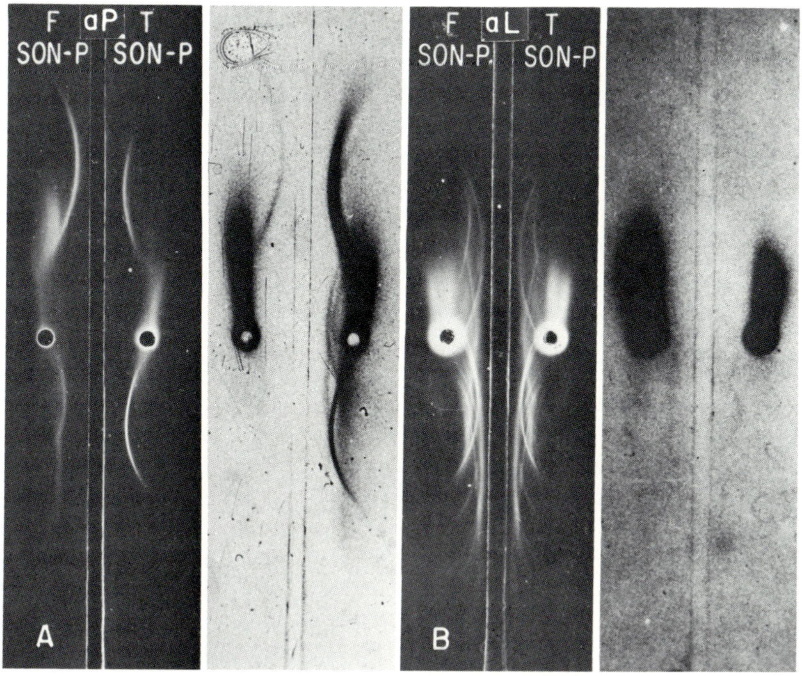

FIG. 6. IEA patterns (A and B) and corresponding autoradiograms of products released from pellets by sonication after incubation (SON-P). Incorporation mixtures were made with free ribosomes prepared on a discontinuous sucrose gradient (F) or with the total microsome–ribosome pellet (T). Plate A was developed with antiplasma protein (aP) and carrier plasma diluted 30 times was added to the F/SON-P; the autoradiogram demonstrates that only a trace of labeled albumin is present in the F/SON-P. Plate B is a pattern of the SON-P fractions with carrier cell sap protein developed with antiliver sap protein (aL); no labeled liver protein is detected in the autoradiogram.

The IEA patterns of F/SON-P and T/SON-P developed by antiliver antibodies (aL) are due entirely to added carrier cell sap. Since very little labeled liver protein was expected, twice as much product concentrate was employed on this plate as on that developed with antiplasma antibodies. Also exposure of the autoradiogram was extended to 15 weeks. Still no specific autoradiographic traces are detected. This means that no finished liver proteins remain loosely bound to free or bound ribosomes and that none are released from microsomes upon sonic rupture.

The supernatant fractions of the incubations are analyzed in Fig. 7. In this series four times as much labeled product was used for development with antiplasma antibodies (aP), since few labeled plasma proteins were expected. No carrier was necessary since the sap fraction used in the incubation mixture

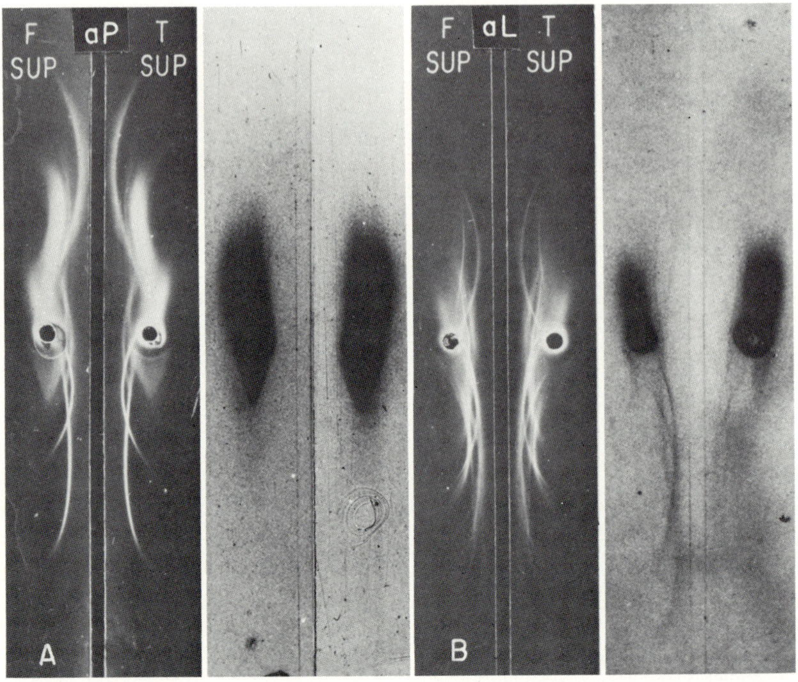

FIG. 7. IEA patterns (A and B) and corresponding autoradiograms of supernatant fraction products (SUP) from incubations of free ribosomes (F) and of the total microsome–ribosome pellet (T). Plate A was developed with antiplasma protein (aP); the autoradiogram demonstrates that neither preparation releases labeled plasma protein into the SUP. Plate B, without carrier added, was developed with antiliver sap (aL); the autoradiogram shows that the F/SUP contains significantly more labeled liver proteins than does the T/SUP.

always has some plasma proteins. No labeled plasma protein was detected in either the supernatant fraction of the free ribosome incubation (F/SUP) or the T/SUP. Development of these fractions by antiliver antibodies (aL) reveals significant labeling of liver proteins by the free ribosomes and their release to the medium upon completion. Appreciably less radioactivity is found in these components in the T/SUP, since the free ribosomes comprise only 15–20% of the total ribosomes. Incubation mixtures are made up on an equal RNA basis so that, assuming equal activity of the free ribosomes in the two preparations, a 90% free ribosome preparation should incorporate amino acids five times more rapidly than the 15–20% preparation.

In two experiments reported by Ganoza and Williams (1969), the activity of the free ribosomal preparation was the same as that of the total ribosomes on an equal RNA basis. This is in agreement with data obtained from *in*

TABLE I

INCORPORATION OF ^{14}C-AMINO ACIDS INTO PROTEIN BY TOTAL, FREE, AND BOUND RIBOSOMES *in Vitro*[a]

Experiment	dpm incorporated per mg ribosomal RNA added to incorporation mixture		
	Total	Free	Bound
1	126,500	130,250	42,500
2	117,900	109,900	50,500

From Ganoza and Williams (1969).

vivo experiments (Redman, 1968). The isolated bound ribosomes were significantly less active, however, probably due to their partial inactivation during preparation. These data are given in Table I. The autoradiograms of IEA patterns obtained with the products of the bound ribosome incubation confirmed that serum proteins were released by sonication of the ribosome–membrane complex, but that no labeled liver component is released. There were very faint traces of labeled nonserum proteins in the incubation supernatant fraction, however, which may have been due to minor contamination by free ribosomes.

These experiments do not exclude the possibility that other nonserum proteins are synthesized on the bound or free ribosomes. At least some membrane proteins appear to be synthesized on bound ribosomes (Dallner *et al.*, 1966; Ragnotti *et al.*, 1969), and some tissues or cell types not known to secrete protein do have ribosomes attached to the endoplasmic reticulum (Andrews and Tata, 1968). In the experiments we have described, over 80% of the labeled peptides synthesized on bound ribosomes could not be released by sonication. In contrast, 50% remained bound to free ribosomes. In both cases part of these products represent unfinished peptides esterified to tRNA; the rest of the bound peptides synthesized seemed to share many characteristic features of structural or membrane proteins (Williams and Ganoza, in preparation).

It might be expected that if the large amount of labeled peptides made in these extracts represented liver proteins, then their rate of synthesis should increase when tissue proliferation is accelerated as, for example, by partial hepatectomy. We had actually performed such experiments several years earlier based on the work of Hultin and von der Decken (1957), but we could not interpret them at the time. Table II summarizes the incorporation data. The distribution of incoporated amino acids into subfractions of the incubation mixtures is much as expected based on the function of free and bound

TABLE II

INCORPORATION OF ^{14}C-LEUCINE INTO DIFFERENT PROTEIN FRACTIONS BY RAT LIVER MICROSOMES *in Vitro*[a]

Group (no. animals)	Average weight (g)	dpm incorporated per mg RNA			
		Incubation supernatant fraction	Sonic release from pellet	Voided sonicated pellet	Total
C (3)	220	29,000	19,500	79,500	146,200
					128,000[b]
R (7)	230	111,300	24,800	336,900	371,100
					473,000[b]
N (4)	220	25,500	15,000	94,200[c]	133,700
NI (4)	150	57,750	31,600	214,000	269,000
					303,500[b]

[a] Cell free incorporation mixtures were prepared with complete microsomal pellet fraction (12.5 mg RNA), the corresponding cell sap fraction, 13 μCi ^{14}C-leucine (223 μCi/μmole), amino acids and an ATP generating system in a volume of 5 ml. Incubation was at 37°C for 50 min. Microsomes from groups C (sham operated) and R (regenerating after median and left lateral lobectomy) were prepared for assay 40 hr after operation. Group NI (normal immature) was assayed with groups C and R. Group N (normal mature) was run separately. Livers of the number of animals indicated were pooled.

[b] Sum of incorporation into fractions.

[c] Fraction not assayed, value by difference.

liver ribosomes and the state of the animals. Major increases in incorporation in the case of regenerating livers are found only in the supernatant fraction and the voided pellet. By contrast, control systems from younger, rapidly growing animals showed a doubling in all the fractions. Analysis of the products by IEA and autoradiography revealed no significant qualitative changes in the synthesis of serum proteins or soluble liver proteins.

IV. Discussion

A cell-free protein synthesizing system derived from rat liver tissues has been devised and shown capable of generating a large number of liver proteins. The proteins synthesized *in vitro* have been equated with specific liver proteins by rigorous immunochemical criteria. That the cell-free system truly reflects the physiological situation in the intact animal is verified by the fact that extracts derived from "acute phase" animals produce a biosynthetic profile of plasma proteins almost identical to that of the intact animal (Williams, Ganoza, and Lipmann, 1965; chapter by Williams, this volume); and in extracts derived from rapidly regenerating liver, there is a strong bias toward the synthesis of soluble liver proteins and probably of structural or membrane proteins.

The experiments described here demonstrate that secretory plasma proteins of hepatic tissue are synthesized on membrane-bound ribosomes, whereas a variety of nonserum liver proteins are made on free ribosomes. The proteins made on bound ribosomes or microsomes remain associated with those structures. These products are detected only following disruption of membranous structures by sonication or detergent lysis. The proteins synthesized on free ribosomes, however, are released spontaneously into the medium, and presumably *in vivo* into the cytoplasm. Thus, we believe that these results conclusively establish separate functional roles for these two classes of ribosomes.

The finding of Hicks *et al.* (1969) and Redman (1969) that ferritin, another nonexport protein, is preferentially labeled by free ribosomes adds to the generality of the phenomenon. The experiments which suggest that membrane generality of the phenomenon. The experiments which suggest that mem-brane-bound ribosomes synthesize structural proteins are consistent with *in vivo* suudies (Dallner *et al.*, 1966); thus, the binding of ribosomes to microtubtles may be not only important in secretion, but it may also play a role in the generation of proteins which comprise those membranes and possibly structural elements of other cellular organelles. We have not established whether free ribosomes synthesize membrane proteins. One such report has appeared (Ragnotti *et al.*, 1969).

The interesting and as yet unanswered questions posed by these experiments concern the mechanism of binding ribosomes to the membranes and the differences between free and bound ribosomes or their complexes with messenger RNA. If there are chemical differences between ribosomes, these might be associated with either membrane binding or messenger recognition. If there are no chemical differences, membrane binding must be directed in part by the messenger, possibly by the nascent peptide chain (Ganoza and Williams, 1969; Bock, 1969). Research into this mechanism might reveal very important factors in the regulation of specific protein synthesis.

Acknowledgments

We would like to acknowledge the expert technical assistance of Mr. George Kuzmycz, and to thank Doctors Fritz Lipmann and Edward L. Tatum for their advice and interest in this work.

This work was supported in part by grants from the National Science Foundation (GB-1383 and G-17436) and the U.S. Public Health Service (AI-07346) and by a U.S. Public Health Service fellowship (5-F2-GM-17,672-02).

Figures 1, 3, and 4 are adapted from a contribution by C. A. Williams to "Protein Metabolism and Biological Function" (R. Hilf and C. P. Bianchi, eds.). Rutgers Univ. Press, New Brunswick New Jersey, 1970.

References

Andrews, T. M., and Tata, J. R. (1968). *Biochem. Biophys. Res. Commun.* **32,** 1050.

Bock, R. M. (1969). *In* "Exploitable Molecular Mechanisms and Neoplasia." p. 217. Williams & Wilkins, Baltimore.

Campbell, P. N., Greengard, O., and Kernaot, B. A. (1960). *Biochem. J.* **74,** 107.

Dallner, G., Siekevitz, P., and Palade, G. E. (1966). *J. Cell Biol.* **30, 73,** 97.

Epstein, W., and Beckwith, J. R. (1968). *Annual Rev. Biochem.* **38,** 673.

Ganoza, M. C., and Williams, C. A. (1969). *Proc. Nat. Acad. Sci. U.S.* **63,** 1370.

Ganoza, M. C., Williams, C. A., and Lipmann, F. (1965). *Proc. Nat. Acad. Sci. U.S.* **53,** 619.

Hicks, L. J., Drysdale, J. W., and Munro, H. N. (1969). *Science* **164,** 584.

Hultin, T., and von der Decken, A. (1957). *Exp. Cell Res.* **13,** 83.

Jamieson, J. D., and Palade, G. E. (1968). *J. Cell. Biol.* **39,** 580.

Lengyel, P., and Söll, D. (1969). *Bacteriological Reviews* **33,** 264.

Peters, T. (1959). *J. Histochem. Cytochem.* **7,** 224.

Ragnotti, G., Lawford, G. R., and Campbell, P. N. (1969). *Biochem. J.* **112,** 139.

Redman, C. M. (1968). *Biochem. Biophys. Res. Commun.* **31,** 845.

Redman, C. M. (1969). *Fed. Proc. Fed. Amer. Soc. Exp. Biol.* **28,** 726.

Redman, C. M., and Sabatini, D. D. (1966). *Proc. Nat. Acad. Sci. U.S.* **56,** 608.

Siekevitz, P., and Palade, G. E. (1960). *J. Biophys. Biochem. Cytol.* **7,** 619.

Thorbecke, R. J., Hochwald, G. M., and Williams, C. A. (1970). *In* "Methods in Immunology and Immunochemistry" (C. A. Williams and M. W. Chase, eds.), Vol. 3. Academic Press, New York.

Wettstein, F. O., Staehelin, T., and Noll, H. (1963). *Nature* (*London*) **197,** 430.

Williams, C. A. (1970). *In* "Methods in Immunology and Immunochemistry" (C. A. Williams and M. S. Chase, eds.), Vol. 3. Academic Press, New York.

Williams, C. A., and Ganoza, M. C. (1969). *Fed. Proc. Fed. Amer. Soc. Exp. Biol.* **28,** 909.

Williams, C. A., Ganoza, M. C., and Lipmann, F. (1965). *Proc. Nat. Acad. Sci. U.S.* **53,** 622.

11

SERUM PROTEIN SYNTHESIS IN RATS FORCE-FED A THREONINE-DEVOID DIET

Jerry Lynn and Herschel Sidransky

DEPARTMENT OF PATHOLOGY, UNIVERSITY OF PITTSBURGH SCHOOL OF MEDICINE, PITTSBURGH, PENNSYLVANIA

I. Introduction

In recent years there have been a number of reports of enhanced hepatic protein synthesis in certain pathologic states of the liver. In altered metabolic states, such as following partial hepatectomy (Cammarano *et al.*, 1965), in experimentally induced antikidney serum nephrosis with albuminuria (Drabkin, 1963), following administration of cortisone and hydrocortisone (Korner, 1960a; Leon *et al.*, 1962; Pena *et al.*, 1966) or growth hormone (Korner, 1959, 1960b; Wagle, 1963), or following whole-body X-irradiation (Hidvégi *et al.*, 1968), there is an increase in hepatic protein synthesis. In studies with hepatotoxic agents, such as turpentine (Zeldis *et al.*, 1957, 1964), methylcholanthrene (Gelboin and Sokoloff, 1961; von der Decken and Hultin, 1960), phenobarbital (Gelboin and Sokoloff, 1961), or *Salmonella* toxin (Guidotti *et al.*, 1963), doses of each have been found to elicit increased incorporation of amino acids into liver protein. In nutritional deficiency studies, increased hepatic protein synthesis has been observed in animals fed diets deficient in or devoid of proteins or amino acids (Sidransky and Farber, 1958b; Garrow, 1959; Waterlow, 1959; Nimni and Bovetta, 1961; Wannemacher, 1961; Sidransky and Verney, 1964; Sidransky *et al.*, 1964).

In our laboratory we have been interested in the acute effect of administering amino acids on hepatic protein synthesis. Also, other experimental studies from different laboratories have been concerned with this problem. The livers of fasted animals respond rapidly after a single feeding of protein or of a complete amino acid mixture with a shift in polyribosomes from lighter to heavier aggregates and with enhanced protein synthesis (Fleck *et al.*, 1965; Webb *et al.*, 1966; Wunner *et al.*, 1966; Sox and Hoagland, 1966; Wilson and Hoagland, 1967; Staehelin *et al.*, 1967; Munro, 1968; Sidransky *et al.*, 1968). Studies have demonstrated this effect when fasted animals (rats or mice) received one tube feeding of a complete amino acid mixture but not when they were tube-fed a complete amino acid mixture devoid of tryptophan (Fleck *et al.*, 1965; Wunner *et al.*, 1966; Sidransky *et al.*, 1968). Subsequently, it has been reported that the administration of tryptophan alone, but not the administration of other single essential amino acids, elicited a stimulatory hepatic response which was similar to that obtained with a complete amino acid mixture (Sidransky *et al.*, 1968; Cammarano *et al.*, 1968; Pronczuk *et al.*, 1968; Rothschild *et al.*, 1969). The acute action of amino acids, including that of tryptophan alone, on the regulation of hepatic polyribosomes appears to be at the translational level rather than at the transcriptional level. Experiments with actinomycin support this concept (Fleck *et al.*, 1965; Wunner *et al.*, 1966; Staehelin *et al.*, 1967; Sidransky *et al.*, 1968).

In addition to studies dealing with the role of amino acids, following one acute administration, on hepatic polyribosomes and protein synthesis, we have been concerned for a number of years with the pathologic alterations induced by multiple tube feedings of purified diets devoid of single essential amino acids for a number of days. In earlier studies (Sidransky and Farber, 1958a,b; Sidransky and Baba, 1960; Sidransky and Verney, 1964), we reported that young rats force-fed for 3–8 days purified diets devoid of any one of several essential amino acids developed pathologic changes that closely resembled many of those found in infants with kwashiorkor (Trowell *et al.*, 1954). The morphologic changes in this kwashiorkor-like model consisted of a periportal fatty liver, excess hepatic glycogen, and atrophy of the pancreas, submaxillary gland, stomach, spleen, and thymus. Among the biochemical changes in these experimental animals were increases in lipid, glycogen, and total RNA in the liver, and an increase in the *in vivo* incorporation of labeled amino acids into hepatic proteins and serum proteins (Sidransky and Farber, 1958b; Sidransky and Verney, 1964). The increase in hepatic protein synthesis was also demonstrated in *in vitro* studies (Sidransky, *et al.* 1964) and was associated with an increase *in vivo* in the number of polyribosomes relative to free monosomes and with a shift of the average polyribosome size toward heavier aggregates (Sidransky *et al.*, 1964; Staehelin *et al.*, 1967). The enhanced *in vivo* hepatic protein synthesis in rats force-fed the experimental diets was not mediated through adrenal (Wagle and Sidransky, 1965) or pituitary (Sidransky and Verney, 1968) hormonal stimulation. While hepatic protein synthesis was increased, skeletal muscle protein synthesis was found to be decreased in studies *in vivo* (Sidransky and Farber, 1958b) and *in vitro* (Sidransky and Verney, 1967).

In an attempt to determine whether the enhanced hepatic protein synthesis in this kwashiorkor-like model involved all or only selected proteins synthesized by the liver, we investigated some aspects of serum protein synthesis in young rats force-fed a complete diet or one devoid of a single essential amino acid, threonine. Since most of the plasma proteins, excluding immunoglobulins, are normally synthesized in the liver (Schultze and Heremans, 1966), it was hoped that from a study of serum proteins we could gain some insight into the overall biochemical changes involving protein metabolism in the liver.

II. Experimental Procedure

Young female Sprague-Dawley rats weighing from 70 to 140 g were force-fed for 1, 3, or 6 days a purified complete diet or one devoid of threonine (Sidransky and Verney, 1964). Each animal was force-fed daily in 3 divided feedings an amount of diet equivalent to 8% body weight. The dry purified diet was mixed with an equal amount of water and was tube-fed using a plastic tube. At the beginning of each experiment the rats were divided into two groups so that the body weights of experimental and control animals were identical. One hour before killing, each rat was given intraperitoneally 2.5–25 μCi of ^{14}C-L-leucine (uniformly labeled, specific activity 10 mCi/mmole)/100 g body weight. Although the doses of ^{14}C-leucine administered to rats varied in different experiments, in each experiment the doses given to control and experimental groups were identical. In two experiments 9 μCi ^{14}C-amino acid mixture (uniformly labeled, specific activity 1.49 mCi/mg) instead of ^{14}C-leucine was used. Since the results obtained were similar regardless of the labeled compounds administered, the results were combined (Tables I and III). One hour after receiving the labeled amino acids, the rats were anesthetized with ether and blood was collected from the abdominal aorta. Sera of animals from the same group were pooled. The total serum proteins were precipitated in 5% trichloroacetic acid. In most experiments albumin and globulins were fractionated from serum using the acid alcohol

TABLE I

DIFFERENCES IN SPECIFIC ACTIVITIES OF SERUM TOTAL PROTEINS, ALBUMIN, AND GLOBULIN OF RATS FORCE-FED COMPLETE (C) OR THREONINE-DEVOID (TD) DIET

Duration (days)	Group	No. of experiments[b]	Specific activity[a] (cpm/mg protein)		
			Total protein	Albumin	Globulin
			Ave. % difference between groups		
1	C	5	(23)[c]		
	TD		(21) + 9.6	−4.4	+17.0[d]
3	C	9	(48)		
	TD		(53) + 31.0[d]	+1.2	+58.0[d]
6	C	1	(6)		
	TD		(7) + 31.0	−5.0	+60.0

[a] All rats received ^{14}C-leucine or ^{14}C-amino acid mixture 1 hr before killing.
[b] In each experiment sera from 2–8 animals of each group were pooled.
[c] Number of rats in parentheses.
[d] $P < 0.01$.

extraction method (Fernandez *et al.*, 1966). Seromucoid was extracted from serum using 0.6 *M* perchloric acid (Winzler, 1955), and the remaining proteins were then treated as a group and arbitrarily designated as nonseromucoid fraction. In three experiments serum was fractionated according to Neuhaus' scheme (Neuhaus *et al.*, 1966) into seromucoid and then further into albumin and globulins, using both 0.6 *M* perchloric acid and acid alcohol. The results obtained with this fractionation procedure were identical to those using the earlier method. Proteins were washed three times with 5% trichloroacetic acid (one wash contained nonradioactive leucine as carrier), once with 1:3 propanol ether, twice with ether, dried in air overnight, weighed, dissolved in hyamine, and counted with toluene base in a Packard liquid scintillation counter. Paper electrophoresis of serum proteins was performed according to Smith (1968). Polyacrylamide gel electrophoresis of serum proteins and seromucoid was done by the method of Ornstein and Davis (Davis, 1964). Determination of protein concentration in serum was done by the biuret method (Layne, 1957).

Since there were differences in the levels of serum proteins, particularly seromucoid, in the control and experimental animals (Table II), it became desirable to correct the specific radioactivity values of the experimental groups force-fed the threonine-devoid diet by a concentration factor in order to obtain a more meaningful comparison of the rate of amino acid

TABLE II

CONCENTRATIONS OF SERUM TOTAL PROTEINS, ALBUMIN, GLOBULIN, AND SEROMUCOID OF RATS FORCE-FED COMPLETE (C) OR THREONINE-DEVOID (TD) DIET

Duration (days)	Group	No. of experiments[a]	No. of rats	Concentrations (g/100 ml) of serum proteins				
				Total	Albumin	Globulin	Seromucoid	Nonseromucoid
1	C	3	18	6.23 ± 0.37^b	3.64 ± 0.19^b	2.95 ± 0.31^b	0.32 ± 0.07^b	6.29 ± 0.25^b
	TD		16	6.26 ± 0.44	3.45 ± 0.27	3.00 ± 0.54	0.46 ± 0.16	5.58 ± 0.12
% Difference				+0.3	−5.3	+4.3	+44.9	−6.9
3	C	3	19	6.41 ± 0.21	3.48 ± 0.30	3.43 ± 0.18	0.35 ± 0.02	6.46 ± 0.45
	TD		21	5.78 ± 0.36	2.96 ± 0.13	3.09 ± 0.19	0.61 ± 0.06^c	5.70 ± 0.69
% Difference				−10.1	−14.2	−9.9	+75.6	−12.3
6	C	1	6	6.91	3.72	3.25	0.43	7.10
	TD		7	5.69	2.39	3.29	1.34	4.93
% Difference				−17.7	−35.8	+1.2	+211.6	−30.6

[a] In each experiment sera from 2 to 8 animals of each group were pooled.
[b] Mean ± standard error of mean.
[c] $0.02 > P > 0.01$.

TABLE III

DIFFERENCES IN SPECIFIC ACTIVITIES OF SERUM SEROMUCOID AND NONSEROMUCOID FRACTIONS OF RATS FORCE-FED COMPLETE (C) OR THREONINE-DEVOID (TD) DIET

Duration (days)	Group	Specific activity[a] (cpm/mg protein)		
		No. of experiments[b]	Seromucoid	Nonseromucoid
			Ave. % difference between groups	
1	C	5	(23)[c]	(18)[c]
	TD		(21) + 45.0[d]	(16) − 0.3
3	C	4	(22)	(17)
	TD		(26) + 108.5[d]	(20) + 30.0
6	C	1	(6)	(6)
	TD		(7) + 305.0	(7) + 7.0

[a] All rats received ^{14}C-leucine or ^{14}C-amino acid mixture 1 hr before killing.
[b] In each experiment sera from 2–8 animals of each group were pooled.
[c] Number of rats in parentheses.
[d] $0.05 > P > 0.01$.

incorporation into serum proteins between the two groups of rats. Therefore, the data in 8 of 15 experiments (Tables I and III) were corrected for concentration changes by the following formula: Corrected cpm/mg protein of threonine-devoid group = uncorrected cpm/mg protein of threonine-devoid group × (concentration in pooled serum of threonine-devoid group)/(concentration in pooled serum of complete diet group). In using the above formula, it was assumed that the blood volumes of the experimental animals and the control animals were equal. In an earlier study, Sidransky and Farber (1958b) reported that blood volumes were equal in rats force-fed the complete or the threonine-devoid diet.

III. Results

In the initial experiments we determined the influence of force-feeding a threonine-devoid diet for 1, 3, or 6 days on the incorporation of labeled amino acid into total serum proteins. After 1 day, there was a 9.6% increase and after 3 and 6 days a 31% increase in incorporation into total serum proteins of the experimental groups in comparison with controls (Table I).

These results are similar to those described in earlier studies (Sidransky and Farber, 1958b; Sidransky *et al.*, 1964). We next decided to determine whether the increase of labeled amino acid incorporation into total serum proteins was uniform in some of the subfractions of serum proteins, particularly those of hepatic origin. Since many studies have shown that serum albumin, the major constituent of serum proteins, is synthesized in the liver (Schultze and Heremans, 1966), the effect of force-feeding the threonine-devoid diet on albumin synthesis was studied after fractionating the serum into albumin and globulins. The results summarized in Table I indicate that the rats force-fed the threonine-devoid diet for 1, 3, or 6 days showed little or no change in the rate of incorporation of labeled amino acids into serum albumin, but in the case of globulin a 17% increase was observed after 1 day and a 58–60% increase after 3 or 6 days. The increases observed in the 1-day and 3-day force-feeding experiments were both statistically significant. It is interesting to note that the percent increases in incorporation into globulins were approximately twice the increases observed in total serum proteins. This is to be expected since albumin which constitutes around half of serum proteins did not show any elevation in incorporation in the experimental groups.

A number of studies (Winzler, 1955, 1965a) have demonstrated increases in several serum glycoproteins in a variety of pathological and experimental conditions, including infection, neoplasm, immunological diseases, experimental scurvy, and after parathyroid hormone administration. In addition, other studies have demonstrated this response after injury (Neuhaus *et al.*, 1966) and turpentine injection (Darcy, 1965; Sarcione and Bogden, 1966) Maung *et al.*, 1968). Therefore, it became of interest to determine whether the levels of seromucoid, a fraction which accounts for approximately 10% of the total glycoprotein hexose (Winzler, 1955), became changed and then to determine whether the rate of incorporation of labeled amino acids into seromucoid was affected by force-feeding the threonine-devoid diet. In several experiments the serum concentrations of seromucoid as well as of total proteins, albumin, and globulin (Table II) were determined by the biuret method. Of these fractions, the only one which showed a significant increase in serum concentration after force-feeding the threonine-devoid diet was seromucoid. In rats force-fed the threonine-devoid diet for 1 day, there was an average increase in seromucoid of 44.9%. In experiments of 3 days duration, serum seromucoid concentration increased by 76.5%, while in one experiment in which the animals were force-fed the deficient diet for 6 days there was a 211.6% increase in seromucoid. In incorporation studies the seromucoid of the rats force-fed the threonine-devoid diet for 1 or 3 days showed significant increases in specific radioactivity, indicating enhanced synthesis of this group of glycoproteins (Table III). In one experiment where rats were force-fed the threonine-devoid diet for 6 days they showed even greater stimulation (+305%) of seromucoid synthesis when compared to the control group. The changes observed in the rate of incorporation of labeled amino acids into the nonseromucoid fraction were not statistically significant.

In addition to the above studies, in one experiment ^{14}C-D-glucosamine (uniformly labeled, specific activity 10.1 mCi/mmole) was used for labeling serum glycoprotein of the seromucoid fraction. The rats force-fed the threonine-devoid diet for 3 days showed a 51% increase in incorporation of ^{14}C-D-glycosamine into seromucoid compared to rats force-fed the complete diet. On the other hand, the incorporation of ^{14}C-D-glucosamine into the other serum proteins did not show differences.

In an attempt to do a more exact analysis of the serum proteins, pooled sera of rats force-fed the complete or the threonine-devoid diet for 3 or 6 days were subjected to electrophoresis on paper and on 7.5% polyacrylamide gel (pH 8.9). In addition, isolated seromucoid from the control and experimental groups was dialized and lyophylized before electrophoresis on polyacrylamide gel. No differences were observed after paper electrophoresis. After polyacrylamide gel electrophoresis, the serum proteins were separated into 17 distinct bands while the seromucoid fraction separated into 10 bands. In general, the proteins of the complete and the threonine-devoid diet groups showed similar electrophoretic patterns.

IV. Discussion

As a follow-up to earlier findings of enhanced synthesis of hepatic and serum proteins of rats force-fed a threonine-devoid diet (Sidransky and Farber, 1958b), this investigation was conducted to determine whether all or only certain selected species of hepatic proteins were increased under these experimental conditions. Our results demonstrate that serum albumin, which is known to be synthesized in the liver (Schultze and Heremans, 1966) and apparently consists of a single polypeptide chain without any prosthetic group (Peters, 1962; Jungblut and Turba, 1963) showed no enhancement of protein synthesis as measured by incorporation of labeled amino acid into albumin in the experimental animals when compared to control animals. On the other hand, serum globulins, most of which are synthesized in the liver, had increased incorporation of labeled amino acid in rats force-fed the threonine-devoid diet in comparison with those fed the complete diet. This indicated that the increase of hepatic protein synthesis in the experimental animals did not affect all hepatic proteins but that this effect was selective.

While there was little or no difference in the specific radioactivity of serum albumin following ^{14}C-leucine administration between control and experimental animals after 1–6 days (Table I), there was a progressive decrease in serum albumin concentration in the experimental rats (Table II). This drop in serum albumin may be due to a small, yet progressive, decrease in synthesis, increased degradation, or a combination of both. Other studies (Weimer, 1961; Platt *et al.*, 1964; Kirsch *et al.*, 1968) have reported diminished levels of serum albumin following dietary protein deprivation.

After observing an increase in the synthesis of serum globulins in rats force-fed the threonine-devoid diet in comparison with rats force-fed the complete diet, it was of interest to determine which fractions of the globulins were involved. In this study we assayed for serum seromucoid, a fraction of the serum glycoproteins. This fraction as defined by Winzler (1955) is not precipitable by 0.6 *M* perchloric acid and accounts for approximately 10% of the total glycoprotein hexose. It is a very heterogeneous fraction and consists of a complex mixture of glycoproteins. In addition, several factors which may affect the amount and composition of this fraction have been reported (Anderson, 1965). Yet since seromucoid has been considered to be most probably of hepatic origin (Sarcione, 1963; Sarcione *et al.*, 1965; Winzler, 1965a,b) and can be readily assayed, we decided to measure this fraction. Our results indicated that there was an increase in the concentration as well as in the incorporation of labeled amino acids and glucosamine into the seromucoid fraction of the experimental animals. In future studies we hope to determine which specific proteins in the seromucoid fraction are involved in the increase.

Since elevations of seromucoid level have been reported in a wide variety of pathological conditions (Winzler, 1955), as well as under our own experimental conditions with the force-feeding of a threonine-devoid diet, this suggests that a rather general systemic response is involved. Although evidence has been demonstrated that a humoral factor, present in plasma during the acute phase after an inflammatory stimulus, may be responsible for the increased hepatic synthesis of seromucoid (Sarcione *et al.*, 1965), the complete mechanism for the enhanced seromucoid synthesis under a variety of different stimuli (Winzler, 1955) is still not known. Likewise, the mechanism responsible for the increased hepatic protein synthesis, including enhanced seromucoid production, in rats force-fed a threonine-devoid diet is not known. Although enhanced seromucoid production can be set into motion by a wide variety of initiating stimuli, it is difficult to conceive that each agent or condition can elicit this common reaction pattern by entirely different mechanisms.

Earlier studies with protein deficiency (Waterlow, 1959; Schreier and Kazassis, 1960) as well as amino acid deficiency (Sidransky and Verney,

1964, 1965, 1967) have demonstrated, using ^{14}C-labeled amino acids, that incorporation into liver protein *in vivo* was increased while that into muscle protein was decreased when compared to control animals. Recent studies (Sidransky and Verney, unpublished data) indicate that skeletal muscle of rats force-fed a threonine-devoid diet undergoes greater breakdown than in control animals. The decrease in muscle protein synthesis and increase in muscle protein catabolism along with the decrease in muscle protein content suggest that a diversion of amino acids from the muscle into the circulation occurs which enables other organs, such as the liver, to maintain normal or even enhanced protein synthesis. This concept is consistent with the hypothesis proposed by Garrow (1959) that during protein depletion, the synthesis of protein is concentrated in the more essential organs in the body at the expense of the less essential ones. However, the results of the present study indicate that the alterations in hepatic protein metabolism in our experimental model are not due to a simple passive response to the increased availability of circulatory amino acids but due to a more complex process. The response of increased hepatic protein synthesis which does not seem to involve serum albumin, a major protein synthesized by the liver, suggests that if the circulating amino acids play a regulatory role, it is probably an indirect one which then can allow for a selective type of response.

Consistent with our findings of enhanced hepatic protein synthesis in rats force-fed a threonine-devoid diet in comparison with control animals force-fed a complete diet are findings in an ultrastructural study (Shinozuka *et al.*, 1968) showing nucleolar enlargement and in biochemical studies (Sidransky *et al.*, 1964, 1969; Staehelin *et al.*, 1967; Kaufmann *et al.*, 1969) indicating increased amounts of nuclear and ribosomal RNA, increased synthesis of nuclear, particularly necleolar, and ribosomal RNA, increased DNA-dependent RNA polymerase activity, and increased average polyribosome size toward heavier aggregates. These findings clearly indicate that the hepatic cells of rats force-fed the threonine-devoid diet undergo significant alterations which lead to an overall enhancement of protein synthesis. At this time the underlying mechanism for these changes is not clear. Also it is not known which proteins are involved in the increased hepatic protein synthesis. Since our findings in this study reveal an increase in serum seromucoid in the experimental animals which is so like the common reaction pattern observed under a variety of conditions, we question the significance of this finding and feel that it is probably of only minor importance in terms of the overall effect on liver protein synthesis. Further studies will be necessary to clarify which hepatic proteins are involved in the increased hepatic protein synthesis in rats force-fed a threonine-devoid diet.

Acknowledgments

This investigation was supported by Public Health Service Grants AM-05908 from the National Institute of Arthritis and Metabolic Diseases, and GM-135 and GM-10269 from the National Institute of General Medical Science. Jerry Lynn is a Sarah Mellon Scaife Fellow, 1966–1970.

References

Anderson, A. J. (1965). *Nature* (*London*) **208,** 491.
Cammarano, P., Giudice, G., and Lukes, B. (1965). *Biochem. Biophys. Res. Commun.* **19,** 487.
Cammarano, P., Chinali, G., Gaetani, S., and Spadoni, M. A. (1968). *Biochim. Biophys. Acta* **155,** 302.
Darcy, D. (1965). *Brit. J. Exp. Pathol.* **46,** 155.
Davis, B. J. (1964). *Ann. NY. Acad. Sci.* **121,** 404.
Drabkin, D. L. (1963). *Ann. NY. Acad. Sci.* **104,** 469.
Fernandez, A., Sobel, C., and Goldenberg, H. (1966). *Clin. Chem.* **12,** 194.
Fleck, A., Shepherd, J., and Munro, H. N. (1965). *Science* **150,** 628.
Garrow, J. S. (1959). *J. Clin. Invest.* **38,** 1241.
Gelboin, H. V., and Sokoloff, L. (1961). *Science* **134,** 611.
Guidotti, G. G., Rossi, C. B., and Ragnotti, G. (1963). *Proc. Soc. Exp. Biol. Med.* **114,** 308.
Hidvégi, E. J., Holland, J., Boloni, E., Laoni, P., Antoni, F., and Varteresz, V. (1968). *Biochem. J.* **109,** 495.
Jungblut, P. W., and Turba, F. (1963). *Biochem. Z.* **337,** 88.
Kaufmann, E., Sidransky, H., and Shinozuka, H. (1969). *Exp. Cell Res.* **55,** 130.
Kirsch, R. E., Brock, J. F., and Saunders, S. J. (1968). *Amer. J. Clin. Nutr.* **21,** 820.
Korner, A. (1959). *Biochem. J.* **73,** 61.
Korner, A. (1960a). *J. Endocrinol.* **21,** 177.
Korner, A. (1960b). *Biochem. J.* **74,** 462.
Layne, E. (1957). *Methods Enzymol.* **3,** 447.
Leon, H. A., Arrhenius, E., and Hultin, T. (1962). *Biochim. Biophys. Acta* **63,** 423.
Maung, M., Baker, D. G., and Murray, R. K. (1968). *Can. J. Biochem.* **46,** 477.
Munro, H. N. (1968). *Fed. Proc. Fed. Amer. Soc. Exp. Biol.* **27,** 1231.
Neuhaus, O., Balengo, H., and Chandler, A. (1966). *Amer. J. Physiol.* **221,** 151.
Nimni, M. E., and Bovetta, L. A. (1961). *Proc. Soc. Exp. Biol. Med.* **108,** 38.
Pena, A., Dvorkin, B., and White, A. (1966). *J. Biol. Chem.* **241,** 2144.
Peters, T., Jr., (1962). *J. Biol. Chem.* **237,** 1181.
Platt, B. S., Heard, C. R. C., and Stewart, R. J. C. (1964). *In* "Mammalian Protein Metabolism" (H. N. Munro and J. B. Allison, eds.), Vol. 2, pp. 446–522. Academic Press, New York.
Pronczuk, A. W., Baliga, B. S., Triant, J. W., and Munro, H. N. (1968). *Biochim. Biophys. Acta* **157,** 204.
Rothschild, M. A., Oratz, M., Mongelli, J., Fishman, L., and Schreiber, S. S. (1969). *J. Nutr.* **98,** 395.
Sarcione, E. (1963). *Arch. Biochem. Biophys.* **100,** 516.
Sarcione, E., and Bogden, A. (1966). *Science* **153,** 547.

Sarcione, E., Bohne, M., and Krauss, S. (1965). *Fed. Proc. Fed. Amer. Soc. Exp. Biol.* **24,** 230.

Schreier, K., and Kazassis, C. (1960). *Nature* (*London*) **187,** 1117.

Schultze, H. E., and Heremans, J. F. (1966). "Molecular Biology of Human Proteins," Vol. 1, pp. 353–365. Elsevier, Amsterdam.

Shinozuka, H., Verney, E., and Sidransky, H. (1968). *Lab. Invest.* **18,** 72.

Sidransky, H., and Baba, T. (1960). *J. Nutr.* **70,** 463.

Sidransky, H., and Farber, E. (1958a). *Arch. Pathol.* **66,** 119.

Sidransky, H., and Farber, E. (1958b). *Arch. Pathol.* **66,** 135.

Sidransky, H., and Verney, E. (1964). *Arch. Pathol.* **78,** 134.

Sidransky, H., and Verney, E. (1965). *J. Nutr.* **86,** 73.

Sidransky, H., and Verney, E. (1967). *Biochim. Biophys. Acta* **138,** 426.

Sidransky, H., and Verney, E. (1968). *J. Nutr.* **96,** 28.

Sidransky, H., Staehelin, T., an d Verney, E. (1964). *Science* **146,** 766.

Sidransky, H., Sarma, D. S. R., Bongiorno, M., and Verney, E. (1968). *J. Biol. Chem.* **243,** 1123.

Sidransky, H., Verney, E., and Shinozuka, H. (1969). *Exp. Cell Res.* **54,** 37.

Smith, I. (1968). "Chromatographic and Electrophoretic Techniques," Vol. 2, pp. 17–83 Wiley (Interscience), New York.

Sox, H. C., Jr., and Hoagland, M. B., (1966). *J. Mol. Biol.* **20,** 113.

Staehelin, T., Verney, E., and Sidransky, H. (1967). *Biochim. Biophys. Acta* **145,** 105.

Trowell, H. C., Davies, J. N. P., and Dean F. R. A. (1954). "Kwashiorkor." Arnold, London.

von der Decken, A., and Hultin, T. (1960). *Arch. Biochem. Biophys.* **90,** 201.

Wagle, D. S., and Sidransky, H. (1965). *Metab. Clin. Exp.* **14,** 932.

Wagle, S. R. (1963). *Arch. Biochem. Biophys.* **102,** 373.

Wannemacher, R. W., Jr. (1961). *Proc. Soc. Exp. Biol. Med.* **107,** 277.

Waterlow, J. S. (1959). *Nature* (*London*) **184,** 1875.

Webb, T. E., Blobel, G., and Potter, V. R. (1966). *Cancer Res.* **26,** 253.

Weimer, H. E. (1961). *Ann. NY. Acad. Sci.* **94,** 225.

Wilson, S. H., and Hoagland, M. B. (1967). *Biochem. J.* **103,** 556.

Winzler, R. J. (1955). *Methods Biochem. Anal.* **2,** 279.

Winzler, R. J. (1965a). *In* "The Amino Sugars" (R. W. Jeanoz and E. A. Balazs, eds.), Vol. 2A, pp. 337–352. Academic Press, New York–London.

Winzler, R. J. (1965b). *Clin. Chem.* **11,** 339.

Wunner, W. H., Bell, J., and Munro, H. N. (1966). *Biochem. J.* **101,** 417.

Zeldis, L. J., Ekegren, J. K., Block, J. H., and Madden, S. C. (1957). *Fed Proc. Fed. Amer. Soc. Exp. Biol.* **16,** 1619.

Zeldis, L. J., Smith, G. G., Moyer, D. L., and Madden, S. C. (1964). *Amer. J. Pathol.* **44,** 29a.

IV

Albumin Metabolism

12

ALBUMIN METABOLISM

Marcus A. Rothschild, Murray Oratz, and Sidney S. Schreiber

RADIOISOTOPE SERVICE, VETERANS ADMINISTRATION HOSPITAL, AND NEW YORK UNIVERSITY SCHOOL OF MEDICINE, NEW YORK, NEW YORK

I. Introduction

The level of serum albumin continues to be considered a bellwether for health and disease even though the amount of circulating albumin is the complex end result of synthesis, degradation, and distribution. The properties of this, usually the most prevalent of the serum proteins, have been known for over 100 years and amazingly little new information has been forthcoming in spite of considerable interest (Ancell, 1839–1840). The physiologic properties of serum albumin which are vital to the organism are its ability to bind and transport hormones, cations, fatty acids, drugs, and degradative and toxic products; its regulation of oncotic pressure; and perhaps its source of nitrogen as a reserve pool (Daughaday, 1956; Schultz and Heremans, 1966; Von Bennhold, 1962; Foster, 1960; Rosner and Deakins, 1968).

Albumin variants or subclasses based primarily upon electrophoretic mobility have been described (Gitlin *et al.*, 1961; Melarlen, 1967), and these variants may have somewhat different actions particularly in terms of the binding properties. Serum albumin is produced in the liver; in the absence of this organ no newly made albumin can be detected (Kukral *et al.*, 1961; Miller and Bale, 1954). However, this is the only fact which is clearly recognized in terms of the metabolic aspects of serum albumin. The regulation of synthesis, the sites of degradation, and the regulation of distribution are not known.

II. Synthesis

A. Rate

The absolute rates of albumin synthesis appear to be quite constant in many species ranging rather closely around 200 mg/kg/day (Schultz and Heremans, 1966). Serum albumin is synthesized on a specific ribosome–mRNA complex which is bound to the endoplasmic reticulum in the hepatocyte (Campbell *et al.*, 1965; Hicks *et al.*, 1969). Approximately 30 min are required for assembly of the albumin molecule and for its release into the systemic circulation, but the time elements in these steps are not known (McFarlane, 1963; Peters, 1962).

B. Regulation

1. *Nutrition*

The most important aspect in the regulation of albumin production is nitrogen intake (Waterlow, 1968; Whipple and Madden, 1944). A short-term fast rapidly depletes the amount of albumin synthesized (Hoffenberg *et al.*, 1966; Kirsch *et al.*, 1968; Rothschild *et al.*, 1968; Staehelin *et al.*, 1967). Chronic malnutrition is also associated with a low level of albumin synthesis, although this disorder is certainly complicated by infection, altered hormonal balance, and frequently elevated globulin levels (Freeman and Gordon, 1964). Upon refeeding, albumin production rapidly increases and may achieve levels much higher than those observed normally. While few studies have been performed using subcellular fractions and isolating albumin per se, the incorporation of labeled amino acids into protein by subcellular fractions removed from livers of fasted animals is markedly inhibited. The decreased synthesizing capacity must involve either an altered polysome, a decreased quantity of messenger RNA, or perhaps decreased synthesis of an inhibitor to RNase with resulting degradation of RNA and disaggregation of the ribosome from mRNA (Brewer *et al.*, 1969). Since albumin is believed to be made and released on an mRNA ribosome membrane complex, fasting might be expected to alter this part of the intracellular mechanism.

However, fasting does not specifically reduce the quantity of endoplasmic membrane-bound polysome; while the total RNA is reduced, the partition between free (nonmembrane bound) RNA and RNA attached to the membrane remains unaltered (Blobel and Potter, 1967). However, both the free or unbound polysome and the membrane-bound polysome are disaggregated (mRNA with a smaller number of attached ribosomes) with degradation to smaller subunits which are relatively inactive in synthesizing protein. A protein inhibitor to RNase is present in the cell sap and deficient protein synthesis within the cell could possibly permit greater RNase activity (Blobel and Potter, 1967), by decreasing the synthesis of this inhibitor.

2. *Amino Acids*

In order to restimulate and stabilize this system to a fed state, it is not necessary to employ a balanced diet. It has been shown that reaggration of ribosomes with messenger RNA occurs within hours following the administration of excess tryptophan and isoleucine, and both amino acids result in increased albumin production (Munro, 1968; Rothschild *et al.*, 1969b; Sidransky *et al.*, 1968). The stimulatory mechanism of action of these two amino acids is not known. However, since albumin synthesis is stimulated so rapidly and to such a high level after the administration of these amino acids, it is possible that some of the mRNA coded for albumin is long-lived and is available for reaggregation. Values for the half-life of the albumin mRNA are widely divergent (John and Miller, 1966; Wilson *et al.*, 1967).

3. *Hormones*

In addition to nitrogen, hormones also play an important role in regulating albumin synthesis. Excess thyroid and cortisone stimulate albumin synthesis. These effects have been observed clinically (Rothschild *et al.*, 1957, 1958). But again the mechanisms remain obscure. Cortisone enhances messenger RNA and ribosomal RNA synthesis and promotes the appearance of a new species of mRNA (Drews and Braverman, 1967; Novikoff and Potter, 1948). Further, cortisone increases the nitrogen delivered to the liver secondary to increased protein degradation of muscle, connective tissue, and lymphatic tissue. These actions, as well as the interconnected alterations in intermediate metabolism resulting from cortisone administration, may all influence albumin synthesis more or less directly. In addition, cortisone produces a change in albumin distribution with a shift in the fraction of the available exchangeable albumin pool located extravascularly into the plasma. The decrease in extravascular albumin is reflected within the liver interstitial space, and it is possible that albumin synthesis may be sensitive to the colloid content within or near the site of albumin production (Rothschild *et al.*, 1966b).

Thyroid may stimulate albumin production by enhancing mRNA production rates (Tata, 1964) though again increased nitrogen turnover could give the same result. The lack of the trophic hormones following hypophysectomy is associated with a decrease in ribosomal RNA which in turn may cause a depression of albumin production (Garren *et al.*, 1957). All of these hormones act quickly, and may act in conjunction with each other, and all their effects may be mediated by nutritional factors.

4. *Environment*

Albumin synthesis is altered by environmental changes (Oratz *et al.*, 1967). Higher serum albumin levels are seen in cold climates; albumin synthesis is depressed during heat acclimatization. This effect is also complex in that both nitrogen intake and thyroid function are depressed in the heat. It is most probable that equivalent changes in albumin synthesis occur with other environmental stresses.

5. *Osmotic content*

A lowering of the serum albumin level is seen in disease states characterized by an elevation of the γ-globulin fraction, and depressions in albumin synthesis have been seen when the colloid content within the fluid spaces of the body has been altered due to elevations in γ-globulin levels or by the use of dextran (Bjorneboe and Schwartz, 1959; Rothschild *et al.*, 1966a). An osmotic regulatory system located perhaps at the site of albumin production has been postulated as one of the factors controlling albumin synthesis. The space available for albumin in the hepatic interstitial water is very small (Rothschild *et al.*, 1966b). Changes in albumin concentration in this area could prove to be an exquisitely sensitive regulator. Osmotic equilibrium is a vital factor in fluid transfer and in the shark family, which has little serum albumin, acute dehydration is prevented by very high urea concentrations. The level of serum albumin or albumin-like proteins is less than 0.1–0.3 g/100 ml and upon a decrease in the urea level after exposure to fresh water an albumin-like protein makes it appearance (Goldstein *et al.*, 1968; Rasmussen and Rasmussen, 1967). These observations support the concept that osmotic regulation is more than a fortuitous byproduct of the albumin level.

III. Albumin Degradation

A. Rates

Albumin degradation is also rather constant in man and in many lower animal species. There are two aspects to degradation which must be separated clearly: the fractional degradative rate which is that part of either the plasma albumin mass or the total exchangeable albumin mass which is metabolized daily; and the total absolute amount of albumin degraded. Experimentally the fractional rate may remain relatively constant in the face of large changes in the absolute rate and albumin degradation may be considered to be unchanged when actually marked alterations have occurred (Rothschild *et al.*, 1966a). In hypoalbuminemia the fractional rate is quite often normal, particularly when the hypoalbuminemia is associated with hypergammaglobulinemia (Rothschild *et al.*, 1966a). Few studies have been performed in hyperalbuminemic states since this is a clinical condition rarely seen except perhaps in acute dehydration. However, after albumin infusions, a marked increase in the absolute degradative rate occurs and is associated with an increase in the fractional rate of degradation. These changes were sufficient to account for most of the excess colloid infused. No alteration in albumin synthesis occurred indicating that albumin synthesis and degradation are not interdependent (Rothschild *et al.*, 1964). While albumin synthesis is affected acutely by a decrease in food intake, albumin degradation falls only after the albumin pool has decreased. In situations where gastrointestinal or renal losses of albumin are excessive, the absolute endogenous rate of albumin degradation is frequently depressed (Bauman *et al.*, 1955; Kerr *et al.*, 1967), perhaps in an attempt to prevent further loss of albumin stores. An increase in synthesis per se would result only in a higher rate of external loss. Hormones also affect albumin degradation and excess thyroid and cortisone result in rapid increments in both the absolute and fractional degradation rates. Alterations in the degradative rates, absolute and fractional, of other serum proteins do not influence albumin degradation, and in hypergammaglobulinemia when more than 1 g of γ-globulin/kg of body weight may be degraded, the fractional rate of albumin degradation was unaltered (Rothschild *et al.*, 1962).

B. Sites

The site of degradation for serum albumin is not known although it is in rapid equilibrium with the plasma compartment. Significant degradation of albumin does not occur in the plasma, in the liver, nor in the kidney, and only a minute fraction is lost normally into the gastrointestinal tract (Cohen and Gordon, 1958; Katz *et al.*, 1964; Kerr *et al.*, 1967). It is clear that either albumin must enter a degradative cell or be attached to it, but no such system has been recognized. Albumin may be lost in excessive amounts into the urine, stool, exudates, or burned areas, but extracts of purulent material do not stimulate albumin degradation (Catsoulis *et al.*, 1965). Nephrosis is associated with an increased renal degradation suggesting that as the albumin moves across an interstitial area it is either degraded or made susceptable to degradation.

IV. Albumin Distribution

Organ Systems

Albumin is distributed between the plasma and extraplasma pools in about a 40:60 ratio in man. Most of the extravascular albumin is located in the skin in the human, and the concentration of albumin in the extravascular extracellular skin water approaches that in the plasma (Rothschild *et al.*, 1955). Besson *etal.,* 1953). This high level may play a role in regulating the osmotic equilibrium of the skin during water exchange. In the liver the amount of albumin located in hepatic interstitial water is small; assuming an even distribution of albumin in hepatic interstitial space, the concentration would be less than one-tenth that in plasma or lymph. However, the hepatic interstitial space is made up of a protein polysaccharide matrix which may exclude albumin from its domain (Rothschild *et al.*, 1966b). Albumin has a similar unequal distribution in the pulmonary interstitial volume, and other tissues will probably show variations of equivalent degree. Albumin has been isolated in tears, sweat, saliva, gastric juice, pleural effusions, ascites, and edema and is most certainly present in every body fluid. The concentrations will vary markedly from less than 0.1 g/100 ml in edema to 2–3 g/100 ml of exudate (Schultz and Heremans, 1966). While the lymph draining the liver has an albumin content of about 90% that of the plasma, recent studies have shown that normally all the newly made albumin entering the lymph can be accounted for as having come from the plasma. Thus, the newly made albumin must reach the systemic circulation directly indicating that the interstitial space and plasma are essentially one freely equilibrating compartment (Smallwood *et al.*, 1968). In cirrhotic subjects with ascites, however, albumin synthesis is elevated in many patients (Rothschild *et al.*, 1969a), and there is some transfer of albumin directly into the lymph. Furthermore a short pathway for albumin transfer directly from the liver into ascites has been demonstrated indicating that the normal method for albumin distribution can be altered in various disease states (Zimmon *et al.*, 1969).

V. Summary

Albumin levels are quite constant in health but are reduced in debilitating disease and trauma. Albumin synthesis should be able to keep pace with any demand and that hypoalbuminemia does exist indicates that though the functional capacity of the liver is still tremendous, other unknown factors must be operative preventing full use of this capacity. Changes in albumin distribution in burns, cirrhosis and nephrosis, loss of vital amino acids in GI disease, and alterations in hormonal balance may pose a pathologic ceiling to a hepatocyte which otherwise could continue to synthesize albumin at an adequate rate. The answers to these questions are not known.

The regulation of degradation, synthesis, and distribution may be interrelated or changed independently. Evaluation of alterations in all these aspects of albumin metabolism during changing states at control levels of nutrition, in isolated systems, where hormonal influence and organ permeability can be controlled, and in subcellular systems, where the intracellular organelles can be examined directly, should all prove to be effective in providing information concerning the regulation of this important protein.

ACKNOWLEDGMENTS

This work was supported in part by Public Health Service Research Grant, AM-02469. This article is reprinted in part from Rothschild *et al.* (1970) *Gastroenterology* **58**, 402.

REFERENCES

Ancell, H. (1839–1940). *Lancet* **1**, 222.

Bauman, A., Rothschild, M. A., Yalow, R. S., and Berson, S. A. (1955). *J. Clin. Invest.* **34**, 1359.

Berson, S. A., Yalow, R. S., Schreiber, S. S., and Post, J. (1953). *J. Clin. Invest.* **32**, 746.

Bjorneboe, M., and Schwartz, M. (1959). *J. Exp. Med.* **110**, 259.

Blobel, G., and Potter, V. R. (1967). *J. Mol. Biol.* **26**, 279.

Brewer, E. N., Foster, L. B., and Sells, B. H. (1969). *J. Biol. Chem.* **244**, 1389.

Campbell, P. N., Guldborg, S. H., and Lowe, E. (1965). *Biochem. J.* **97**, 422.

Catsoulis, E. A., Rothschild, M. A., Oratz, M., and Franklin, E. C. (1965). *Arthritis Rheum.* **8**, 38.

Cohen, S., and Gordon, A. H. (1958). *Biochem. J.* **70**, 544.

Daughaday, W. H. (1956). *J. Lab. Clin. Med.* **48**, 799.

Drews, J., and Braverman, G. (1967). *J. Biol. Chem.* **242**, 801.

Foster, J. F. (1960). *Plasma Proteins* **1**, 179.

Freeman, T., and Gordon, A. H. (1964). *Clin. Sci.* **26**, 17.

Garren, L. D., Richardson, A. P., and Crocco, R. M., (1957). *J. Biol. Chem.* **242**, 650.

Gitlin, D., Schmid, K., Earle, D. P., and Givelber, H. (1961). *J. Clin. Invest.* **40**, 820.

Goldstein, L., Oppelt, W. W., and Maren, T. H. (1968). *Amer. J. Physiol.* **215**, 1493.

Hicks, S. J., Drysdale, J. W., and Munro, H. N. (1969). *Science* **164**, 584.

Hoffenberg, R., Black, E., and Brock, J. F. (1966). *J. Clin. Invest.* **45**, 143.

John, D. W., and Miller, L. L. (1966). *J. Biol. Chem.* **241**, 4817.

Katz, J., Sellers, A. L., and Bonorris, G. (1964). *J. Lab. Clin. Med.* **63**, 680.

Kerr, R. M., Dubois, J. J., and Holt, P. R. (1967). *J. Clin. Invest.* **46**, 2064.

Kirsch, R., Frith, L., Black, E., and Hoffenberg, R. (1968). *Nature* (*London*) **217**, 578.

Kukral, J. C., Kerth, J. D., Pancner, R. J., Cromer, D. W., and Henegar, G. C. (1961). *Surg. Gynecol. Obstet.* **113**, 360.

McFarlane, A. S. (1963). *Biochem. J.* **89**, 277.

Melarlen, L. (1967). *Acta Pathol. Microbiol. Scand. Suppl.* **191**, 10.

Miller L. L., and Bale, W. F. (1954). *J. Exp. Med.* **99**, 125.

Munro, H. N. (1968). *Fed. Proc. Fed. Amer. Soc. Exp. Biol.* **27,** 1231.

Novikoff, A. B., and Potter, V. R. (1948). *J. Biol. Chem.* **173,** 223.

Oratz, M., Walker, C., Schreiber, S. S., Gross, S., and Rothschild, M. A. (1967). *Amer. J. Physiol.* **213,** 1341.

Peters, T. (1962). *J. Biol. Chem.* **237,** 1181.

Rasmussen, L. E., and Rasmussen, R. A. (1967). *In* "Sharks, Skates, and Rays" (P. Gilbert, R. F. Mathewson, and D. P. Dall, eds.), Ch. 25. Johns Hopkins Press, Baltimore, Maryland.

Rosner, W., and Deakins, S. M. (1968). *J. Clin. Invest.* **47,** 2109.

Rothschild, M. A., Bauman, A., Yalow, R. S., and Berson, S. A. (1955). *J. Clin. Invest.* **34,** 1334.

Rothschild, M. A., Bauman, A., Yalow, R. S., and Berson, S. A. (1957). *J. Clin. Invest.* **36,** 422.

Rothschild, M. A., Schreiber, S. S., Oratz, M., and McGee, H. L. (1958). *J. Clin. Invest.* **37,** 1129.

Rothschild, M. A., Oratz, M., Franklin, E. C., and Schreiber, S. S. (1962). *J. Clin. Invest.* **41,** 1564.

Rothschild, M. A., Oratz, M., Evans, C., and Schreiber, S. S. (1964). *J. Clin. Invest.* **43,** 1874.

Rothschild, M. A., Oratz, M., and Schreiber, S. S. (1966a). *Protides Fluids Proc. Colloq.* **14,** 267.

Rothschild, M. A., Oratz, M., Evans, C., and Schreiber, S. S. (1966b). *Amer. J. Physiol.* **210,** 57.

Rothschild, M. A., Oratz, M., Mongelli, J., and Schreiber, S. S. (1968). *J. Clin. Invest.* **47,** 2591.

Rothschild, M. A., Oratz, M., Zimmon, D., Schreiber, S. S., Weiner, I., and Van Canegham, X. (1969a). *J. Clin. Invest.* **48,** 344.

Rothschild, M. A., Oratz, M., Mongelli, J., Fishman, L., and Schreiber, S. S. (1969b). *J. Nutr.* **98,** 395.

Rothschild, U. A., Oratz, M., and Schreiber, S. S. (1970). *Gastroenterology,* **58,** 402.

Schultze, H. E., and Heremans, J. F. (1966). "Molecular Biology of Human Proteins," Vol. 1. Elsevier, New York.

Sidransky, H., Sarma, D. S. R., Bongiorno, M., and Verney, E. (1968). *J. Biol. Chem.* **243,** 1123.

Smallwood, R. A., Jones, E. A., Craigie, A., Raia, S., and Rosenoer, V. M. (1968). *Clin. Sci.* **35,** 35.

Staehelin, T., Verney, E., and Sidransky, H. (1967). *Biochim. Biophys. Acta* **145,** 105.

Sterling, K. (1964). *J. Clin. Invest.* **43,** 1721.

Tata, J. R. (1964). *Biochim. Biophys. Acta* **87,** 528.

Von Bennhold, H. (1962). *Protides Biol. Fluids Proc. Colloq.* **9,** 58.

Waterlow, J. C. (1968). *Lancet* **11,** 1091.

Whipple, G. H., and Madden, S. C. (1964). *Medicine* (*Baltimore*) **23,** 215.

Wilson, S. H., Hill, H. Z., and Hoagland, M. D. (1967). *Biochem. J.* **103,** 567.

Zimmon, D., Oratz, M., Kessler, R., Schreiber, S. S., and Rothschild, M. A. (1969). *J. Clin. Invest.* **48,** 2074.

13

NUTRITIONAL, HORMONAL, AND TEMPORAL FACTORS REGULATING NET PLASMA PROTEIN BIOSYNTHESIS IN THE ISOLATED PERFUSED RAT LIVER

Effects of Feeding or Fasting Liver Donors and of Supplementation with Amino Acids, Insulin, Cortisol, and Growth Hormone

Leon L. Miller and David W. John

DEPARTMENT OF RADIATION BIOLOGY AND BIOPHYSICS,
UNIVERSITY OF ROCHESTER, SCHOOL OF MEDICINE AND DENTISTRY,
ROCHESTER, NEW YORK

I. Introduction

It was just about 20 years ago when we first became convinced that the role of the liver in plasma protein synthesis might be defined by a study of the isolated perfused liver. With the collaboration of Bly, Watson, and Bale (Miller *et al.*, 1951), we developed the technique of isolated rat liver perfusion to a point where experiments 6 hours in duration were readily carried out. With the aid of 6-^{14}C-lysine (Miller *et al.*, 1951) and later of preparative-zone electrophoresis (Miller *et al.*, 1954) we were able to demonstrate qualitatively the dominant role of the liver in the synthesis of all plasma proteins, excepting the γ-globulins; the converse—that the nonhepatic tissues synthesize only γ-globulins—was also concluded from parallel studies of 6-^{14}C-lysine incorporation into the γ-globulins by the eviscerated surviving rat (Miller and Bale, 1954) and by the hepatectomized rat (Miller *et al.*, 1964).

The incorporation of isotopically labeled amino acids into plasma proteins is at best a presumptive quantitative measure of net synthesis; observations *in vivo* (Neuhaus *et al.*, 1966) have indicated that more than 4–6 hours was necessary before the acute phase response in plasma protein synthesis was manifest in response to injury. Accordingly, we developed the rat liver perfusion technique (John and Miller, 1969a) to permit routine perfusions of 12-hr duration and adapted known highly specific serological methods (Fahey and McKelvey, 1965; Mancini *et al.*, 1965) to the direct quantitative measurement of net increases in synthesis of rat plasma proteins during such 12-hr intervals.

The use of the isolated rat liver perfused with heterospecies blood for 12 hr affords a model system for direct evaluation of effects of various factors on the net biosynthesis of plasma proteins by the liver. This report describes: (1) effects of feeding or fasting rat liver donors on the net synthesis of rat serum albumin (RSA), fibrinogen, α_1-acid glycoprotein, α_2-(acute phase) globulin, and haptoglobin; (2) effects of supplementation *in vitro* with glucose, a nutritionally complete amino acid mixture, and the hormones insulin, hydrocortisone, and growth hormone. This report also extends the novel demonstration (John and Miller, 1969b) that the combination of amino acids, insulin, and glucocorticoid (cortisol) can induce *in vitro* increased rates of synthesis of the acute-phase proteins. Additional data are included on significant effects of the combination of amino acids and hormones on the incorporation of 1-^{14}C-L-lysine into hepatic proteins, oxidation of 1-^{14}C-L-lysine to $^{14}CO_2$, production of bile, utilization of glucose, net urea synthesis, and net changes in amino acid nitrogen and on nitrogen balance as defined for the liver perfusion system (Miller *et al.*, 1956; John and Miller, 1969b).

II. Methods

A. Perfusion Technique

The perfusion apparatus, operative technique, and general conditions were similar to those previously described by us for 12-hr experiments (John and Miller, 1969a); however, in these studies defibrinated rabbit blood was replaced by either bovine red blood cells or rabbit red blood cells which were washed three times in sterile Ringer solution (to remove the buffy coat) and then suspended in Krebs-Ringer bicarbonate containing bovine serum albumin (Armour Fraction V) 3 g/100 ml.

Liver donors were adult male rats of Sprague-Dawley (Holtzman) strain and weighed between 320 and 460 g. All rats were maintained on a diet of commercial rat food (Purina Checkers) and tap water ad libitum. Fed rats, allowed access to food until sacrifice, had full stomachs and were actively assimilating as indicated by the prominence of the blood vasculature of the gastrointestinal tract, pancreas, and mesenteric structures. Rats fasted 18 hr or 6 days were allowed access to drinking water at all times. At operation their abdominal viscera were comparatively inactive and avascular. Prominent in the 6-day fasted liver donors was the atrophy of the gastrointestinal tract and pancreas and the relatively small size of the liver.

In so-called refeeding experiments, after 6 days fast, rats were allowed unlimited food for periods of 12 hr, 24 hr, or 37 hr when livers were removed for perfusion.

B. Hormones and Amino Acids

Insulin, generously supplied by the Eli Lilly Company, was a glucagon-free, porcine, crystalline zinc insulin dissolved with the aid of 0.1 *N* HCl and diluted to volume with Ringer's solution. Growth hormone was a purified, lyophilized preparation of bovine growth hormone generously supplied by Armour and Company (Lot No. M10810); it was dissolved in Ringer's solution, 1 mg/ml. Cortisol was the sodium succinate derivitive (Solucortef, Upjohn Company). L-Amino acids were the purest commercially available, (General Biochemicals). For each perfusion, the amino acids added consisted of 7.7 mg cysteine-HCl plus 312.3 mg of a mixture prepared by pulverizing the following weights in g of L-amino acids: asparagine, 1.710; threonine, 0.820; serine, 0.985; glutamine, 2.160; proline, 0.503; alanine, 0.472; glycine, 0.960; valine, 0.633; methionine, 0.179; isoleucine, 0.813; leucine, 1.085; tyrosine, 0.565; phenylalanine, 0.623; lysine·HCl, 1.850; histidine·HCl, 0.546; arginine · HCl, 1770; and tryptophan, 0.400. This mixture reflects the amino acid composition we have found by analysis of purified rat fibrinogen

(John and Miller, 1966), by standard methods (Spackman *et al.*, 1958); asparagine and glutamine are substituted for aspartic and glutamic acids.

C. "Full Supplementation" Conditions

For experiments with addition of amino acids plus all three hormones (designated in this report as "full supplementation" experiments), the perfusate consisted of red cells, 38 ml; Krebs–Ringer–bicarbonate–albumin solution (adjusted to pH 7.4 with 1.0 *N* NaOH), 50 ml; glucose, 100 mg; heparin, 10,000 units; penicillin, 3000 units; streptomycin, 3.0 mg; insulin, 5.1 units; bovine growth hormone, 0.5 mg; cortisol, 5.0 mg; and enough Ringer's to bring the volume to 100 ml. For these full supplementation experiments, 18 ml of Ringer's solution was infused for 12 hr at a constant rate (1.5 ml/hr) and contained: glucose, 500 mg; penicillin, 300 units; streptomycin, 3.0 mg; 1-^{14}C-L-lysine·HCl, 15 μCi (0.62 mg); insulin, 6.8 units; bovine growth hormone, 1.0 mg; cortisol, 5.0 mg; and the complete amino acid mixture, 320 mg. Control experiments were set up exactly as detailed above except hormones and amino acids were omitted (from both perfusates and infusions). Other experiments involved omitting the amino acid mixture and/or certain of the hormones. Throughout this chapter when these supplements are mentioned, the above doses and methods of administration are implied.

D. Serological Measurement of Individual Plasma Proteins

Specific plasma proteins synthesized by the rat liver and released into the perfusion medium were quantitatively measured in aliquots of perfusion plasma by the single-radial diffusion technique of Mancini, Carbonara, and Heremans (Mancini *et al.*, 1965) as modified by Fahey and McKelvey (Fahey and McKelvey, 1965). Details of the method and of the preparation of the pure rat antigens and rabbit antisera to RSA, fibrinogen, α_1-acid glycoprotein, and α_2-(acute phase) globulin have been described (John and Miller, 1968). Rat haptoglobin was prepared by the method of Lombart, Moretti, and Jayle (Lombart *et al.*, 1965) and further purified, as we have described for RSA (John and Miller, 1966), by elution from preparative acrylamide gels after electrophoresis. This haptoglobin was used to immunize rabbits by a technique similar to that we have described for preparation of anti-α_1-acid glycoprotein (John and Miller, 1968). The antihaptoglobin sera were pooled and tested by immunoelectrophoresis (Grabar and Williams, 1953) against whole rat serum. Because the antiserum reacted weakly with α_1-acid glycoprotein as well as strongly with haptoglobin, it was absorbed with α_1-acid glycoprotein (John and Miller, 1968). Subsequent tests of the absorbed antiserum showed reaction only against haptoglobin. The antiserum was stored frozen in small glass ampules until used.

E. Other Methods

Procedures for measuring $^{14}CO_2$ production, incorporation of 1-^{14}C-lysine into plasma total protein, liver protein, and bile, and for determining net changes in glucose, urea, and amino acid nitrogen have been described (John and Miller, 1966, 1968). Net changes in glucose and amino acid nitrogen include the amounts infused during the 12-hr duration of these experiments. The ninhydrin method used for measurement of amino acids determines α-amino acid nitrogen with leucine as a standard. Total amino acid nitrogen was approximated by multiplying the latter by a factor (1.59) obtained by dividing the total calculated amino acid nitrogen of the mixture by the total calculated α-amino (ninhydrin) nitrogen of the mixture.

Nitrogen balance for the perfusion system is computed as previously reported (John and Miller, 1969b; Miller *et al.*, 1956) by subtracting net change in urea nitrogen from net change in amino acid nitrogen in perfusate. Thus, positive nitrogen balance occurs when net loss of amino acid nitrogen exceeds net gain of urea nitrogen.

III. Results

A. Net Biosynthesis of Rat Serum Albumin

Figure 1a reveals the capacity of the rat liver to elaborate serum albumin in the absence of amino acid and hormone supplement. Synthesis by livers

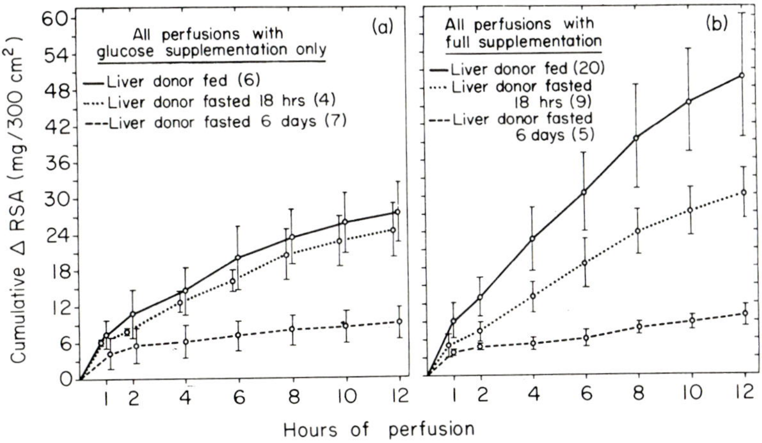

FIG. 1. (a) and (b): Cumulative average net change in rat serum albumin (RSA) in perfusate per 300 cm² of body surface area of liver donor rat (John and Miller, 1966). Numbers in parentheses refer to numbers of perfusions in each group. Bars indicate the standard deviation for each group. In this and in Figs. 2, 3, 4, and 5, data are from the same experiments for which additional data are included in Table I.

from fed rats is substantial but is only about 60% of that from livers of fed rats fully supplemented (Fig. 1b).

In contrast to results seen in response of livers from 18 hr fasted rats with synthesis of fibrinogen, α_1-acid glycoprotein, and haptoglobin, the liver of the 18-hr fasted rat does not respond to full supplementation with significantly increased albumin synthesis. The 6-day fast greatly impairs the capacity of the rat liver to synthesize albumin (Fig. 1a); even with full supplementation, synthesis does not exceed 10–15% of the maximum reached by livers from fed rats with full supplementation (Fig. 1b).

B. Net Synthesis of Fibrinogen

Figure 2a depicts the capacity of the isolated liver to elaborate fibrinogen without the support of amino acid or hormone supplementation; even after 6 days fast we see significant, if small, synthesis of fibrinogen. Figure 2b reveals that full supplementation elicits substantial increase in the net rate of fibrinogen synthesis but only after 4–6 hr. The increased rate of synthesis induced in the last 6 hr of perfusion of livers from both fed and 18-hr fasted rats is about three times that seen in the first 4 hours. In contrast to albumin, the synthesis of fibrinogen by livers from 18-hr fasted rats was substantially increased by full supplementation. Even the livers of 6-day fasted rats respond to full supplementation with a qualitatively similar pattern in induction, and a relative increase in the rate of fibrinogen synthesis.

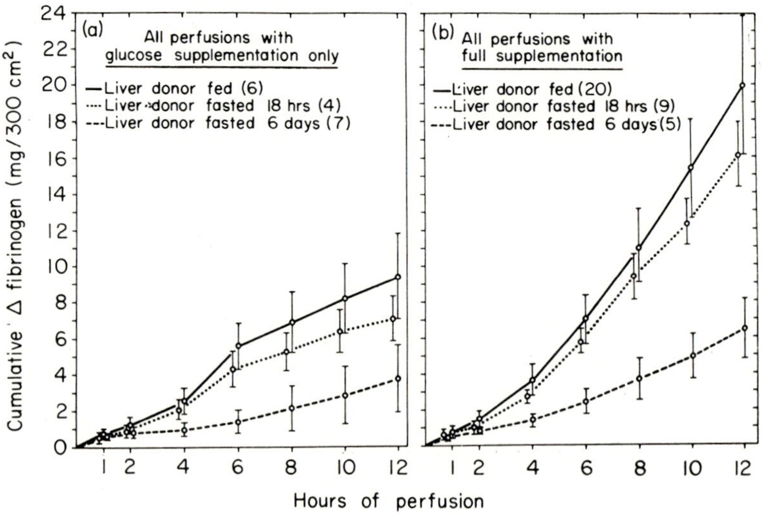

FIG. 2. (a) and (b): Cumulative average net change in rat fibrinogen in perfusate per 300 cm² body surface area of liver donor rat.

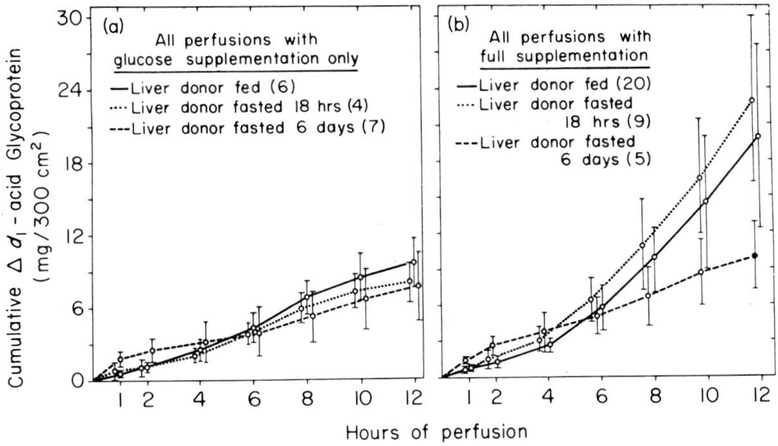

FIG. 3. (a) and (b): Cumulative average net change in rat α_1-acid glycoprotein in perfusate per 300 cm^2 body surface area of liver donor rat.

C. Net Synthesis of α_1-Acid Glycoprotein

In the absence of full supplementation, the perfused liver possesses a remarkable capacity to elaborate and secrete α_1-acid glycoprotein with no significant difference between the livers from rats fed or even fasted 6 days (Fig. 3a); the addition of full supplementation leads to about 100% increase in net synthesis compared with controls. Noteworthy is the approximately threefold increase in rate of synthesis after the fourth to sixth hour. As also in the case of fibrinogen and haptoglobin, the induction seen after the fourth to sixth hour (Fig. 3b) is dependent on the presence of cortisol in full supplementation (John and Miller, 1969b).

D. Net Synthesis of α_2-(Acute Phase) Globulin

In control perfusions of livers from fed rats (Fig. 4a), there is a roughly linear mean synthesis of α_2-(acute phase) globulin during the last 11 hr; an 18-hr fast is associated with somewhat decreased early mean rate of synthesis, which is closely similar to that of fed controls in the last 4 hr. Full supplementation elicits a marked increase in rate of synthesis only after 8–10 hr (Fig. 4b) and only in the livers of fed rats. The response, if any, of the livers of 18-hr fasted rats is seen only after 10 hr. None of the perfusions of livers from 6-day fasted rats showed significant synthesis of α_2-(acute phase) globulin regardless of supplementation; accordingly, the data are omitted from Fig. 4.

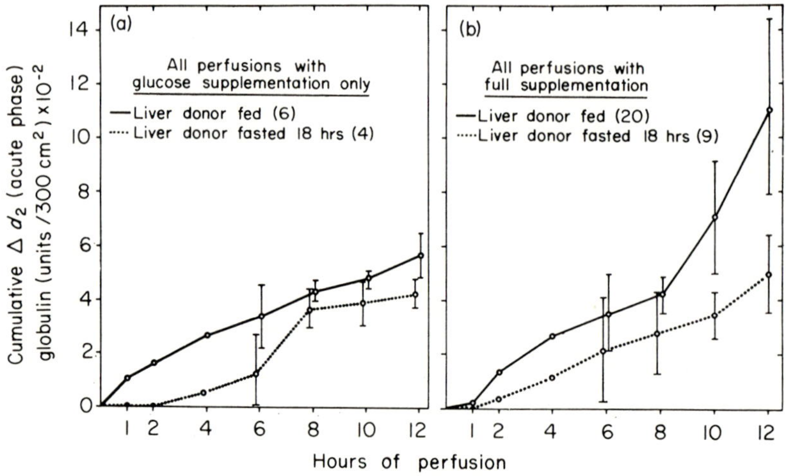

FIG. 4. (a) and (b): Cumulative average net change in rat α_2-(acute phase) globulin in perfusate per 300 cm^2 of body surface area of rat liver donor. Because no measurable change occurred in perfusions of livers from 6-day fasted liver donors, data from these experiments are not included here.

E. NET SYNTHESIS OF HAPTOGLOBIN

Figure 5a shows that neither the absence of amino acid and hormone supplementation nor fasting 18 hrs or even 6 days totally impairs the capacity of the perfused liver to effect net synthesis of haptoglobin. However, as

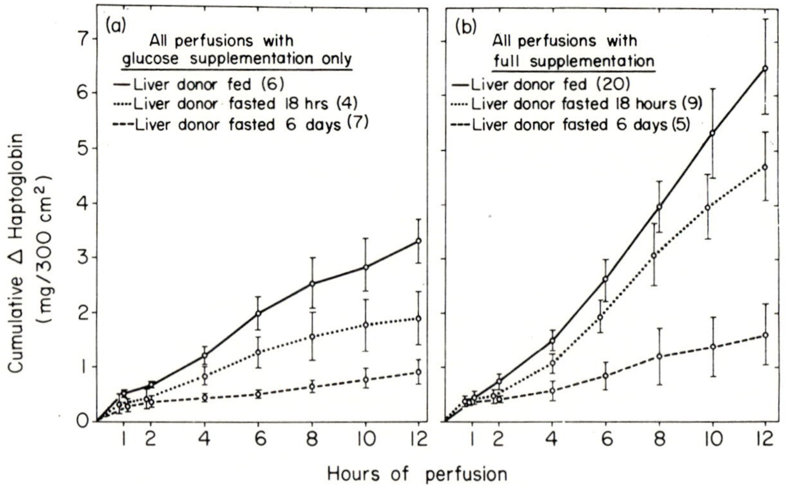

FIG. 5. (a) and (b): Cumulative average net change in rat haptoglobin in perfusate/300 cm^2 of body surface area of liver donor rat.

revealed in Fig. 5b, full supplementation substantially improves the rate of synthesis, but only after an induction interval of 4–6 hr, after which the rate of synthesis is about three times that seen in controls and in early hours of perfusions with full supplementation.

F. Effects of Refeeding Liver Donors after 6-Day Fast

The variably profound impairment of a 6-day fast on the biosynthesis of albumin and fibrinogen, α_2-(acute phase) globulin, and haptoglobin led us to examine the effects of refeeding with the stock diet. As the results of perfusion experiments depicted in Figs. 6 and 7 show, 12-hr refeeding has minimal effect on albumin or fibrinogen synthesis even with full supplementation, but 37 or 47 hr refeeding and full supplementation of the perfused liver is already associated with significant restoration of impaired function. Comparison with Fig. 1b and Fig. 2b reveals that 47 hr refeeding has restored net synthetic capacity for albumin and for fibrinogen to about 40% of that seen with full supplementation of livers from normal fed rats.

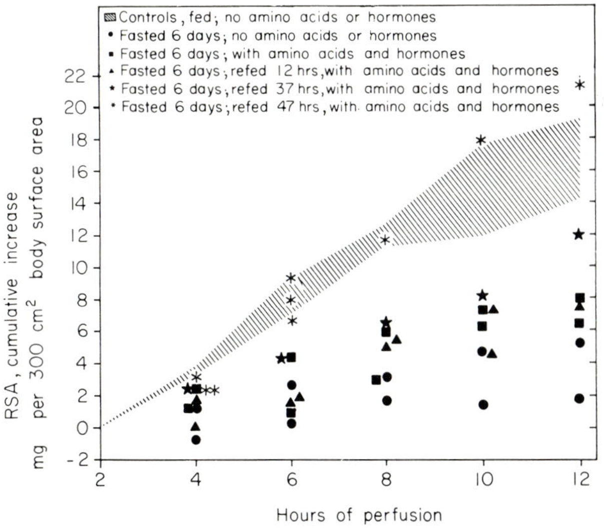

FIG. 6. The effect of a 6-day fast of the rat liver donor on the net synthesis of serum albumin by the isolated perfused liver. The terms amino acids and hormones indicate the addition of amino acids and hormone as described for "full supplementation" in Section II, C. Since preformed albumin is largely released in the first 2 hr (John and Miller, 1966, 1969b) of perfusion, the figure shows net synthesis from 2 hr on. The cross-hatched area represents the range of values of four livers from normal fed rats; only the 1-^{14}C-L-lysine and 500 mg of glucose were given in the continuous infusion. Where refed, liver donors were allowed access to their normal diet of commercial rat food (Purina Checkers) for the times indicated before perfusion.

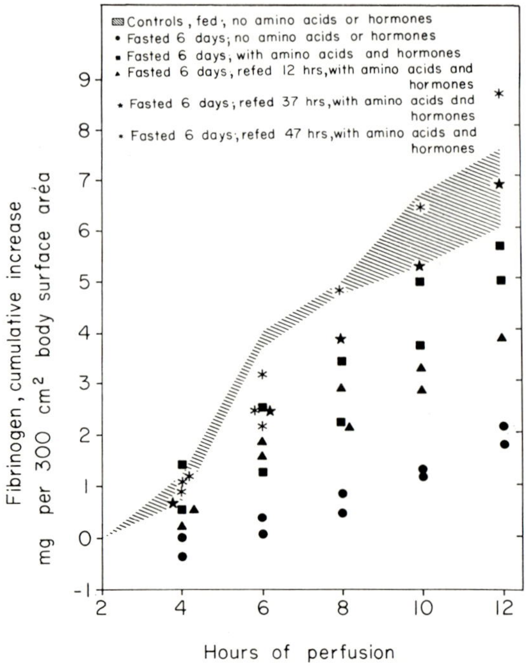

FIG. 7. The effect of a 6-day fast on the net biosynthesis of rat fibrinogen by the isolated perfused rat liver. Experiments and details are the same as those in legend of Fig. 6.

G. EFFECTS OF FULL SUPPLEMENTATION ON NITROGEN BALANCE

Table I summarizes additional data obtained from the perfusions of Figs. 1–5. These data are perhaps best interpreted along with those of a similar table in a paper (John and Miller, 1969b) in which we have recently examined not only the effects of the amino acid mixture but also the added separate contribution of insulin, cortisol, and growth hormone singly or together. The latter table made clear that: (1) the combination of glucose plus amino acids and insulin was necessary for positive nitrogen balance (i.e., more amino acid nitrogen removed from the perfusion system than urea nitrogen produced); (2) induction of increased synthesis of acute phase proteins *in vitro* required the presence of cortisol and was maximal only in the presence of both cortisol and insulin and that growth hormone per se, or in addition to insulin and cortisol, was without effect on plasma or liver protein synthesis in perfusion of livers from normal fed rats.

Table I reveals that: (1) in the absence of added amino acids the perfusion medium actually shows a net gain of amino acid N which is largest for livers of fed rats and smallest for those of 6-day fasted rats; (2) with full supplementation, approximately equal amounts of amino acid are removed; (3) the net effect of full supplementation is positive nitrogen balance; (4) net glucose

TABLE I

COMPOSITE SUMMARY OF HEPATIC PROTEIN SYNTHETIC AND METABOLIC ACTIVITY NET CHANGES IN 12 HR OF PERFUSION[a]

Liver Donor	Full supplementation			Controls		
	Fed (20)	18 Hr fasted (9)	6 Days fasted (5)	Fed (6)	18 Hr fasted (4)	6 Days fasted (7)
^{14}C in total hepatic protein (% total ^{14}C dose)	11.5 ± 1.6	12.3 ± 1.6	8.6 ± 0.5	8.8 ± 2.0	13.6 ± 2.0	9.9 ± 4.9
Perfusate amino acid nitrogen (mg)	−42.5 ± 2.4	−45.0 ± 2.7	43.4− ± 1.5	14.8 ± 2.1	7.6 ± 0.8	6.3 ± 1.0
Bile amino acid nitrogen (mg)	2.2 ± 0.7	1.9 ± 0.4	0.5 ± 0.2	1.5 ± 0.4	1.2 ± 0.3	0.4 ± 0.1
Perfusate urea nitrogen (mg)	28.7 ± 3.1	29.9 ± 3.7	32.2 ± 5.7	31.1 ± 3.0	24.4 ± 4.0	21.2 ± 2.1
Bile urea nitrogen (mg)	3.8 ± 1.1	5.0 ± 1.5	1.1 ± 0.6	2.0 ± 0.7	2.1 ± 0.6	0.7 ± 0.1
Overall nitrogen balance (mg)	7.8 ± 4.2	8.2 ± 5.4	9.6 ± 5.2	−49.4 ± 2.8	−35.3 ± 5.2	−28.6 ± 2.1
Glucose (mg)	−352 ± 96	−420 ± 39	−160 ± 23	−126 ± 133	−385 ± 53	−196 ± 56
Bile volume (ml)	15.0 ± 2.8	14.6 ± 2.3	4.9 ± 1.4	8.1 ± 1.5	9.3 ± 1.7	4.3 ± 1.6
^{14}C in bile (% total ^{14}C dose)	1.8 ± 0.4	1.4 ± 0.2	0.6 ± 0.2	1.3 ± 0.4	1.0 ± 0.2	0.5 ± 0.2

[a] All values are averages and standard deviations.

disappearance from the perfusate is drastically reduced by 6-day fast; and (5) full supplementation enhances bile production by livers of fed and 18-hr fasted but not by livers of 6-day fasted rats.

IV. Discussion

Much previous work on effects of nutritional factors or hormones on plasma protein synthesis has used incorporation of ^{14}C-labeled metabolites into relatively crude protein fractions *in vivo* or in subcellular systems as an index of protein synthesis. By their very nature, *in vivo* studies preclude unequivocal definition of direct effects on the liver from indirect effects. In the absence of unequivocal data on precursor pool size and ^{14}C specific activity, ^{14}C incorporation is at best a presumptive index of net protein synthesis.

By exploiting the advantages of the isolated liver perfused with hetero-species blood, the studies described in this chapter and in our closely related recent paper (John and Miller, 1969b) represent a distinct advance with regard to definition of the role of the liver in the net synthesis of specific plasma proteins. These studies reveal that the net biosynthesis by the isolated perfused rat liver of five specific plasma proteins varies with the nutritional state of the liver donor with respect to feeding, fasting, or fasting followed by refeeding of the liver donor. In addition, these studies confirm and extend those of our recent report (John and Miller, 1969b); taken together they make clear that optimal conditions for maximum net synthesis of albumin, fibrinogen, α_1-acid glycoprotein, α_2-(acute phase) globulin, and haptoglobin entail not only that the liver be from a fed donor, but also that it receive a nutritionally complete amino acid mixture, and at least the combination of hormones insulin and cortisol. These studies reaffirm the need that 3–8 hr intervals elapse before induction of increased synthesis of the acute phase proteins occurs.

We have used the term induction in the sense discussed in detail by Greengard (1967) to mean an increase in the amount of a specific protein as a consequence of an increase in the ratio of the rate of synthesis to the rate of degradation; the term implies no specific mechanism. However, it is likely that the increases observed in the experiments of Figs. 2–5 are primarily the result of more rapid synthesis, since suppression of the relatively low rate of plasma protein catabolism known to occur in isolated liver perfusions (John and Miller, 1968) would be inadequate to account for the large net increases in synthesis observed.

Several authors have reported apparent increase in rate of synthesis of acute phase plasma proteins when livers isolated from normal rats were perfused with blood from acutely injured rats (Gordon and Darcy, 1967; Sarcione *et al.*, 1965); however, subsequent investigations (Gordon and Koj, 1968; Schumer *et al.*, 1967) including unpublished reports from our own laboratory have failed to confirm such an effect of blood from acutely injured rats.

Impressive is the fact that even without exogenous supplementation or amino acids or hormones, livers from fed, 18-hr fasted, and even 6-day fasted rats can effect significant synthesis of specific rat plasma proteins presumably at the expense of rat liver proteins or of the heterospecies erythrocyte or albumin proteins present in the perfusion system. Under the extreme conditions of a prior 6-day fast albumin synthesis is proportionately more depressed than that of fibrinogen, α_1-acid glycoprotein, or haptoglobin; unlike the synthesis of the latter three acute phase proteins, synthesis of α_2-(acute phase) globulin is totally suppressed in livers from 6-day fasted rats and for that reason is not included in Fig. 4.

Equally impressive is the prompt twofold enhancement of net albumin synthesis seen with full supplementation by the liver from fed rats (Fig. 1). The failure of full supplementation to elicit enhanced albumin or α_2-(acute phase) globulin synthesis by livers from 18-hr fasted rats is puzzling, since the synthesis of fibrinogen, α_1-acid glycoprotein, and haptoglobin is quantitatively only slightly less than that seen with livers from fed donors. That the depressing effect of an 18-hour fast of the liver donor on albumin synthesis by the isolated liver is not peculiar to the rat is suggested by observations of Rothschild *et al.* (1968) on the synthesis of albumin by the isolated rabbit liver. Using $^{14}CO_2$ incorporation into arginine of serum albumin and urea, they estimated that an 18-hr fast of the liver donor led to about a 50% decrease in albumin synthesis in experiments 4 hr in duration. One may conjecture that this reflects a biological priority for the synthesis of fibrinogen, α_1-acid glycoprotein, and haptoglobin; or one may speculate that the processes of protein digestion and absorption supply the liver with a factor or factors necessary for maximal albumin synthesis. It is noteworthy that these observations on albumin and fibrinogen synthesis by the isolated rat liver are qualitatively similar to those made in intact dogs studied with the technique of plasmapheresis (Madden *et al.*, 1938, 1939). Whatever the mechanisms, one may conclude that the rate of synthesis of albumin *in vivo* may best be described by some type of oscillating function, with maxima corresponding to periods of active protein digestion and absorption and associated hormonal responses (particularly that of insulin), and minima corresponding to the intervening intervals of inanition and decreased blood levels of insulin.

Effects of hormones on plasma protein levels of intact animals have been studied by ablation of endocrine glands and by hormone treatment (Leathme,

1964); most of the results have been difficult to interpret because changes observed could have been secondary to altered nutrition, blood volume, radioactive precursor specific activity, or other factors and not due to effects of the specific hormones studied.

Observations on direct effects of hormones on plasma protein synthesis by rat liver are few in number. Korner (1962; Jefferson and Korner, 1967) has reported that the ability of rat liver microsomes to incorporate activated ^{14}C-amino acids in an *in vitro* system into protein in general and serum albumin in particular is regulated by growth hormone *in vivo*. Our recent study (John and Miller, 1969b) fails to show more than a minimal effect of added bovine growth hormone on albumin synthesis by perfused livers from normal fed rats, and no apparent effect on synthesis of the four other plasma proteins studied or on incorporation of 1-^{14}C-L-lysine into liver protein.

A role for cortisol in maintaining serum albumin levels in hypophysectomized rats has been suggested by the work of Levin and Leathem (1942); liver perfusion studies with ^{14}C-leucine incorporation have led Gordon (1964) to conclude that cortisol significantly reduces synthesis of serum albumin, fibrinogen, and transferrin.

Because this report presents effects of hormones given together as a part of "full supplementation," the reader must refer to the detailed data of our recent report (John and Miller, 1969b) to see the experimental basis for our conclusion: The nutritionally complete amino acid mixture and glucose are by themselves incapable of supporting nitrogen equilibrium or positive nitrogen balance without added insulin. The same table reveals that cortisol is essential for the induction of enhanced synthesis of acute phase proteins, and that this induction is maximal only in the presence of insulin.

Although "full supplementation" affects the net synthesis of specific plasma proteins by the perfused liver from 6-day fasted rats only slightly (albumin) or minimally (fibrinogen, α_1-acid glycoprotein, haptoglobin), the net positive nitrogen balance (Table I) suggests that full supplementation first promotes synthesis of liver ribonucleic acids and protein before the capacity for plasma protein synthesis is restored toward normal. This view is supported by the fact that after 6-day fast, refeeding must go on for at least 24–37 hr before substantial increase in response to "full supplementation" is seen (Figs. 6 and 7).

The mechanism of the remarkable effect on plasma protein synthesis of cortisol combined with insulin remains to be elucidated. Inductive changes produced by cortisol in synthesis of some liver proteins such as tryptophan pyrollase (Barnabei and Sereni, 1964; Goldstein *et al.*, 1962) and tyrosine transaminase (Goldstein *et al.*, 1962) have been thought by some to be secondary to increased synthesis of messenger RNA (Barnabei and Sereni, 1964). Wool has suggested that insulin has a role in somehow facilitating the association of ribosomes with messenger RNA (Wool *et al.*, 1968). Whether a combination of such effects is involved in induction of acute phase plasma protein synthesis is conjectural at present.

V. Summary

Livers from normal, male, fed, 18-hr fasted, 6-day fasted, and 6-day fasted refed rats were isolated and perfused for 12 hr with bovine or rabbit red blood cells suspended in Krebs–Ringer–bicarbonate solution containing bovine serum albumin (3.0 g/100 ml) with 1-^{14}C-L-lysine and 500 mg of glucose continuously infused. Net synthesis of five specific plasma proteins was estimated by serological measurements of changes in their concentration in perfusate. The influence of "full supplementation," namely the addition of porcine, insulin, cortisol, bovine growth hormone, and a nutritionally complete mixture of amino acids on net synthesis of plasma proteins, was evaluated and compared with results of control perfusions (no hormones or amino acids added).

Control perfusions with livers from fed donors displayed relatively linear net synthesis of the specific plasma proteins throughout the experiments; in experiments with full supplementation, albumin synthesis increased twofold during the first hr of perfusion, fibrinogen and haptoglobin synthesis increased threefold between 2 and 6 hr, α_1-acid glycoprotein synthesis increased 3.5-fold between 4 and 8 hr, and α_2-(acute phase) globulin synthesis increased 11-fold between 8 and 10 hr.

In control perfusions with livers from 18-hr fasted donors, plasma protein synthesis was not significantly different from that seen with livers from fed donors. However, response to full supplementation was different; livers from 18-hr fasted donors failed to show increased synthesis of albumin and induction of α_2-(acute phase) globulin synthesis, but did show significant increase and induction of synthesis of fibrinogen, haptoglobin, and α_1-acid glycoprotein.

In control perfusions, perfusions with livers from 6-day fasted donors, net synthesis of albumin was only 10–20% of that seen with fed controls, and synthesis of fibrinogen, α_1-acid glycoprotein, and haptoglobin was decreased from 20 to 40% of that from fed controls. The response of livers from 6-day fasted controls to full supplementation was absent for albumin and α_2-(acute phase) globulin, and was perceptible, but small, for the other three proteins studied. Partial recovery of this response was perceptible only after 6-day fasted liver donors had been refed for 37–47 hr.

Positive nitrogen balance seen with full supplementation of livers from 6-day fasted donors points to the need for hepatic parenchymal protein and RNA synthesis to precede renewed plasma protein synthesis.

Acknowledgments

This work is supported in part by the United States Atomic Energy Commission. Report No. UR-49-1183, at the University of Rochester Atomic Energy Project, and in part by Grant 1-ROI-AM 11029 MET from the National Institutes of Health, United States Public Health Service.

We thank Donna M. Eddy, Louise Finch, Catherine S. Plane, Drusilla Wemett, and Janice White for their invaluable technical assistance, and Leon Schwartz and Gerald Cooper for their preparation of the illustrations.

References

Barnabei, O., and Sereni, F. (1964). *Biochim. Biophys. Acta* **91**, 239.

Fahey, J. L., and McKelvey, E. M. (1965). *J. Immunol.* **94**, 84.

Goldstein, L., Stella, E. J., and Knox, W. E. (1962). *J. Biol. Chem.* **237**, 1723.

Gordon, A. H. (1964). *Biochem. J.* **90**, 18P.

Gordon, A. H., and Darcy, D. A. (1967). *Brit. J. Exp. Pathol.* **48**, 81.

Gordon, A. H., and Koj, A. (1968). *Brit. J. Exp. Pathol.* **49**, 436.

Grabar, P., and Williams, C. A. (1953). *Biochim. Biophys. Acta* **10**, 193.

Greengard, O. (1967). *Enzymol. Biol. Clin.* **8**, 81.

Jefferson, L. S., and Korner, A. (1967). *Biochem. J.* **104**, 826.

John, D. W., and Miller, L. L. (1966). *J. Biol. Chem.* **241**, 4817.

John, D. W., and Miller, L. L. (1968). *J. Biol. Chem.* **243**, 268.

John, D. W., and Miller, L. L. (1969a). *Biochem. Pharmacol.* **18**, 1135.

John, D. W., and Miller, L. L. (1969b). *J. Biol. Chem.* **244**, 6134.

Korner, A. (1962). *Protein Metab.: Influence Growth Horm. Anabolic Steroids Nutr. Health Dis. Int. Symp., Leiden* p. 8.

Leathem, J. H. (1964). *In* "Mammalian Protein Metabolism" (H. N. Munro, and J. B. Allison, eds.), Vol. 1, p. 354. Academic Press, New York.

Levin, L.. and Leathem, J. H. (1942). *Amer. J. Physiol.* **136**, 306.

Lombart, C., Moretti, J., and Jayle, M.-F. (1965). *Biochim. Biophys. Acta* **97**, 262.

Madden, S. C., George, W. E., Waraich, G. S., and Whipple, G. H. (1938). *J. Exp. Med.* **67**, 675.

Madden, S. C., Noehren, W. A., Waraich, G. S., and Whipple, G. H. (1939). *J. Exp. Med.* **69**, 721.

Mancini, G., Carbonara, A. O., and Heremans, J. F. (1965). *Immunochemistry* **2**, 235.

Miller, L. L., and Bale, W. F. (1954). *J. Exp. Med.* **99**, 125.

Miller, L. L., Bly, C. G., Watson, M. L., and Bale, W. F. (1951). *J. Exp. Med.* **95**, 531.

Miller, L. L., Bly, C. G., and Bale, W. F. (1954). *J. Exp. Med.* **99**, 133.

Miller, L. L., Burke, W. T., and Haft, D. E. (1956). "Some Aspects of Amino Acid Supplementation," p. 44. Rutgers Univ. Press, New Brunswick, New Jersey.

Miller, L. L., Hanavan, H. R., Titthasiri, N., and Chowdhury, A. (1964). *Advan. Chem. Ser.* **44**, 17.

Neuhaus, O. W., Bolegno, H. F., and Chandler, A. M. (1966). *Amer. J. Physiol.* **211**, 151.

Rothschild, M. A., Oratz, M., Mongelli, J., and Schreiber, S. S. (1968). *J. Clin. Invest.* **47**, 2591.

Sarcione, E. J., Bohne, M., and Krauss, S. (1965). *Fed Proc. Fed. Amer. Soc. Exp. Biol.* **24**, 230,

Schumer, W., Molnar, J., Dowling, J. N., and Winzler, R. (1967). *Amer. J. Physiol.* **212**, 184.

Spackman, D. H., Stein, W. H., and Moore, S. (1958). *Anal. Chem.* **30**, 1185.

Wool, I. G., Stirewalt, W. S., Kurihara, K., Low, R. B., Bailey, P., and Oyer D. (1968). *Recent Progr. Horm. Res.* **24**, 139.

14

ONCOTIC PRESSURE AND ALBUMIN SYNTHESIS

Murray Oratz

RADIOISOTOPE SERVICE, VETERANS ADMINISTRATION HOSPITAL,
NEW YORK, NEW YORK, AND DEPARTMENT OF BIOCHEMISTRY,
NEW YORK UNIVERSITY COLLEGE OF DENTISTRY, NEW YORK, NEW YORK

I. Introduction

Bjorneboe (1946) postulated that the reciprocal changes in the serum albumin and globulin levels which occurred in the presence of hypergammaglobulinemia might be related to a mechanism affecting control over oncotic pressure. Further, this depression of serum albumin associated with elevated globulin levels is not necessarily associated with hepatic dysfunction or excessive albumin loss, again suggesting the presence of an oncotic system regulating albumin synthesis. This concept has been tested *in vivo* and *in vitro* and much evidence now exists to support its presence.

II. Oncotic Pressure and Albumin Synthesis

Effect of High Molecular Weight Substances on Albumin Synthesis

1. *Infusions of Dextran*

Initially, animal studies were conducted by infusing dextran (average MW 186,000) daily for 2–4 weeks and measuring albumin metabolism with ^{131}I-labeled albumin before, during, and after the dextran infusions. As the plasma dextran levels rose, the albumin levels in the plasma decreased. Dextran produced the expected increase in the plasma volume, albumin degradation decreased as the albumin levels, and pool size decreased and albumin synthesis fell by over one-third during the period of dextran infusions (Rothschild *et al.*, 1961).

2. *Hypergammaglobulinemia*

These observations were extended in animals whose γ-globulin synthesis was stimulated to a marked degree by immunizing rabbits with a polyvalent pneumococcal antigen. These animals produced as much as 1 g of γ-globulin/kg of body weight/day; again the same sequence of events was observed. As the plasma γ-globulin levels rose, albumin levels fell, albumin degradation fell, and the albumin pool size decreased. Since both albumin degradation and pool size fell, albumin synthesis must have decreased and a 23% decrement was measured (Rothschild *et al.*, 1962).

3. *Infusion of γ-Globulin*

In evaluating this decrease in albumin synthesis the available nitrogen pool for all serum protein production must be considered. It is possible that the rapid production of one protein may interfere with the synthesis of another by competing for the available substrates. This concept was examined by infusing γ-globulin daily in amounts sufficient to raise the γ-globulin pool size to the level seen in the hyperimmunized animals and for the third time the same changes as described with dextran, and during the anamnestic response occurred, and albumin production fell by 31% during the 3 weeks (Rothschild *et al.*, 1965) (Table I).

In these studies there was little change in the calculated oncotic pressure of the serum. The infused colloids or the synthesized γ-globulin essentially

TABLE I

In Vivo

Colloid	Albumin synthesis (% change)
Infused dextran	−35
Endogenous γ-globulin	−23
Exogenous γ-globulin	−31
Exogenous albumin	0
In vitro—perfused liver	
Hypoalbuminemia	+50
Hyperalbuminemia	−40
Sucrose	−45

compensated for the loss of intravascular albumin. However, in a variety of disease states low serum oncotic pressure is not effective in stimulating albumin synthesis. In fact, the lowest rates of albumin production are often seen associated with the lowest levels of serum albumin. Thus the level of serum albumin does not appear to be an effective signal for albumin production.

These clinical observations are made in well established conditions whereas changes in albumin production occur rather rapidly as the experimental situation changes until a new steady state is achieved.

III. Site of Oncotic Regulation

A. Extravascular Location

Plasma levels do not play a significant role in regulating albumin synthesis, rather albumin synthesis appears to bear an inverse relationship to extravascular albumin and colloid content (Table II). In hypothyroidism and after dextran infusions (Rothschild *et al.*, 1961), there is a marked shift of albumin from plasma to extravascular sites and albumin synthesis is depressed. After excess cortisone (Rothschild *et al.*, 1958; Grossman *et al.*, 1960; Sterling, 1960) and thyroid (Rothschild *et al.*, 1957) and following plasmaphoresis (Whipple and Madden, 1944; Mathews, 1961), the plasma albumin content is maintained at the expense of the extravascular albumin pool and albumin production rises. Plasma oncotic pressure does not correlate with albumin synthesis, and it is possible that if an oncotic regulator exists, it is situated not in the plasma but adjacent to or within the site of albumin synthesis, namely, within the liver itself.

TABLE II

ALBUMIN DISTRIBUTION[a]

	Plasma level	IV%	EV%	Synthesis
Human				
Myxedema	±	−	+	−
Excess thyroid	±	+	−	+
Excess cortisone	±	+	−	+
Nephrosis	−	+	−	±
Rabbit				
Plasmaphoresis	−	±	−	+
Cortisone	−	+	−	+
Dextran	−	−	+	−
Hypergammaglobulinemia	−	±	±	−
Dextran + cortisone	—	+	−	+

[a] (+), elevated; (−), depressed; IV, intravascular; EV, extravascular.

These concepts prompted a study of hepatic intravascular and extravascular–extracellular (interstitial) albumin distribution.

B. Oncotic Control in Hepatic Interstitial Fluid

Normally the hepatic interstitial fluid has an albumin content equivalent to 0.3 g/100 ml of interstitial fluid. This is doubled during the administration of dextran and reduced to about 0.1 g/100 ml after the administration of cortisone (Rothschild *et al.*, 1966). Since newly made albumin is transferred directly from the hepatocyte to the plasma (Woolley and Courtice, 1962; Smallwood *et al.*, 1968), the interstitial fluid probably plays only a very insignificant part in this transfer. Otherwise this low albumin content would not be compatible with the concentration found in either lymph or plasma. However, the hepatic interstitial fluid is made up of a protein polysaccharide matrix which probably excludes other large molecules from its aqueous domain (Trowell, 1946; Ogston and Phelps, 1960; Laurent, 1963). Thus the interstitial volume measured with a small molecule is much larger than the true volume available for albumin distribution. If available water is used as the volume for albumin solution, then the albumin concentration may approach the levels found in plasma and lymph.

In either event the hepatocyte would be exposed to a very low quantity of albumin and colloid. If albumin production within the hepatocyte responds to changes in oncotic pressure in the interstitial space, then small changes in the amount of albumin or colloid in this area could serve as a very sensitive regulator for albumin production. All of these changes have been studied *in vivo* where stress, hormonal effects, and nutrition could alter albumin production. It now remained to confirm these findings *in vitro*.

IV. Isolated Rabbit Liver Perfusions

A. TECHNIQUE

Male rabbits were employed as donors and were fed ad lib until operation. The portal vein was cannulated, and perfusion, from a reservoir at a height of 30–40 cm above the liver, was started with a flow of 10–15 ml/min. The liver was cut off from portal circulation for less than 1 min. The inferior vena cava was cannulated from below the liver and ligated above the diaphragm; the liver was removed and mounted on a platform in a humidified 37°C box. The inflow to the portal vein was transferred to a pump recirculation system where the outflow was delivered to a reservoir leading to a disk oxygenator. The oxygenated blood was pumped through a blood and silk cloth filter into the portal vein at a constant rate.

The perfusion solution contained two parts heparinized rabbit blood, one part oxygenated Krebs–Henseleit solution containing 3–4 μmoles/ml amino acids, equivalent to rabbit plasma, including the essential amino acids, and, in addition, 0.69 mM L-glutamine and 0.24 mM L-arginine. Rabbit albumin (Pentex) was added to adjust the final albumin level. The oxygenated perfusate was pumped into the portal vein at flow rates of 0.3–1.4 ml/min/g wet liver (Rothschild *et al.*, 1969a). The ^{14}C-carbonate technique was used to measure albumin synthesis (McFarlane, 1963). The ^{14}C labels the intracellular arginine pool. The carbon of urea and the precursor arginine guanido carbon have the same specific activity. A direct product precursor relationship exists and the specific activity of newly made urea can be used as an index of intracellular arginine specific activity. The perfusion was continued for 2.5 hr at which time the total perfusate was collected, albumin isolated, and synthesis calculated. It should be emphasized that with this technique it is most important to obtain a pure albumin preparation (Fig. 1) and to be certain that the albumin isolated is not contaminated with radioactive urea.

B. ONCOTIC CONTENT OF PERFUSATE AND ALBUMIN SYNTHESIS

In these studies, albumin synthesis was inversely related to the osmotically active content of the perfusate; and these changes occurred within the short experimental period. The perfusate was varied as to the colloid content. With a hypoalbuminemic perfusate (0.4–0.6 g/100 ml) albumin synthesis increased by 49% during 2½ hr. When the perfusate was made hyperosmotic

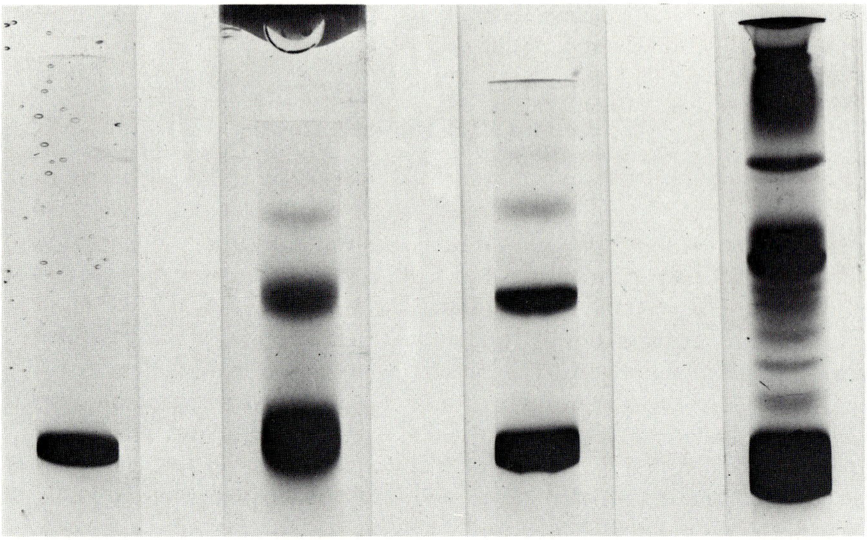

FIG. 1. Qualitative acrylamide gel electrophoresis. Reading from left to right: Tube 1, the perfusate; tube 2, commercial albumin preparation; tube 3, albumin isolated by TCA–alcohol–ether technique; and tube 4, albumin isolated by preparative acrylamide gel electrophoresis. The specific activity of albumin isolated in tubes 3 and 4 were identical.

with either excess albumin (7–9 g/100 ml) or sucrose (0.5–1.0 g/100 ml), albumin synthesis decreased to less than one-half of the control value (Table I) (Rothschild *et al.*, 1969b).

V. Other Studies with Isolated Livers

In the perfused liver, dextran has been shown to inhibit the increased ^{14}C-leucine incorporation into albumin associated with hypoalbuminemic perfusates (Tracht *et al.*, 1967). Ryoo and Tarver (1968) noted a decrease in albumin synthesis in the perfused rat liver when the perfusate contained polyvinylpyrrolidone.

In the present studies an increase in albumin synthesis and/or release was seen with hypoalbuminemic perfusates. Whereas hyponocotic states are present in cirrhosis, nephrosis, and analbuminemia, the absolute synthetic rate is not consistently elevated *in vivo* (Blahd *et al.*, 1955; Dykes, 1968; Hasch *et al.*, 1967) though high rates have been observed (Hasch *et al.*, 1967; Rothschild *et al.*, 1969c). In these states other operative factors which might inhibit the albumin-producing capacity are present, such as nitrogen deficiency and liver cell damage. These factors are not present in the perfused system nor do they seem to have altered the subcellular systems once the latter are removed from their own environment. Marsh and associates (1966) have shown that hepatic microsomes removed from nephrotic animals are more active in their albumin synthesizing ability than controls.

VI. Hyperalbuminemia

The decrease in albumin synthesis produced by high albumin levels in the perfusate does not occur *in vivo*. Clinically, hyperalbuminemia is a condition which rarely occurs except perhaps in acute dehydration. However, by infusing large quantities of albumin, elevated albumin levels and pool size can be produced in a few days. In this situation the excess albumin did not inhibit albumin synthesis, since the additional colloid was rapidly removed from the body by a considerable increase in albumin degradation (Rothschild *et al.*, 1964). This marked increase in albumin degradation *in vivo* could have provided the liver with an excess amount of amino acids which in turn might have stimulated albumin synthesis. In the perfused liver no significant degradation of albumin occurs and this opposing action to the effects of excess colloid would not be present (Cohen and Gordon, 1958). That albumin synthesis was neither increased nor decreased in these *in vivo* studies does show that albumin synthesis and degradation are not interdependent.

Thus, there are many pieces of evidence to support the existence of an osmotic regulatory system effecting control over albumin production. The data also support the concept that this type of regulatory system is located within the liver itself.

VII. Incorporation of Isotopic Amino Acids into Protein in *in Vitro* Systems

It now became important to examine the effects of excess colloid on the basic cellular mechanism for protein synthesis. The genetic code is transmitted by mRNA, and this nucleic acid is associated with another RNA containing organelle, the ribosome. The combination ribosome–mRNA (polysome) is the functional unit in forming protein. Amino acids are activated by enzymes and esterified to specific transfer ribonucleic acids (tRNA). These tRNAs contain an anticodon triplet which hydrogen bonds to a codon triplet in mRNA. Thus the amino acid is placed in proper sequence on the growing peptide chain. The polysome may be free within the cell or attached to the endoplasmic reticulum (Schultze and Heremans, 1966).

Albumin is produced on a ribosome–mRNA–endoplasmic reticulum complex (Campbell *et al.*, 1965, 1967; Takagi and Ogata, 1968; Hicks *et al.*, 1969). Altered oncotic pressure would have to interfere with this mechanism at some step and do it rapidly, certainly within hours, as has been shown by the observation that 2½ hr of hepatic perfusion with a hyperoncotic perfusate inhibits albumin synthesis. There are many possibilities or combination of possibilities in this complex process: The ribosome–mRNA complex would be disengaged from the membrane; the ribosome–mRNA could be disaggregated with release and subsequent degradation of mRNA; the polysome could be altered qualitatively, mRNA synthesis could be blocked; and transfer of amino acid by RNA blocked so that the amino acid could not be delivered to the growing peptide chain. If knowledge of the effects of altered colloid content on albumin production is to be furthered, studies of these facets of cellular activity must be undertaken.

A. Effect of Dextran Pretreatment

Some progress has been made and some questions can be answered and some possibilities eliminated. First of all it is necessary to determine if the complete subcellular system (isolated from livers removed from animals whose albumin production has been slowed by excess colloid) is reflective of this decreased albumin production.

These studies were conducted in control animals and animals treated with dextran. After 2 weeks of dextran infusions, albumin synthesis was measured. The livers were removed from these dextran treated rabbits and microsomes were isolated from a postmitochondrial supernatant. The tissue homogenate was centrifuged at 15,000 *g* for 10 min and the resulting postmitochondrial supernatant centrifuged for 1 hr at 105,000 *g* in a Model L ultracentrifuge.

TABLE III

Medium for Protein Biosynthesis by Hepatic Microsomes[a,b]

Microsomes in buffered medium	7–10 mg protein
Cell sap in buffered medium	10 mg protein
ATP	2 μmoles
GTP	0.5 μmoles
Phosphoenolpyruvate	20 μmoles
Pyruvate kinase	50 μg
^{14}C-amino acid	2.5 μCi
Amino acid mixture minus ^{14}C-amino acid	0.1 μmoles each

[a] The additions are present in a total volume of 1.0 ml.

[b] Incubation was carried out at 37°C for 30 min. The buffered medium contained 0.15 *M* sucrose, 25 m*M* KCl_2, 10 m*M* $MgCl_2$, 6 m*M* 2-mercaptoethanol, and 35 m*M* tris-HCl buffer, pH 7.8.

The sedimented pellet was suspended in buffer and all incubations were carried out in the presence of control cell sap passed through a column of Sephadex G25 to assure constant specific activity of the tracer amino acid. The rate and quantity of incorporation of the label into total protein was measured over a 30-min time period. The system used in these experiments are detailed in Table III.

It is important to point out that only animals who had been eating well and steadily were used in both groups, even a 12-hr fast can alter not only albumin synthesis but also microsomal ability to incorporate activity into protein (Rothschild *et al.*, 1968).

1. *Incorporation of ^{14}C-Leucine or ^{14}C-Phenylalanine by Microsomes*

Using either ^{14}C-phenylalanine or ^{14}C-leucine microsomes obtained from the dextran treated rabbits incorporated less activity into total protein than did microsomes obtained from livers of control animals (Table IV). Both

TABLE IV

PRIOR EFFECT OF DEXTRAN TREATMENT ON ^{14}C-LEUCINE INCORPORATION OF MICROSOMES

Microsomes[a]	Specific activity[b]	% Incorporation
Control	1840	100
Dextran	1140	62
Dextran (washed)	1233	67
Control + 1% dextran	1800	98
	^{14}C-Phenylalanine	
Control	4600	100
Dextran	2900	63

[a] All incubations were in the presence of control cell sap previously passed through Sephadex G25.

[b] Counts/min/mg microsomal protein.

groups of microsomes incorporated the tracer into total protein at the same rate and both reached a plateau at the same time, but the level of incorporation by the dextran microsomes was 38% lower than that observed with the control preparation.

2. *Lysosomal Inhibition*

This defect in the *in vitro* repressed microsomes could be related to increased lysosomal breakdown with the release of hydrolyases resulting in destruction of these subcellular components. Dextran, when given parenterally to mice, was found to accumulate in hepatic lysosomes, and it was presumed that these larger dextran loaded lysosomes were easily disrupted during homogenization (Baudhuin *et al.*, 1965). It has also been shown that one of the protein synthesis inhibitors, by microsomes, originated in lysosomes (Scornick *et al.*, 1967). Lysosomal contamination with the microsomes was studied by measuring acid phosphatase activity. Total acid phosphatase of control and dextran microsomes was measured as namonoles of *p*-nitrophenol released from *p*-nitrophenylphosphate/min/mg of microsomal protein. Control and dextran microsomes were nearly identical averaging 214 and 223 nanomoles/min/mg. Lysosomal acid phosphatase activity determined by F^- inhibition was only 10–12% of the total acid phosphatase activity in both preparations. Thus the presence of dextran within the Kupffer cell and within the hepatocyte did not result in either an increase of lysosomes in the cell or a disruption of the lysosomal membrane during homogenization.

3. *RNase Activity*

The possibility of an increase in "active" RNase in dextran livers due to a decrease in RNase inhibitor (Roth, 1956) levels was also studied. Livers from animals given dextran were assayed by the method of Shortman (1961) for total RNase after inhibiting the inhibitor with *p*-chloromercuribenzoate, active RNase (in the absence of *p*-chloromercuribenzoate), and the degree of inhibition of added ribonuclease. Total RNase in dextran livers averaged 394 ng/g liver and controls averaged 389 ng/g liver. No significant active RNase activity was found in any of the livers, and inhibition of the added 10 ng of pancreatic ribonuclease was nearly 100%.

Thus neither lysosomal fragility nor an increase in RNase activity, or a decrease in RNase inhibitor can account for the dextran induced depression in the albumin synthesis or the lower incorporating activity of the dextran microsomes.

4. *Polysome Activity*

The possibility that the decreased incorporation of amino acids into protein by the dextran microsomes might be due to a decreased quantity of functioning polysomes was examined by sucrose density gradient analysis. The dextran microsomes contained twice as many free ribosomes as did the control microsomes. Ribosomes detached from mRNA are relatively inactive in manufacturing protein. The activity of these ribosomes was studied with ^{14}C-phenylalanine and polyuridylic acid (poly U). The latter acts as an artificial mRNA and results in the synthesis of polyphenylalanine. In the absence of poly U dextran microsomes incorporated 62% of the

^{14}C-phenylalanine compared to control microsomes. The addition of poly U increased the incorporation of phenylalanine by control microsomes 26% but had no effect on the dextran microsomes. Removal of endogenous messenger from the microsomes by preincubation in the amino acid incorporating medium abolished the incorporation of phenylalanine (Weksler and Gelboin, 1967). The addition of poly U stimulated the incorporation of phenylalanine in both groups of microsomes (Table V). In contrast to the results obtained

TABLE V

EFFECT OF POLYURIDYLIC ACID ON ^{14}C-PHENYLALANINE INCORPORATION

	Specific activity (counts/min/mg microsomal protein)	
	Control	Dextran
− Poly U	4600 ± 100[a]	2900 ± 50
+ Poly U	5800 ± 90	2700 ± 50
Preincubation		
− Poly U	90 ± 30	16 ± 10
+ Poly U	7100 ± 100	3900 ± 60

[a] ± Standard error of the mean.

with microsomal preparations, polysomes isolated from control and dextran livers incorporated ^{14}C amino acids to the same degree. These observations suggest that the available polysomes within the dextran liver cell can function well; if protein synthesis is proportional to the number of polysomes, then the quantity of polysomes in the dextran microsome preparation is less than in control. Further, since the free ribosome in the dextran preparation failed to be stimulated by poly U, a functional defect in this RNA moiety exists as well (Oratz *et al.*, 1970).

B. EFFECT OF SUCROSE ON THE ISOLATED PERFUSED LIVER

These studies were continued employing the isolated perfused liver and increasing the colloid content by the addition of 1% sucrose to the perfusate as described above.

1. *Incorporation of ^{14}C-Leucine or ^{14}C-Phenylalanine by Microsomes*

Using ^{14}C-phenylalanine and ^{14}C-leucine for the incorporation of activity into total protein by microsomes isolated from these perfused livers was

TABLE VI

INCORPORATION BY MICROSOMES FROM LIVERS PERFUSED WITH 3% ALBUMIN PLUS 1% SUCROSE

Expt. no.	Microsomes	^{3}H-leuceine (% incorporation)	^{14}C-phenylalanine	% Stimulation with poly U
1	Control	100		
	Sucrose	100		
2	Control	100		
	Sucrose	72		
3	Control	100		
	Sucrose	82		
5	Control		100	250
	Sucrose		72	290
6	Control		100	246
	Sucrose		100	245

depressed in three studies by 18, 28, and 32% and unchanged in two even though albumin production in the perfused liver was inhibited by 50% in all five studies (Table VI). In contrast to the lack of stimulation of phenylalanine incorporation by the addition of poly U to dextran microsomes, sucrose microsomes were stimulated to the same extent as were control microsomes. Thus the 2½-hr perfusion with 1% sucrose did not produce an inactive ribosome population as did the 2 week infusion of dextran.

In these studies using subcellular organelles the incorporation of a tracer amino acid into protein should reflect a cell's total protein synthesizing polysome population. These polysomes may exist free or bound to the endoplasmic reticulum and it has been suggested that polysomes bound to endoplasmic reticulum synthesize proteins that the cell secretes. Further, it has been shown that serum albumin was synthesized exclusively by polysomes bound to the endoplasmic reticulum (Campbell *et al.*, 1965, 1967; Takagi and Ogata, 1868; Hicks *et al.*, 1969). Thus the elevated oncotic activity might effect only those polysomes concerned with protein synthesis for export—the endoplasmic membrane polysomes.

2. *Distribution of Polysomes in the Postmitochondrial Supernatant—Bound and Free*

The rapid effect of sucrose of hyperoncotic perfusates on albumin production might be due to a dissociation of the polysome from the endoplasmic

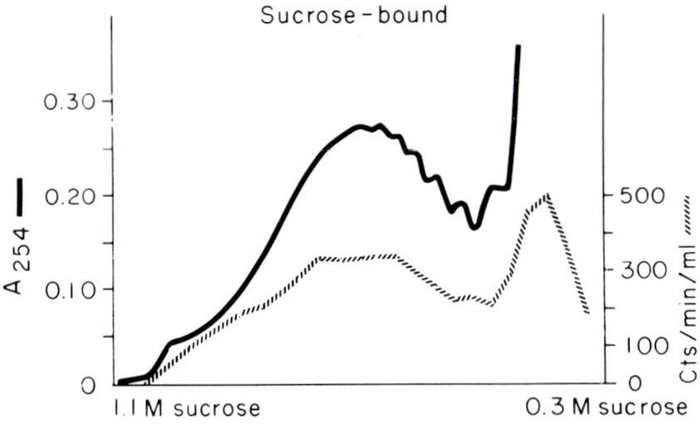

FIG. 2. Sucrose density gradient analysis of membrane-bound ribosomes. Membrane-bound ribosomes were isolated at the interface between 1.38 *M* sucrose and 2.0 *M* sucrose and treated with 20% Triton–1% deoxycholate (Blobel and Potter, 1967). Preparations were layered on top of a linear 0.3–1.0 *M* sucrose gradient. Following centrifugation the contents of the tube were pumped through an ISCO uv analyzer. The hatched line represents the distribution of ^{3}H-uridine following its addition to the perfusate.

The sucrose solutions were prepared from RNase-free sucrose obtained from Mann Research Laboratories.

reticulum, or to a disaggregation of the polysome to ribosomes and unattached mRNA. These possibilities were examined. Free and membrane-bound ribosomes were isolated from the postmitochondrial supernatant of sucrose perfused livers, by sedimentation through a discontinuous sucrose gradient (Blobel and Potter, 1967). RNA analysis (Fleck and Begg, 1965) indicated that free ribosomes accounted for 55–60% of the total. Similar results were obtained with control livers. Sucrose density gradient analysis (Fig. 2) did not demonstrate any significant degree of disaggregation of the polysome in the sucrose livers as compared to control livers.

In these short-term experiments, where the liver is perfused for $2\frac{1}{2}$ hr with 3% albumin–1% sucrose, albumin synthesis is decreased by 50%. This decrease does not appear to be as a result of polysome dissociation from the membrane, or polysome disaggregation. It is worth mentioning an obvious fact that all amino acid incorporation studies are usually carried out in 5–8% sucrose (0.15–0.25 *M*) so that toxicity of sucrose is not a factor.

3. *Incorporation of ^{3}H-Uridine into RNA*

The previous studies have demonstrated that while albumin synthesis is depressed, neither the incorporating ability of microsomes nor polysomal profiles are altered in these studies using sucrose perfused livers. Another possibility concerns the influences of hyperoncotic solutions on the ability of these perfused livers to incorporate uridine into various forms of RNA: transfer, messenger, and ribosomal. Total RNA was isolated from liver homogenates obtained from the perfused livers by phenol extraction in the presence of sodium dodecylsulfate and bentonite. Tritiated uridine had been added to the perfusates used in the studies with livers from fed rabbits as controls and fed preparations whose perfusates were enriched with sucrose. The isolated RNA was resolved into 4, 18, and 28 S RNA's by centrifugation in a 5–20% linear sucrose gradient. Again no significant difference in the specific activity, counts/min/mg RNA, was obtained for any of the RNAs isolated on the sucrose gradients. These results suggest that the hyperoncotic solution does not influence the synthesis of messenger or transfer RNA. The results suggest that another mechanism must be interfered with to result in decreased albumin synthesis. A defect in the binding properties of tRNA and altered mRNA or a defective synthesis of some intermediate enzmyic step may be present. Excessive sucrose does not simply inhibit release of albumin in the isolated perfused liver for the specific activity of the newly made albumin released into the perfusate which reaches a peak in the sucrose perfusates at the same rate as in the control or in fasting preparations (Rothschild *et al.*, 1969b). Different total levels of incorporation are achieved to account for the different synthesis rates.

VIII. Conclusion

Albumin synthesis is influenced by diet, hormones, stress, trauma, environment, and disease; but ultimate regulation must depend on day to day fluctuations. Certainly diurnal variations in hormonal secretion and the variation in food intake would result in changes of the rate of albumin synthesis during the day. Alterations in colloid content and effective oncotic pressure also must vary as thirst, urinary loss, and sweating change the fluid distribution within the body. Albumin production probably responds to these changes but as yet the mechanism must remain speculative. That the oncotic effect exists is strongly suggested, but it is not known where albumin synthesis is inhibited. Some basic questions must be answered. For example, what is the state of the endoplasmic reticulum on exposure to hyperosmotic perfusates? Pitot (1969) has suggested a functional structure called a membron, which is a membrane-associated RNA template whose activity may be regulated by alterations in the cellular environment. Haldar and Freeman (1969) found the incorporation of radioactive leucine into protein by mitochondria to be inhibited in hyperosmolar media (0.45 *M* sucrose). Decreasing the osmolarity to a value equivalent to 0.25 *M* sucrose, allowed incorporation to proceed at a substantial rate. Their studies indicated that the inhibition of protein synthesis occurred at a stage beyond activation and coupling of amino acid to tRNA. Since mitochondria isolated in 0.44 *M* sucrose were elongated and had a dense matrix with few and narrow cristae, while mitochondria isolated in 0.25 *M* sucrose were round and swollen (Whittaker, 1966), they suggested that the contracted state in some way inhibits protein synthesis. Further, not only has the work employing altered colloidal solutions shown an inhibition of albumin production *in vivo* and in the perfused livers, but also studies in lower animals have suggested osmotic control. Changes in the urea concentration in sharks as they are removed from salt to fresh water are accompanied by the appearance of an albumin-like protein suggesting that perhaps both urea and albumin production are under osmotic regulation (Rasmussen and Rasmussen, 1967). It is interesting to speculate that the control of albumin synthesis by oncotic pressure could be effected by a deformation of the endoplasmic reticulum–mRNA–ribosome complex. This deformation may cause a decrease in albumin synthesis because of a sterically hindered template. Thus the subcellular synthesizing mechanisms are still unclear. Albumin synthesis does respond to changes in colloid content. The exact mechanism remains to be uncovered.

ACKNOWLEDGMENTS

This work was supported, in part, by Public Health Service Research Grant AM-02489.

REFERENCES

Baudhuin, P., Beoufay, H., and de Duve, D. (1965). *J. Cell. Biol.* **25**, 219.
Bjorneboe, M. (1946). *Acta Med. Scand.* **123**, 393.
Blahd, W. H., Fields, M., and Goldman, R. (1955). *J. Lab. Clin. Med.* **46**, 747.
Blobel, G., and Potter, V. R. (1967). *J. Mol. Biol.* **26**, 279.
Campbell, P. N., Serck-Hanssen, G., and Lowe, E. (1965). *Biochem. J.* **97**, 422.
Campbell, P. N., Lowe, E., and Serck-Hanssen, G. (1967). *Biochem. J.* **103**, 280.
Cohen, S., and Gordon, A. H. (1958). *Biochem. J.* **70**, 544.
Dykes, P. W. (1968). *Clin. Sci.* **34**, 161.
Fleck, A., and Begg, D. (1965). *Biochimt Biophys. Acta* **108**, 333.
Grossman, J., Yalow, A. A., and Weston, R. E. (1960). *Metab. Clin. Exp.* **9**, 528.
Haldar, D., and Freeman, K. B. (1969). *Biochem. J.* **111**, 653.
Hasch, E., Jarnum, S., and Tygstrup, N. (1967). *Acta Med. Scand.* **182**, 83.
Hicks, S. J., Drysdale, J. W., and Munro, H. N. (1969). *Science* **164**, 584.
Laurent, T. C. (1963). *Biochem. J.* **89**, 253.
McFarlane, A. S. (1963). *Biochem. J.* **89**, 277.
Marsh, J. B., Drabkin, D. L., Broun, G. A., and Parks, J. S. (1966). *J. Biol. Chem.* **241**, 4168.
Mathews, C. M. E. (1961). *J. Clin. Invest.* **40**, 603.
Ogston, A. G., and Phelps, C. F. (1960). *Biochem. J.* **78**, 827.
Oratz, M., Rothschild, M. A., and Schreiber, S. S. (1970). *Am. S. Physiol.* **218**, 1108.
Pitot, H. C. (1969). *Arch. Pathol.* **87**, 212.
Rasmussen, L. E. and Rasmussen, R. A. (1967). *In* "Sharks, Skates, and Rays" (P. Gilbert, R. F. Mathewson, and D. P. Dall, eds.), Ch. 25. Johns Hopkins Press, Baltimore.

Roth, J. S. (1956). *Biochim. Biophys. Acta* **21**, 34.
Rothschild, M. A., Bauman, A., Yalow, R. S., and Berson, S. A. (1957). *J. Clin. Invest.* **36**, 422.
Rothschild, M. A., Schreiber, S. S., Oratz, M., and McGee, H. L. (1958). *J. Clin. Invest.* **37**, 1229.
Rothschild, M. A., Oratz, M., Wimer, E., and Schreiber, S. S. (1961). *J. Clin. Invest.* **40**, 545.
Rothschild, M. A., Oratz, M., Franklin, E. C., and Schreiber, S. S. (1962). *J. Clin. Invest.* **41**, 1564.
Rothschild, M. A., Oratz, M., Evans, C., and Schreiber, S. S. (1964). *J. Clin. Invest.* **43**, 1874.
Rothschild, M. A., Oratz, M., Mongelli, J., and Schreiber, S. S. (1965). *J. Lab. Clin. Med.* **66**, 733.
Rothschild, M. A., Oratz, M., and Schreiber, S. S. (1966). *Protides Biol. Fluids* **14**, 267.
Rothschild, M. A., Oratz, M., Mongelli, J., and Schreiber, S. S. (1968). *J, Clin. Invest.* **47**, 2591.
Rothschild, M. A., Oratz, M., Mongelli, J., Fishman, L., and Schreiber, S. S. (1969a). *J. Nutr.* **98**, 395.
Rothschild, M. A., Oratz, M., Mongelli, J., and Schreiber, S. S. (1969b). *Amer. J. Physiol.* **216**, 1127.
Rothschild, M. A., Oratz, M., Zimmon, D., Schreiber, S. S., Weiner, I., and Van Cangeham, A. (1969c). *J. Clin. Invest.* **48**, 344.
Ryoo, H., and Tarver, H. (1968). *Proc. Soc. Exp. Biol. Med.* **117**, 730.
Schultze, H. E., and Heremans, J. F. (1966). "Molecular Biology of Human Proteins," Vol. 1, pp. 321–337. Elsevier, Amsterdam.
Scornick, O. A., Hoagland, M. B., Pfefferkorn, L. C., and Bishop, E. A. (1967). *J. Biol. Chem.* **242**, 131.
Shortman, K. (1961). *Biochim. Biophys. Acta* **51**, 37.
Smallwood, R. A., Jones, E. A., Craigie, A., Raia, S., and Rosenoer, V. M. (1968). *Clin. Sci.* **35**, 35.
Sterling, K. (1960). *J. Clin. Invest.* **39**, 1900.
Takagi, M., and Ogata, K. (1968). *Biochem. Biophys. Res. Commun.* **33**, 55.
Tracht, M. E., Tollol, L., and Tracht, D. G. (1967). *Life Sci.* **6**, 2621.
Trowell, O. A. (1946). *J. Physiol.* (*London*) **105**, 268.
Weksler, M. E., and Gelboin, H. V. (1967). *J. Biol. Chem.* **242**, 727.
Whipple, G. H., and Madden, S. C. (1944). *Medicine* (*Baltimore*) **23**, 215.
Whittaker, V. P. (1966). *In* Biochim. Biophys. Acta Library, Vol. 7: "Regulations of Metabolic Processes in Mitochondria" (Tager, J. M., Papa, S., Quagliariello, E., and Slater, E. C. eds.), p. 1. Elsevier, Amsterdam.
Wolley, G., and Courtice, F. C. (1962). *Aust. J. Exp. Biol. Med. Sci.* **40**, 121.

15

CONTROL OF ALBUMIN DEGRADATION IN VIVO AND IN THE PERFUSED LIVER

R. Hoffenberg

M.R.C. CLINICAL RESEARCH CENTRE, HARROW, MIDDLESEX
ATTACHED TO: DIVISION OF BIOPHYSICS,
NATIONAL INSTITUTE FOR MEDICAL RESEARCH,
LONDON, N.W.1, ENGLAND

I. Introduction

The level of plasma albumin may be altered in many disease processes, and attempts are promptly made by the body to restore it to normal and to maintain it within the relatively narrow range found in health.

There are obviously a limited number of ways in which these compensatory adjustments can be made. As a matter of urgency, transfer can take place between the plasma and extravascular compartments, but this can only

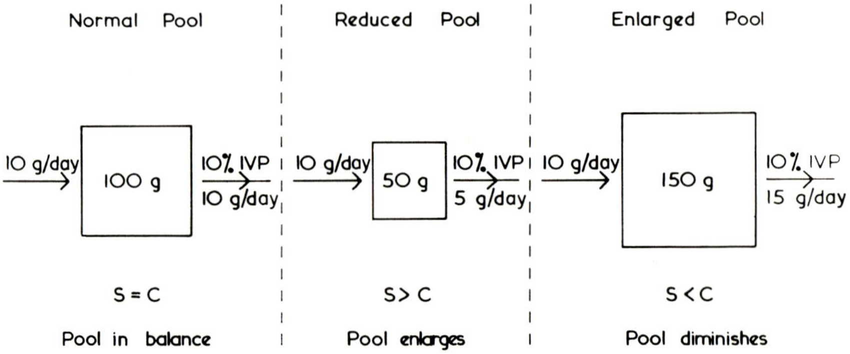

FIG. 1. Effect of catabolism on pool size. If synthesis is maintained, a constant fractional breakdown rate corrects reduction or enlargement of the pool. IVP, intravascular pool; S, synthesis; C, catabolism.

be a temporary measure and other, more long-lasting changes must take place. In the face of significant loss of albumin, e.g., renal or gastrointestinal, the liver seems capable of increasing its production, but little is known of its absolute capacity to respond to extra demands. Evidence suggests that this capacity is limited and increased synthesis cannot be expected to play a major role in processes of restoration. The third, and perhaps most expedient mechanism for adjustment exists in the catabolic rate.

It is frequently held that catabolism of albumin is a first-order process, i.e., that the fractional rate of catabolism is constant, the absolute amount degraded per day varying according to the plasma content (McFarlane, 1964; Birke, 1967). This would indeed provide a powerful means of correcting any abnormality (Fig. 1).

While such a mechanism can usually be shown to exist, significant deviations of fractional catabolic rate may be found under certain circumstances. It is the purpose of this chapter to consider the responses of catabolism and, where appropriate, synthesis to alterations in plasma albumin. It should perhaps be stressed that no satisfactory method exists for short-term (hourly) assessment of catabolic rates, so that immediate acute responses cannot readily be determined.

II. Reduction of Plasma Albumin

Diminution of the plasma albumin pool can be produced through three mechanisms: (1) reduced synthesis; (2) loss; and (3) increased degradation.

A. Reduced Albumin Synthesis

1. *Liver Disease*

Since albumin synthesis is confined to this organ, lowered production would be expected in diseases which affect its function, and reduction of plasma albumin levels is certainly found in most cases of advanced, diffuse hepatocellular disease. But the concentration of albumin does not always correlate with other tests of liver function, and normal or even high rates of synthesis have been found in severe liver disease, even where plasma albumin is reduced (Hasch *et al.*, 1967; Rothschild *et al.*, 1969). This unexpected finding probably reflects the complexity of factors which might influence synthesis in cirrhosis (food intake, especially of protein, absorption, infection) or which might affect the level of plasma albumin (rate of catabolism, plasma volume, distribution through abnormal extravascular pools, or loss through concomitant proteinuria).

The liver is far less concerned with albumin catabolism than with synthesis. Disease of this organ would not therefore be expected to have a major direct influence on the overall catabolic rate, and the almost invariable lowering of catabolism found in cirrhosis must be secondary to changes in pool size (Dykes, 1961, 1968; Tavill *et al.*, 1968; Rothschild *et al.*, 1969). This dependency of catabolism on pool size in cirrhosis is supported by the findings of Dykes (1961) who demonstrated considerable increases in catabolic rate after albumin infusions. Dykes and Wilkinson and Sherlock (1962) found that far greater amounts were needed to boost the plasma levels to normal than had been expected. While this could partly be explained by the presence of inordinately large extravascular pools, hypercatabolism was also shown during infusion periods, in one instance 34 g of albumin being degraded per day (Dykes, 1961).

2. *Reduced Amino Acid Supply*

The effects of *dietary deprivation* on albumin metabolism have been studied in disease and under experimental conditions (Cohen and Hansen, 1962; Picou and Waterlow, 1962; Freeman and Gordon, 1964; Hoffenberg *et al.*, 1966; Kirsch *et al.*, 1968a; Davis and Richmond, 1968), and a correlation can be shown between the level of dietary protein and the plasma albumin concentration. Kirsch *et al.* (1968b) studied the sequence of events following exposure of rats to a nonprotein diet and showed that catabolism was maintained at a normal level for 6–7 days and then began to fall. This fall affected both absolute and fractional rates of breakdown. On refeeding, a prompt increase in synthesis led to restoration of the pool, and catabolism gradually returned to normal (Fig. 2).

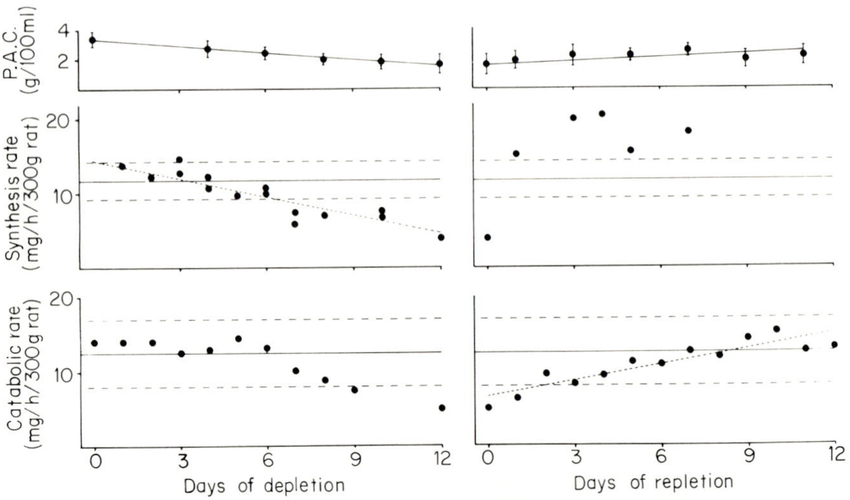

FIG. 2. Serial measurement of plasma albumin concentration (PAC), synthesis and catabolic rates in rats during depletion and repletion of dietary protein. Mean control values ±2 SD are shown.

This sequence provides the answer to the question once asked by Waterlow (1962)—whether the reduced albumin breakdown seen in states of protein malnutrition reflected the lowered protein content of the diet or the state of depletion at the time of the test. Synthesis of albumin, or of any other protein, inherently depends on an adequate supply of amino acids: If, under experimental conditions, this supply is limited, reduced synthesis leads gradually to a fall in plasma and, presumably, total body albumin; only then does a fall in catabolism become apparent, as a reflection of the reduced pool, not directly as a consequence of altered protein intake. This effect was shown in an earlier study (Hoffenberg *et al.*, 1966) in which a reduced catabolic rate followed low protein feeding and persisted after provision of an adequate diet. James and Hay (1968) have largely substantiated these conclusions in malnourished children, who were shown to have significantly lower albumin catabolic rates than recovered children on the same protein intake. Refeeding of a malnourished child was associated with a return to normal catabolic rate only after a period of several weeks.

This change in catabolic rate provides an impressive means of conserving protein in states of dietary deprivation. In kwashiorkor (Cohen and Hansen, 1962) and in experimental protein depletion (Hoffenberg *et al.*, 1966) a fall in fractional catabolic rate enhances this conservation.

Good support for these observations is received from *in vitro* studies. Amino acid activating enzymes in the liver actually increase on low protein feeding (Gaetani *et al.*, 1964; Mariani *et al.*, 1966; Stephen and Waterlow,

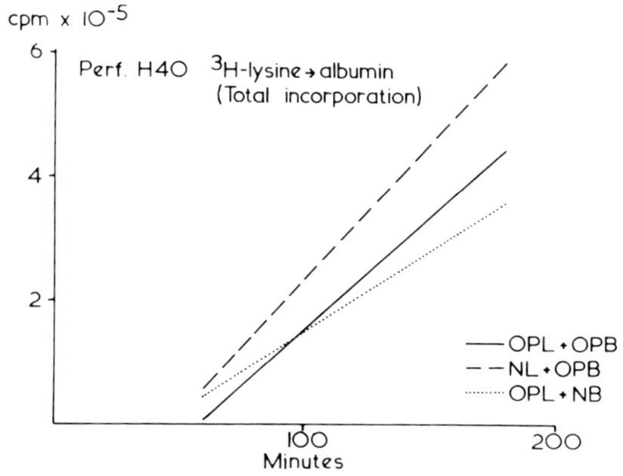

FIG. 3. Incorporation of ^{3}H-lysine into albumin by isolated perfused livers. NL and OPL refer to livers taken from rats fed normal diet and nonprotein diet for two weeks, respectively. NB and OPB refer to perfusate blood taken from either group.

1968), as though the organ is primed to revert promptly to normal protein synthesis when adequate amino acids are supplied. The associated changes in polysome profile and reduced ability of ribosomes to synthesize protein are rapidly reversed by refeeding (Munro, 1966, 1968; Sidransky *et al.*, 1967; Wannemacher *et al.*, 1968). In Fig. 3 the isolated perfused liver, taken from a rat fed a nonprotein diet for 2 weeks, is shown to be capable of incorporating labeled amino acid into protein. After 2 days of refeeding the liver shows an above-normal rate of incorporation (Fig. 4). While the liver thus retains its synthetic potential, there is a fall of enzymes of the urea cycle, which reflects both the quantity and quality of the protein in the diet, and these return more gradually to normal when protein supplies are restored (Schimke, 1962; Muramatsu and Nakata, 1967; Stephen and Waterlow, 1968).

Malabsorptive syndromes similarly reduce the available amino acid supply, and lowered albumin synthesis rates were reported in a series of patients with gastrointestinal disease (Jeejeebhoy, 1964). In this group, in contrast to the finding in protein-losing enteropathy, catabolism was very low. Restoration of the catabolic rate to normal followed high protein feeding, albumin infusions or provision of a gluten-free diet.

B. LOSS OF ALBUMIN

1. *Renal Disease*

Normally a small amount of albumin is filtered through the glomeruli, reabsorbed by the tubules, and possibly degraded within the tubular cells

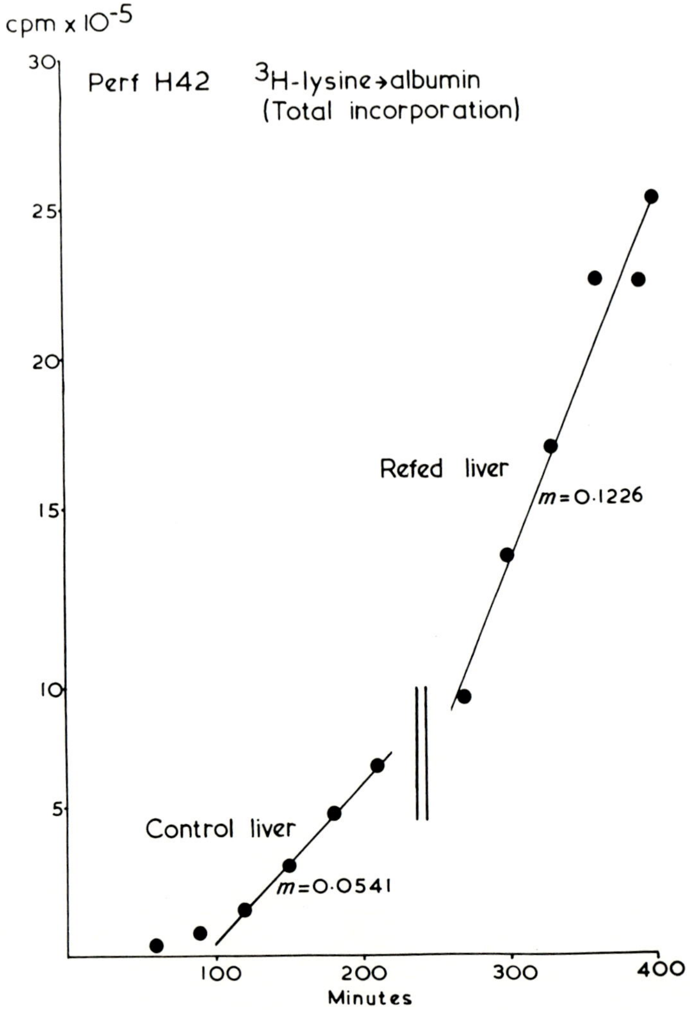

FIG. 4. Incorporation of ³H-lysine into albumin by isolated perfused control liver, replaced by liver taken from rat fed a normal diet for 2 days after 2 weeks of protein deprivation. m = Slope (significance of difference of slopes, $p < 0.001$).

(Rather, 1952; Sellers, 1956; Sussman *et al.*, 1968). This contribution to the overall body catabolism of albumin is probably negligible in health, but may reach significant proportions in albuminuric states, in which marked hypercatabolism has been reported (Gitlin *et al.*, 1956; Kaitz, 1959; Katz *et al.*, 1964, 1967; Jensen *et al.*, 1967; Wochner *et al.*, 1968; Jensen, 1969). It is probable that this increased breakdown takes place within the kidney, as a reduction

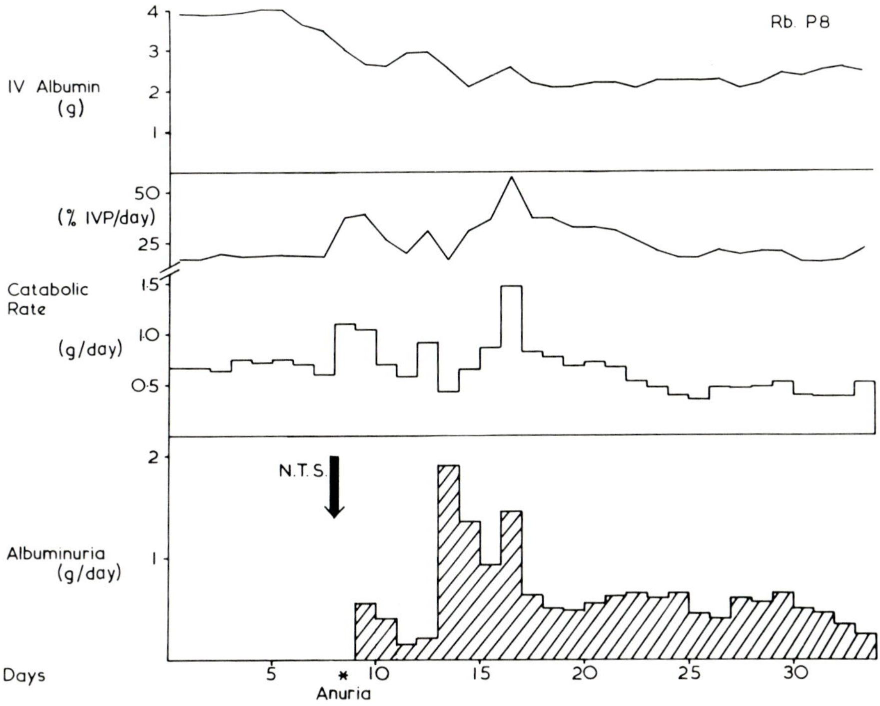

FIG. 5. Serial changes in intravascular albumin pool (IVP) and catabolic rate in rabbit made albuminuric by injection of nephrotoxic serum (NTS). Note considerable increase in fractional and absolute catabolic rates during initial phase of albuminuria with subsequent reduction of absolute amount degraded per day.

has been shown following nephrectomy (Sellers *et al.*, 1961). A direct correlation between the level of albuminuria and the fractional or absolute catabolic rates has not always been demonstrated (Bianchi *et al.*, 1969), possibly because of variations in renal tubular function. Very high fractional catabolic rates may be found despite small pool sizes, and absolute amounts degraded are quite variable. Increased breakdown of albumin may add considerably to the overall depletion that is found. Figure 5 shows serial changes in a rabbit made albuminuric by injection of nephrotoxic serum. For a short time very high fractional and absolute albumin catabolic rates are seen, but these fall gradually as the plasma pool becomes depleted. If part of this terminal breakdown is renal, as a consequence of continued albuminuria, then degradation elsewhere in the body must have been severely depressed.

It should perhaps be stressed that care is needed in the interpretation of catabolic rate data in renal disease, as excretion of iodide released from protein breakdown may be impaired, leading to spuriously high figures for

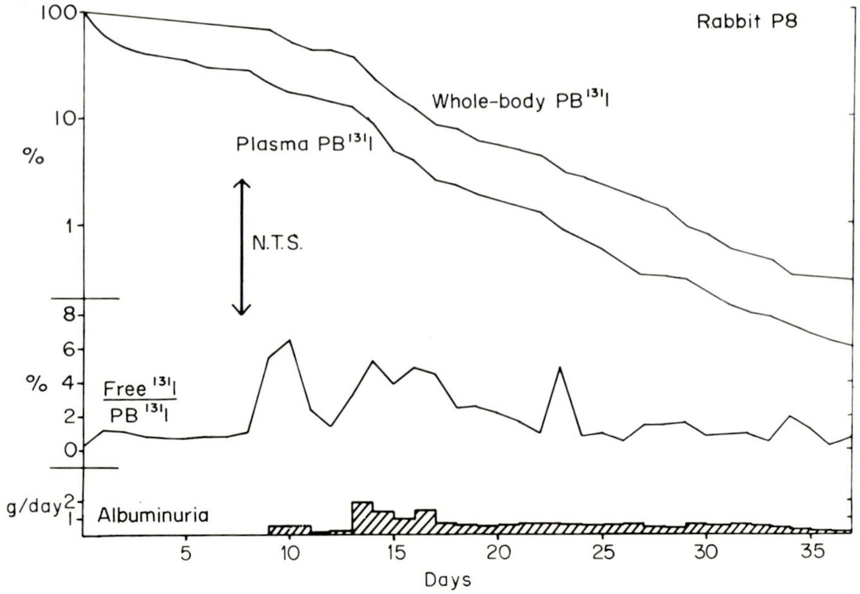

FIG. 6. Changes in whole-body and plasma protein-bound radioactivity following injection of ^{131}I-albumin in a rabbit made albuminuric by nephrotoxic serum (N.T.S.).

whole-body and plasma radioactivity (Vitek *et al.*, 1967). Figure 6 shows that free iodide can account for a significant part of plasma radioactivity and, because of differences in albumin and iodide distribution spaces, for an even higher fraction of total body activity. As urine iodide is correspondingly diminished, misleadingly low catabolic rates will be determined, with apparent large increases when the kidneys finally dispose of the iodide. This error is obviated if fluctuations in total body water iodide are taken into account.

2. *Gastrointestinal Loss*

Several authors have suggested that the gastrointestinal tract accounts for a major part of the total daily albumin degradation in health. While estimates range up to 60% (Jeejeebhoy and Coghill, 1961; Tarver *et al.*, 1961; Glenert *et al.*, 1962; Wetterfors, 1965), others using ^{51}Cr-albumin felt that this mechanism accounted for less than 20% of the overall turnover (Waldmann, 1966). When protein leaks through the gut wall in abnormal amounts, greatly increased catabolic rates are found (Waldmann *et al.*, 1961; Wochner *et al.*, 1968), the effects of which severely aggravate the rate of fall of total body and plasma albumin pools.

3. *Plasmapheresis*

Plasmapheresis provides an experimental means of simulating the protein loss seen in the nephrotic syndrome of protein-losing enteropathy, without the complications arising from enhanced local catabolic activity. Matthews (1961), Hoffenberg *et al.* (1966), and Andersen and Rossing (1967) have all demonstrated a fall in albumin catabolism, including the fractional rate following depletion of body albumin stores as a result of repeated plasmapheresis. An abrupt fall in catabolism is also seen as an acute response to a single bleed (Table I) and much of the albumin removed will be restored by this means. This prompt reaction contrasts with the slower adaptation seen after exposure to a low protein diet.

TABLE I

EFFECT OF SINGLE PLASMAPHERESIS[a] ON CATABOLISM OF ^{131}I-ALBUMIN AND SYNTHESIS[b]

Rabbit 227		Control period (day 1–6)	Day of plasmapheresis (day 7)	Post plasmapheresis period (day 8–12)
Catabolic rate	% IVP/day	19.9[c] (16.5–26.8)[d]	12.1	19.9[c] (17.1–23.3)
	mg/day	795[c]	459	756[c]
Synthesis rate	% IVP/day		24.0	
	mg/day		907	

[a] Removal of 1200 mg albumin in 35 ml plasma.
[b] Measured by $^{14}CO_2$ method.
[c] Mean.
[d] Range.

4. *Adaptation to Loss or Removal of Albumin*

These contributions to the normal catabolic process in states of loss via the kidney or gut obscure the relationship between plasma albumin pool and degradation rate that would be expected were the latter a first-order process. The body is thus deprived of a simple and theoretically ideal method of conserving protein, which is still seen to operate after removal of albumin by plasmapheresis. In all three states—nephrotic syndrome, protein-losing enteropathy, and after plasmapheresis—it seems that synthesis of albumin may increase, although the extent to which it does is apparently limited and may vary with factors such as anorexia, malabsorption, and associated liver

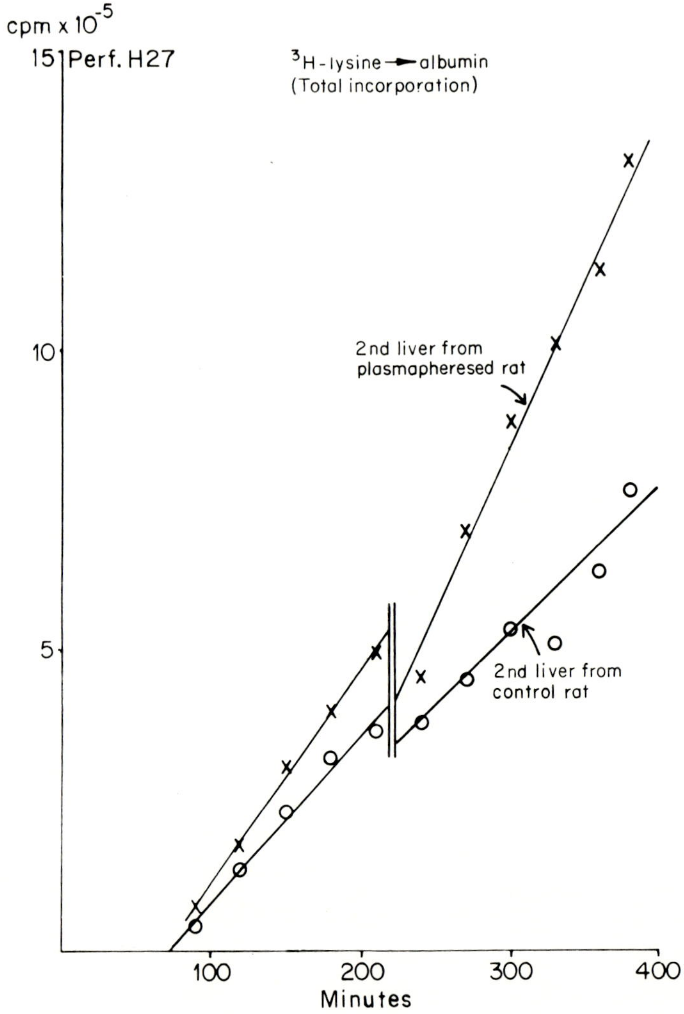

FIG. 7. Incorporation of ^{3}H-lysine into albumin by isolated perfused liver, control livers being replaced after 200 min by livers from a plasmapheresed and a normal rat. Plasmapheresis consisted of removal of 6 ml plasma on day preceding perfusion.

dysfunction (Bauman *et al.*, 1955; Blahd *et al.*, 1955; Gitlin *et al.*, 1956; Freeman and Cohen, 1960; Matthews, 1961, 1965; Hoffenberg *et al.*, 1966; Andersen and Rossing, 1967; Wochner *et al.*, 1968). Rats from which about 6 ml plasma had been removed on the previous day showed mean albumin synthesis rates, measured by the $^{14}CO_2$ method (McFarlane, 1963; McFarlane *et al.*, 1965), of 15.9 mg/hr compared to a control figure of 10.7 mg/hr (Hoffenberg and Black, unpublished). A greater-than-normal capacity of the isolated perfused liver to synthesize albumin has been shown in experimental nephrosis (Marsh and Drabkin, 1958, 1960; Katz *et al.*, 1967), and may also be seen after plasmapheresis (Fig. 7).

In summary, a fall in catabolic rate would ideally compensate for losses of albumin and is seen to play a significant role after plasmapheresis. When protein is lost via the kidneys or gut, however, local hypercatabolism obscures or eclipses this compensation and increased synthesis is the major means of restoring the body pool. Since this appears to have a limited capacity, a rapid lowering of albumin pool is observed.

C. Increased Degradation

It is unlikely that a primary increase in catabolism ever causes severe reduction of the body albumin pool. In thyrotoxicosis and Cushing's disease or after ACTH or cortisone treatment, albumin breakdown is slightly increased, but this is readily compensated by synthesis, which keeps the pool more-or-less constant (Rothschild *et al.*, 1957, 1961). An increase in catabolism seen in pregnancy is countered by an increase in synthesis, although a slight fall in plasma albumin may be found terminally (Masters *et al.*, 1969). In acute or active chronic disease, synthesis may fail to compensate for an increase in degradation and a slightly lower plasma albumin pool may result (Jarnum and Lassen, 1961). Hypercatabolism and a low serum albumin are reported in some women with idiopathic edema, but the relationship is not fully established (Gill *et al.*, 1969).

D. Hypo- or Analbuminaemia

In this rare condition it is interesting to note that extremely low catabolic rates have been reported. Freeman (1969) quotes a figure of 1.4% of the intravascular pool degraded per day, which was not altered by acute infusion of albumin. After 2 years of infusion, an increase in pool size, concentration, and fractional catabolic rate was observed. This lack of immediate response suggests a delayed adaptation of catalytic enzymes, analogous perhaps to the refeeding of protein-deprived children or animals discussed earlier.

III. Increased Albumin Pool

Except in states of hemoconcentration or as a result of albumin infusion, hyperalbuminaemia is not known to occur. Theoretically, it could arise from increased synthesis, decreased catabolism, or net shifts of albumin from the extravascular to the intravascular pool. In practice, these changes are found only as reactions to other disturbances, and a primary increase in albumin pool has not been described.

The response to two circumstances is worth considering.

A. High Protein Feeding

One might anticipate an increase in albumin synthesis on a high protein diet. This has in fact been shown by a number of authors (Iber *et al.*, 1958; Jeffay and Winzler, 1958; Yuile *et al.*, 1959; Holt *et al.*, 1962). In all cases the overproduction has been met by increased breakdown, usually by maintenance of a constant fractional rate, but in some cases by a small increase in this, as well (Iber *et al.*, 1958; Hoffenberg *et al.*, 1966). Schimke (1962) showed adaptive changes of all urea-cycle enzymes, which increased considerably in activity when a high protein diet was exhibited.

It seems that overproduction of albumin as a result of feeding is readily compensated by an increase in breakdown.

B. Albumin or Plasma Infusions

Infusion of albumin into human subjects or experimental animals is associated with increased absolute or fractional rates of albumin breakdown

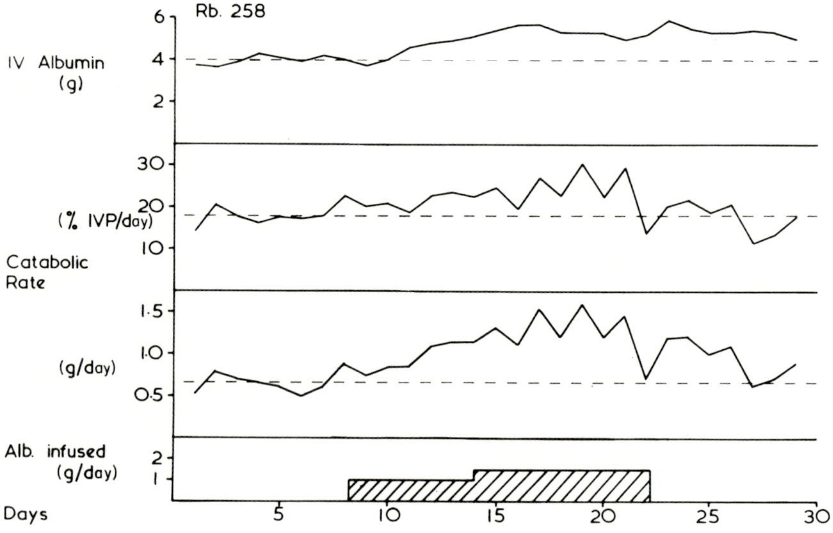

Fig. 8. Serial changes in intravascular albumin and catabolic rate in a rabbit infused with albumin for 14 days. Broken line indicates mean values taken from initial control period.

(Friedberg, 1963; Rothschild *et al.*, 1964; Andersen and Rossing, 1967). Figure 8 shows the calculated breakdown rates for albumin in a rabbit infused over a 14-day period. During this time, "surplus" catabolism could account for not more than one-third of the amount infused, but maintenance of a high breakdown rate after infusions had stopped further compensated for the additional load. It is significant in this animal that the plasma pool was not much increased. While deposition in extravascular sites will partly explain this, it is likely that synthesis was reduced during the infusion period. Albuminuria did not occur (see below).

The isolated rat liver has been used to study catabolism of plasma albumin (Cohen and Gordon, 1958; Gordon, 1962; McFarlane, 1964). Although only about 10–15% of normal rat albumin catabolism can be accounted for by the liver under these conditions, the system can be used to examine the effect of acute changes in the perfusate pool. Figure 9 demonstrates that constancy of fractional breakdown rate is maintained after addition of albumin in an amount sufficient to double the concentration. This helps to dispose of the extra load, although synthesis also appears to be acutely suppressed.

In the dog, albuminuria may be induced by continued infusion of plasma (Terry *et al.*, 1948) and in man by infusion of albumin (Waterhouse and Holler, 1948). Figure 10 shows the effect of massive plasma infusions given to a hemophiliac patient in an attempt to boost his antihemophilic-globulin levels prior to essential surgery. An increase in catabolism was promptly seen, which rose to extreme heights after a few days when he developed an "overflow" albuminuria (Mibashan and Hoffenberg, unpublished data). This response in humans and dogs appears to provide an additional safeguard against the development of hyperalbuminaemia.

From the human and rabbit data referred to, and in the report of Andersen

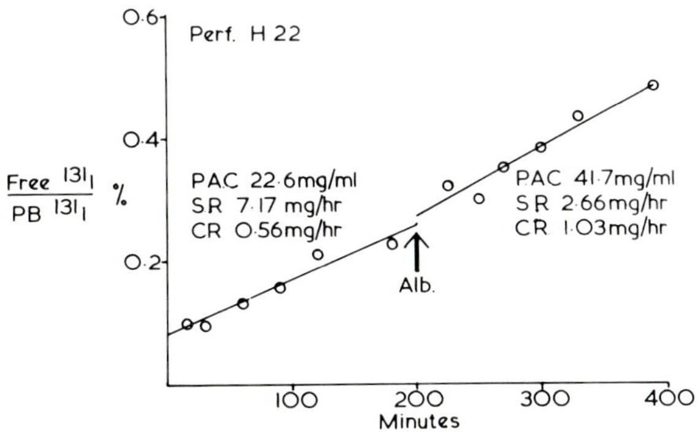

FIG. 9. Catabolism of ^{131}I-albumin by isolated perfused liver before and after addition of albumin to perfusate. PAC, Plasma albumin concentration; SR, synthesis rate (measured by $^{14}CO_2$ method); CR, catabolic rate.

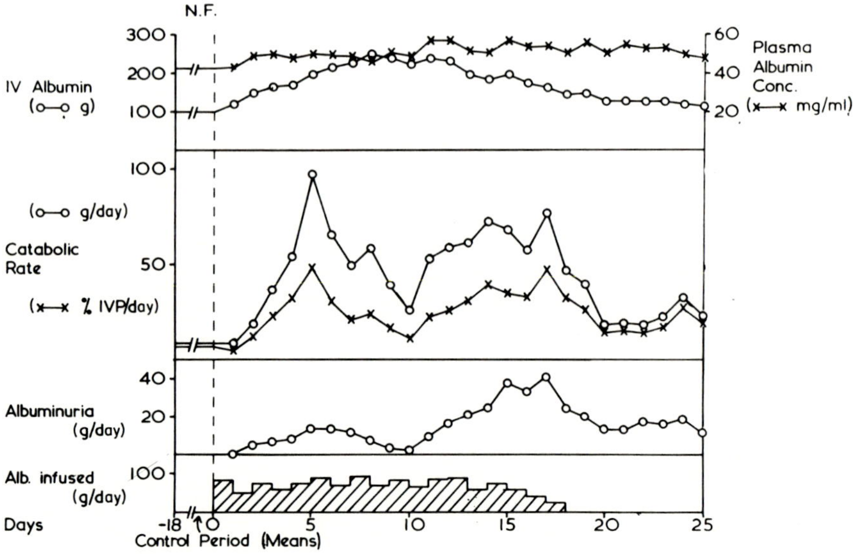

FIG. 10. Effect of massive plasma infusions in human subject. Control values are shown for 18 days prior to start of infusions.

and Rossing (1967), albumin synthesis appeared to be depressed. Yet, neither Rothschild *et al.* (1964) in rabbits, nor Tavill *et al.* (1968) in human subjects were able to confirm a fall after albumin infusion. Even if synthesis were to be totally suppressed, the major process for disposing of massive additional loads of albumin is catabolism, and the ability of the catabolic rate to respond to this demand is remarkable.

IV. Conclusions

In most cases of perturbation of the plasma albumin pool, prompt compensation is achieved by alterations in catabolism, dependent largely on maintenance of a fixed fractional rate. In extreme depletion or overload this fractional rate can also alter to provide further, more profound adjustments. A deleterious change is seen in renal or gastrointestinal protein loss, in which very high rates of catabolism may obscure the expected response to reduction of the albumin pool. Variations in synthesis rate may also contribute to the compensatory process, increasing in the face of depletion and being suppressed by albumin overloads; but many other factors may influence the liver's capacity to produce albumin, and it cannot match the ability of the catabolic sites to compensate for change. In this chapter reference has been made to daily albumin breakdown ranging from a few milligrams to 97 g, and of fractional rates varying between 1.4% of the plasma pool per day and 57%. It is this versatility which makes adjustment of catabolism such a powerful regulatory mechanism for albumin homeostasis.

Much has been written about the site of albumin catabolism, the identification of which has eluded investigators. From this survey it seems that the kidney, the gut, and the isolated perfused rat liver are all capable of participating in the normal process and their individual contributions can be enhanced by increasing the flow of albumin through them. The overall rate of catabolism is augmented by adding albumin to the plasma, perhaps by increasing the flow through widely-distributed sites of breakdown throughout the body.

ACKNOWLEDGMENTS

I wish to thank the Editors of *Nature* for permission to reproduce Fig. 2 (Kirsch *et al.*, 1968b).

REFERENCES

Andersen, S. B., and Rossing, N. (1967). *Scand. J. Clin. Lab. Invest.* **20**, 183.
Bauman, A., Rothschild, M. A., Yalow, R. S., and Berson, S. A. (1955). *J. Clin. Invest.* **34**, 1359.
Bianchi, R., Federighi, G., Giagnoni, P., Giordani, R., Navalesi, R., and Donato, L. (1969). *In* "Physiology and Pathophysiology of Plasma Protein Metabolism" (G. Birke, R. Norberg, and L.-O. Plantin, eds.), p. 195. Macmillan (Pergamon), New York.
Birke, G. (1967). *Protides Biol. Fluids Proc. Colloq.* **14**, 253.
Blahd, W. H., Fields, M., and Goldman, R. (1955). *J. Lab. Clin. Med.* **46**, 747.
Cohen, S., and Gordon, A. H. (1958). *Biochem. J.* **70**, 544.
Cohen, S., and Hansen, J. D. L. (1962). *Clin. Sci.* **23**, 351.
Davis, M. M., and Richmond, J. E. (1968). *Amer. J. Physiol.* **215**, 366.
Dykes, P. W. (1961). *Quart. J. Med.* **30**, 297.
Dykes, P. W. (1968). *Clin. Sci.* **34**, 161.
Freeman, T. (1969). *In* "Physiology and Pathophysiology of Plasma Protein Metabolism" (G. Birke, R. Norberg, and L.-O. Plantin, eds.), p. 75. Macmillan (Pergamon), New York.
Freeman, T., and Cohen, S. (1960). *Protides Biol. Fluids Proc. Colloq.* **8**, 230.
Freeman, T., and Gordon, A. H. (1964). *Clin. Sci.* **26**, 17.
Friedberg, W. (1963). *Amer. J. Physiol.* **204**, 501.
Gaetani, S., Paolucci, A. M., Spadoni, M. A., and Tomassi, G. (1964). *J. Nutr.* **84**, 173.
Gill, J. R., Waldmann, T. A., Delea, C. S., and Bartter, F. C. (1969). *J. Clin. Invest.* **48**, 28a
Gitlin, D., Janeway, C. A., and Farr, L. E. (1956). *J. Clin. Invest.* **35**, 44.
Glenert, J., Jarnum, S., and Riemers, S. (1962). *Acta Chir. Scand.* **124**, 63.
Gordon, A. H. (1962). *Biochem. J.* **82**, 531.
Hasch, E., Jarnum, S., and Tygstrup, N. (1967). *Acta Med. Scand.* **182**, 83.
Hoffenberg, R., Black, E., and Brock, J. F. (1966). *J. Clin. Invest.* **45**, 143.

Holt, L. E., Halac, E., and Kajdi, C. N. (1962). *J. Amer. Med. Ass.* **181**, 699.
Iber, F. C., Nassau, K., Plough, I. C., Berger, F. M., Meroney, W. H., and Fremont-Smith, K. (1958). *J. Clin. Invest.* **37**, 1442.
James, W. P. T., and Hay, A. M. (1968). *J. Clin. Invest.* **47**, 1958.
Jarnum, S., and Lassen, N. A. (1951). *Scand. J. Clin. Lab. Invest.* **13**, 357.
Jeejeebhoy, K. N. (1964). *Brit. Med. J.* **1**, 30.
Jeejeebhoy, K. N., and Coghill, N. F. (1961). *Gut* **2**, 123.
Jeffay, H., and Winzler, R. J. (1958). *J. Biol. Chem.* **231**, 111.
Jensen, H. (1969). *In* "Physiology and Pathophysiology of Plasma Protein Metabolism" (G. Birke, R. Norberg, and L.-O. Planton, eds.), p. 213. Macmillan (Pergamon), New York.
Jensen, N., Rossing, N., Andersen, S. B., and Jarnum, S. (1967). *Clin. Sci.* **33**, 445.
Kaitz, A. L. (1959). *J. Lab. Clin. Med.* **53**, 186.
Katz, J., Sellers, A. L., and Bonorris, G. (1964). *J. Lab. Clin. Med.* **64**, 709.
Katz, J., Bonorris, G., Okuyama, S., and Sellers, A. L. (1967). *Amer. J. Physiol.* **212**, 1255.
Kirsch, R. E., Brock, J. F., and Saunders, S. J. (1968a). *Amer. J. Clin. Nutr.* **21**, 820.
Kirsch, R., Frith, L. Black, E., and Hoffenberg, R. (1968b). *Nature (London)* **217**, 578.
McFarlane, A. S. (1963). *Biochem. J.* **89**, 277.
McFarlane, A. S. (1964). *In* "Mammalian Protein Metabolism" (H. N. Munro and J. B. Allison, eds.), Vol. 1, p. 297. Academic Press, New York.
McFarlane, A. S., Irons, L., Koj, A., and Regoeczi, E. (1965). *Biochem. J.* **95**, 536.
Mariani, A., Migliaccio, P. A., Spadoni, M. A., and Ticca, M. (1966). *J. Nutr.* **90**, 25.
Marsh, J. B., and Drabkin, D. L. (1958). *J. Biol. Chem.* **230**, 1073.
Marsh, J. B., and Drabkin, D. L. (1960). *Metab.* **9**, 946.
Masters, C. L., Bignold, L. P., and Morgan, E. H. (1969). *Amer. J. Physiol.* **216**, 876.
Matthews, C. M. E. (1961). *J. Clin. Invest.* **40**, 603.
Matthews, C. M. E. (1965). *In* "Radioaktive Isotope in Klinik und Forschung" (K. Kellinger and R. Höfer, eds.), p. 240. Urban & Schwarzenberg, Munick.
Munro, H. N. (1966). *Nutr. Diet.* **8**, 179.
Munro, H. N. (1968). Fed. Proc. *Fed. Amer. Soc. Exp. Biol.* **27**, 1231.
Muramatsu, K., and Nakata, I. (1967). *Agr. Biol. Chem.* **31**, 958.
Picou, D., and Waterlow, J. C. (1962). *Clin. Sci.* **22**. 459.
Rather, L. J. (1952). *Medicine* (Baltimore) **31**, 357.
Rothschild, M. A., Bauman, A., Yalow, R. S., and Berson, S. A. (1957). *J. Clin. Invest.* **36**, 422.
Rothschild, M. A., Oratz, M., Wimer, E., and Schreiber, S. S. (1961). *J. Clin. Invest.* **40**, 545.
Rothschild, M. A., Oratz, M., Evans, C., and Schreiber, S. S. (1964). *J. Clin. Invest.* **43**, 1874.
Rothschild, M. A., Oratz, M., Zimmon, D., Schreiber, S. S., and Weiner, I. (1969). *J. Clin. Invest.* **48**, 344.
Schimke, R. T. (1962). *J. Biol. Chem.* **237**, 459.
Sellers, A. L. (1956). *Arch. Intern. Med.* **98**, 801.
Sellers, A. L., Katz, J., and Rosenfeld, S. (1961). *Nature (London)* **192**, 562.
Sidransky, H., Bongiorno, M., Sarma, D. S., and Verney, E. (1967). *Biochem. Biophys. Res. Commun.* **27**, 242.
Stephen, J. M. L., and Waterlow, J. C. (1968). *Lancet* **i**, 118.
Sussman, H. C., Fowler, B., and Findley, T. (1968). *Metab. Clin. Exp.* **17**, 336.
Tarver, H., Armstrong, F. B., Debro, J. R., and Margen, S. (1961). *Ann. N.Y. Acad. Sci.* **94**, 23.
Tavill, A. S., Craigie, A., and Rosenoer, V. M. (1968). *Clin. Sci.* **34**, 1.

Terry, R., Hawkins, D. R., Church, E. H., and Whipple, G. H. (1948). *J. Exp. Med.* **87**, 561.
Vitek, F., Bianchi, R., and Federighi, G. (1967). *J. Nucl. Biol. Med.* **11**, 139.
Waldmann, T. A. (1966). *Gastroenterology* **50**, 422.
Waldmann, T. A., Steinfeld, J. L., Dutcher, T. F., Davidson, J. D., and Gordon, R. S. (1961). *Gastroenterology* **41**, 197.
Wannemacher, R. W., Cooper, W. K., and Yatvin, M. B. (1968). *Biochem. J.* **107**, 615.
Waterhouse, C., and Holler, J. (1948). *J. Clin. Invest.* **27**, 560.
Waterlow, J. C. (1962). *Lancet* **ii**, 1279.
Wetterfors, J. (1965). *Acta Med. Scand. Suppl.* **430**, 1.
Wilkinson, P., and Sherlock, S. (1962). *Lancet* **ii**, 1127.
Wochner, R. D., Weissman, S. M., Waldmann, T. A., Houston, D., and Berlin, N. I. (1968). *J. Clin. Invest.* **47**, 971.
Yuile, C. L., Lucas, F. V., Olson, J. P., and Shapiro, A. B. (1959). *J. Exp. Med.* **109**, 173.

V

Immunoglobulin Metabolism

16

GENETIC REGULATION OF IMMUNOGLOBULIN LEVELS IN MAN

Edward C. Franklin

DEPARTMENT OF MEDICINE, RHEUMATIC DISEASES STUDY GROUP,
NEW YORK UNIVERSITY SCHOOL OF MEDICINE,
NEW YORK, NEW YORK

I. Introduction

The immunoglobulins are a group of structurally related proteins produced by plasma cells and lymphocytes. They provide antibody activity against an almost infinite number of antigens to which the individual is normally exposed. Since it is now generally accepted that all antibodies differ from one another in their primary amino acid sequence, the immunoglobulins require a degree of heterogeneity not encountered in any other biologic system to adequately perform their biologic function (Cohen and Milstein, 1967).

Studies during the past 10 or 15 years have clearly resolved some of the questions related to the mechanisms responsible for generating such diversity. Thus, there is little evidence in support of "instructive" theories of antibody formation which endow antigen or some derivative thereof with the capacity of inducing in the antibody-forming cell the synthesis of a new and unique antibody molecule. Instead, the prevailing opinion now favors the idea that, in every individual, there are cells potentially capable of synthesizing all antibody molecules that will be required during the life of the individual. The role of the antigen then is simply to "select" the appropriate cells and to cause them to divide and to produce large amounts of antibody specifically reactive with the inciting antigen (Burnet, 1969). While the role of antigen then is largely "selective," it remains to be determined whether the necessary diversity is generated at the level of the germ line which may contain a very large number of genes which code for specific antibodies, or whether there is a small number of genes which is subject to somatic mutation.

Since most, if not all, γ-globulin consists of antibody, it is obvious that the type of antibody protein that the individual is able to produce is genetically controlled, but that the amounts actually synthesized are largely determined by environmental factors, especially exposure to antigen. This is best illustrated by comparing γ-globulin levels in normal animals with those in germ-free lines; the latter are almost completely agammaglobulinemic as long as they exist in a sheltered environment. What little γ-globulin they produce is probably stimulated by antigens transmitted by non-infectious agents, such as dietary constituents. Once exposed to a natural environment, these animals then respond with a vigorous immune response which results in a marked increase in γ-globulin production (Gustafsson and Laurell, 1959). Further support for this modulation by environmental factors is given by a comparison of the γ-globulin levels of normal healthy men and those recovering from infections or who are exposed to an unusually contaminated environment. In the latter, for example, in individuals residing in regions where malaria is endemic, there are often marked elevations of γ-globulin, most of which is antibody directed to the offending organisms to which the individual has been exposed.

II. Genetic Control of the Immune Response

In recent years, as a result of studies of inbred strains of mice and guinea pigs immunized with a number of synthetic polypeptide antigens and a simple small protein, insulin, it has become obvious that, under these well controlled conditions, the immune response to a given antigen is genetically controlled (McDevitt and Benacerraf, 1969). These studies have demonstrated distinct differences in both the qualitative and quantitative responses to these well defined antigens. Thus, the response to conjugates of a hapten to poly-L-lysine or a copolymer of Lys—Ala with either TyrGlu or HisGlu appear to be under the control of a separate autosomal dominant gene, now known as *Ir–1* (Immune Response 1) gene. It was somewhat surprising in the case of poly-L-lysine antigen that a "nonresponder" strain can produce antibody to the conjugates if they are coupled to an immunogenic carrier like albumin, thus indicating that, in this case, the specificity of the genetic difference is for the carrier. Similarly, in the case of mice, the ability to produce antibody to certain branched multichain synthetic polypetides is under the control of an autosomal dominant gene linked to, and possibly identical to an allele at the H_2 locus. Similar studies with a natural protein (insulin) demonstrated that two strains of guinea pigs responded to different determinants on the molecule.

III. Structure of Human Immunoglobins

Before considering the genetic control of the immunoglobulins, we must first look at their structure. In man there exist five well defined immunoglobulin classes, known as γG, A, M, D, and E in decreasing order of concentration (Cohen and Milstein, 1967) (Table I). While all of them possess

TABLE I

PROPERTIES OF FIVE CLASSES OF IMMUNOGLOBULINS IN MAN

	γG	γA	γM	γD	γE
Molecular weight	145,000	±160,000	±900,000	±160,000	±200,000
Sedimentation coefficient (S)	7	7 (polymers)	19	7	8
Approximate concentration (mg %)	1200	200	100	3	0.1
Subclasses	γ1, 2, 3, 4	α1, 2	μ1, 2	?	?
Genetic factors	*Gm*	*Am*	?	?	?
"Paraprotein"	G. myeloma	A myeloma	Macroglobulin	D myeloma	E myeloma

antibody activity, two of these occupy unique positions: One, the γA globulin, provides antibody activity in the external secretions (Franklin and Frangione, 1969); the other, the recently discovered γE fraction, contains the reaginic antibodies associated with a variety of allergic conditions (Ishizaki *et al.*, 1966). In the case of γG, γM, and γA globulins, subclasses (4, 2, and 2, respectively) are now recognized on the basis of their chemical and antigenic properties (Franklin and Frangione, 1969). It seems likely, however, as will become apparent later, that these subclasses will eventually be considered to be closely related classes of immunoglobulins on the basis of their structural differences and unique genetic control. In time, therefore, there will probably be an even larger number of recognized immunoglobulin classes. All of the immunoglobulins are synthesized by plasma cells and lymphocytes (Zucker-Franklin, 1969). There is now abundant evidence to suggest that a plasma cell can synthesize only a single class of γ-globulin at any point in its existence (Pernis, 1967). Since the basic structure of all immunoglobulins is similar, it seems likely that the time needed for the synthesis and secretion of each immunoglobulin class must be roughly the same. Therefore, the concentration of a particular immunoglobulin class is directly proportional to the number of cells synthesizing it and inversely proportional to its catabolic rate. Thus, γG-, A-, and M-producing cells are much more common than D- and E-producing cells (Pernei, 1967).

A great deal has been learned about the structure of the immunoglobulins during the past 15 years, and several of these proteins have now been completely sequenced (Cohen and Milstein, 1967). Much of the work has been done with myeloma proteins which are now generally considered to serve as models of antibodies and to possess antibody activity when properly tested (Eisen *et al.*, 1967). The major feature to note is that each class of γ-globulin, like hemoglobin, is a complex molecule composed of two types of polypeptide chains, each under separate genetic control. One of these, the light chain, is common to all classes of immunoglobulins. There are at least two types of light chains, known as κ and λ, and any γ-globulin molecule has two identical light chains in its basic 7 S, 150,000–200,000 molecular weight subunit (Cohen and Milstein, 1967). The other, known as the heavy chain, differs for each immunoglobulin class and, thus, is responsible for the chemical and biological properties characteristic of each class. Because of these structural differences, the heavy chains are denoted by the Greek letter corresponding to the class: γ, α, μ, δ, and ε for γG, A, M, D, and E, respectively. Here again each 7 S subunit is composed of two identical heavy chains (Cohen and Milstein, 1967).

As a result of sequence studies of a number of light and heavy chains in recent years, a rather unique structural arrangement of the immunoglobulin chains has been established (Fig. 1). It would appear that each type of chain is composed of a region of constant composition which is the same for all proteins of that type and accounts for the class specific properties, and another region of variable composition which is largely involved in providing antibody specificity (Cohen and Milstein, 1967). In the case of the light chain, the variable region constitutes the amino terminal half of the molecule; for the heavy chain, on the other hand, it makes up the amino terminal quarter (Edelman *et al.*, 1967). A major question facing immunologists at the moment is to determine a means whereby this limited variability can be achieved in a structure that appears to be a single chain. The possibility has

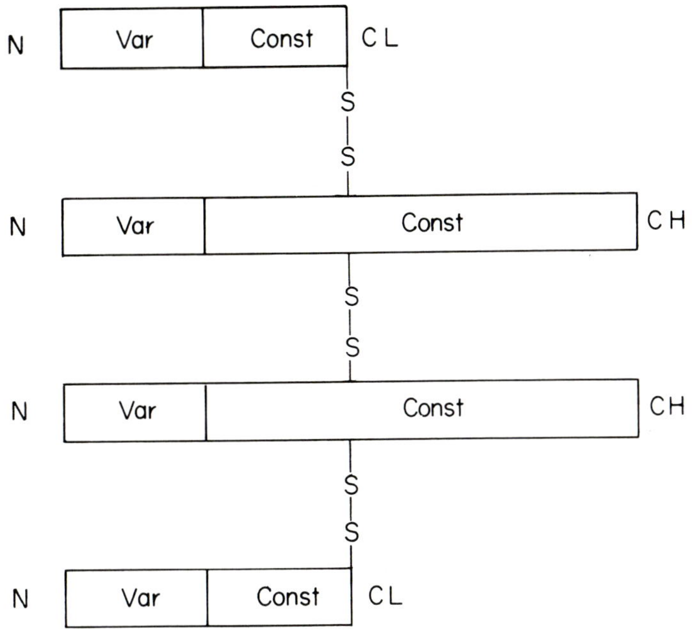

FIG. 1. Schematic diagram of immunoglobulin molecule. H, heavy chain; L, light chain; N, amino terminal; C, carboxy terminal; Var, variable; Const, constant; S—S—, disulfide bond.

been seriously raised that the variable and constant regions are under separate genetic control, and there is abundant genetic evidence in support of this view (Milstein, 1967). However, all chemical and biosynthetic studies to date favor the idea that both the light and heavy chains are a single continuous polypeptide chain (Fleischman, 1967). Thus, if fusion of two separate chains were to occur, it would have to be at the level of the DNA.

IV. Genetic Polymorphism of Immunoglobulins

The concentration in serum of each class and subclass of immunoglobulin is approximately the same for all normal individuals in a given population. However, as a consequence of the discovery of genetic polymorphism of the γ chains by Grubb in 1956, it became apparent that the distribution of different genetic Gm factors varied in different populations, indicating the existence of subtle qualitative differences which are under genetic control among individuals with grossly similar immunoglobulins (Grubb and Laurell, 1956). Since then, polymorphism and similar variability among individuals has been best delineated for the γ chains (Gm factors) and the κ chains (Inv factors) (Franklin and Fudenberg, 1969). Recently, the first allotypic specificity of the α_2 chains (Am) has also been described (Vyas and Fudenberg, 1969; Kunhel *et al.*, 1969). While allotypic factors for the μ, δ, ε, and λ chains remain to be discovered, it seems probable that they also exist and that they will soon be found; in fact, some of these have been delineated in mice and rabbits.

The method originally used to detect these factors depended on the ability of a γ-globulin of a given allotypic specificity to inhibit the agglutination by an anti-γ-globulin of a human Rh^+ cell coated by an incomplete γG antibody (Grubb and Laurell, 1956). More recently, the system has been modified and simplified by coating cells with myeloma proteins of a given specificity and, in the case of the Am system, the first such factor was discovered with an antiserum from a multitransfused individual and cells coated with a γA_2 myeloma protein. For some of the factors, precipitating systems employing antisera prepared in rabbits or primates are now available and greatly simplify genetic typing and quantitation (Franklin and Fudenberg, 1969).

Much has been learned about these genetic factors in the past 10 years. The Inv locus, which controls the κ chain, has two major alleles known as a and b. These differ by a single amino acid residue at position 191 [leucine for Inv(a) and valine for Inv(b)] (Franklin and Fudenberg, 1969). They express themselves in the carboxy terminal half of the light chain which is also known as the constant half. In a recent study, light chains prepared from γG-globulin from a number of normal heterozygous subjects contained approximately equal quantities of these two genetic factors (Terry *et al.*, 1969). The *Oz* factors on the λ chain, which at first glance appeared analogous to the Inv factors, proved to be different. They did not represent an example of genetic polymorphism, since all normal individuals studied were heterozygous (Ein, 1968).

The situation in the case of the γ chains proved to be somewhat more complex since at least 23 serologic factors are now known (Franklin and Fudenberg, 1969). Initially, it was assumed that these represented multiple alleles at a single γ-chain locus. However, with the delineation of four γ-chain subclasses and the realization that the different genetic *Gm* factors were associated with different subclasses, it was soon shown that there exists a series of at least three and probably four closely linked γ-chain loci, one for each γ-chain subclass. The major *Gm* factors associated with the γ1 locus are *az*, *non-a*, and *f*; at the γ2 locus, we have $Gm(n^+)$ and (n^-) and at the γ3 locus, *Gm b* and *g*. The γ4 locus has not yet been delineated (Kunkel, 1969). While the precise structural basis for all of these factors is not yet known, there are known to be at least two amino acid differences in a single pentapeptide between the *Gm*(*a*) and "*non-a*" molecules (Franklin and Fudenberg, 1969). As a result of some elegant genetic studies, Kunkel and his group have recently postulated that the three γ-chain loci are located in close proximity on a single chromosome with $\gamma 1$ next to $\gamma 3$ followed by $\gamma 2$ and probably 4 (Kunkel, 1968).

Studies with homogeneous myeloma proteins leave little doubt that a single protein carries only the factors controlled by one allele at the appropriate locus, and that in the heterozygous subject there must then be two types of molecules, each carrying the product of one allele (Franklin and Fudenberg, 1969). Similarly, it seems certain that a cell can synthesize only a single allotypic specificity at each locus, i.e., that only one allele at a particular locus is expressed (Curtain and Baumgarten, 1966). This is unusual for proteins controlled by autosomal loci and more clearly resembles products under the control of X-linked genes since one of the X chromosomes is always inactivated, as first pointed out by Lyon (Lyon, 1962). The reason for this unusual behavior of the *Gm* factors, as well as certain other features involved in the genetic control of the immunoglobulin chains, remains to be elucidated.

In the normal population there is little evidence to suggest that the genetic makeup of an individual can in any way effect the concentration of immunoglobulins or his ability to produce antibodies. The only exception to date appears to be in the $\gamma 3$ class where individuals who were $Gm(b^{+})$ appeared to have higher concentrations of $\gamma 3$ molecules than those who were $Gm(b^{-})$, a finding suggesting that they may be more responsive to antigenic and other stimuli resulting in increased production of $\gamma 3$ globulin antibodies (Yount, *et al.*, 1967). Of some interest in this regard are several families with low $\gamma 3$ levels as a result of a heterozygous state with deletion of the $\gamma 3$ cistron, a mechanism which, especially in the homozygous state, may lead to low levels of certain immunoglobulins (Yount *et al.*, 1969). Whether the same features will also be found for some other immunoglobulin classes remains to be determined. However, if it does occur, such a mechanism may be important in determining an individual's resistance to infection. It would appear that a few cases of hypogammaglobulinemia may have selective subclass deficiencies of this nature (Terry, 1968).

V. Control of Immunoglobulin Synthesis in Health and Disease

What factors then regulate immunoglobulin synthesis in normal man? The genetic capacity to form antibodies exists in all at birth, but remains dormant until challenged by antigen. It is too early to say whether qualitative differences like those in the $\gamma 3$ locus exist for other specificities as well, and as yet it is not possible to determine whether quantitative differences among different ethnic groups are purely environmental or also in part genetic. Answers to these questions will have to wait quantitative studies of different allotypic specificities and controlled studies of members of different ethnic groups exposed to different environments. It is conceivable that the nature of the antigenic stimulus an individual is exposed to may determine his immunoglobulin composition and that the apparent overall similarity in most populations represents a mean of a very large number of antigenic stimuli which thus tends to obscure differences caused by selective responses to certain antigens. There is now abundant evidence that antibodies to certain antigens (only a few have been carefully studied to date) preferentially or exclusively occur in only one subclass of immunoglobulin, and not infrequently a minor one (Yount *et al.*, 1968). Thus, if an individual's exposure to some of these is excessive, his immunoglobulin composition may deviate from the mean.

Little is known about factors regulating the magnitude of the immune response under normal circumstances. As currently visualized, antigen interacts with macrophages and, after undergoing some change, combines with a specific receptor (possibly preformed antibody) on the surface of the lymphocyte to stimulate it to proliferate. Thus, it does not come as a surprise that circulating antibody is able to inhibit additional antibody synthesis probably by neutralizing antigen and that it can act as a feedback inhibitor (Uhr and Moller, 1968).

There are a number of instances where a naturally occurring disease or an experimental procedure can drastically affect the regulation of γ-globulin production. The best example of the latter is the allotype suppression of long duration induced in newborn rabbits born of mothers who had antibody to one of the fetal allotypes in their circulation (Mage and Dray, 1965). Such offspring have a prolonged suppression of synthesis of γ-globulin with that allotypic specificity, but do show a compensatory increase of the product of its allele. This defect in synthesis is probably due to destruction of the cells involved, since fluorescent antibodies fail to detect cells producing that particular allotypic specificity.

Two types of disorders of γ-globulin synthesis might conceivably provide insight into factors regulating γ-globulin synthesis if we were to know the nature of the defect. These diseases are the agammaglobulinemic syndromes and the neoplastic disorders of plasma cells and lymphocytes accompanied by the excessive production of a homogeneous immunoglobulin or fraction thereof. In the former, there is generally a deficiency of plasma cells and lymphocytes (Franklin, 1968). Whether failure of maturation of this cell line is the primary defect or whether it reflects a defect in immunoglobulin synthesis remains to be determined. The selective absence of some γG subclasses in certain individuals provides the first clearcut support for the previously mentioned possibility that these disorders may be the result of structural or possibly regulator gene abnormalities (Yount *et al.*, 1969).

In the proliferative disorders of plasma cells and lymphocytes, the opposite situation occurs (Engle and Wallis, 1969). Here a clone of cells proliferates in an uncontrolled fashion, possibly as a result of a viral infection or a mutation. The abnormal cell and its progeny may continue to produce the same immunoglobulin as before. The large amounts probably reflect the increase in cell number rather than the increased production per cell. Not infrequently, however, the control of γ-globulin synthesis and assembly becomes distorted, and such cells may then produce only light chains, excessive amounts of light chains, or parts of heavy chains. The mechanism responsible for these aberrations in synthesis remains to be determined.

VI. Summary and Conclusions

In summary, it would appear that the qualitative features and the required heterogeneity of the immunoglobulins are under the direct control of genetic mechanisms, some of which remain to be elucidated. The quantitative expression of these factors, however, appears to be determined by antigenic stimuli derived from the environment. It is the balance of these two forces that determines the immunologic state of competence of an individual at any point in time, and aberrations in both of these can interfere with normal immunoglobulin synthesis.

Acknowledgments

This study supported by U.S.P.H.S. Grants AM 1431, 2594, 5064 and N.Y.C. Health Research Council I-274.

References

Burnet, Sir M. (1969). "Cellular Immunology." Melbourne Univ. Press, Melbourne, Australia.

Cohen, S., and Milstein, C. (1967). *Advan. Immunol.* **7**, 1.

Curtain, C. C., and Baumgarten, A. (1966). *Immunology* **10**, 499.

Edelman, G. M., Cunningham, B., Gall, W. E., Gottlieb, P. D., Rutishauser, U., and Waxdal, M. (1969). *Proc. Nat. Acad. Sci. U.S.* **63**, 78.

Ein, D. (1968). *Proc. Nat. Acad. Sci. U.S.* **60**, 982.

Eisen, H., Little, J. R., Osterland, C. K., and Simms, E. C. (1967). *Cold Spring Harbor Symp. Quant. Biol.* **32**, 75.

Engle, R. L., and Wallis, L. A. (1969). *In* "Immunoglobulinopathies" (I. N. Kugelmass, ed.). Thomas, Springfield, Illinois.

Fleischman, J. B. (1967). *Biochemistry* **6**, 1311.

Franklin, E. C. (1968). *Amer. J. Med.* **45**, 817.

Franklin, E. C., and Frangione, B. (1969). *Annu. Rev. Med.* **20**, 155.

Franklin, E. C., and Fudenberg, H. H. (1969). *Rheumatology* **2**, 62.
Grubb, R., and Laurell, A. B. (1956). *Acta Pathol. Microbiol. Scand.* **39**, 390.
Gustafsson, B. E., and Laurell, C. B. (1959). *J. Exp. Med.* **110**, 675.
Ishizaka, K., Ishizaka, T., and Hornbrook, M. M. (1966). *J. Immunol.* **97**, 75.
Kunkel, H. G. (1968). *J. Exp. Med.* **128**, 763.
Kunkel, H. G. (1969). *J. Immunol.* **102**, 611.
Kunhel, H. G., Smith, W. K., Joslin, J. B. Natvig, J. B., and Litwin, S. D. (1969). *Nature* **223**, 1247.
Lyon, M. F. (1962). *Amer. J. Hum. Genet.* **14**, 135.
McDevit, H., and Benacerraf, B. (1969). *Advan. Immunol.* **11**, 31.
Mage, R., and Dray, S. (1965). *J. Immunol.* **95**, 525.
Milstein, C. (1967). *Nature* **216**, 330.
Pernis, B. (1967). *Cold Spring Harbor Symp. Quant. Biol.* **32**, 333.
Terry, W. D., Hood, L. E., and Steinberg, A. (1969). *Proc. Nat. Acad. Sci. .U.S.* **63**, 71.
Terry, W. D. (1968). *Birth Defects. Orig. Artic. Ser.* **4**, 357.
Uhr, J., and Moller, G. (1968). *Advan. Immunol.* **8**, 81.
Vyas, G. N., and Fudenberg, H. H. (1969). *Proc. Nat. Ac. Sci.* (1969). **64**, 1211.
Yount, W. J., Kunkel, H. G., and Litwin, S. D. (1967). *J. Exp. Med.* **125**, 177.
Yount, W. J., Dorner, M. M., Kunkel, H. G., and Kabat, E. D. (1968). *J. Exp. Med.* **127**, 633.
Yount, W. J., Hong, R., Seligmann, M., and Kunkel, H. G. (1969). *J. Clin. Invest.* **48**, 92a.
Zucker-Franklin, D. (1969). *Semin. Hematol.* **6**, 4.

17

PHYSIOLOGICAL FACTORS CONTROLLING IMMUNOGLOBULIN METABOLISM

T. A. Waldmann, R. M. Blaese, and W. Strober

IMMUNOPHYSIOLOGY SECTION OF THE METABOLISM BRANCH,
NATIONAL CANCER INSTITUTE, NATIONAL INSTITUTES OF HEALTH,
BETHESDA, MARYLAND

I. Introduction

One of the major developments in our understanding of the circulating immune response was the recognition that there are five major classes of immunoglobulin molecules (IgG, IgA, IgM, IgD, and IgE) that differ in chemical structure and in biological function. With the parallel development of techniques for isolating and radiolabeling of these proteins, the study of their metabolism in normal individuals, as well as in patients with immunological deficiencies, has become possible.

The serum concentrations of the different classes of immunoglobulins and antibody molecules reflect a balance among the rates of synthesis, rates of catabolism, and patterns of distribution of these molecules. Each of the major classes of immunoglobulins has a unique pathway and rate of catabolism and synthesis, and, in some cases, of transport as well.

A study of these pathways using radioabeled immunoglobulins have provided major insights into the physiological and pathophysiological factors controlling immunoglobulin metabolism.

II. The Metabolism of the Different Classes of Immunoglobulin in Normal Man

The results of studies of the metabolism of the five major classes of immunoglobulins in normal human subjects are summarized in Table I. It can be seen that there is a wide range in the protein concentrations of the different immunoglobulins from over 1200 mg/100 ml for IgG to 0.05 mg/100 ml for IgE. There is a similar range in the total circulating and total body pools of these proteins. The fraction of the intravascular pool catabolized per day ranges from 6.3% for IgG to 89% for IgE. The synthetic rate for IgG is 32 mg/kg/day, quite similar to that of IgA, but about five times that of IgM, 100 times that of IgD, and about 2000 times that of IgE.

In addition to the differences in catabolic rates between major classes of immunoglobulins, different subclasses of these proteins have different rates of catabolism. Recently, four subclasses of human IgG, G-1, G-2, G-3, and G-4, with different antigenic and structural characteristics have been recognized (Terry and Fahey, 1964; Frangione *et al.*, 1966). IgG-1, -2, and -4 have fractional catabolic rates similar to that of IgG isolated from normal serum by chromatography (Spiegelberg *et al.*, 1968; Morell *et al.*, 1970). IgG-3, however, has a much higher fractional catabolic rate with 17% of the intravascular pool catabolized per day compared to 6.7% for the other subclasses (Morell *et al.*, 1970).

It is known that IgG-3 molecules are more labile *in vitro* than the other subclasses of IgG. It was necessary, therefore, to demonstrate that the high fractional catabolic rate of this protein was an inherent property of the molecule and was not due to denaturation during isolation and radiolabeling. The survival of the protein, therefore, was determined in a patient with a G-3 myeloma protein using *in vivo* labeling with ^{14}C-guanido arginine. In this patient, ^{14}C-guanido arginine was administered intravenously and the rate of decline of specific activity of the purified G-3 myeloma protein was determined with time. A turnover study with the patient's IgG-3 labeled *in vitro* with ^{125}I was then performed. The survival of the protein labeled

TABLE I

SURVIVAL OF IMMUNOGLOBULINS IN NORMAL MAN[a]

Immuno-globulin class	No. of controls studied	Serum concentration (mg/ml)	% Intra-vascular	Total circulating pool (mg/kg)	$T\frac{1}{2}$ (days)	Fractional catabolic rate fraction of IV (pool/day)	Synthetic rate (mg/kg/day)	Reference
IgG	23	12.1	45	494	23	0.067	34	Solomon *et al.*, (1963)
IgG	6	12.0	45	530	18	0.046	25	Birke *et al.* (1963)
IgG	21	12.6	52	527	19	0.069	36	Andersen (1964)
IgG	21	11.3	49	483	—	0.058	28	Freeman (1965)
IgM	7	0.93	76	37	5.1	0.187	6.9	Barth *et al.* (1964)
IgM	4	0.78	—	35	—	0.11	4.5	Birke *et al.* (1967)
IgA	3	—	40	80	5.0–6.5	—	30	Solomon and Tomasi (1964)
IgA	12	2.5	42	95	5.8	0.252	24	Strober *et al.* (1968)
IgD	27	0.023	75	1.1	2.8	0.37	0.4	Rogentine *et al.* (1966)
IgE	23	0.0005	51	0.019	2.4	0.89	0.017	Waldmann *et al.* (1969)

[a] All proteins were prepared by DEAE cellulose chromatography and gel filtration and were iodinated by the iodine monochloride method

in vivo with ^{14}C and of the protein labeled *in vitro* with ^{125}I were identical, indicating that the short survival of G-3 is an inherent property of these molecules.

III. Physiological Factors Regulating Immunoglobulin Synthesis

A number of physiological factors can be shown to control immunoglobulin synthesis. Central lymphoid organs, such as the bursa of Fabricius of birds, have been shown to be of major importance in the development of immunological capabilities. If the development of the bursa of Fabricius is prevented hormonally, using androgens *in ovo*, or if the bursa is surgically removed *in ovo* or immediately following hatching with subsequent irradiation, plasma cells, germinal centers, and circulating antibody are not produced (Glick *et al.*, 1956; Cooper *et al.*, 1966). These animals may have an almost complete absence of immunoglobulins in the serum. In preliminary studies in our laboratory we have shown that the survival of IgG is normal, or slightly reduced, in busectomized birds, while the rate of IgG synthesis is reduced to less than 1% of normal in such animals.

The primary factor controlling immunoglobulin synthesis in immunologically competent animals is antigenic stimulation. Immunoglobulin levels are exceedingly low in germ-free animals. The rate of IgG, IgA, and IgM synthesis is one-fiftieth to less than one-three-hundredth of normal in mice raised in a germ-free environment (Fahey and Sell, 1965; Sell and Fahey, 1964). In contrast, mice hyperimmunized with hemocyanin, or raised in an environment with high bacterial counts, have rates of synthesis of immunoglobulins 5–10 times that seen in normal animals. Thus, there is a 500- to 1500-fold range in the rate of immunoglobulin synthesis between germ-free and high pathogen mice. The rates of synthesis of immunoglobulins are thus not controlled by their serum concentration or pool size but are determined by the bacterial environment and the magnitude of antigenic exposure.

Another factor influencing immunoglobulin synthesis is the age of the animal. The rate of immunoglobulin synthesis especially IgG and IgA synthesis is very low in newborn animals. The rate of IgG synthesis in neonatal rabbits, for example, is only one-thirty-sixth that of adult animals (Deichmiller and Dixon, 1960). Other physiological factors influencing immunoglobulin synthesis include antibody acting as a self-regulator of antibody synthesis (Uhr and Möller, 1968) and the genetic background of the host (McDevitt and Benacerraf, 1969). The administration of certain drugs, such as cortocosteroids, nitrogen mustard, and 6-mercaptopurine, are associated with decreased immunoglobulin synthesis.

IV. Physiological Factors Regulating Immunoglobulin Catabolism

A. The Relationship between the Metabolic Rate and Immunoglobulin Catabolism

The catabolism of immunoglobulins is controlled in part by factors that affect all serum proteins, and in part by factors that are specific for one or more of the immunoglobulin classes. The rate of body metabolism influences the catabolism of many serum proteins, including γ-globulins, and this presumably accounts for the difference in the catabolic rates among the various species. A plot of the logarithm of the IgG fractional catabolic rate against the logarithm of the body weight is shown in Fig. 1 for horses, cows, man, dogs, baboons, monkeys, rabbits, guinea pigs, rats, and mice. The regression line obtained from this plot is defined by the equation.

$$\% \text{ IV IgG catabolized/day} = 74 \text{ body weight in grams}^{-0.19} \qquad (1)$$

Similar kinds of relationships, although with different exponents, can be found between many other physiological functions, including oxygen

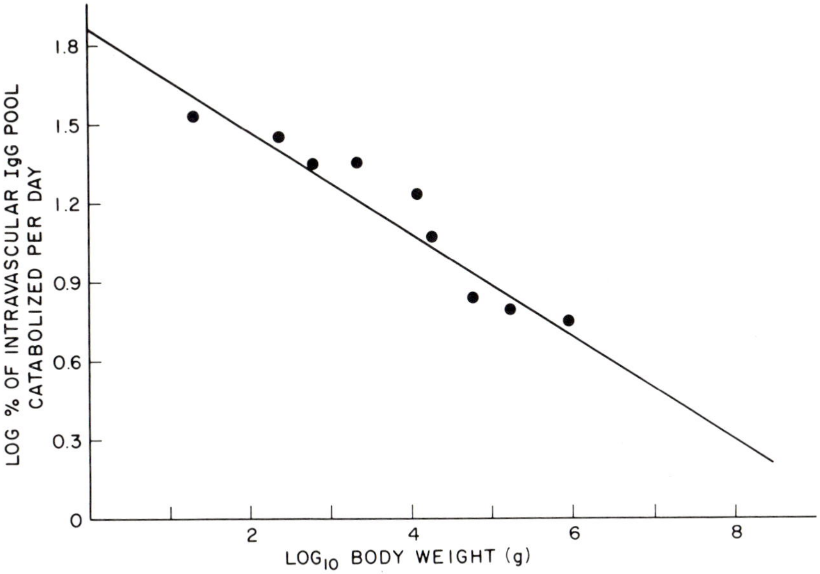

FIG. 1. Fractional catabolic rate (*Fcr*) in various species in relation to body weight in grams. In ascending order of body weight: mouse, rat, guinea pig, rabbit, dog, baboon, man, cow, and horse. The regression line obtained from this plot is defined by the equation: $Fcr = 74$ body weight$^{-0.19}$.

consumption and body weight. The relationship between oxygen consumption and body weight for mammals, for example is given by the expression.

$$0_2 \text{ ml/g/min} = 0.0633 \text{ body weight}^{-0.27} \qquad (2)$$

Functions related to a common function (e.g., body weight) can be related to each other (Adolph, 1949). A quantitative relationship between the fractional catabolic rate for IgG and oxygen consumption/g body weight/min in different species may be calculated from Eqs. (1) and (2) and expressed as follows.

$$\begin{aligned} &\text{FCR } (\% \text{ IV IgG catabolized/day}) \\ &= 513\ 0_2 \text{ consumption ml/g/min}^{0.7} \end{aligned} \qquad (3)$$

This relationship suggests, but does not prove, the possibility that differences in the metabolic rate may be one of the factors producing the wide range in fractional catabolic rate for IgG seen in different species. The effect of the metabolic rate on IgG catabolism has been more directly supported by the observation that there is a significant shortening of IgG survival in hypothyroid or euthyroid guinea pigs, rats, and rabbits following the administration of thyroxin (Dixon *et al.*, 1952; Farthring *et al.*, 1960).

B. The Effect of Steroid Administration on Immunoglobulin Catabolism

A second method of affecting the rate of catabolism of many proteins is by steroid administration. Rothschild and co-workers (1958) and Flick and Steinfeld (1958) had previously shown that the rate of albumin catabolism is increased in animals receiving corticosteroids. However, most investigations on the effect of corticosteroids on immunoglobulin survival have shown that the reduced immunoglobulin levels observed were due solely to decreased synthesis and that the survival was normal (Bjorneboe *et al.*, 1951; Germuth *et al.*, 1951). We decided to reinvestigate the effect of steroid hormones on the rate of catabolism of IgG molecules in the mouse and to determine whether the resultant hypogammaglobulinemia could be explained by decreased synthesis alone, or by a combination of decreased synthesis and hypercatabolism. The change in immunoglobulin concentration produced by the administration of 2.5 mg of cortison daily over a 3-week period to low pathogen mice weighing 18–22 g is shown in Table II. It can be seen that the mice receiving steroids had a significantly reduced serum level of IgM, IgA, and the three subtypes of IgG. A similar reduction of immunoglobulin levels was noted in normal pathogen mice following steroid administration. The effect of steroids on the catabolism of IgG molecules of the γ_{2A} subtype was then studied. Both low pathogen and normal pathogen

TABLE II

THE EFFECT OF HYDROCORTISONE ADMINISTRATION ON THE IMMUNOGLOBULIN CONCENTRATION OF PATHOGEN-FREE MICE

	IgG subclasses (mg/ml)			IgM	IgA
	γ_1	γ_2A	γ_2B		
Control	0.626 ± 0.181	0.365 ± 0.216	0.156 ± 0.082	0.787 ± 0.152	0.275 ± 0.104
Hydrocortisone 2.5 (mg/day)	0.106 ± 0.042	0.140 ± 0.092	0.050 ± 0.008	0.164 ± 0.021	0.177 ± 0.047

mice had a significant shortening of IgG survival and a marked increase in the fractional catabolic rate of IgG when receiving high doses of hydrocortisone daily. Thus, it is clear that corticosteroids affect not only immunoglobulin synthesis, but immunoglobulin catabolism.

C. The Relationship between Serum Immunoglobulin Concentration and Immunoglobulin Catabolism

Another major physiological factor that affects immunoglobulin catabolism is the serum immunoglobulin concentration. The different classes of immunoglobulins differ significantly when the relationship between the fractional catabolic rate and the concentration of the immunoglobulin in the plasma is considered. In the first pattern, exemplified by IgM (Barth *et al.*, 1964) and IgA (Strober *et al.*, 1968), the fractional catabolic rate is independent of the serum concentration. The fractional catabolic rate of IgM is essentially the same in normal subjects, as in patients with agammaglobulinemia with markedly reduced IgM concentrations, or in patients with macroglobulinemia of Waldenström, with markedly increased serum IgM concentrations. The lack of relationship between serum immunoglobulin concentration and protein survival observed with IgA and IgM implies that a constant volume of plasma was cleared of protein per unit time by some mechanism independent of immunoglobulin concentration.

A second pattern of the relationship between serum concentration and catabolic rate is seen with haptoglobin (Freeman, 1965), transferrin, and to a lesser extent with IgD (Rogentine *et al.*, 1966) and IgE (Waldmann *et al.*, 1970). Here, there is an inverse relationship between the fractional catabolic rate and the protein concentration. The fractional catabolic rate of the protein tends to be low in patients with a high serum concentration of that protein.

A third pattern of the concentration–catabolism relationship is observed with IgG. With this protein the fractional rate of catabolism varies in direct proportion to the concentration of IgG in the plasma. Thus, the fractional catabolic rate of IgG is significantly reduced (the survival prolonged) in patients who have low serum γ-globulin concentration secondary to decreased synthesis (Freeman, 1965; Solomon *et al.*, 1963; Waldmann and Schwab, 1965; Waldmann and Strober, 1969). Conversely, with increasing serum-IgG concentration, the survival half-time decreases from the normal of 23 days progressively until a limit is approached of 10 days (Lippencott *et al.*, 1960; Solomon *et al.*, 1963; Birke *et al.*, 1963; Waldmann and Strober, 1969). This effect on IgG catabolism is specifically related to the serum IgG concentration; the survival of IgG is not affected by the concentration of the other immunoglobulins or albumin in the serum. Thus, patients with macroglobulinemia of Waldenström with elevated IgM levels or multiple myeloma associated with the production of IgA or Bence-Jones proteins, have a normal or prolonged survival of IgG when the serum concentration of IgG is normal or reduced. Similarly, patients with analbuminemia and a prolonged albumin survival have a normal IgG survival.

It should be noted that the concentration of the different subclasses of IgG affect the survival of the other subclasses. Thus, an elevated concentration of one subclass, for example IgG-1, results in a short survival of G-1, G-2, G-3, and G-4 subclasses (Morell *et al.*, 1970). The survival curve of G-3 is similar to those of the other subclasses with the exception that the curve is shifted to the left, that is, the G-3 survival is shorter at all IgG concentrations. This indicates that although G-3 proteins share metabolic pathways with the other subclasses they are also acted upon by other catabolic mechanisms.

A concentration–catabolism effect for IgG similar to that noted in man has been reported in mice with a range in survival half-times of from 10 days in hypogammaglobulinemic germ-free mice to 1.8 days in hypergammaglobulinemic mice (Sell and Fahey, 1964; Fahey and Robinson, 1963). In rabbits, the relationship between γ-globulin concentration and survival is less clear with conflicting data reported from different laboratories. In clear contrast to the situation in the man and the mouse, there is no apparent relationship between γ-globulin concentration and γ-globulin survival in the guinea pig. The survival of γ-globulin is the same, approximately 8–9 days, in newborn guinea pigs with low γ-globulin level, and in germ-free guinea pigs with low γ-globulin concentration, as it is in adult guinea pigs (Sell, 1964a). The survival of homologoùs and heterologous γ-globulin in the guinea pig is not affected by infusions of guinea pig, human, or mouse γ-globulin or human serum albumin (Sell, 1964b). Thus, over a wide range of γ-globulin concentrations in the guinea pig there is no change in γ-globulin survival.

In those animals where there is a correlation between the IgG concentration and the IgG survival, there are a number of factors that may be affecting the survival, including the serum γ-globulin concentration and the number of plasma cells synthesizing γ-globulin. It is now clear that the concentration of IgG is the significant factor. The survival of human or mouse IgG in low pathogen mice is markedly shortened when human or mouse IgG is infused into these animals (Fahey and Robinson, 1963). This effect of concentration on γ-globulin survival was shown to be specific and selective since infusions of IgA, IgM, or albumin did not cause an acceleration of IgG catabolism.

The structural configuration of the γ-globulin molecule determining the effect of concentration on catabolism appears to be restricted to the Fc piece of the molecule. Infusions of Fc piece obtained from mouse myeloma protein or from human IgG accelerate the catabolism of mouse γ-globulin while the infusions of Fab piece either prolonged the survival of IgG or did not affect it (Fahey and Robinson, 1963). In a related study the survival of Fc piece in the plasma was affected by the γ-globulin concentration. There was a long survival of Fc piece in hypogammaglobulinemic pathogen-free and germ-free mice and a markedly shortened survival when intact IgG molecules were infused (Waldmann and Strober, 1969). The association of the Fc piece with the concentration–catabolism effect is in keeping with the fact that this effect is specific for the immunoglobulin class.

A number of mechanisms have been considered to explain the concentration–catabolism relationship observed with IgG molecules in humans and certain rodents. One possibility is that with increasing serum IgG concentrations a greater proportion of molecules are present in the catabolic compartment, i.e., the fraction intravascular increases. This would have the effect of increasing the number of molecules in the whole body at risk of degradation and of shortening the survival half-time but would not affect the fraction of the intravascular pool catabolized per day. Andersen and Bjorneboe (1964) studying hyperimmunized rabbits reported such an effect. However, in the majority of studies on the effect of concentration on catabolism in a variety of species, the distribution of IgG molecules remained constant with changes in serum concentration, while at the same time there was a seven to tenfold change in the fractional catabolic rate (Sell, 1964b; Catsoulis *et al.*, 1964; Waldmann and Strober, 1969). At present one must conclude that the effect of serum concentration is on the catabolic mechanism itself and cannot be simply explained by shifts of proteins between compartments.

A second hypothesis is that enzyme induction explains the increased catabolism found with increasing concentration of the IgG substrate. Our preliminary studies, however, indicate that the concentration–catabolism effect is still demonstrable in animals receiving large doses of actinomycin-D or cyclohexamide, materials that would inhibit the protein synthesis required for enzyme induction.

Brambell *et al.* (1964) have proposed an interesting hypothesis to explain the concentration–catabolism effect that postulates a mechanism similar to that proposed for the selective transport of IgG molecules across the placenta, yolk sac, or intestinal epithelium that is a saturable protein protection system specific for IgG molecules. In this model a constant fraction of the plasma protein molecules, including IgG molecules, are isolated from the circulating protein pool into a catabolic pool. Some of the IgG molecules are attached to a limited number of specific protection receptors and are ultimately returned to the circulation while all of the remaining IgG molecules that are isolated are degraded. Such isolation could occur in pinocytotic vacuoles and the receptors could be on the walls of the vacuoles. At a low serum concentration, the majority of the isolated IgG molecules would be protected and returned to the circulation, producing a long survival of the protein, whereas at a high concentration of IgG only a small proportion of the isolated IgG molecules would be protected and a shorter IgG survival would be observed. To obtain a mathematical definition of the proposed model, Brambell *et al.* (1964) made two simplifying assumptions: (1) The fraction of the body pool of protein isolated per day (termed a) is unaffected by the protein concentration and is constant. (2) There is a minimum concentration of IgG molcules (b) that will saturate the receptor sites. Until the b concentration is achieved, the survival of the protein would be infinite. After this concentration is achieved, the number of molecules protected would be constant.

The milligrams of IgG isolated into the catabolic pool was represented by Vac, if V is the volume of the distribution of the molecules, c the concentration of the IgG molecules, and a the fraction of the body pool isolated per day. The concentration of proteins exposed to catabolism will be $c - b$, that is, the concentration isolated minus the concentration required to saturate the receptor protection sites. The milligrams of IgG catabolized would then be $Va(c - b)$ and the fraction (f) of the whole body IgG pool (Vc) catabolized would be:

$$f = \frac{Va(c - b)}{Vc} = \frac{a(c - b)}{c} \tag{4}$$

Since

$$T_{\frac{1}{2}} \text{ of survival} = \frac{\ln 2}{f} = \frac{0.693}{f} \tag{5}$$

then by substitution

$$T_{\frac{1}{2}} = \frac{0.693c}{a(c - b)} \tag{6}$$

Brambell *et al.* (1964) obtained values for a and b for a species by graphical analysis of the half-time of IgG survival vs. the serum IgG concentration, using the data of Fahey and Robinson (1963) for the mouse and Andersen and Bjorneboe (1964) for the rabbit. A similar plot of half-time of IgG survival as a function of serum IgG concentration for patients we have studied with a wide range of IgG concentration is shown in Fig. 2. It should be noted that when the concentration c is very great, the proportion of molecules protected is negligible and $a = 0.693/T_{\frac{1}{2}}$, and that when the $T_{\frac{1}{2}}$ approaches infinity, $c = b$. Values determined in this way by Brambell for

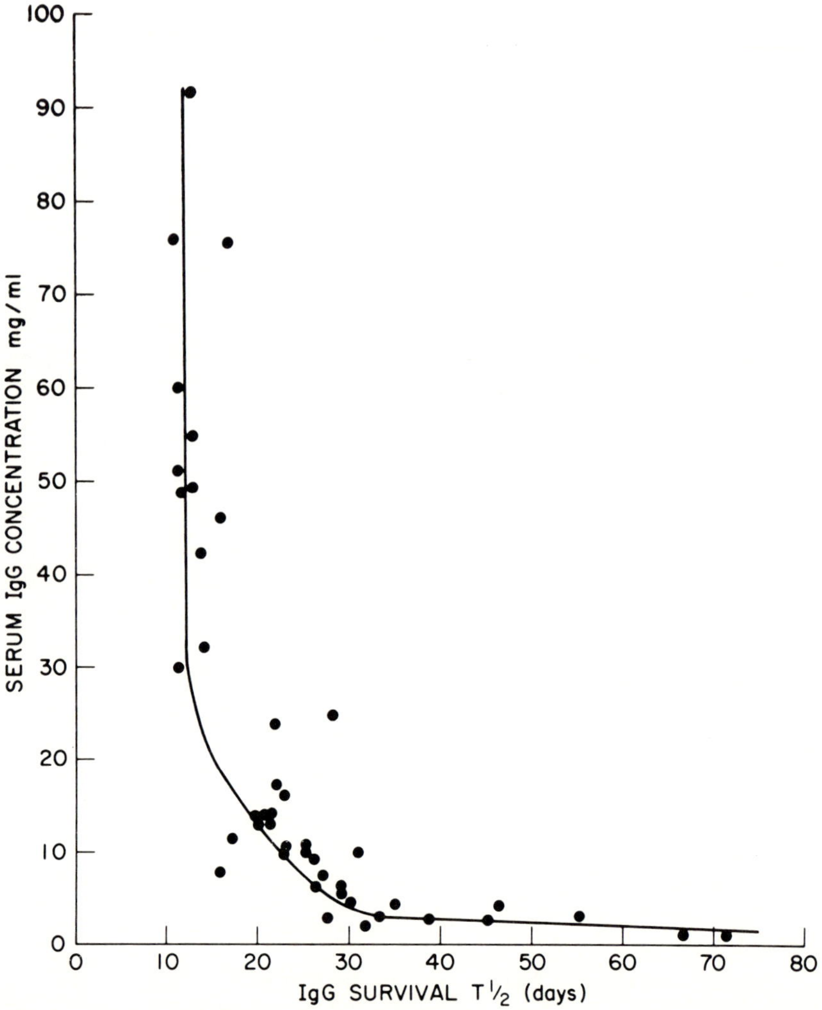

FIG. 2. The relationship between the survival $T_{\frac{1}{2}}$ of IgG and the serum IgG concentration obtained from turnover studies performed in patients with a wide range of IgG concentrations.

the mouse were: a (fraction of the body pool isolated/day) = 0.34 and $b = 2.5$ mg/ml of plasma required to saturate protection sites. The observed values for the half-time of survival of IgG in the mouse agreed well with those predicted from a curve using these values for a and b and varying concentrations of c.

Although this hypothesis in general provides the best explanation for the concentration catabolism effect, it has certain limitations. The use of the half-time of survival or the fraction of the whole-body pool catabolized per day in the mathematical definition of the model was unfortunate since these parameters to not directly estimate the more significant fraction of the intravascular pool catabolized per day. An examination of the data of Andersen and Bjorneboe (1964) used by Brambell *et al.* (1964) in the description of their model provides an example of the difficulties that may occur when the half-time of survival is used in this analysis. In the studies of Andersen and Bjorneboe (1964) in rabbits, the fractional catabolic rate remained constant over a wide range of γ-globulin concentrations in spite of changes in the half-time of survival. Brambell *et al.* (1964) are thus substantiating their analysis with some experiments which do not, in fact, demonstrate the concentration–catabolism effect.

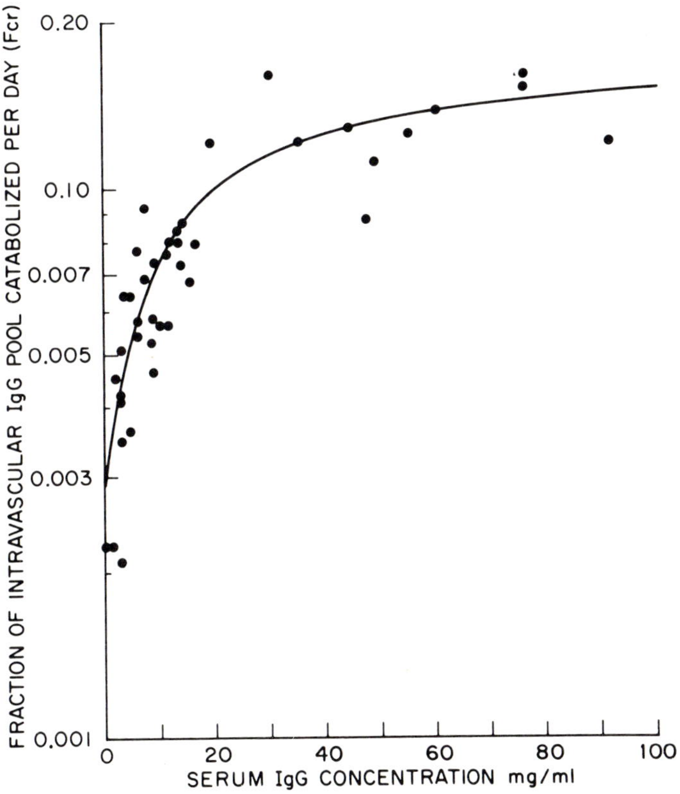

FIG. 3. The relationship between the fraction of the intravascular pool of IgG catabolized per day and the serum IgG concentration obtained from patients with a wide range of IgG concentrations. The dots represent the experimentally determined values and the solid line indicates the values predicted from Eq. (8) in the text.

Another criticism of the model in its present form is that the second simplifying assumption does not appear to be fulfilled. It was assumed that after a minimum concentration, *b*, was achieved the receptor sites would be saturated and there would be no increase in the number of molecules protected, that is, there should be no increase in the number of milligrams of IgG protected as the IgG concentration increases and the IgG survival half-time declines. In order to test this assumption we have plotted the milligrams of IgG protected by receptors vs. the total circulating IgG pool, *Vc*. The data for these determinations were obtained from IgG turnover studies in patients with a wide range of IgG concentrations. The milligrams of IgG

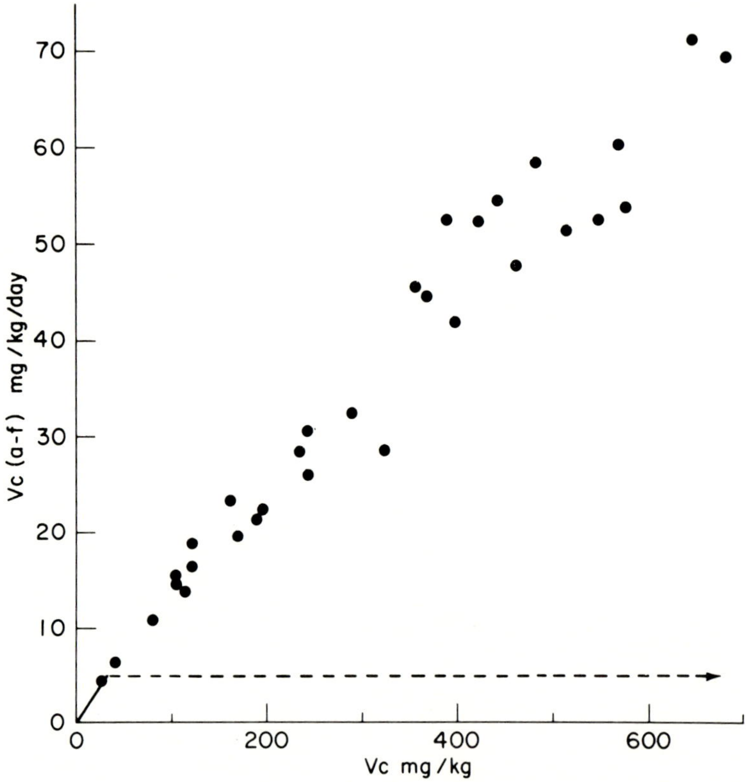

FIG. 4. The relationship between the milligrams of IgG protected per day and the plasma IgG pool size. The milligrams of IgG protected is equal to $Vc(a-f)$ where V is the plasma volume, c is the plasma IgG concentration, f is the observed fractional catabolic rate, and a is the fraction of the plasma pool isolated each day. The number of milligrams of IgG protected $Vc(a-f)$ is not constant over a wide range of plasma IgG pool sizes, Vc, as predicted by the assumption of Brambell but increases with increasing concentration of IgG. The values predicted using the Brambell hypothesis are indicated by the dashed line.

protected by the receptors is defined by Vab which can be shown to be equal to $Vc\,(a-f)$ by transposition of Eq. (4). Thus, the amount protected at each concentration may be calculated from the parameters of the system. The serum concentration of IgG, c, and the fractional catabolic rate observed, f, are determined experimentally for each individual. The fraction of the body pool isolated per day, a, is the equivalent to the IgG fractional catabolic rate in that species at infinite IgG concentration. It is obtained from the asymptotic value of a plot of the IgG fractional catabolic rate as a function of the concentration (Fig. 3). It can be seen from Fig. 4 that the number of milligrams of IgG protected, $Vc(a-f)$, is not constant over a wide range of body IgG pools, Vc, as predicted by the second assumption of Brambell (predicted values are indicated by the dashed line) but increases with increasing concentrations of IgG.

One might expect such an increase in the number of molecules protected with increasing IgG concentration if the relationship between IgG and the receptor site was similar to an enzyme and substrate or a carrier transport system with a reversible interaction between the carrier and the substrate. To show the parallel of this system with carrier mediated transport systems or enzyme kinetics, we have plotted the reciprocal of the amount protected from catabolism $1/Vc(a-f)$ against the reciprocal of the pool size of IgG, $1/Vc$, as in a Lineweaver-Burke plot (Fig. 5). A straight line function is obtained with the number of molecules protected increasing until infinite pool size is reached. This plot may be used to determine the number of molecules protected per unit time at full saturation, termed d, from the reciprocal of the value of the function at its intercept with the ordinate. One may also predict the fractional catabolic rate for any protein concentration using the slope m and ordinate intercept $1/d$ of the reciprocal plot in Fig. 5. From this figure

$$\frac{1}{Vc(a-f)} = \frac{1}{d} + \frac{m}{Vc} \tag{7}$$

where $d =$ maximum number of milligrams of IgG protected per day at full saturation, and m is the slope of the line. Solving for the fractional catabolic rate f we obtain

$$f = a - \frac{d}{Vc + md} \tag{8}$$

if a is defined as the fraction of the intravascular rather than body pool isolated, then f will be the fraction of the intravascular pool catabolized per day. For man, a, the fraction of the plasma pool isolated per day, is 0.18 (the fraction of the intravascular pool catabolized per day at infinite IgG

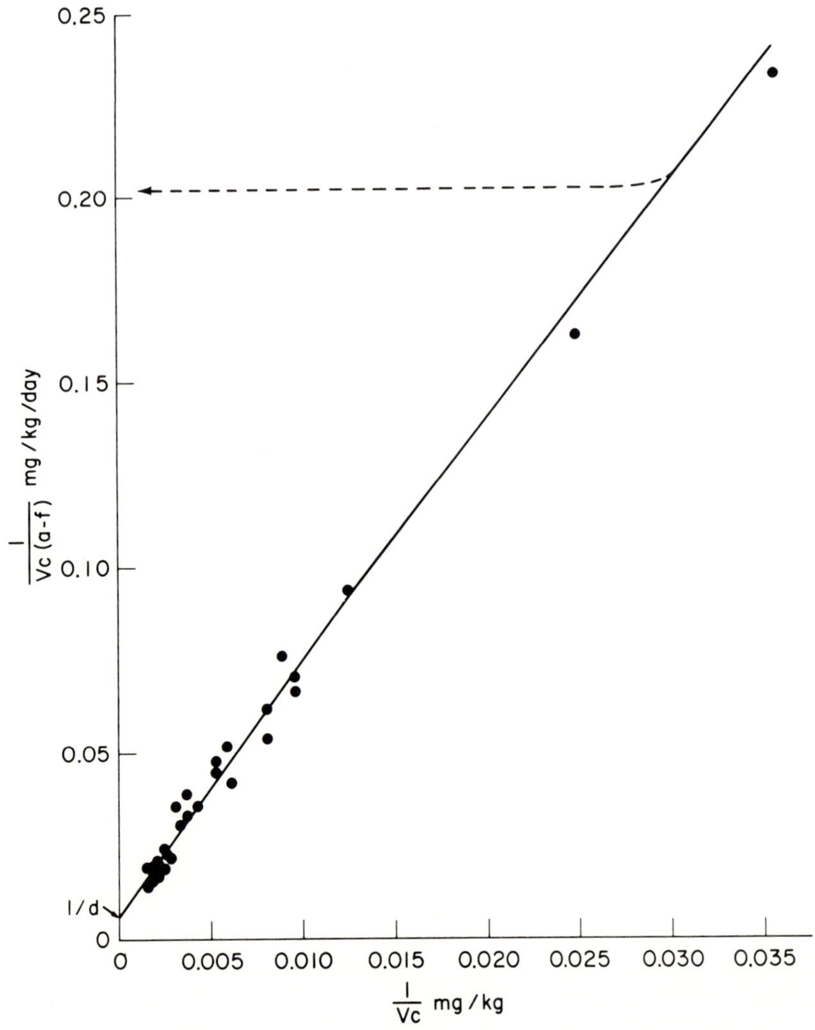

FIG. 5. A plot of the reciprocal of the milligrams of IgG protected $1/Vc(a\text{-}f)$ against the reciprocal of the plasma pool of IgG, $1/Vc$. A straight line function is obtained with a slope m and an ordinate intercept $1/d$. The values predicted by the Brambell formulation are indicated by the dashed lines. The number of milligrams of IgG protected per day/kg body weight at full saturation is given by d. For man, the value obtained was approximately 147 mg/kg/day.

concentration). The value for d, the number of milligrams of IgG protected per day at full saturation, determined graphically from the previous figure is roughly 147 mg of IgG/kg/day (equivalent to 5.5×10^{17} molecules of IgG/kg/day). A plot of f vs. the IgG concentration obtained from patients with a variety of diseases affecting IgG level is shown in Fig. 3 with the values predicted from Eq. (8) indicated by the solid line. There is a close correlation demonstrated between the experimentally obtained and the predicted values.

Further work will be necessary using IgG infusions in animals to determine if this model gives valid estimates for IgG fractional catabolic rates at different IgG concentrations and to define further the nature of the postulated saturable protection system.

V. Summary

The five major classes of immunoglobulins have different rates of synthesis, patterns of distribution, and rates of catabolism in normal man. In addition, the subclasses of IgG have different survival half-times and fractional catabolic rates. The rate of immunoglobulin synthesis is normally determined by the nature and magnitude of antigenic stimuli, the age and genetic background of the host, and the functional status of the central lymphoid organs, such as the bursa of Fabricius of birds. The rate of immunoglobulin catabolism is determined in part by factors such as the metabolic rate and the level of steroids which affect all serum proteins, as well as by factors specific for one immunoglobulin class. The effect of IgG concentration on the IgG catabolic rate is an example of a factor specifically affecting the metabolism of one immunoglobulin class. A direct relationship between the serum IgG concentration and fractional catabolic rate has been demonstrated in man and the mouse. In man the fraction of intravascular pool of IgG catabolized daily rises from 2% in patients with extreme hypogammaglobulinemia to a limit of 16–18% in patients with markedly elevated serum IgG levels. It has been suggested that this concentration–catabolism effect could be explained by postulating a saturable carrier mediated protection system for IgG molecules similar to that postulated for the transport of IgG molecules from the mother to the fetus or newborn.

References

Adolph, E. F. (1949). *Science* **109**, 579.

Andersen, S. B. (1964). Metabolism of Human Gamma Globulin (γ_{ss}-Globulin). Blackwell, Oxford.

Andersen, S. B., and Bjorneboe, M. (1964). *J. Exp. Med.* **119**, 537.

Barth, W. F., Wochner, R. D., Waldmann, T. A., and Fahey, J. L. (1964). *J. Clin. Invest.* **43**, 1036.

Birke, G., Liljedahl, S. O., Olhagen, B., Plantin, L. O., and Ahlinder, S. (1963). *Acta Med. Scand.* **173**, 589.

Birke, G., Norberg, R., Olhagen, B., and Plantin, L. O. (1967). *Scand. J. Clin. Lab. Invest.* **19**, 171.

Bjorneboe, M., Fischel, E. E., and Stoerk, H. C. (1951). *J. Exp. Med.* **93**, 37.
Brambell, F. W. R., Hemmings, W. A., and Morris, I. G. (1964). *Nature* (*London*) **203**, 1352.
Catsoulis, E. A., Franklin, E. C., Oratz, M., and Rothschild, M. A. (1964). *J. Exp. Med.* **119**, 615.
Cooper, M. D., Peterson, R. D. A., South, M. A., and Good, R. A. (1966). *J. Exp. Med.* **123**, 75.
Deichmiller, M. P., and Dixon, F. J. (1960). *J. Gen. Physiol.* **43**, 1047.
Dixon, F. J., Talmage, D. W., Maurer, P. H., and Deichmiller, M. (1952). *J. Exp. Med.* **96**, 313.
Fahey, J. L., and Robinson, A. G. (1963). *J. Exp. Med.* **118**, 845.
Fahey, J. L., and Sell, S. (1965). *J. Exp. Med.* **122**, 41.
Farthring, C. P., Gerwing, J., and Shewell, J. (1960). *J. Endocrinol.* **21**, 83.
Flick, A. L., and Steinfeld, J. L. (1958). *Amer. J. Med. Sci.* **236**, 65.
Frangione, B., Franklin, E. C., Fudenberg, H. H., and Koshland, M. E. (1966). *J. Exp. Med.* **124**, 715.
Freeman, T. (1965). *Ser. Hematol.* **4**, 76.
Germuth, F. G., Jr., Oyama, J., and Ottinger, B. (1951). *J. Exp. Med.* **94**, 139.
Glick, B., Chang, T. S., and Jaap, R. G. (1956). *Poultry Sci.* **35**, 224.
Lippincott, S. W., Korman, S., Fong, C., Stickley, E., Wolins, W., and Hughes, W. L. (1960). *J. Clin. Invest.* **39**, 565.
McDevitt, H. O., and Benacerraf, B. (1969). *Advan. Immunol.* **11**, 31.
Morell, A., Terry, W. D., and Waldmann, T. A. (1970). *J. Clin. Invest.* **49**, 673.
Rogentine, G. N., Jr., Rowe, D. S., Bradley, J., Waldmann, T. A., and Fahey, J. L. (1966). *J. Clin. Invest.* **45**, 1467.
Rothschild, M. A., Schrieber, S. S., Oratz, M., and McGee, H. L. (1958). *J. Clin. Invest.* **37**, 1229.
Sell, S. (1964a). *J. Immunol.* **92**, 559.
Sell, S. (1964b). *J. Exp. Med.* **120**, 967.
Sell, S., and Fahey, J. (1964). *J. Immunol.* **93**, 81.
Solomon, A., and Tomasi, T. B., Jr. (1964). *Clin. Res.* **12**, 452.
Solomon, A., Waldmann, T. A., and Fahey, J. (1963). *J. Lab. Clin. Med.* **62**, 1.
Spiegelberg, H., Fishkin, B., and Grey, H. (1968). *Fed. Proc. Fed. Amer. Soc. Exp. Biol.* **27**, 731.
Strober, W., Wochner, R. D., Barlow, M. H., McFarlin, D. E., and Waldmann, T. A. (1968). *J. Clin. Invest.* **47**, 1905.
Terry, W. D., and Fahey, J. L. (1964). *Science* **146**, 400.
Uhr, J. W., and Möller, G. (1968). *Advan. Immunol.* **8**, 81.
Waldmann, T. A., and Schwab, P. J. (1965). *J. Clin. Invest.* **44**, 1523.
Waldmann, T. A., and Strober, W. (1969). *Progr. Allergy* **13**, 1.
Waldmann, T. A., Strober, W., Johansson, S. G. O., Bennech, H., Ogawa, M., McIntyre, O. R., Morell, A., and Terry, W. D. (1970). In preparation.

18

ABNORMALITIES OF IMMUNOGLOBULIN METABOLISM

Warren Strober, R. Michael Blaese, and Thomas A. Waldmann

IMMUNOPHYSIOLOGY SECTION OF THE METABOLISM BRANCH,
NATIONAL CANCER INSTITUTE, NATIONAL INSTITUTES OF HEALTH,
BETHESDA, MARYLAND

I. Introduction

Recent studies of immunoglobulin metabolism in patients with immunological deficiency have led to major insights into the pathophysiology of abnormal immunoglobulin levels. In addition, such studies have led to the discovery of a number of new types of immunological deficiency diseases. It is clear from this work that abnormalities of immunoglobulin levels may be secondary to defective synthesis of immunoglobulins, excessive loss of immunoglobulins, or to a short survival of the proteins associated with abnormalities of the endogenous degradative mechanisms.

II. Hypogammaglobulinemia Associated with Decreased Immunoglobulin Synthetic Rate

The first immune deficiency states described were those disorders that have a decrease in immunoglobulin synthesis as their chief abnormality. Table I is a compendium of these disorders. In the agammaglobulinemias

TABLE I

DISORDERS OF IMMUNOGLOBULIN SYNTHESIS

I. Deficiency of all immunoglobulin classes
  A. Autosomal recessive lymphopenic Agammaglobulinemia (Swiss-type agammaglobulinemia)
  B. Sex-linked recessive lymphopenic agammaglobulinemia (thymic alymphoplasia)
  C. Sex-linked agammaglobulinemia (Bruton-type agammaglobulinemia)
  D. Primary immunoglobulin deficiency
    1. Acquired agammaglobulinemia
    2. Congenital agammaglobulinemia (sporadic, nonsex-linked)
  E. Secondary immunoglobulin deficiency—chronic lymphocytic leukemia
  F. Acquired agammaglobulinemia with thymoma
II. Selectic immunoglobulin deficiencies
  A. Lymphopenic immune deficiency with dysgammaglobulinemia
  B. Dysgammaglobulinemia Type I—↑IgM, ↓IgG, ↓IgA
  C. Other dysgammaglobulinemias with varying immunoglobulin patterns
  D. Ataxia telangiectasia
  E. Isolated absence of IgA associated with malabsorption
  F. Asymptomatic absence of IgA

involving a deficiency in all immunoglobulin classes (Category I in Table I), the number of plasma cells is markedly reduced and the machinery for production of immunoglobulins is absent. Metabolic studies performed in these disorders have amply shown that the defect is primarily one of decreased immunoglobulin synthesis. The survivals of IgM, IgA, IgD, and IgE were normal in such patients (Barth *et al.*, 1964; Stiehm *et al.*, 1966; Strober *et al.*, 1968a; Rogentine *et al.*, 1966; Waldmann *et al.*, 1969a). The survival of IgG in these patients was long with a mean half-time of survival of 38 days and a fractional catabolic rate of 3.9%/day as compared to a survival $T_{\frac{1}{2}}$ of 23 days and a fractional catabolic rate of 6.8%/day in normal individuals (Waldmann and Schwab, 1965). The prolonged survival of IgG is in accord with the physiological concentration–catabolism effect governing IgG (discussed in the preceding chapter) which leads to a long IgG survival when the serum IgG levels are low. When the serum IgG concentration of patients with agammaglobulinemia is raised to normal by IgG infusions the survival of IgG returns to normal (von Koblet *et al.*, 1964).

The defect of immunoglobulin synthesis may not affect all classes of immunoglobulin molecules but may be restricted to one or two classes of proteins in a group of disorders that have been called dysgammaglobulinemias by some authors (Category II in Table I). In these disorders plasma cells are generally present and the basic mechanism for the decreased immunoglobulin synthesis is poorly understood. For the dysgammaglobulinemia characterized by reduction in the serum concentration of IgG and IgA and an increased concentration of IgM (dysgammaglobulinemia Type I) the mechanism may involve a maturation arrest of lymphoid tissues. This is suggested by the studies of Cain *et al.* (1968) wherein chickens bursectomized *in ovo* just before hatching had an immunoglobulin pattern similar to the patients with reduced IgG synthesis but normal IgM synthesis; animals bursectomized at an earlier period had agammaglobulinemia. Another hypothesis that has been advanced is that this type of dysgammaglobulinemia results from a failure of controller or regulatory genes to derepress protein synthetic mechanisms for α- and γ-polypeptide chains (Barth *et al.*, 1965). In each of the patients with this form of dysgammaglobulinemia studied with radiolabeled immunoglobulins there was a marked decrease in IgA and IgG synthesis with an increase in IgM synthesis (Barth *et al.*, 1965; Gleich *etal.*, 1965; Stiehm *et al.*, 1966).

Another major form of dysgammaglobulinemia is the absence of IgA either as an isolated defect or in association with the features of ataxia telangiectasia. In these patients the primary immunoglobulin metabolic defect is the decreased or absent IgA synthesis (Strober *et al.*, 1968a). In addition, as discussed below these patients may have an exceedingly short survival of IgA molecules due to the development of circulating antibodies to this immunoglobulin.

Patients with hypogammaglobulinemia or dysgammaglobulinemia may develop gastrointestinal disorders associated with excessive protein loss into the gastrointestinal tract (von Barandun *et al.*, 1960; Waldmann and Laster, 1964). In this case the primary defect in γ-globulin synthesis may coexist with decreased immunoglobulin survival secondary to loss of proteins into the gastrointestinal tract. When the gastrointestinal disorder is successfully treated, the survival of the proteins returns to normal but the underlying defect in immunoglobulin synthesis persists.

III. Hypogammaglobulinemia Associated with Decreased Immunoglobulin Survival

In addition to the diseases associated with synthetic defects it has been possible in the past few years to delineate groups of abnormalities due primarily to decreased protein survival. This may involve either abnormal loss of protein to the external environment or changes in endogenous catabolic rates. In the latter instances, pathologic processes may involve only a single immunoglobulin or may involve all the immunoglobulin classes as well as other serum proteins. The diseases associated with decreased protein survival are listed in Table II and will now be considered in greater detail.

TABLE II

DISORDERS OF IMMUNOGLOBULIN CATABOLISM

I. Hypercatabolism
  A. Excessive loss of immunoglobulins
    1. Nephrotic syndrome
    2. Protein-losing gastroenteropathy
  B. Endogenous hypercatabolism
    1. Hypercatabolism affecting a single immunoglobulin
      a. Myotonic dystrophy—hypercatabolism of IgG
      b. Dysgammaglobulinemia and antiimmunoglobulin antibodies
    2. Hypercatabolism affecting more than one immunoglobulin (as well as other serum proteins)
      a. Wiskott-Aldrich syndrome
      b. Familial hypercatabolic hypoproteinemia
      c. Hypermetabolic states
      d. Nephrotic syndrome
II. Hypocatabolism—nephron loss disease with decreased endogenous catabolism of L-chains and other low molecular weight serum proteins
III. Abnormal pathway of metabolism—renal tubular disease with ↑protein excretion and ↓endogenous catabolism of protein

A. URINARY PROTEIN LOSS

Urinary protein loss may occur as a result of renal glomerular defects or renal tubular defects or a combination of these abnormalities. In disorders with glomerular defects such as nephrosis there is damage to the glomerular basement membrane increasing its permeability to serum proteins. Immunoglobulins such as IgG and IgA which are ordinarily retained by the basement membrane are lost more readily than are a larger immunoglobulin molecule such as IgM (Joachim *et al.*, 1964). Through the experimental work of Katz and Sellers and their associates in rodents (Katz *et al.*, 1963, 1964) as well as studies in some children with nephrosis (Gitlin *et al.*, 1956), we now recognize that the glomerular protein loss in the urine may be associated with an increase in endogenous fractional catabolic rates probably due to increased exposure of these proteins to catabolic mechanisms in the renal tubule. Thus, in glomerular disease hypogammaglobulinemia and decreased IgG survival is due primarily to loss of this protein into the urine with increased IgG catabolism by the kidney representing a contributing factor.

Tubular proteinuria occurs in a variety of diseases affecting the proximal tubular cells primarily, such as Fanconi syndrome, cystinosis, and cadmium poisoning (Butler and Flynn, 1958). In these conditions low molecular proteins in the 20,000–40,000 molecular weight class are found in the urine while relatively small amounts of intact immunoglobulin molecules are found. The urinary protein of tubular proteinuria form characteristic patterns on two-dimensional starch block electrophoresis and acrylamide gel electrophoresis (Butler *et al.*, 1962). Harrison *et al.* (1968) have shown that in tubular proteinuria several serum enzymes of low molecular weight have increased excretion rates, i.e., increased clearance of low molecular weight protein into the urine. These authors postulated that low molecular weight proteins normally filtered and reabsorbed by the renal tubules are not reabsorbed in diseases associated with tubular proteinuria and a characteristic urinary protein pattern results.

Further insight into the mechanism of tubular proteinuria has resulted from studies on the role of the kidney in the catabolism of low molecular weight serum proteins. Initial studies with Bence-Jones proteins demonstrated that patients with renal disease catabolized Bence-Jones proteins less well

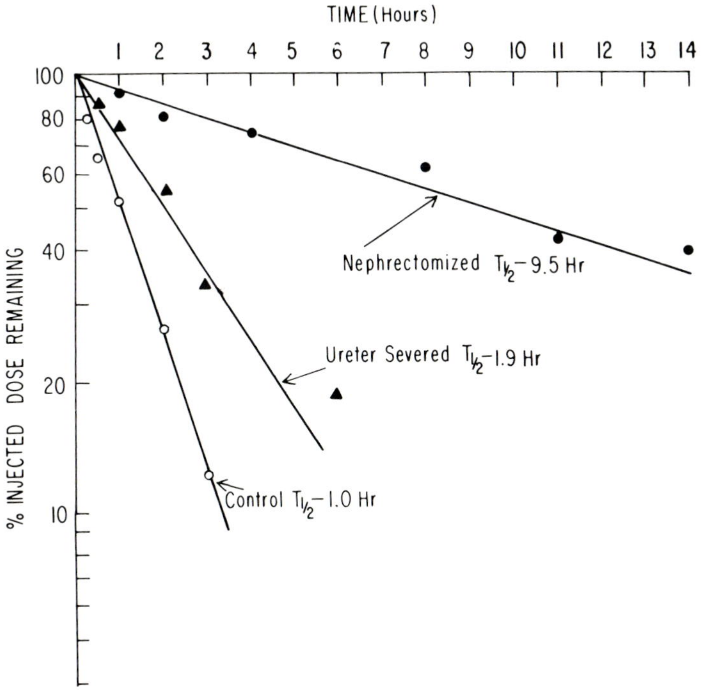

FIG. 1. Survival of human L-chains (derived from IgG by reduction and alkylation) in mice. The catabolism of L-chains was markedly impaired in nephrectomized animals.

than patients without renal disease (Solomon *et al.*, 1964). The reason for this phenomenon was made apparent by turnover studies in nephrectomized mice. In such animals the rate of catabolism of Bence-Jones proteins or L-chain fragments of IgG is greatly decreased (Fig. 1) while in ureter-severed mice (uremic controls) the rate of catabolism is close to that observed in unoperated animals. Thus, the kidney is an important site of Bence-Jones protein catabolism and the prolonged survival of this protein observed in severe renal disease is due to the loss of the renal catabolic site. Bence-Jones proteins are not the only molecules affected by the renal catabolic mechanism; it can be shown with similar nephrectomy experiments in mice that the rate of catabolism of the enzymes, ribonuclease and lysozyme are also markedly decreased when the kidney is removed. It is likely that the metabolism of most proteins small enough to pass through the glomerular basement membrane is subject to renal metabolism (Mogielnicki *et al.*, 1969).

The effect of tubular disease on the renal catabolism of protein was explored in animals using the renal toxin, maleic acid (Mogielnicki and Strober, 1969). This material had previously been shown to produce experimental Fanconi syndrome in rodents. Both treated animals and control animals underwent prior ureter-severing in order to control for the effect of uremia on L-chain metabolism. Maleic acid treatment retarded L-chain catabolism to about the same extent as did total nephrectomy suggesting that renal protein catabolism is occurring in the kidney tubules (Fig. 2).

The findings observed in maleic acid treated mice have been extended to humans in that studies of L-chain metabolism in patients with tubular proteinuria have been performed (Waldmann *et al.*, 1969b). It was found that while the survival of labeled protein in the serum of patients with pure tubular proteinuria (i.e., normal glomerular filtration) is quite normal, the fractional amount of injected protein appearing intact in the urine is greatly increased. Thus, the fractional disappearance rate which includes proteinuria and endogenous catabolism is unchanged, but proteinuria is increased 10- to 60-fold and endogenous catabolism is correspondingly decreased. Since the increase in the proteinuric rate is exactly balanced by the decrease in endogenous catabolism, it is reasonable to assume that in tubular proteinuria low molecular weight protein normally filtered, reabsorbed, and catabolized by the kidney tubule is being filtered and lost intact in the urine.

These concepts may be summarized in the following way. in pure glomerular diseases associated with large basement membrane pores and little or no tubular dysfunction there is both increased loss of intact immunoglobulin molecules into the urine and increased endogenous catabolism. It is possible that in this class of disease those small proteins which in the normal state pass through the basement membrane of the glomerulus do not undergo a change in metabolism. In tubular diseases there is no change in the cata-

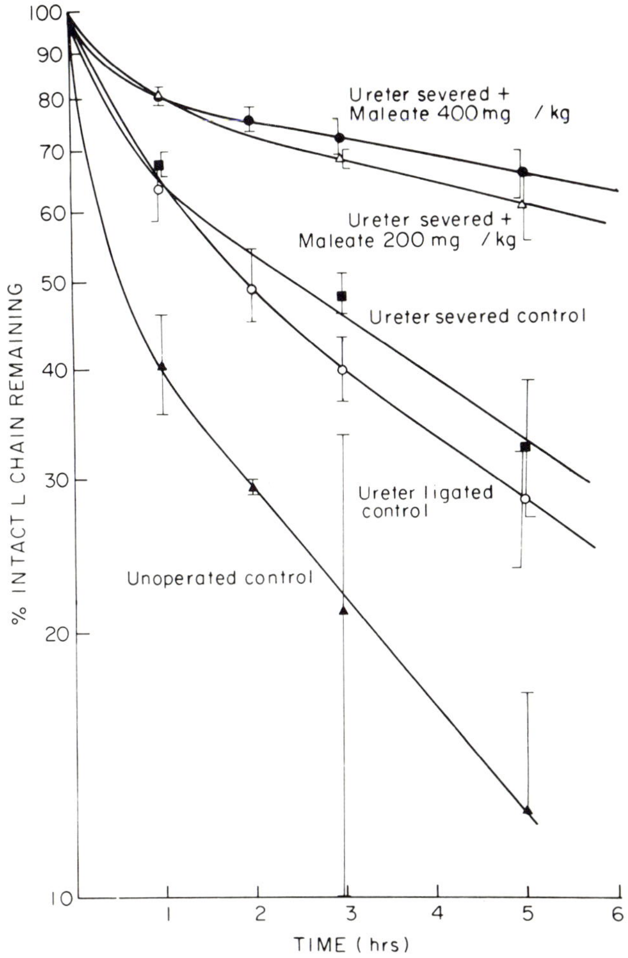

FIG. 2. Survival of mouse Bence-Jones protein in mice pretreated with maleic acid to produce experimental Fanconi syndrome. Survival of B-J protein is prolonged in ureter-severed animals treated with maleic acid as compared to either ureter-severed or ureter-ligated control animals.

bolism of large proteins which are normally retained by the basement membrane. However, in these diseases there is increased proteinuria because there is decreased uptake and decreased endogenous catabolism of small proteins by the renal tubule. Because the total metabolism is unchanged, serum protein levels of these proteins remain normal. In chronic renal disease associated with destroyed nephrons and uremia there is normal endogenous catabolism and little or no proteinuria of large proteins. In this situation small proteins which pass into the glomerular filtrate are lost into the urine in normal amounts but endogenous catabolism is decreased because there are less glomeruli and tubules to filter and take up these proteins. Here the survival of low molecular weight proteins is prolonged and serum levels of these proteins may increase. For example, increased levels of L-chains are observed in patients with lupus erythematosis associated with severe renal disease (Epstein and Tan, 1966). Thus, renal disease may produce both decreased or increased levels of serum immunoglobulins and their fragments by pathophysiological mechanisms involving increased glomerular permeability to proteins or decreased renal tubular catabolism of proteins. The renal catabolic mechanisms may play an important role in the rapid inactivation of biologically active low molecular weight proteins and the loss of this catabolic site in chronic renal disease may lead to the accumulation in the circulation of such substances with adverse clinical consequences.

B. Protein Loss into the Gastrointestinal Tract

A second major route of protein loss is into the gastrointestinal tract (protein-losing enteropathy). This condition may appear in association with over 70 intestinal diseases as well as several diseases not limited to the gastrointestinal tract such as cardiac disease and lymphoma (Waldmann, 1966). In contrast to the protein loss seen in glomerular proteinuria, the protein loss associated with protein-losing enteropathy (as studied in intestinal lymphangiectasia patients) appears to be a bulk process unrelated to protein size. This is evidenced by the fact that the increase in the fractional catabolic rates was approximately the same for all proteins studied (Fig. 3).

Protein-losing enteropathy may occur secondary to a variety of pathological mechanisms. One such mechanism is a disorder of lymphatics leading to leakage of lymph into the gastrointestinal lumen. Intestinal lymphangiectasia, cardiac enteropathies, and Whipple's disease exhibit this kind of disorder. A second mechanism is that of inflammation leading to leakage of inflammatory exudate. Protein-losing enteropathy associated with ulcerative colitis and transient protein-losing enteropathy associated with bacterial enteritis would be in this class. The mechanism of protein-losing enteropathy in agammaglobulinemia, gluten-induced enteropathy (sprue), allergic gastroenteropathy, and many other conditions has not been defined.

The prototype of lymphatic disorders is intestinal lymphangiectasia, a generalized disorder of lymphatic channels characterized by extreme hypoalbuminemia, hypogammaglobulinemia, edema which may be asymmetrical, and, in many cases, chylous effusions. The hallmark morphologic lesion of this disease is dilated lymphatic channels of the small bowel due to a primary abnormality in the development of the central lymphatic channels or the secondary blockage of these channels by an unrelated pathologic process.

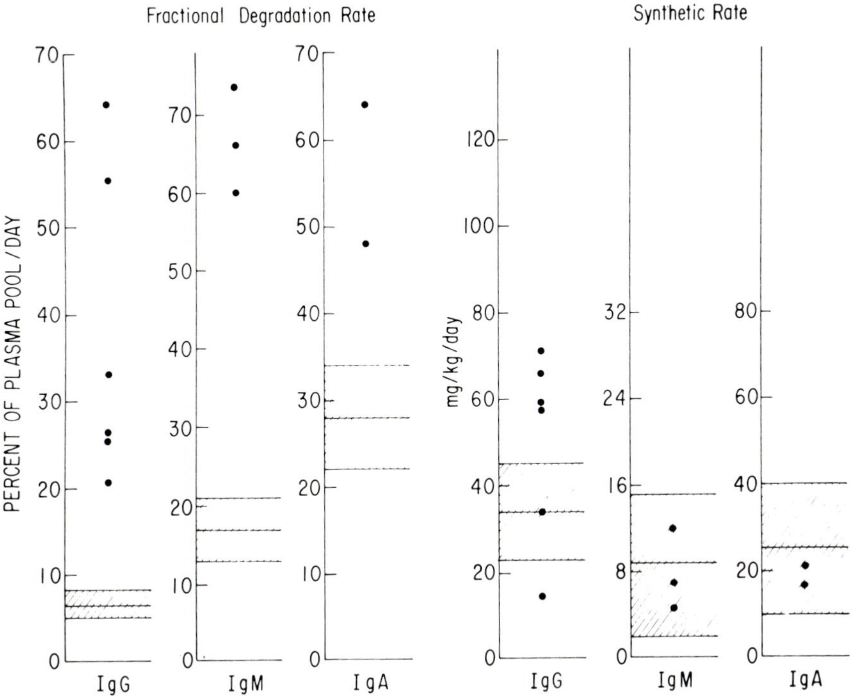

FIG. 3. Metabolism of the major immunoglobulins in intestinal lymphangiectasia. Synthetic rates are normal while fractional catabolic rates are increased due to loss of protein into the gastrointestinal tract.

In this case, the protein-losing enteropathy with short immunoglobulin survival is associated with striking lymphocytopenia (Strober *et al.*, 1967). Patients with intestinal lymphangiectasia may have the functional equivalent of a thoracic duct fistulae into the bowel with intestinal loss of lymphocytes as well as protein. Such patients, who have fairly normal antibody production, have profoundly depressed delayed hypersensitivity responses and are unable to reject skin grafts from unrelated donors. This delayed hypersensitivity defect is the consequence of their profound lymphocytopenia.

A second class of protein-losing enteropathy associated with lymphatic abnormalities is protein-losing enteropathy occurring with cardiac disease. Constrictive pericarditis is the most common cardiac disorder associated with gastrointestinal protein loss but other cardiac diseases such as myocardiopathy and tricuspid regurgitation can lead to protein loss as well (Strober *et al.*, 1968b; Davidson *et al.*, 1961). The physiological factor common to these cardiac conditions is a marked elevation of central venous pressure; this can lead, on the one hand, to a functional obstruction of lymph flow from the main thoracic lymph channels into the venous system and, on the other hand, to increased production of lymph at the capillary level. However, not every patient with an elevation in the central venous pressure develops protein losing enteropathy. Such factors as chronicity of the disorder and local lymphatic conditions may also be important in the development of protein loss of cardiac origin.

Patients with constrictive pericarditis or tricuspid regurgitation with associated protein-losing enteropathy are lymphocytopenic out of proportion to the expansion of the plasma volume occurring in congestive heart failure. It has been shown that just as in intestinal lymphangiectasia, cardiac patients have delayed hypersensitivity abnormalities similar to those of intestinal lymphangiectasia (Strober *et al.*, 1968b). It is probable that these immunologic abnormalities are similar in origin to those found in intestinal lymphangiectasia, that is, there is a loss of both immunoglobulins and lymphocytes directly into the gastrointestinal tract. It is significant that in gastrointestinal protein loss associated with cardiac abnormalities, the short serum protein survival as well as the immunologic deficiency are eliminated by surgical and/or medical correction of the cardiac disease.

In summary, we have termed the protein-losing enteropathy associated with lymphatic abnormalities and lymphocytopenia, the lymphocytopenic gastroenteropathies. In this clinical category a single pathophysiologic

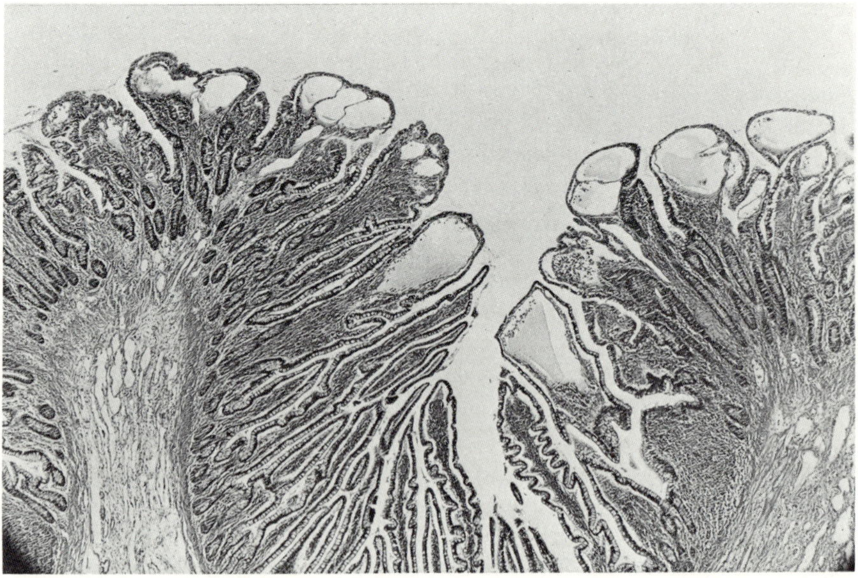

FIG. 4. Intestinal biopsy from a patient with intestinal lymphangiectasia. Dilated lymphatic channels are present in mucosa and in submucosa.

mechanism results in hypoproteinemia as well as deficient delayed hypersensitivity responses.

C. Endogenous Hypercatabolism Affecting a Single Protein

The next category of disorders are those diseases with a short survival affecting a single class of immunoglobulin. The first representative of this group is myotonic dystrophy. Myotonic dystrophy is an autosomal dominant disorder characterized by myotonia, muscle wasting, balding, testicular atrophy, and bone abnormalities. In addition, it has been shown that such patients have low IgG levels due to a defect in the catabolism of the IgG immunoglobulin molecules (Wochner *et al.*, 1966). The survival of ^{125}I-labeled IgG was markedly shortened in patients with myotonic dystrophy with IgG survival half-times averaging 11.4 days in patients compared to 23 days in controls. A mean of 14% of the circulating IgG was catabolized per day in the patients compared to 6.8%/day in controls. In contrast, the serum level and metabolism of albumin, IgA, IgM, IgD, and IgE was normal.

The metabolism of L-chains as well as Fc fragments of IgG was also studied in patients with myotonic dystrophy (Ein and Waldmann, 1969; Strober and Waldmann, 1969). The metabolism of L-chains was normal in these patients while the survival of IgG-Fc was short indicating that the catabolic abnormality is related to the Fc portion of the molecule.

The pathogenesis of the short IgG survival in patients with myotonic dystrophy has not been defined. Host factors appear to be responsible for the accelerated breakdown of IgG. IgG isolated from the serum of myotonic patients behaved normally in control patients while either normal IgG or IgG from myotonic patients had a short survival in myotonic dystrophy patients (Fig. 5). No antibodies to IgG molecules have been detected in the serum of these patients and preincubation of labeled IgG with myotonic serum did not affect IgG survival in normal individuals. Since in myotonic dystrophy only IgG metabolism is affected and further since only IgG demonstrates the concentration–catabolism effect, it is possible that the short survival of IgG in myotonic dystrophy results from loss of the IgG protective mechanism which has been postulated to decrease the fractional catabolic rate for IgG. At this time there is no evidence for or against this hypothesis.

Another condition in which the catabolism of a single protein is affected is ataxia telangiectasia. Ataxia telangiectasia is a multisystem disease taking its name from the major clinical symptoms occurring in this condition: progressive cerebellar ataxia and cutaneous telangiectasia (Boder and Sedgwick, 1958). In addition, abnormalities of the liver, carbohydrate metabolism, and immunologic function can also be found in this disorder (Peterson *et al.*, 1966; McFarlin *et al.*, 1970). This disease has aroused interest in that it is

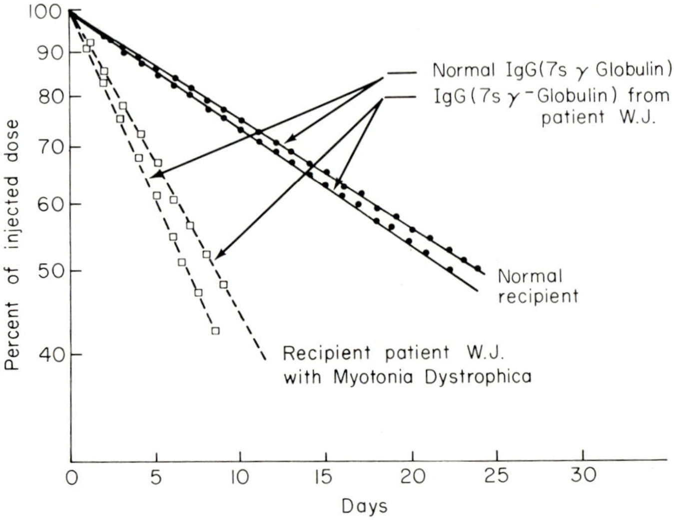

FIG. 5. IgG metabolism in mytotonic dystrophy. Normal IgG or IgG from a patient with myotonic dystrophy has a normal survival in a normal recipient. These same proteins have a reduced survival in a patient recipient. Host factors are, therefore, responsible for the defect in IgG metabolism.

a provocative model of immune dysfunction: These patients have thymic hypoplasia, poor delayed hypersensitivity responses, and, in many cases, lack serum IgA.

To determine if the reduced serum IgA level was due to a synthetic or catabolic abnormality the metabolism of IgA immunoglobulin has been studied in patients with ataxia telangiectasia (Strober *et al.*, 1968a). It was shown that the IgA deficiency is primarily due to profoundly decreased IgA synthesis. In addition, however, several patients had markedly decreased survival of radiolabeled IgA protein. The fractional catabolic rates in these instances were 4–20 times normal. Using a variety of techniques including radioimmunoelectrophoresis, immunoprecipitation and the C′1a fixation and transfer test (Borsos and Rapp, 1965) this increased catabolism was shown to be associated with the presence of circulating anti-IgA antibodies. That such antibodies are responsible for the shortened survival is strongly indicated by the fact that labeled IgA preincubated *in vitro* with the serum of patients with ataxia telangiectasia and then injected into a normal recipient was very rapidly catabolized.

Recently it was found that anti-IgA antibodies may occur in any patient with IgA deficiency and even in some individuals with serum IgA who develop anti-IgA antibodies against genetic determinants not found on their own IgA (Vyas *et al.*, 1968). In addition we have demonstrated an exceedingly short survival in a patient with isolated IgA deficiency (in the absence of ataxia telangiectasia) who had anti-IgA antibodies (Fig. 6). This patient had severe anaphylactoid reactions when blood products containing even very small amounts of IgA were administered. This experience emphasizes the point that transfusion of plasma must be given with care to patients who lack one or another serum protein component because such patients may develop antibodies to the missing component. In summary, it should be stressed that the patients who lack serum IgA and have anti-IgA antibodies have as their primary defect, decreased or absent IgA synthesis. Presumably as a result of antigenic exposure they may develop antibodies against IgA. Hypercatabolism of exogenously injected IgA may then be observed as a secondary phenomenon.

Hypercatabolism of a single class of immunoglobulin may occur in other protein–protein interactions. Hypercatabolism of IgG due to cryoprecipitation of the normal IgG with circulating monoclonal IgM has been observed

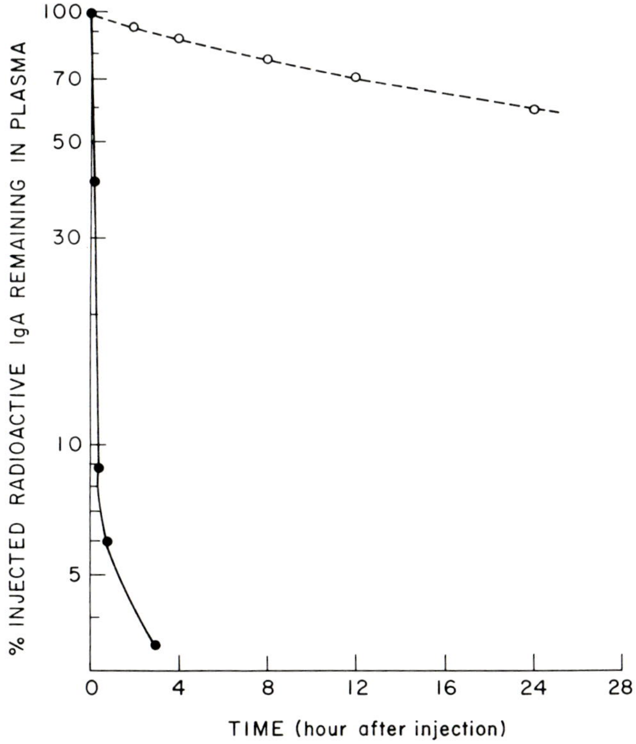

FIG. 6. Survival of IgA in a normal recipient (dashed line) and in a recipient with isolated absence of IgA and anti-IgA antibody (solid line).

in a patient with Sjögren's disease (Waldmann *et al.*, 1968a). In this case the monoclonal IgM interacted at body temperature with IgG molecules to form a complex gel containing both IgM and IgG molecules. These complexes were rapidly catabolized when formed. The monoclonal IgM did not interact with IgA or IgM and the metabolism of these proteins was normal.

Another instance where one would expect protein–protein interaction as the cause of hypercatabolism is in rheumatoid arthritis and other collagen diseases in which the survival of IgG is shortened (Wochner, 1968; Levy *et al.*, 1969). However, Wochner (1968) found no evidence of protein–protein interaction in these instances and the cause of altered protein survival in connective tissue diseases remains unexplained.

D. Endogenous Hypercatabolism Affecting Several Serum Proteins

A second category of disorders of endogenous catabolism encompasses those diseases in which more than one immunoglobulin is affected. The first example is familial idiopathic hypercatabolic hypoproteinemia (Waldmann *et al.*, 1968b). This name has been applied to the condition observed in two siblings with low albumin and γ-globulin levels. The patients had a marked reduction in the serum concentration of IgG and albumin while the concentration of IgM was only slightly reduced, and the concentration of IgA was normal or even slightly elevated. In addition to these abnormal protein levels, the patients had other abnormalities including bilateral bowing of the radii bones and abnormal glucose metabolism. The latter was manifest as diabetic glucose tolerance curves associated in one of the patients with necrobiosis lipoidica diabeticorum.

A series of metabolic studies were performed to identify the pathologic processes operant. The two patients had total circulating albumin pools that were less than 30% of the mean of normal and total body IgG pools that were less than 15% of normal. The rates of IgG and albumin synthesis were essentially normal, but the fractional catabolic rates were markedly increased. The fractional catabolic rates for IgG were 34 and 35% of the IV pool/day, five times the normal mean of 6.8%/day, while fractional catabolic rates for albumin were 22 and 19% of the IV pool/day as compared to the normal mean of 10.3%/day. Thus, the low levels of IgG and albumin in the serum were due to short survivals of these proteins.

More recently, the metabolic behavior of IgA was determined in one of the two siblings. In this case, also, there was an increased catabolism of the protein; furthermore, the IgA synthetic rate was elevated.

When two or more diverse proteins such as albumin and IgG are present in reduced concentration and when this is associated with increased fractional catabolic rates for both proteins, one is generally correct in assuming that the defect is excessive loss of serum proteins, either through the gastrointestinal tract or into the urine. In this case, however, the patients had normal ^{51}Cr-albumin excretion rates, clearing less than 20 ml of plasma daily into the gastrointestinal tract or an amount of serum representing less than 3% of the overall metabolism of albumin and γ-globulin. These values are within the range seen in normal individuals. In addition, the patients had no proteinuria. The presence of circulating antibodies to proteins was unlikely since the catabolic effect involved more than one protein, nevertheless, antibodies were sought by radioimmunoelectrophoresis and immunoprecipitation and were not found. Other causes of generalized hypercatabolism such as thyrotoxicosis and hypersecretion of adrenal corticoids were investigated and were not present. Finally, the patients did not have the stigmata of another disease associated with hypercatabolism of several serum proteins, namely Wiskott-Aldrich syndrome which is discussed in greater detail below. It was concluded that the hypogammaglobulinemia and hypoalbuminemia could not be explained on the basis of a mechanism heretofore described and that these patients represent a new disease syndrome.

A second catabolic disorder involving several immunoglobulins as well as albumin has been observed in patients with the Wiskott-Aldrich syndrome. The Wiskott-Aldrich syndrome is a sex-linked recessive disorder characterized by the triad of thrombocytopenia, eczema, and multiple infections with all types of organisms (Cooper *et al.*, 1968). Recently evidence has been adduced indicating that the immunologic abnormality in this disorder involves the afferent limb or antigen processing limb of the immune response, in contrast to the efferent limb disorders noted in most other forms of immunologic deficiency (Blaese *et al.*, 1968; Cooper *et al.*, 1968). In the area of cellular immune responses the patients with Wiskott-Aldrich syndrome have near normal levels of lymphocytes and these cells can undergo blast transformation when exposed *in vitro* to phytohemagglutinin. In spite of this, the patients have a marked inability to manifest delayed hypersensitivity responses in skin tests and have abnormal homograft rejection. On the humoral side of immunity, the patients have normal total amounts of serum immunoglobulin yet have markedly depressed antibody responses to antigenic challenge, Thus, patients with Wiskott-Aldrich syndrome have a broad immunological defect associated with normal effector organs: functional lymphocytes and plasma cells capable of producing adequate immunoglobulin levels. This suggests that the Wiskott-Aldrich syndrome is a disorder of antigen recognition or processing, perhaps involving malfunction of the reticuloendothelial system or other cellular systems participating in this part of immune response.

Metabolic turnover studies of purified immunoglobulins proved to be quite interesting in patients with the Wiskott-Aldrich syndrome. The synthetic rates of all classes of immunoglobulins were elevated in these patients thus confirming that the effector mechanism of immunoglobulin synthesis is intact. On the other hand, the fractional catabolic rate of all immunoglobulins and albumin were increased with fractional catabolic rates of 25%/day for IgG (normal 6.8%/day), 48%/day for IgA (normal 25%/day), 20%/day for IgM (normal 9%/day), and 19%/day for albumin (normal 11%/day) (Fig. 7).

As yet there is no adequate explanation for these increased fractional catabolic rates. ^{51}Cr-albumin excretion studies were performed in several of these patients to determine the presence or absence of gastrointestinal protein loss. These studies have shown that Wiskott-Aldrich patients have a slightly increased gastrointestinal clearance of serum proteins with 1.5 to 2.5% of the intravascular protein pool cleared into the gastrointestinal tract daily compared to 1% in normals. Such minor increases in gastrointestinal protein loss cannot explain the increased fractional catabolic rates observed. These patients do not have proteinuria and elevated corticosteroid excretion or increased thyroid function have been excluded by appropriate tests. Patients with the Wiskott-Aldrich syndrome thus appear to have a form of shortened serum protein survival due to a mechanism not previously described.

The cause of this metabolic defect may be associated with the cause of

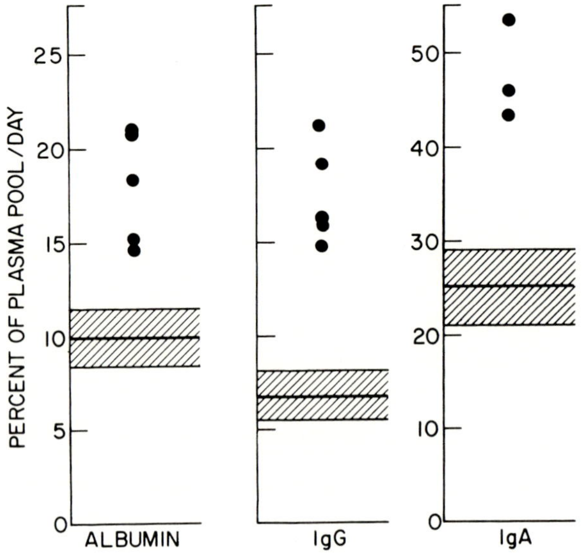

FIG. 7. Protein catabolism in patients with Wiskott-Aldrich syndrome. The fractional catabolic rates of immunoglobulins as well as albumin are increased.

the immunologic defect discussed above. As noted, patients with the Wiskott-Aldrich syndrome may have a defect in the recognition or processing of antigens. In addition, they have markedly hyperplastic reticuloendothelial systems and have rapid R-E clearance of colloidal gold and iodinated aggregated albumin. In view of these findings it is possible, but by no means proven that reticuloendothelial malfunction may be instrumental in the short protein survival. In this regard previous studies on the role of the reticuloendothelial system on the normal metabolism of serum proteins do not exclude this system as a possible site of protein catabolism. These previous studies showed that the survival of albumin was not prolonged in animals with reticuloendothelial systems blocked by carbon or thorium dioxide particles (Benacerraf *et al.*, 1959; Thorbecke *et al.*, 1958). However, it is known that one macrophage function may be blocked by X rays while another is unaffected by the same dose of X rays (Gallily and Feldman, 1967). It is, therefore, quite possible that reticuloendothelial system blockade may affect the uptake of denatured or aggregated proteins while the uptake and catabolism of normal proteins remains intact. One independent piece of evidence suggesting that the reticuloendothelial system may play some role in protein catabolism is that hyperimmunized animals with hyperplastic reticuloendothelial systems have a shortened survival of albumin and γ-globulin. Thus, Sell (1964) found that hyperimmunized guinea pigs have increased fractional catabolism of albumin and γ-globulin. The demonstration of short survival of many serum proteins in guinea pigs with reticuloendothelial hyperplasia and in the Wiskott-Aldrich syndrome suggest that further examination of the role of this system in protein catabolism would be of value.

IV. Summary

Disorders of immunoglobulin pool size and concentration may occur secondary to a variety of pathological mechanisms.

At one end of the metabolic process are the disorders characterized by decreased immunoglobulin synthesis. This occurs in the most common immune deficiency states and may involve all immunoglobulins or only one or two classes as in the dysgammaglobulinemias.

At the other end of the metabolic process are the disorders characterized by increased immunoglobulin loss or catabolism. In this category we have first, those diseases involving loss of proteins to the external environment. Renal disease may result in loss of intermediate-size serum proteins due to increased glomerular permeability or of smaller proteins in patients with tubular disease who have decreased tubular uptake and catabolism of these proteins. Immunoglobulins may also be lost into the gastrointestinal tract in association with over 70 disorders. One mechanism for such loss is a disorder of lymphatic channels of the bowel in patients with intestinal lymphangiectasia or cardiac disease. Patients with lymphatic disorders have lymphocytopenia and abnormalities of delayed hypersensitivity.

Next we have diseases involving endogenous catabolic mechanisms of a single immunoglobulin. In myotonic dystrophy there is an isolated hypercatabolism of IgG. In ataxia telangiectasia decreased synthesis of IgA is frequently found; in addition, some patients have a very short survival of IgA due to the development of anti-IgA antibodies.

Finally, we have two diseases associated with increased endogenous catabolism of many serum proteins including immunoglobulins. Familial idiopathic hypercatabolic hypoproteinemia is a hereditary disease in which the protein abnormality is associated with disorders of bone and carbohydrate catabolism. The mechanism of hypercatabolism in this disease is unknown. The Wiskott-Aldrich syndrome is another hereditary disease associated with immunologic deficiency. The immunologic abnormality may be due to defective antigen processing and it is suggested, but not proven, that in this case some abnormality of the reticuloendothelial system explains the hypercatabolism.

References

Barth, W. F., Wochner, R. D., Waldmann, T. A., and Fahey, J. L. (1964). *J. Clin. Invest.* **43**, 1036.

Barth, W. F., Asofsky, R., Liddy, T. J., Tanaka, Y., Rowe, D. S., and Fahey, J. L. (1965). *Amer. J. Med.* **39**, 319.

Benacerraf, B., Sebestyen, M., and Cooper, N. A. (1959). *J. Immunol.* **82**, 131.

Blaese, R. M., Strober, W., Brown, R. S., and Waldmann, T. A. (1968). *Lancet* **i**, 1056.

Boder, E., and Sedgwick, R. P. (1958). *Pediatrics* **21**, 526.

Borsos, T., and Rapp, H. J. (1965). *J. Immunol.* **95**, 559.

Butler, E. A., and Flynn, F. V. (1958). *Lancet* **ii**, 978.

Butler, E. A., Flynn, F. V., Harris, H., and Robson, E. B. (1962). *Clin. Chim. Acta* **7**, 34.

Cain, W. A., van Alten, P. J., Good, R. A., and Cooper, M. D. (1968). *Fed. Proc. Fed. Amer. Soc. Exp. Biol.* **27**, 493.

Cooper, M. D., Chase, H. P., Lowman, J. T., Krivit, W., and Good, R. A. (1968). *Amer. J. Med.* **44**, 499.

Davidson, J. D., Waldmann, T. A., Goodman, D. S., and Gordon, R. S., Jr. (1961). *Lancet* **i**, 899.

Ein, D., and Waldmann, T. A. (1969). *J. Immunol.* **103**, 345.

Epstein, W. V., and Tan, W. (1966). *Arthritis Rheum.* **9**, 713.

Gallily, R., and Feldman, M. (1967). *Immunology* **12**, 197.

Gitlin, D., Janeway, C. A., and Farr, L. E. (1956). *J. Clin. Invest.* **35**, 44.

Gleich, G. J., Condemi, J. J., and Vaughan, J. H. (1965). *New Engl. J. Med.* **272**, 331.

Harrison, J. F., Lunt, G. S., Scott, P., and Blainey, J. D. (1968). *Lancet* **i**, 371.

Joachim, G. R., Cameron, J. S., Schwartz, M., and Becker, E. L. (1964). *J. Clin. Invest.* **43**, 2332.
Katz, J., Bonorris, G., and Sellers, A. L. (1963). *J. Lab. Clin. Med.* **62**, 910.
Katz, J., Sellers, A. L., and Bonorris, G. (1964). *J. Lab. Clin. Med.* **64**, 709.
Levy. J., Barnett, E. V., MacDonald, N. S., and Klinenberg, J. R. (1969). *Clin. Res.* **17**, 118.
McFarlin, D. E., Strober, W., and Waldmann, T. A. (1970). Unpublished observations.
Mogielnicki, R. P., and Strober, W. (1969). Unpublished observations.
Mogielnicki, R. P., Strober, W., and Waldmann, T. A. (1969). Unpublished observations.
Peterson, R. D. A., Cooper, M. D., and Good, R. A. (1966). *Amer. J. Med.* **41**, 342.
Rogentine, G. N., Jr., Rowe, D. S., Bradley, J., Waldmann, T. A., and Fahey, J. L. (1966). *J. Clin. Invest.* **45**, 1467.
Sell, S. (1964). *J. Exp. Med.* **120**, 967.
Solomon, A., Waldmann, T. A., Fahey, J. L., and McFarlane, A. S. (1964). *J. Clin. Invest.* **48**, 103.
Stiehm, E. R., Vaerman, J. P., and Fudenberg, H. H. (1966). *Blood* **28**, 918.
Strober, W., and Waldmann, T. A. (1969). Unpublished observations.
Strober, W., Wochner, R. D., Carbone, P. P., and Waldmann, T. A. (1967). *J. Clin. Invest.* **46**, 1643.
Strober, W., Wochner, R. D., Barlow, M. H., McFarlin, D. E., and Waldmann, T. A. (1968a). *J. Clin. Invest.* **47**, 1905.
Strober, W., Cohen, L. S., Waldmann, T. A., and Braunwald, E. (1968b). *Amer. J. Med.* **44**, 842.
Thorbecke, G. J., Sebestyen, M., Benacerraf, B., and Green, H. (1958). *Proc. Soc. Exp. Biol. Med.* **99**, 439.
von Barandun, S., Aebersold, J., Bianchi, R., Kluthe, R., von Muralt, G., Poretti, G., and Riva, G. (1960). *Schweiz. Med. Woschenschr.* **90**, 1458.
von Koblet, H., Diggelmann, H., Barandun, S., Kaser, H., and Riva, K. (1964). *Helv. Med. Acta* **31**, 618.
Vyas, G. N., Perkins, H. A., and Fudenberg, H. H. (1968). *Lancet* **ii**, 312.
Waldmann, T. A., (1966). *Gastroenterology* **50**, 422.
Waldmann, T. A., and Laster, L. (1964). *J. Clin. Invest.* **43**, 1025.
Waldmann, T. A., and Schwab, P. J. (1965). *J. Clin. Invest.* **44**, 1523.
Waldmann, T. A., Johnson, J., and Talal, N. (1968a). Unpublished observations.
Waldmann, T. A., Miller, E. J., and Terry, W. D. (1968b). *Clin. Res.* **16**, 45.
Waldmann, T. A., Strober, W., Johansson, S. G. O., Bennech, H., Ogawa, M., McIntyre, O. R., Morell, A., and Terry, W. D. (1969a). Unpublished observations.
Waldmann, T. A., Strober, W., and Mogielnicki, R. P. (1969b). Unpublished observations.
Wochner, R. D. (1968). *J. Lab. Clin. Med.* **72**, 1028.
Wochner, R. D., Drews, G., Strober, W., and Waldmann, T. A. (1966). *J. Clin. Invest.* **45**, 321.
Wochner, R. D., Strober, W., and Waldmann, T. A. (1967). *J. Exp. Med.* **126**, 207.

19

THE SUBMOLECULAR SITE RELATED TO THE RATE OF CATABOLISM OF IgG IMMUNOGLOBULINS

Hans L. Spiegelberg

DEPARTMENT OF EXPERIMENTAL PATHOLOGY,
SCRIPPS CLINIC AND RESEARCH FOUNDATION,
LAJOLLA. CALIFORNIA

I. Introduction

It has been shown in a number of species that different classes of immunoglobulins have a characteristic catabolic rate. In man, IgG immunoglobulins are relatively slowly catabolized with an average half-life of about 20 days (Anderson, 1964), whereas IgA, IgM, IgD, and IgE immunoglobulins are more rapidly catabolized with average half-lives of 6 (Solomon and Tomasi, 1964), 5 (Barth *et al.*, 1964), 3 (Rogentine *et al.*, 1966), and 2 days (Johannson and Bennich, 1969), respectively. The mechanism controlling the catabolic rate of immunoglobulins, however, is still poorly understood. Brambell proposed a theory on the control of catabolism of immunoglobulins (Brambell *et al.*, 1964; Brambell, 1966) based on the finding that the rate of IgG catabolism in mice depends on the IgG plasma concentration (Fahey and Robinson, 1963). This theory proposes that immunoglobulin molecules are pinocytosed at a constant rate and that a fraction of the pinocytosed molecules attaches via a specific site onto receptors which protect this fraction from proteolysis and subsequently returns it to the circulation; in contrast, pinocytosed molecules which are not attached to the receptors are catabolized. Since the number of protective receptors would be constant, a smaller portion of the pinocytosed molecules could be protected at a high IgG plasma concentration resulting in an increase of the catabolic rate, whereas at a low IgG concentration the rate of catabolism would be decreased since most of the molecules pinocytosed could attach to protective receptors. The present experiments were undertaken in order to attempt to define the site or sites on the IgG molecule which are related to the rate of IgG catabolism and which would, in Brambell's theory, react with the protective receptor. In order to define this submolecular site, the rate of elimination from the circulation and the excretion into the urine of fragments (Spiegelberg and Weigle, 1965) and polypeptide chains (Spiegelberg and Weigle, 1966) of normal IgG immunoglobulins and of IgG myeloma proteins (Spiegelberg *et al.*, 1968; Spiegelberg and Grey, 1968) and heavy chain disease proteins (Spiegelberg and Fishkin, 1970) of different heavy chain subclasses was studied.

II. Structure of IgG Immunoglobulins

In the past ten years much has been learned about the structure of IgG immunoglobulins. Figure 1 shows the model of the IgG molecule proposed by Porter (Cohen and Porter, 1964). IgG immunoglobulins are composed of two light and two heavy polypeptide chains linked by disulfide bonds. Both types of polypeptide chain have a variable portion at the N-terminus, probably related to the antibody combining site, and a constant portion at the C-terminal end. Digestion of IgG with papain results in the formation of three fragments of about equal size, two Fab fragments each containing one antibody combining site and one Fc fragment which does not interact with antigen but is responsible for other biological functions of IgG such as complement fixation (Taranta and Franklin, 1961), transmission to the fetal circulation (Brambell *et al.*, 1960), and fixation to the skin (Franklin and Ovary, 1963). Digestion of IgG with pepsin results in the formation of a large fragment called $F(ab')_2$ fragment which consists of two Fab-like fragments linked by the inter-heavy chain disulfide bonds (Nisonoff *et al.*, 1960). The Fc fragment is degraded by pepsin into small peptides. The heavy and light polypeptide chains can be isolated from IgG following reduction and

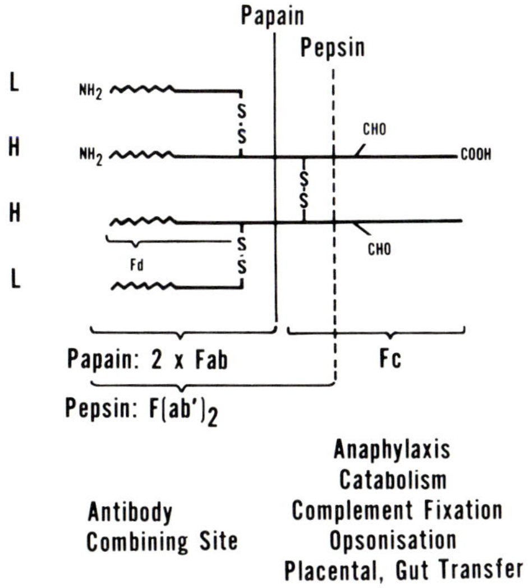

FIG. 1. Model of structure of γG immunoglobulins and biological activities mediated by either the Fab fragment or Fc fragment.

alkylation by Sephadex gel filtration in propionic or acetic acid (Fleischman *et al.*, 1962). It has also been shown that in certain neoplastic diseases of the lymphocyte-plasma cell series, subunits of IgG can be present in the serum and urine. In some cases of multiple myeloma, the patients excrete into the urine light chains which have previously been known as Bence-Jones proteins (Edelman and Gally, 1962). In a recently discovered disease, called heavy chain disease (Franklin *et al.*, 1964), a protein is present in the serum and in some cases in the urine which in structure closely resembles the Fc fragment of normal IgG and this protein has been called H chain disease protein. In the present work, both the fragments obtained by proteolytic digestion from normal IgG and the subunits formed in neoplastic diseases, were labeled with ^{131}I and/or ^{125}I (Weigle and Dixon, 1959; McConahey and Dixon, 1966), injected intravenously into either man or experimental animals and their elimination from the circulation and excretion into the urine were studied. In some experiments, the disappearance of the ^{131}I radioactivity was also determined by whole body counting.

III. The Catabolism of IgG Fragments and Polypeptide Chains in Experimental Animals

The elimination of rabbit IgG fragments in rabbits is shown in Fig. 2. The F(ab')$_2$ and Fab fragments, having molecular weights of about 109,000 and 50,000, respectively, were very rapidly eliminated from the circulation.

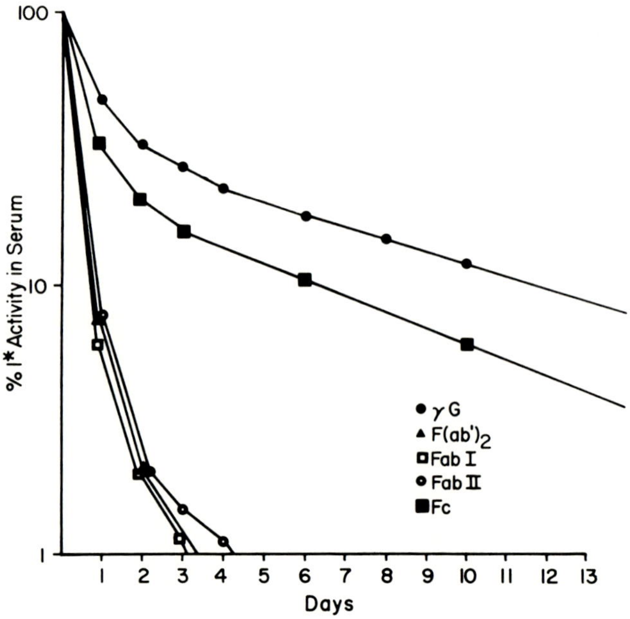

FIG. 2. Elimination of rabbit γG and rabbit γG fragments from the circulation in rabbits (Spiegelberg and Weigle, 1965).

During the first 3 days following injection, about 20% of the injected radioactivity was recovered in the urine as protein bound radioactivity and about 50% as nonprotein-bound radioactivity, indicating that most of the $F(ab')_2$ and Fab fragments were catabolized rather than excreted intact into the urine. In contrast to the $F(ab')_2$ and Fab fragments, the Fc fragments having a molecular weight of about 50,000 were slowly eliminated from circulation. During the first 3 days following injection, only about 2% of the injected radioactivity of the Fc fragments was recovered as protein-bound radioactivity in the urine, as compared to 50% as nonprotein bound. Similar findings were made in guinea pigs and mice with both homologous and heterologous IgG fragments (Spiegelberg and Weigle, 1965). These experiments suggested that the site responsible for the rate of catabolism of IgG immunoglobulins is located in the papain Fc fragment.

In all species tested, the *in vivo* behavior of the Fc fragments, but not Fab and $F(ab')_2$ fragments, depended on the duration of digestion with the enzyme. As can be seen in Fig. 3, a portion of the Fc fragments was rapidly eliminated within the first few days following injection and another portion was more slowly eliminated. The rapidly eliminated portion increased with longer periods of papain digestion. Similar biphasic elimination curves of different Fc fragment preparations were observed by whole-body counting.

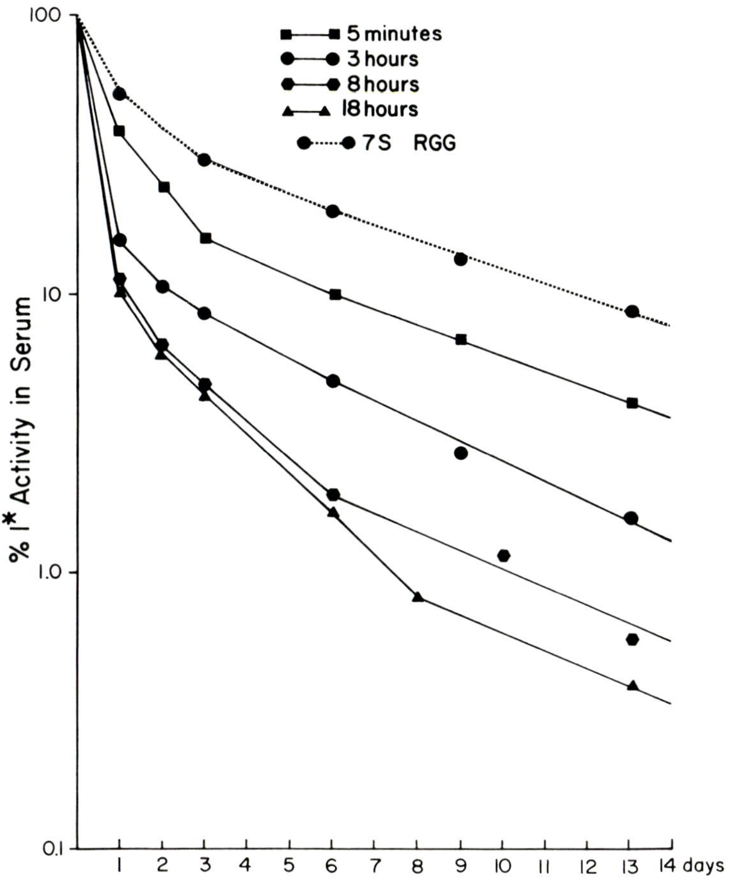

FIG. 3. Elimination of rabbit γG and rabbit γG Fc fragments (isolated after digestion for 5 min, 3, 8, and 18 hr) from the circulation of rabbits (Spiegelberg and Weigle, 1965).

These experiments indicated that a progressive alteration of the Fc fragments occurred as the period of digestion was extended. It has been shown that prolonged digestion of human IgG with papain results in the formation of a heterogenous population of Fc fragments lacking structures both at the N-terminus as well as at the C-terminus (Grey and Abel, 1967). It appears likely that prolonged papain digestion either removed or partially destroyed the site on the Fc fragment which is related to its ability to persist in the circulation. Since prolonged digestion with papain removes first peptides at the N-terminal end of the Fc fragment, the N-terminus of the Fc fragment might be particularly important in forming the site responsible for the rate of catabolism of IgG; however, more experiments are needed in order to establish this point. It was attempted to prepare a homogenous preparation of Fc fragment following very brief digestion of IgG with papain; however, these experiments did not succeed. It was therefore decided to study the catabolism of heavy chain disease proteins (Spiegelberg and Weigle, 1966) which have been shown to be relatively homogenous and to be structurally closely related to the IgG Fc fragments (Franklin, 1964). The plasma and whole-body elimination curves of H chain disease proteins and of intact human IgG in rabbits are shown in Fig. 4. During the first 6 days following injection, the whole-body elimination curves of two H chain disease proteins (Cr, Zu) were very similar to that of intact IgG. The plasma elimination curves, however, differed. About 20 to 25% of the H chain disease proteins remained in the intravascular compartment as compared to 45–50% of the intact IgG. Less than 2% of the injected radioactivity of the H chain disease proteins and less than 0.5% of the radioactivity of the IgG were recovered in the urine as protein-bound radioactivity in the urine. As expected, all rabbits showed an immune elimination of the human IgG 6–9 days following injection. None of the rabbits showed an immune elimination of the H chain disease proteins and some of these rabbits even became tolerant to the H

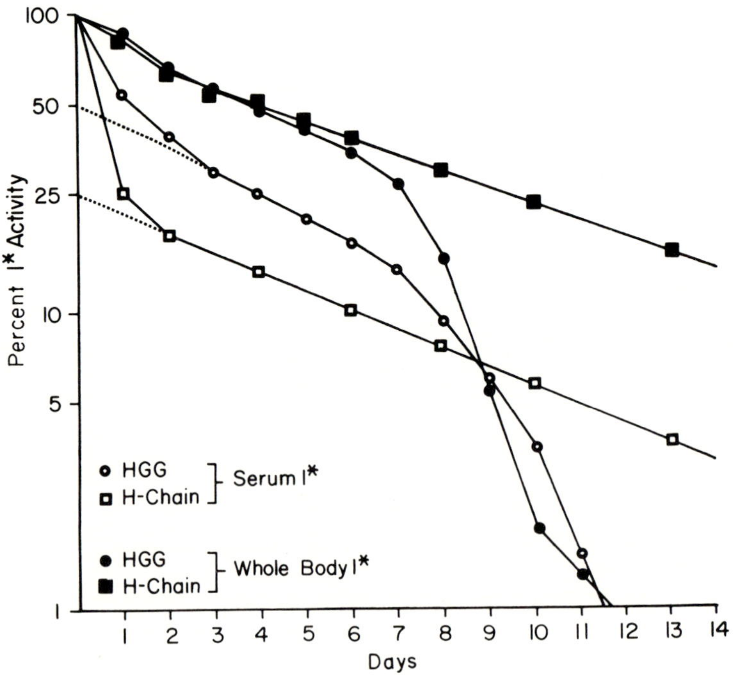

FIG. 4. Elimination of H chain disease protein and human γG in rabbits (Spiegelberg and Weigle, 1966).

chain disease protein (Spiegelberg and Weigle, 1967), probably because the H chain disease proteins have a very low tendency to aggregate. The catabolic rate of rabbit light chains and of human Bence-Jones proteins was also studied in rabbits. Both the rabbit light chains and human Bence-Jones protein of K and λ type were rapidly eliminated from the circulation. Most of the injected radioactivity was recovered in the urine as nonprotein-bound radioactivity indicating that the light chains were catabolized and not excreted intact into the urine. The similar half-lives of IgG and H chain disease proteins confirmed the findings made with the Fab and Fc fragments and suggested that the entire site related to the rate of catabolism might be present in the H chain disease proteins. However, as will be discussed below, the different intra- and extravascular equilibration of IgG and H chain disease proteins indicated a different fractional turnover rate (fraction of the intravascular pool catabolized/day) of these proteins, suggesting that structures in addition to the site in the Fc fragment might also play a role in determining the catabolic rate of IgG.

IV. Catabolism of Human IgG Myeloma Proteins and of H Chain Disease Proteins in Man

Human IgG immunoglobulins can be divided into four subclasses on the basis of antigenic differences in the Fc fragment (Grey and Kunkel, 1964; Terry and Fahey, 1964). Since IgG immunoglobulins of different subclasses differ in certain biological activities such as skin fixation (Terry, 1965) and complement fixation (Ishizaka *et al.*, 1967), it was of interest to determine if they also differ in their catabolic rates. Myeloma proteins are at the present time the only available source of relatively pure immunoglobulins of a particular subclass; therefore, the rate of elimination from the circulation of myeloma proteins was studied in man (Spiegelberg *et al.*, 1968). In order to demonstrate differences in half-lives, pairs of ^{131}I and ^{125}I-labeled myeloma proteins were injected. As can be seen in Table I, the average plasma half-lives of IgG-1, IgG-2, and IgG-3 myeloma proteins were similar and did not differ significantly from normal IgG. In contrast, IgG-3 myeloma proteins were more rapidly catabolized than myeloma proteins of the other three subclasses. The average half-life of normal IgG was shorter than that reported by other investigators (Anderson, 1964). The explanation of this difference probably lies in technical differences; the longer half-lives were obtained by measurement of the retention of radioactivity in the body. In contrast to the average half-lives, individual myeloma proteins within a given subclass were heterogeneous with respect to their catabolic rates. The heterogeneity in the rate of catabolism of different myeloma proteins existed

TABLE I

AVERAGE HALF-LIVES OF NORMAL AND MYELOMA γG IMMUNOGLOBULINS IN HUMANS

	Number of proteins	Number of patients	Average half-life
Normal γG		11	12.3 ± 3.4[a]
Myeloma γG-1	8	44	11.6
Myeloma γG-2	5	16	12.4
Myeloma γG-3	4	15	8.2 ± 3.2
Myeloma γG-4	3	10	11.3

[a] Standard deviation.

within all four heavy chain subclasses. The differences in half-lives of two myeloma proteins was expressed in percent, as shown in Fig. 5. Of 33 different pairs of myeloma proteins injected, 24 pairs differed significantly with differences varying between 10 and 49.5%. The differences in half-lives within the IgG-1 and IgG-2 subclass are shown in Table II. In different patients, a particular myeloma protein consistently either had a relatively long, medium, or short half-life. The half-life of normal IgG was intermediate to the half-lives of myeloma proteins. No correlation existed between half-lives of the myeloma proteins and their light chain type, Gm factor, carbohydrate content, and electrophoretic mobility. The differences in half-lives of myeloma proteins belonging to the same subclass was surprising

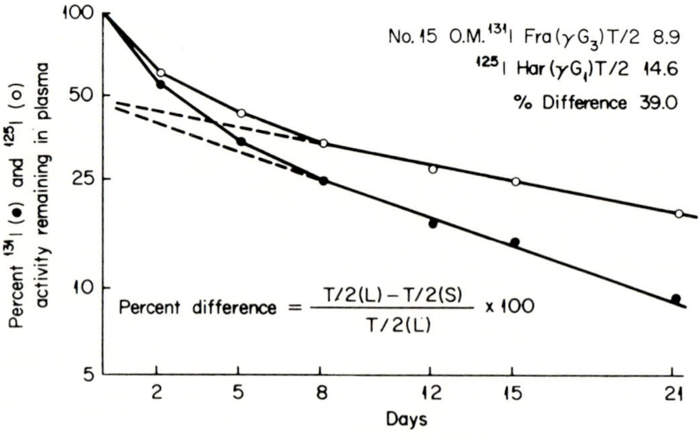

FIG. 5. Plasma elimination curves of the ^{131}I trace-labeled γG-3 myeloma protein Fra and the ^{125}I trace-labeled γG-1 myeloma protein Har in patient O.M. (Spiegelberg *et al.*, 1968).

TABLE II

DIFFERENCES IN HALF-LIVES OF HUMAN γG MYELOMA PROTEINS IN HUMANS

Patient	γG (g/100 ml)		$T/2$		$T/2$	Percent difference
			γG-1 vs. γG-1			
S. J.	2.1	Ing	13.1	Den	13.1	0.0
R. L.	0.8	Car	7.2	Jac	7.6	5.3
W. W.	0.7	Cut	13.0	Cov	12.1	6.9
M. F.	0.5	Car	11.5	Den	10.3	10.4
C. G.	0.6	Car	10.0	Ing	8.6	14.0
W. S.	1.5	Car	9.8	Har	12.0	18.3
F. K.	1.8	Cov	7.2	Har	9.2	21.6
C. M.	0.6	Jon	18.4	Jac	14.0	24.0
			γG-2 vs. γG-2			
M. H.	1.1	Tsc	23.0	Dah	23.0	0.0
M. Y.	0.8	Tsc	10.7	Lig	13.4	20.1
J. B.	1.2	Spa	10.2	Dah	7.7	24.6
L. B.	1.2	Lig	12.0	Dom	7.2	40.0
L. M.	0.3	Spa	12.8	Dom	6.8	46.9

since it had been shown that the Fc fragments of myeloma proteins of the same subclass are structurally very similar, except for allotypic differences (Frangione *et al.*, 1966; Grey and Kunkel, 1967) and probably have an identical amino acid sequence. In order to investigate if the differences in catabolic rates of myeloma proteins reflected differences in the Fc fragments, the rate of catabolism of Fc fragment obtained from myeloma proteins was studied in monkeys and that of H chain disease proteins in man (Spiegelberg and Fishkin, 1970). It had previously been shown that the differences in catabolic rates of individual myeloma proteins in monkeys were similar to those in man (Spiegelberg and Grey, 1968). The catabolism of 16 pairs of Fc fragments obtained from 6 IgG-1, 2 IgG-2, 2 IgG-3, and 2 IgG-4 myeloma proteins was studied. No significant differences in half-lives were demonstrated for any pair of Fc fragments obtained from IgG-1, IgG-3, and IgG-4 myeloma proteins. Fc fragments obtained from IgG-2 myeloma proteins did not persist in the circulation, probably because IgG-2 myeloma proteins can only be digested by papain in the presence of strong reducing agents (0.1 *M* mercaptoethanol) and this treatment might have abolished the ability of the IgG-2 Fc fragment to persist in the circulation. The average half-life in monkeys of the Fc fragments was 4.3 days in contrast to 7.2 days for intact myeloma proteins. The lack of a difference in the catabolic rate of Fc fragments of different heavy chain subclasses suggested that structures of IgG immunoglobulins not present on the Fc fragment are probably responsible for the heterogeneity in catabolic rates of individual myeloma proteins and for the more rapid catabolic rate of IgG myeloma proteins. Whether or not the Fab fragment contributes to the site related to the rate of catabolism or if other features of IgG, such as susceptibility to proteolysis, and tendency to aggregate, are responsible for the differences in catabolic rates of individual myeloma proteins is presently unknown. Additional evidence demonstrating the lack of catabolic heterogeneity of the Fc portion of the heavy chain was obtained through studies of the catabolism of H chain disease proteins in man. As can be seen in Table III, the plasma half-lives and fractional turnover

TABLE III

CATABOLISM OF H CHAIN DISEASE PROTEINS
Cra (γG-1), Cli (γG-2), and Zuc (γG-3) IN MAN

Recipient	Injected protein	Plasma ($T/2$)	% of Plasma pool		% Intravascular
			catabolized/day	excreted/day	
Burg.	Cra	16.5	6.2	0.65	24
	Cli	16.0	6.4	0.7	24
Page	Cra	13.2	8.8	0.8	24
	Zuc	13.2	8.5	0.8	24
Bare.	Cra	19.6	4.0	0.3	22
	Zuc	21.6	4.1	0.4	22

rates of three pairs of H chain disease proteins involving proteins of subclasses IgG-1, IgG-2, and IgG-3 did not differ significantly. In different patients the fractional turnover rate of the H chain disease proteins varied from 4.0 to 8.9% and 0.3 to 0.8% of the intravascular pool was recovered in the urine as protein-bound radioactivity. Following equilibration, 22–24% of the H chain disease proteins remained in the intravascular compartment. The catabolism of H chain disease proteins was also studied in four patients paired with normal IgG (Table IV, Fig. 6). The whole-body half-life (determined by plotting the radioactivity retained in the body) of the H chain disease proteins was only slightly shorter than that of intact IgG. The fractional turnover rate (percent of plasma pool catabolized per day), however, was on the average 1.6 times higher for the H chain disease proteins. It varied from 7.0 to 10.5% for the H chain disease proteins and from 4.5 to 6.0%

TABLE IV

CATABOLISM OF γG AND H CHAIN DISEASE PROTEINS Cra (γG-1), Cli (γG-2), AND Zuc (γG-3) IN MAN

Recipient	Injected protein	Whole-body ($T/2$)	% Plasma pool catabolized/day	% Plasma pool excreted/day	% Intravascular
Nuttl.	γG	26.5	5.0	<0.1	40
	Cra	26.0	9.0	0.4	22
Cart.	γG	28.0	5.3	<0.1	46
	Cli	27.0	7.0	0.75	28
Par.	γG	33.0	4.5	<0.1	42
	Zu	28.0	10.5	0.35	28
Buttl.[a]	γG	35.0	6.0	0.2	37
	Cli	22.0	8.0	1.5	24

[a] Diabetes insipidus.

for IgG. Less than 0.1% of the intravascular pool of IgG and 0.35 to 1.5% of the H chain disease proteins was recovered in the urine as protein-bound radioactivity. Following equilibration, 37–48% of the IgG and 22–28% of the H chain disease proteins remained in the intravascular compartment. About 6–10% of the H chain disease proteins were more rapidly eliminated than the remaining portion during the first few days following injection as shown by the whole-body elimination curve (Fig. 6). The reasons for the

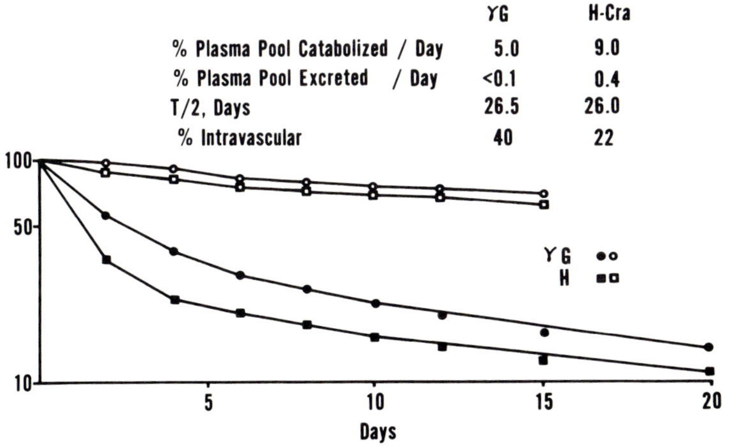

FIG. 6. Elimination of H chain disease protein Cra and normal human γG in patient Nuttl. (Spiegelberg and Fishkin, 1970).

higher fractional turnover rate of H chain disease proteins are presently unknown. It could be the result of the different molecular weights and diffusion coefficients, which appear to be responsible for the difference in the intra- and extravascular equilibration of the two proteins. It has been shown for a number of proteins that the ratio of intra- and extravascular equilibration is related to the diffusion coefficient (Nakamura *et al.*, 1968). The whole-body half-life of the H chain disease proteins was only slightly shorter than that of IgG, probably because more of the H chain disease proteins equilibrated into the extravascular compartment. It has been shown that little if any of the IgG present in the extravascular compartment is being catabolized (Anderson, 1964). Since a larger proportion of the H chain disease proteins was present in the extravascular pool and therefore not available for catabolism, the higher catabolic rate of the intravascular pool of the H chain disease proteins did not result in a significantly shorter whole-body half-life.

V. Summary

The present data demonstrate that the Fc portion of the heavy chain is largely responsible for the persistence in the circulation and rate of catabolism of IgG immunoglobulins. Fc fragments and H chain disease proteins were slowly catabolized at a rate similar to that of intact IgG, whereas $F(ab')_2$ fragments, Fab fragments, and light chains were rapidly eliminated from the circulation and catabolized. These findings are in agreement with those reported by Fahey and Robinson (1963) who demonstrated that passive administration of Fc fragments but not Fab fragments resulted in an increased turnover rate of IgG in mice and support Brambell's theory (1966) suggesting that the IgG catabolism is controlled by a specific submolecular site. However, individual myeloma proteins, which have been shown to represent a single homogeneous population of IgG molecules (Kunkel, 1965; Edelman *et al.*, 1969) differed significantly in their catabolic rates, whereas the Fc fragment obtained from these myeloma proteins were all eliminated at a similar rate. Furthermore, IgG-3 myeloma proteins were on the average catabolized more rapidly than myeloma proteins of the other subclasses, but the catabolic rate of the IgG-3 Fc fragments did not differ from that of Fc fragments of the other subclasses. Three H chain disease proteins of subclasses IgG-1, IgG-2, and IgG-3 also were catabolized at a similar rate. These data indicate that though the Fc fragment is of major importance for determining the catabolic rate of IgG immunoglobulins, some other sites within the molecule must also contribute to the submolecular structure influencing this parameter.

Acknowledgments

Supported by research grants from the American Cancer Society T-457 and the American Heart Association 67-796.

References

Andersen, S. B. (1964). "Metabolism of Human Gamma Globulin." Davis, Philadelphia, Pennsylvania.

Barth, W. F., Wochner, R. D., Waldmann, T. A., and Fahey, J. L. (1964). *J. Clin. Invest.* **43**, 1036.

Brambell, F. W. R. (1966). *Lancet* **ii**, 1087.

Brambell, F. W. R., Hemmings, W. A., Oackley, C. L., and Porter, R. R. (1960). *Proc. Roy. Soc. Ser. B* **151**, 478.

Brambell, F. W. R., Hemmings, W. A., and Morris, I. G. (1964). *Nature (London)* **203**, 1352.

Cohen, S., and Porter, R. R. (1964). *Advan. Immunol.* **4**, 287.

Edelman, G. M., and Gally, J. A. (1962). *J. Exp. Med.* **116**, 207.

Edelman, G. M., Cunningham, B. A., Gall, W. E., Gottlieb, P. D., Rutishauser, U., and Waxdal, M. J. (1969). *Proc. Nat. Acad. Sci. U.S.* **63**, 78.

Fahey, J. L., and Robinson, A. G. (1963). *J. Exp. Med.* **118**, 845.

Fleischman, J. B., Pain, R. H., and Porter, R. R. (1962). *Arch. Biochem. Biophys. Suppl.* **1**, 174.

Frangione, B., Franklin, E. C., Fudenberg, H. H., and Koshland, M. E. (1966). *J. Exp. Med.* **124**, 715.

Franklin, E. C. (1964). *J. Exp. Med.* **120**, 691.

Franklin, E. C., and Ovary, Z. (1963). *Immunology* **6**, 434.

Franklin, E. C., Lowenstein, J., Bigelow, B., and Meltzer, M. (1964). *Amer. J. Med.* **37**, 332.

Grey, H. M., and Abel, C. A. (1967). *Immunochemistry* **4**, 315.

Grey, H. M., and Kunkel, H. G. (1964). *J. Exp. Med.* **120**, 253.

Grey, H. M., and Kunkel, H. G. (1967). *Biochemistry* **6**, 2326.

Ishizaka, T., Ishizaka, K., Salmon, S., and Fudenberg, H. (1967). *J. Immunol.* **99**, 82.

Johansson, S. G. O., and Bennich, H. (1969). Personal communication.

Kunkel, H. G. (1965). *Harvey Lect.* **59**.

McConahey, P. J., and Dixon, F. J. (1966). *Int. Arch. Allergy Appl. Immunol.* **29**, 185.

Nakamura, R. M., Spiegelberg, H. L., Lee, S., and Weigle, W. O. (1968). *J. Immunol.* **100**, 376.

Nisonoff, A., Wissler, F. C., Lipman, L. N., and Woernly, D. L. (1960). *Arch. Biochem. Biophys.* **89**, 230.

Rogentine, G. N., Rowe, D. S., Bradley, J., Waldmann, T. A., and Fahey, J. L. (1966). *J. Clin. Invest.* **45**, 1467.

Solomon, A., and Tomasi, T. B. (1964). *Clin. Res.* **12**, 42.

Spiegelberg, H. L., and Fishkin, B. G. (1970). In preparation.

Spiegelberg, H. L., and Grey, H. M. (1968). *J. Immunol.* **101**, 711.

Spiegelberg, H. L., and Weigle, W. O. (1965). *J. Exp. Med.* **121**, 323.

Spiegelberg, H. L., and Weigle, W. O. (1966). *J. Immunol.* **95**, 1034.

Spiegelberg, H. L., and Weigle, W. O. (1967). *J. Immunol.* **98**, 1020.

Spiegelberg, H. L., Fishkin, B. G., and Grey, H. M. (1968). *J. Clin. Invest.* **47**, 2323

Taranta, A., and Franklin, E. C. (1961). *Science* **134**, 1981.

Terry, W. D. (1965). *J. Immunol.* **95**, 1041.

Terry, W. D., and Fahey, J. L. (1964). *Science* **146**, 400.

Weigle, W. O., and Dixon, F. J. (1959). *J. Immunol.* **82**, 516.

20

THE INFLUENCE OF GASTROINTESTINAL TRACT ON γ-GLOBULIN METABOLISM

S. Jarnum, G. Bendixen, K. B. Jensen, J. Soltoft, B. Weeke, and H. Westergaard

MEDICAL DEPARTMENT P, DIVISION OF GASTROENTEROLOGY, RIGSHOSPITALET, COPENHAGEN Ø, AND DEPARTMENT OF CLINICAL CHEMISTRY BISPEBJERG HOSPITAL, COPENHAGEN NV, DENMARK

I. Introduction

The gastrointestinal tract has been shown to play a major role in the homeostasis of immunoglobulins in both normal and pathological states. It is of significance in the synthesis of immunoglobulins, in the absorption and transport of intact γ-globulins in some newborn animals, and in the degradation of plasma proteins. Excessive loss of immunoglobulins and other serum proteins into the gastrointestinal tract is an important pathological phenomenon seen in association with a variety of disorders.

II. The Role of the Gastrointestinal Tract in Immunoglobulin Synthesis

The gastrointestinal tract is of importance both as the organ from which central lymphoid organs are derived, as well as a site of immunoglobulin synthesis.

A. Central Lymphoid Organs Derived from the Gastrointestinal Tract

The central lymphoid organs are lymphoepithelial tissues that do not participate directly in immune responses, but exert a profound influence over the development of immune capabilities by the host.

The thymus derived from the third and fourth pharyngeal pouch plays a major role in the development of capacity to manifest celular (delayed tuberculin type) immune responses, but plays a lesser role in immunoglobulin metabolism (Good *et al.*, 1962; Miller *et al.*, 1962). Thus, although thymectomized animals have a reduced ability to respond with humoral antibodies to certain antigens, they have at least normal rates of immunoglobulin synthesis (Fahey *et al.*, 1965).

Another central lymphoid organ, the bursa of Fabricius of birds, plays a major role in immunoglobulin synthesis. If the bursa is removed *in ovo*, birds cannot produce plasma cells or synthesize immunoglobulins (Cooper *et al.*, 1966a). Although the mammalian equivalent of the bursa of Fabricius has not been fully defined, it has been suggested that Peyer's patches and appendix play a comparable role. In studies by Cooper and co-workers (1966b), removal of the appendix, sacculus rotundus, and all of the Peyer's patches in young rabbits, together with total body irradiation, produced a lasting and selective suppression of humoral immunity while leaving cellular immune mechanisms intact.

B. Synthesis of Immunoglobulins by the Gastrointestinal Tract

In recent years, considerable evidence has been accumulated to indicate that cells of the gastrointestinal tract synthesize immunoglobulin molecules, and that a distinct immunological system characteristic of external secretion exists. In this regard, animals and man with cholera infection or oral polio immunization may have antibodies detectable in gastrointestinal secretions alone, or have antibody titers in these secretions exceeding those in serum.

The plasma cells of the gastrointestinal tract have been shown to contain all classes of immunoglobulin molecules as demonstrated by immunofluorescent studies. IgA containing cells predominate (Crabbe *et al.*, 1965) with IgM, IgG, IgD, and IgE containing cells present in descending order of frequency. Studies using incorporation of amino acids by small bowel biopsies into immunoglobulins *in vitro* have also shown that all major classes of immunoglobulins are synthesized by the gastrointestinal tract with IgA predominating.

In accord with the immunofluorescent studies and the biosynthesis studies, IgA is the major immunoglobulin in gastrointestinal secretions with an IgA: IgG ratio in these secretions as high as 15 : 1 compared to 1 : 6 in serum (Tomasi and Bienenstock, 1968). Studies using simultaneously administered radiolabeled albumin and IgA have shown that there is no preferential transport of IgA from the serum to gastrointestinal secretions, but that over 90% of IgA in secretions is synthesized locally.

There are a number of differences between serum IgA and that produced by the gastrointestinal secretory system (Tomasi and Bienenstock, 1968). Serum IgA is a 7 S molecule with a molecular weight of 170,000, while secretory IgA has a sedimentation coefficient of 11 S and a molecular weight of 390,000. The secretory IgA has a unique antigenic determinant, a glycoprotein with a molecular weight of 50,000–60,000, that is not present in serum IgA. This determinant, referred to as secretory piece, is probably synthesized by the epithelial cells of the gut. It seems likely that 11 S secretory IgA is a dimer of 7 S IgA coupled with the secretory piece. The secretory immunoglobulin system plays a role as the first line of defense against invading organisms.

III. Physiological Factors Regulating Immunoglobulin Transport by the Gastrointestinal Tract

Although young animals can produce some immunoglobulins, they normally depend on transmission of immunoglobulins from the mother for protection against infections in the newborn period. In many species the transmission involves transcellular transport by a saturable process that shows specificity for IgG molecules, and that may be competitively inhibited by other homologous and heterologous IgG molecules. There is considerable variation between species in terms of the time of transmission (Brambell, 1958). In rabbits, guinea pigs, monkeys, and man, transmission occurs before birth. Transmission of antibodies in the dog, mouse, and rat occurs *in utero*, as well as across the gastrointestinal tract, after birth. Predominantly postnatal transport of γ-globulins has been observed in the mink and in ungulates such as sheep, goat, cow, horse, and pig. The postulated intestinal transport appears to occur by two somewhat different mechanisms in ungulates and rodents. In ungulates there is relatively nonspecific mucosal uptake and subsequent transport to lymphatics of massive amounts of colostral protein during the first 24–36 hr of life (Payne and Marsh, 1962). The oral administration of any protein or the macromolecule polyvinylpyrrolidone will shut off mucosal uptake and transport of proteins within 5 hr of administration (Payne and Marsh, 1962). Once the luminal surface of the mucosal cell is exposed to macromolecules, the "blockade" to intestinal protein transport is complete and permanent.

In contrast to the situation in ungulates, the postnatal transmission of immunoglobulins in mice and rats occurs over a longer period (16–19 days) and is a more selective transport of protein (Halliday, 1955; Brambell, 1958; Morris, 1964). In the rat, for example, rat IgG is much more effectively transported than rat albumin, IgA, or IgM. Autologous γ-globulin is most readily transmitted; while rabbit, bovine, and sheep γ-globulin follow with descending transmissibility. Within the autologous IgG group, intact molecules are transmitted more easily then the Fc fragment of IgG which, in turn, is far more easily transported than the Fab fragment.

Related to the phenomenon of transport selectivity is the ability of certain heterologous IgG molecules or their Fc piece to competitively interfere with the transport of homologous or other heterologous IgG proteins (Brambell *et al.*, 1958; Morris, 1964). In contrast to the situation in ungulates, the transport cannot be inhibited by introduction of nonimmunoglobulin molecules into the gastrointestinal lumen. The inhibition of transport is not permanent, but persists for only a few hours following administration of the interfering molecules.

Specific transport of IgG molecules across the newborn gut of rodents appears to involve at least two steps. The initial step in the transport of protein molecules is the nonspecific uptake of protein and other macromolecules into pinocytotic vacuoles at the base of the microvilli. The specificity of transport in the face of nonspecific uptake is explained by catabolism of most proteins within the mucosal cell with protection of some IgG molecule from this catabolic process. Brambell (1966) proposed that saturable protection receptors exist on the walls of pinocytotic vacuoles. The attachment of IgG molevules to the receptor protects it from catabolism and permits transport. Transmission of heterologous γ-globulin could be accounted for by the extent to which it fits receptors. Competition for protection receptors by heterologous IgG molecules could also account for interference. The Brambell hypothesis is in accord with most of the experimental data concerning γ-globulin transport. It should be noted, however, that it lacks such direct verification as the demonstration of IgG molecules specifically bound to the membrane of the pinocytotic vacuoles.

IV. The Role of the Gastrointestinal Tract in Immunoglobulin Catabolism

A. The Normal Gastrointestinal Tract in Immunoglobulin Catabolism

γ-Globulin and other serum proteins have been demonstrated in the gastrointestinal tract of normal individuals by electrophoretic and immunological techniques (Gullberg and Olhagen, 1950; von Barandun *et al.*, 1960). Thus, the gastrointestinal tract has been shown to play a role in the physiological degradation of serum proteins. The magnitude of this process, however, has been the subject of considerable controversy. A number of workers (Wetterfors, 1965; Andersen *et al.*, 1963) have estimated the magnitude of loss of albumin and γ-globulin into the gastrointestinal tract by administering iodinated proteins intravenously, and aspirating the gastric and small intestinal secretions. These workers have estimated that from 38 to over 60% of the degradation of these proteins in normal subjects could be explained by gastrointestinal protein loss. However, minor trauma to the intestinal mucosa during the intubation and surgical procedures might have given these high estimates for gastrointestinal catabolism of serum proteins. Other studies suggest that the normal gastrointestinal tract plays a less significant role in the catabolism of serum proteins. Removal of the subdiaphragmatic gastrointestinal tracts of rodents resulted in only 0 to 10% prolongation of the survival of iodinated albumin (Gitlin *et al.*, 1956). Clearance studies using ^{51}Cr-albumin (Waldmann *et al.*, 1969), ^{67}Cu-ceruloplasmin (Waldmann *et al.*, 1967), ^{59}Fe-dextran (Jarnum *et al.*, 1968), or ^{95}Nb-albumin (Jeejeehbhoy *et al.*, 1968) suggest that only 5–15% of the normal turnover of albumin and γ-globulin could be accounted for by enteric protein loss. Thus, bulk loss of serum into the gastrointestinal tract appears to be of only minor importance in the normal metabolism of the serum proteins.

B. The Effect of Gastrointestinal Disease on γ-Globulin Metabolism

When a depressed serum albumin occurs in gastrointestinal disorders it is usually due to an abnormal protein loss, since the capacity of the liver to increase albumin synthesis is limited. In contrast, the capacity of immunoglobulin producing plasma cells is so great that synthesis can be increased manifold. Consequently, the resulting serum concentration of immunoglobulins may be normal or elevated even in the face of severe gastrointestinal protein loss. The estimation of the serum levels of immunoglobulins is, therefore, insufficient to evaluate the state of immunoglobulin metabolism in gastrointestinal disease. It can only be done by means of studies with labeled immunoglobulins.

We have studied IgG and IgM turnover in some gastrointestinal disorders known to be associated with abnormal protein loss and an altered functional state of immunologically reactive tissue (Bendixen *et al.*, 1968; Jensen *et al.*, 1969; Bendixen *et al.*, 1970). Twenty-two cases of Crohn's disease, 14 cases of ulcerative colitis, and 7 cases of intestinal lymphangiectasia were studied.

Figure 1 shows the serum concentrations of albumin and IgG, IgM and IgA in 12 cases of Crohn's disease in whom simultaneous studies were made of IgG and IgM turnover. There was no correlation between the concentration of the various immunoglobulins, nor were they related to serum albumin or the size of intestinal protein loss as estimated by means of ^{59}Fe-labeled iron dextran.

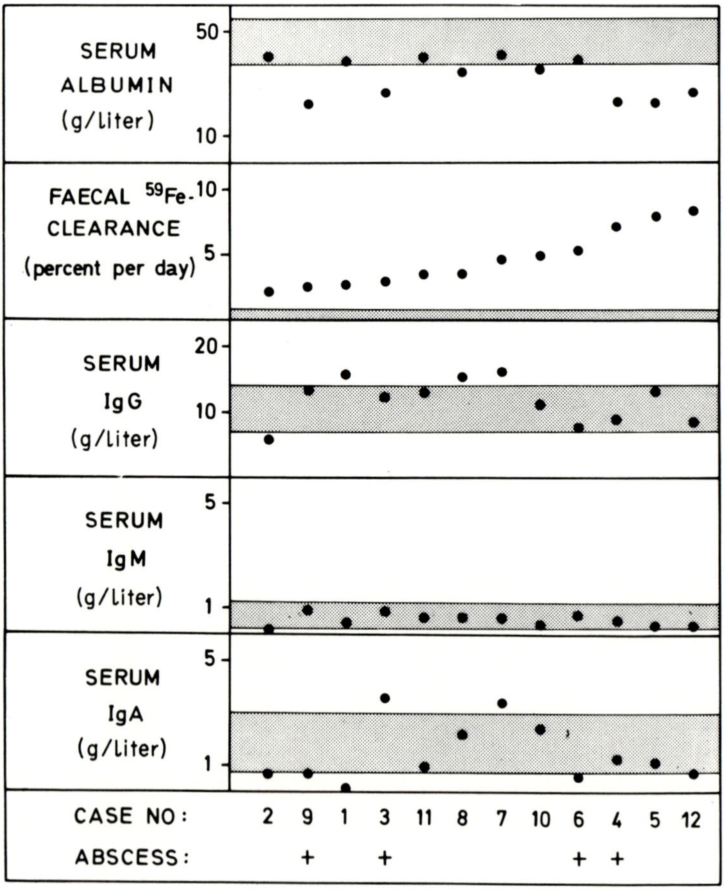

FIG. 1. Serum concentrations of albumin, IgG, IgM, and IgA in Crohn's disease (12 cases).

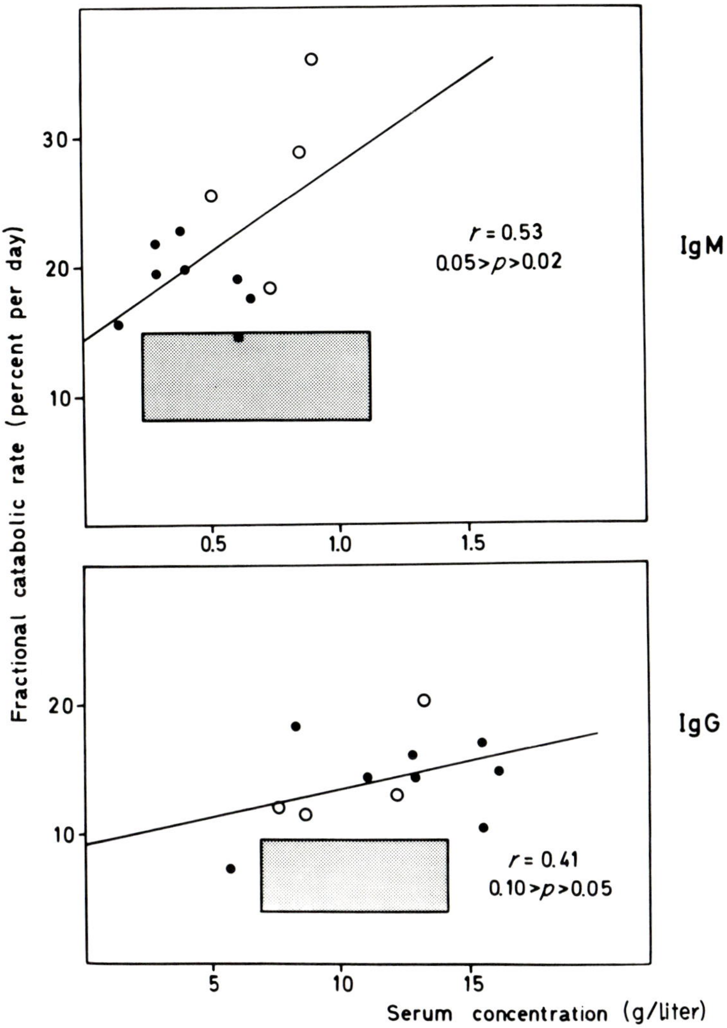

FIG. 2. Crohn's disease: ○, patient with an abscess.

In Fig. 2, it is shown that a slight positive correlation was present between serum concentration of IgM and its fractional catabolic rate. This is in contrast to the general pattern of IgM metabolism according to which catabolism of IgM is independent of its serum concentration.

For IgG the same trend of correlation was present. It was more obvious in another series studied with labeled IgG and albumin (Fig. 3) in which it is seen that, whereas an inverse relationship existed between serum albumin

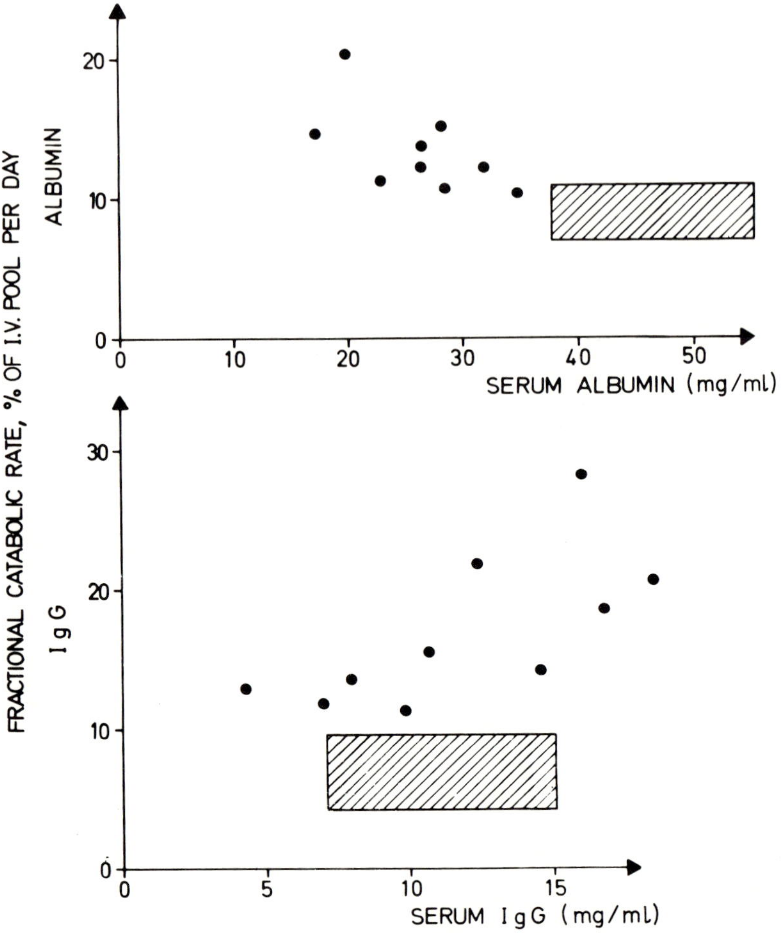

FIG. 3. Serum level and catabolic rate of IgG and albumin in Crohn's disease.

and its degradation rate, a positive relation was present as far as IgG is concerned.

A comparison between the degradation rate of IgG and albumin (Fig. 4) showed a high degree of correlation between these two functions. The regression line ($y = 1.46x - 2.8$) indicated that a certain increase of albumin catabolism was associated with an even higher (about 50%) increase of IgG catabolism. The inverse observation was made in 7 cases of intestinal lymphangiectasia in whom albumin catabolism was consistently higher than IgG catabolism ($y = 0.75x + 2.2$).

It has been claimed that the protein loss in protein-losing gastroenterop-

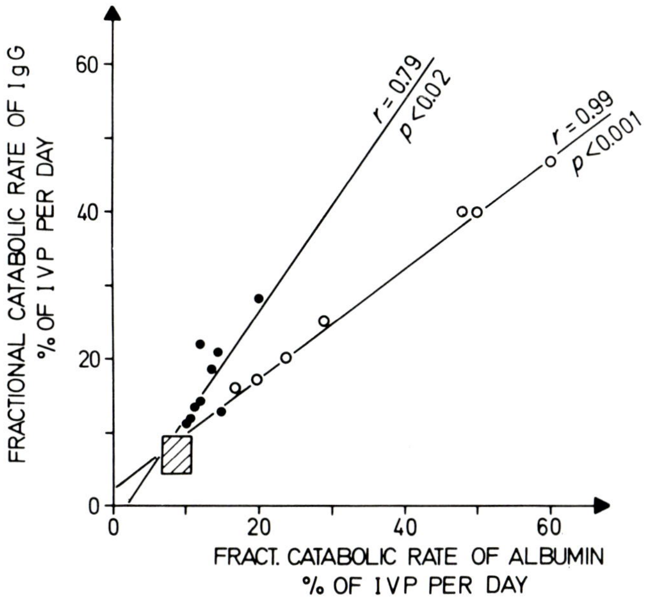

FIG. 4. Catabolic rate of albumin and IgG in Crohn's disease (●) and intestinal lymphangiestasia(○).

athy—and in particular in intestinal lymphangiectasia—is a "bulk loss" in the sense that small and large protein molecules are lost to the gut to the same extent. The finding that albumin catabolism is more increased than IgG catabolism is not incompatible with this conception. It is reasonable to believe that intestinal protein loss proceeds from the interstitial fluid of the mucosa. The interstitial concentration of IgG is lower than that of albumin, and, relative to the serum concentrations, IgG is probably also lower than albumin. Indirect evidence for this difference is furnished by turnover studies with labeled IgG and albumin which always show that the intravascular fraction of IgG is appreciably higher than that of albumin. Consequently, if a bulk loss takes place, the increase of IgG catabolism will be less than that of albumin, because catabolism is expressed as fraction of the intravascular mass catabolized per day.

Now, returning to IgG turnover in Crohn's disease (Fig. 4), it is highly improbable that the increased catabolism can be due only to intestinal protein loss. If it were so, then one must assume a selective protein loss of IgG exceeding that of albumin, either due to some hypothetic secretory process or due to a much increased interstitial IgG concentration. However, the distribution of IgG and albumin between intra- and extravascular pools in Crohn's

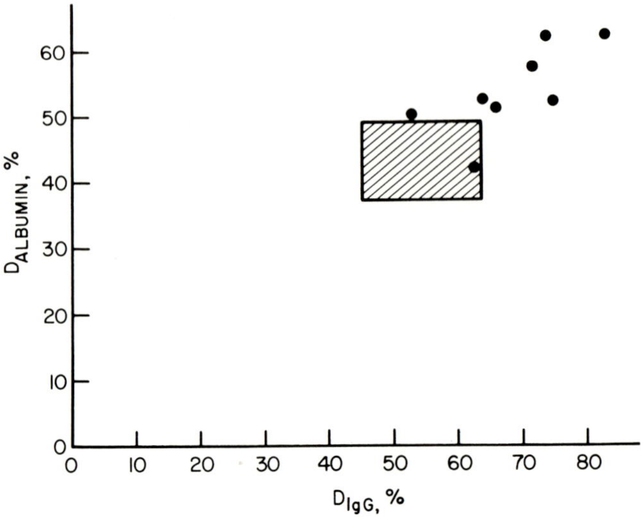

FIG. 5. Relationship of increased catabolism and selective protein loss of IgG exceeding that of albumin. D = Degradation percent of IV pool per day.

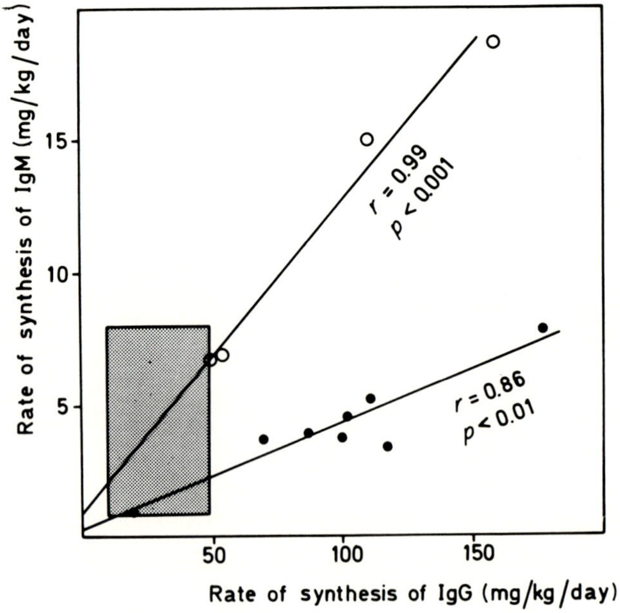

FIG 6. Crohn's disease. ○, patients with an abscess.

disease does not favor the latter possibility. The intravascular fraction of both proteins is even higher than normal (Fig. 5).

A much more likely, although still insufficient, explanation of increased IgG catabolism is that Crohn's disease affects the turnover of this immunoglobulin in the same way as other chronic inflammatory conditions like rheumatoid arthritis and chronic hepatitis. Both of these diseases result in a marked stimulation of IgG-producing cells with an abnormally high synthetic rate of IgG and, for reasons still unknown, an increased degradation rate of the IgG produced.

The synthetic rate of IgG in our case material turned out to be markedly increased, up to 9 times the normal mean, in the majority of cases.

Concerning IgM, we found that the synthetic rate was significantly correlated to that of IgG (Fig. 6). Quite unexpectedly, it turned out that patients with a complicating intraabdominal abscess presented a specific stimulation

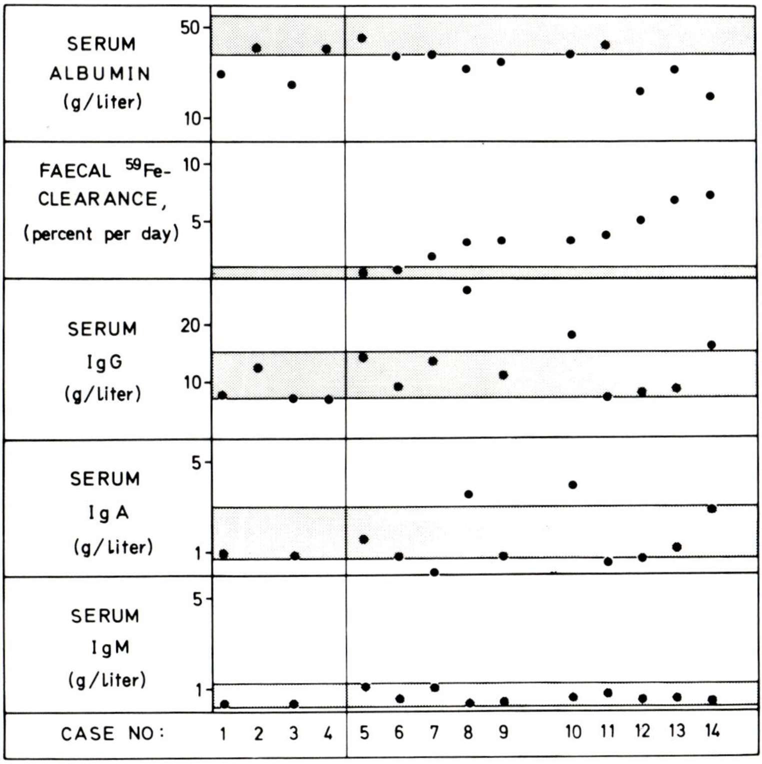

FIG 7. Ulcerative colitis (14 cases).

of IgM production, which was evidently higher than in the remaining patients in whom it fell within normal range.

V. Ulcerative Colitis

Our studies in ulcerative colitis comprised 14 patients. The results were in several respects similar to those obtained in Crohn's disease.

No characteristic change of immunoglobulin concentrations in serum was observed (Fig. 7). Abnormal intestinal protein loss estimated from fecal ^{59}Fe clearance was moderate.

An inverse relation between serum albumin and albumin catabolism was present like in Crohn's disease (Fig. 8), but no significant relation existed between serum IgG and IgG catabolism.

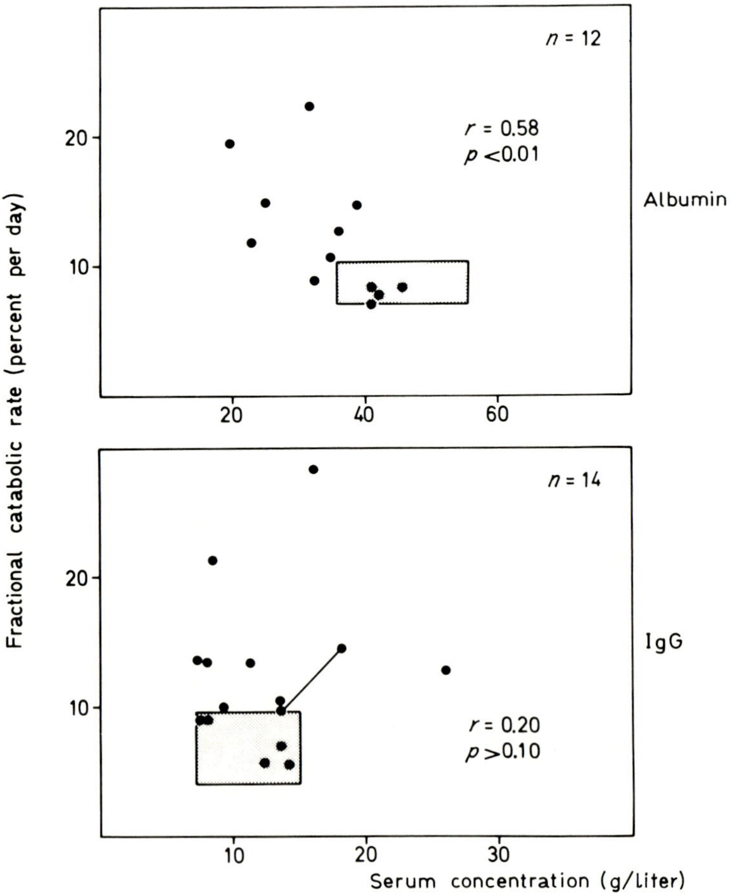

FIG. 8. Ulcerative colitis.

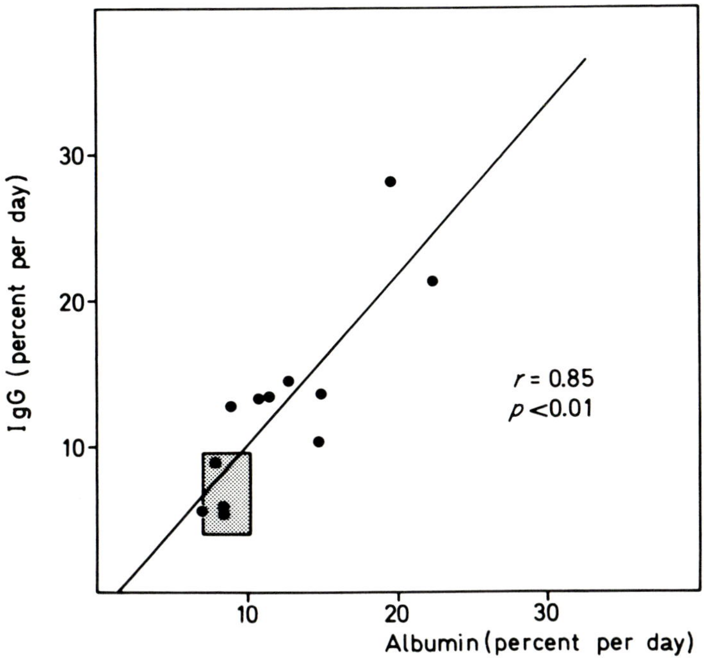

FIG. 9. Ulcerative colitis ($n = 12$): fractional catabolic rate of albumin and IgG.

Like in Crohn's disease, a significant correlation was present between the catabolic rates of IgG and albumin (Fig. 9). The regression line ($y = 1.16x - 1.42$, $r = 0.85$, $p < 0.01$) showed that the increase of catabolic rate was almost identical for the two proteins.

The synthetic rate of IgG was elevated in only half the cases (7 of 14), and the elevation was usually small.

IgM studies were made in three patients. The catabolic rate was increased in two; the synthetic rate was normal.

When the results of the studies in Crohn's disease and ulcerative colitis are compared, one may conclude: (1) The abnormal protein loss is moderate in both conditions; (2) the catabolic rate of IgG is more consistently elevated in Crohn's disease (in 21 or 22 cases) than in ulcerative colitis (9 of 14 cases). (3) The synthetic rate of IgG is also more frequently elevated in Crohn's disease (18 of 22 cases) than in ulcerative colitis (7 of 14 cases). (4) The catabolic rate of IgM is usually increased in both conditions (11 of 12 cases with Crohn's disease, 2 of 3 cases with ulcerative colitis). Whereas, (5) the synthetic rate of IgM is normal in both conditions except in patients with Crohn's disease and an intraabdominal abscess. In these patients, IgM synthesis is elevated.

Thus, in Crohn's disease the ability to produce circulating immunoglobulins seems to be more efficient than in ulcerative colitis, which is in contrast to these patients' ability to develop a state of cellular hypersentitivity. In this respect, patients with Crohn's disease are apparently less efficient than patients with ulcerative colitis. Crohn patients have no hypersensitivity of their leukocytes to colon or small intestine antigen, as opposed to ulcerative colitis where about 50% of the patients do show a cellular hypersensitivity to these antigens as judged from the leukocyte migration test (Bendixen, 1967). Furthermore, Crohn patients are unable or less able to develop cutaneous delayed hypersensitivity to dinitrochlorobenzene (Jones *et al.*, 1969).

Acknowledgments

This work was supported by grants from P. Carl Petersens Foundation, King Christian Xth's Foundation, and the Danish State Foundation of Science.

References

Andersen, S. B., Glenert, J., and Wallevik, K. (1963). *J. Clin. Invest.* **42,** 1873.

Bendixen, G. (1967). *Scand. J. Gastroenterol.* **2,** 214.

Bendixen, G., Jarnum, S., Soltoft, J., Westergaard, H., Weeks, B., and Yssing, M. (1968). *Scand. J. Gastroenterol.* **3,** 481.

Bendixen, G., Goltermann, N., Jarnum, S., Jensen, K. B., Weeks, B., and Westergaard, H. (1970). In press.

Brambell, F. W. R. (1958). *Biol. Rev. Cambridge Phil. Soc.* **33,** 488.

Brambell, F. W. R. (1966). *Lancet* **11,** 1087.

Brambell, F. W. R., Halliday, R., and Morris, I. G. (1958). *Proc. Roy. Soc. Ser. B* **149,** 1.

Cooper, M. D., Peterson, R. D. A., South, M. A., and Good, R. A. (1966a). *J. Exp. Med.* **123,** 75.

Cooper, M. D., Petey, D. Y., McKneally, M. F., Gabrielsen, A. E., Sutherland, D. E. R., and Good, R. A. (1966b). *Lancet* **1,** 1388.

Crabbe, P. A., Carbonara, A. O., and Heremans, J. F. (1965). *Lab. Invest.* **14,** 235.

Fahey, J. L., Barth, W. F., and Law, L. W. (1965). *J. Nat. Cancer Inst.* **35,** 663.

Gitlin, D., Janeway, C. A., and Farr, L. E. (1956). *J. Clin. Invest.* **35,** 44.

Good, R. A., Dalmasso, A. P., Martinez, C., Archer, O. K., Pierce, J. C., and Papermaster, B. W. (1962). *J. Exp. Med.* **116,** 773.

Gullberg, R., and Olhagen, B. (1950). *Nature* (*London*) **184,** 1848.

Halliday, R. (1955). *Proc. Roy. Soc. Ser. B* **143,** 408.

Jarnum, S., Westergaard, H., Yssing, M., and Jensen, H. (1968). *Gastroenterology* **55,** 229.

Jeejeebhoy, K. N., Jarnum, S., Singh, B., Nadkarni, G. D., and Westergaard, H. (1968). *Scand. J. Gastroenterol.* **3,** 449.

Jensen, K. B., Goltermann, N., Jarnum, S., Weeks, B., and Westergaard, H. (1969). *Gut* **10.**

Jones, J. V., Housby, J., Ashurst, P. M., and Hawkins, C. F. (1969). *Gut* **10,** 52.

Miller, J. F. A. P., Marshall, A. H. E., and White, R. G. (1962). *Advan. Immunol.* **2,** 111.
Morris, I. G. (1964). *Proc. Roy. Soc. Ser. B* **160,** 276.
Payne, L. C., and Marsh, C. L. (1962). *J. Nutr.* **76,** 151.
Tomasi, T. B., and Bienenstock, J. (1968). *Advan. Immunol.* **9,** 2.
von Barandun, S., Aebersold, J., Bianchi, R., Kluthe, R., von Muralt, G., Poretti, G., and Riva, G. (1960). *Schweiz. Med. Wochenschr.* **90,** 1458.
Waldmann, T. A., Morell, A. G., Wochner, R. D., Strober, W., and Sternlieb, I. (1967). *J. Clin. Invest.* **46,** 10.
Waldmann, T. A., Wochner, R. D., and Strober, W. (1969). *Amer. J. Med.* **46,** 275.
Wetterfors, J. (1965). *Acta Med. Scand. Suppl.* **430,** 1.

21

PROTEIN-LOSING ENTEROPATHIES OF LYMPHATIC ORIGIN

P. Vesin and Cl. Marche

DEPARTMENT OF MEDICINE, ARGENTEUIL HOSPITAL, ARGENTEUIL, FRANCE;
PATHOLOGY LABORATORY, DEPARTMENT OF MEDICINE,
BICHAT UNIVERSITY HOSPITAL, PARIS, FRANCE

I. Introduction

Since the identification by Gordon (1959) and Schwartz and Jarnum (1959) of protein-losing enteropathy, the role of the lymphatic system both in the mechanism and etiology of this syndrome has been the topic of many deep-searching investigations. Waldmann *et al.* (1962) described the major pathological lymphatic anomaly of the small intestine, "intestinal lymphangiectasia," and isolated several of the responsible etiologies. The main clinical, biological, and pathological anomalies and mechanisms found in the affected patients will be reviewed. Further the various etiologies of the protein-losing enteropathies of lymphatic origin known so far, and the management, will be considered. In the course of this chapter we shall critically assess the concept of "intestinal lymphangiectasia," both from the pathological and nosological point of view.

II. Clinical Picture

It has been possible on the basis of the many papers published in the early sixties and reviewed by Waldmann (1968) to delineate the main clinical features of this syndrome. However, as the clinical manifestations of protein-losing enteropathy and those of the lymphatic disorder responsible for protein leakage into the gut are closely intermingled, it is important to try to recognize those clinical findings that point to a lymphatic origin of the protein-losing enteropathy. (1) The gastrointestinal tract disorders, which may be very marked or mild, consisting of bouts of diarrhea, more seldom of abdominal cramps and vomiting, are the same in all forms of protein-losing enteropathy. (2) As well, the relatively rare occurrence of an increased tendency to respiratory or intestinal infections is not specific of a lymphatic etiology. (3) Manifestations of calcium deficiency such as paresthesias or attacks of tetany are not pathognomonic, but are quite common in protein-losing enteropathies of lymphatic origin. (4) Edemas of the lower limbs are usually conspicuous but a great value must be ascribed to the finding of asymmetrical, unilateral edemas localized to a leg or foot or to an upper limb with in typical cases the classic aspect of *lymphedema*, as in the cases reported by Marchal (1965) and Mistilis and Skyring (1966), and Herskovic *et al.*(1967) among others. The finding of a chylous peritoneal or pleural effusion is one of the best clinical evidences of lymphatic impairment. The fluid is milky and rich in fat, proteins, and lymphocytes. The early development of these disorders in adolescents, children, or even neonates also points to a lymphatic origin, as well as the familial occurrence reported by several authors (Waldmann, 1968).

III. Biological Findings

The main serum changes occur in the blood proteins. Marked hypoproteinemia is common and several serum proteins are usually decreased as demonstrated by chemical, electrophoretic, immunoelectrophoretic, and immunodiffusion studies: albumin, IgG, IgA, and IgM, caeruloplasmin, and transferrin. The decrease in innumoglobulins is only due to the loss of these proteins into the gut and not a decreased synthesis. Waldmann and Strober (1968) have shown that in response to stimulation with antigens the patients respond with a normal or even increased antibody synthesis. These observations explain the relative rarity of severe infections in these patients.

Blood cholesterol and total lipids are either normal or decreased.

Blood calcium is usually decreased as well as calciuria. The serum iron level is frequently decreased. The serum magnesium level is also decreased. The levels of the other serum electrolytes, Cl, Na, K, are usually normal or only moderately decreased, apart from a few cases with severe diarrhea.

There is no proteinuria, but an increased aminoaciduria is frequently discovered, as reported by Nussle *et al.* (1962).

In a few cases normochromic anemia has been reported but the differential count usually shows a major anomaly. A significant or very marked decrease in the lymphocyte count is present. This lymphocytopenia points to a lymphatic origin of the protein-losing enteropathy as emphasized by Waldmann (1968) and Vesin and Cattan (1969). Deep lymphocytopenia is responsible for the loss of delayed hypersensitivity (tuberculin, candidin tests, etc.) and lack of rejection of skin homografts as reported by Waldmann and Strober (1968).

To sum up, the observation in a patient with a picture suggestive of protein-losing enterpathy (edemas, diarrhea, unexplained hypoproteinemia) of findings such as lymphedema, chylous effusions, and lymphocytopenia are in favor of the lymphatic origin of this enteropathy. A final diagnosis will require the use of several investigative procedures designed to assess intestinal function impairment, demonstrate the loss of plasma proteins and eventually of calcium into the gut, and, at last, identify a lymphatic origin.

IV. Diagnostic Studies

Routine workup should include X rays which usually show a "malabsorption pattern." Further steatorrhea is common, but usually moderate. The absorption of carbohydrates (D-xylose test, oral glucose tolerance test) will as a rule be normal. The observation of steatorrhea *and* normal carbohydrate absorption is suggestive of a lymphatic origin of the enteropathy as is well known. The absorption of vitamin A and carotene is usually normal, but vitamin B_{12} absorption may be impaired as reported by Herskovic *et al.* (1967). Endogenous calcium loss into the gut occurs and is an important finding (Milhaud and Vesin, 1961). Finally, protein leakage into the gut should be demonstrated.

To ascertain the lymphatic origin of a protein-losing enteropathy, several procedures can be used.

(1) Duodenal intubation in some cases may, as reported by Vesin *et al.* (1965) and Stoelinga *et al.* (1963), show the presence in the duodenum of a large amount of milky fluid containing an increased amount of proteins and, eventually, of fat. The calcium level of the fluid is normal (around 50–60 mg/liter) but the amount of fluid lost may be considerable, eventually resulting in a net loss of calcium.

(2) Peritoneoscopy may demonstrate dilated lymphatic channels on intestinal loops; a very unusual finding in a fasting subject. Such observations have been made by de Ruyters *et al.* (1964), Roberti *et al.* (1965), and Vesin *et al.* (1965).

(3) Although lymphography only studies the deep prevertebral lymphatic channels and lymph nodes, the cysterna chili and the thoracic duct, it may prove useful. The existence of lymphatic anomalies of the lower limbs, or pelvic and abdominal lymphatic system, such as agenesis or hypogenesis, agenesis, stenosis, or compression of the thoracic duct, or leakage of the injected dye into the small intestine may be found.

(4) Exploratory laparotomy may be necessary for diagnostic purposes. In typical cases dilated lymphatic channels on the small intestine will be seen. A dark pigmentation of the intestine may also be noted. Mesenteric anomalies (enlarged lymph nodes, malformation, tumors) can be recognized.

(5) A major examination should be small intestinal biopsy. It will bring the ultimate proof of the lymphatic origin of a protein-losing enteropathy: the existence in a grossly normal small intestine of an abnormal accumulation of lymph, which may assume several aspects, lymphangiectasia being only one of them.

V. Pathological Study

We shall consider here only the anomalies of the small intestine that have been carefully studied by many authors following the initial studies of Waldmann *et al.* (1962).

A. Macroscopic Findings

Whatever the etiology, the small intestinal findings are very much the same. The jejunum is always involved, thickened, edematous, sometimes with a dark brown pigmentation (due to the unexplained presence of a lipochrome pigment in the smooth muscle cells of the muscularis). Also very typical is the existence of dilated serosal lymphatic channels, with the aspect of contorted whitish vessels; along these dilated channels, nodules may be found; lymph accumulation may also assume the aspect of white areas or bands.

B. Microscopic Findings

The mucosa is practically always normal, the villi are of normal height, and usually no abnormal cellular infiltration is found. The shape of the villi may be somewhat distorted by distension of the lamina propria by edema. The lymphatic channels are dilated, in the lamina propria and especially at the tip of the villus. The dilatation of the central lymphatic channel of the villus is very typical. The lymphatic channels can be also dilated in the muscularis mucosa and in the submucosa. Foamy fat-loaded macrophages may be seen inside the lumen of the dilated lymphatics. These dilated lymphatics have been usually given the name of "lymphangiectasias."

However, associated with these lymphangiectasias, other aspects of lymph accumulation can be observed. Vesin *et al.* (1970) have shown the existence inside the villus, directly under the mucosal cells, of a cell-poor edema, sometimes very conspicuous and resulting in a detachment of the mucosa. The presence of such lymphatic bullae is not always associated with a dilatation of the central lymphatic of the mucosa and these authors emphasize the value of this aspect which, despite the absence of mucosal lymphangiectasia, is clearly in favor of the existence of protein-losing enteropathy of lymphatic origin. The accumulation of lymph just under the mucosal cells is obviously related to the dilatation by fluid of the intercellular spaces of the villous epithelium seen at electron microscopy. At last, in some cases, blood capillaries may also be more or less dilated.

C. Electron Microscopy

Dobbins (1966) made a very interesting electron microscopic study of the dilated lymphatic channels in the intestinal mucosa of a patient with lymphatic malformations and protein-losing enteropathy. This author confirmed the integrity of the absorptive cells and also noted the widened intercellular spaces at the base of intestinal cells. He found no ruptures of the dilated lymphatic channels, but only an increased number of endothelial filaments in the lymphangiectasia, suggesting the existence of an increased hydrostatic pressure within the lymphatic channel. Inside the lumen of the dilated lymphatics, there were lipophages and chylomicrons. As no intrinsic lesion of the lymphatic vessel was found, Dobbins (1966) considered these aspects as resulting from a mechanical block to lymphatic outflow.

Bierring *et al.* (1965) were the first to describe these dilated intercellular spaces, as lacunae filled with a cell-poor fluid containing occasionally lymphocytes and a precipitated, floccular material, also seen by Dobbins and considered to be made of fat and proteins. The lacunae between the cells are mainly found at the base of the bells, but also close to the lumenal surface; however, these dilated spaces have never been observed to open on the lumenal surface of the epithelium.

Laumonnier *et al.* (1970) have made similar observations and also consider, like Bierring *et al.* (1965) that the loss of lymph and proteins takes place between the epithelial cells and not through them.

VI. Mechanisms

Three problems will be discussed: (1) the cause of the accumulation of lymph in the small intestine; (2) the mechanism of lymph leakage into the gut; and (3) the interpretation of small intestinal functional disorders.

A. Intestinal Lymph Accumulation

The accumulation of lymph is primarily the consequence of an impaired lymph flow from the small intestine due to various lesions that impair the flow of lymph along the main lymphatic channels in the abdomen and chest. Experimental studies in animals have confirmed the observations made in patients. Marshall *et al.* (1965) ligated the thoracic duct in dogs and Milhaud and Guillien (1966) performed the same experiment in rats. These authors observed that thoracic duct ligation was accompanied by protein leakage into the gut. If impaired lymph flow is the major cause of protein-losing enteropathy of lymphatic origin, the question may be raised (Wagner *et al.*, 1965) whether venous compression resulting in hyperproduction of lymph in the small intestinal mucosa cannot induce the same picture of lymph accumulation and leakage into the small intestine. Along with lymphatic vessel dilatation, venous and capillary dilatations have been noted (Vesin *et al.*, 1964). Although venous compression in the portal system is not accompanied by protein loss into the gut, and experimental compression in the upper vena cava system [as in the experiments of Milhaud and Guillien (1966)] failed to induce protein loss into the gut, other studies made in constrictive pericarditis are worth mentioning. Venous congestion in the lower vena cava system results in increased lymph production in the small intestine and it is conceivable that if the amount of lymph produced is greater than the capacity of transport of the lymphatic system, even if the latter is anatomically normal, intestinal lymph stasis may develop. Surgical treatment of constrictive pericarditis will then relieve venous hyperpression and lymph and protein leakage will disappear, as often is the case in these patients. On the contrary, if there are lymphatic lesions, such as thickening of the lymph channels and fibrosis of the abdominal nodes in the case of

Petersen and Hastrup (1963) or definitive obstruction of the origin of the thoracic duct in the case reported by Vesin and Cattan (1969), the suppression of venous hyperpression will not be sufficient to suppress lymph leakage into the gut, as an impaired lymph flow persists. Lymphography, as stressed by Vesin and Cattan (1969) has proven of great value to differentiate these cases in which venous hyperpression is the major and primary disorder, the impairment of lymph transport being functional and secondary, from these cases in which, along with increased venous pressure, there are lesions of the lymphatic vessels with lasting lymph flow impairment.

B. Leakage

The convincing demonstration by Bierring *et al.* (1965) and Laumonnier *et al.* (1970) that lymph leakage into the gut lumen between the intestinal cells, but not through the cells, accounts for the bulk loss of all components of lymph: proteins, cholesterol and lipids, calcium, lymphocytes. It should be recalled here that the isotopic studies of Waldmann (1965) in patients with intestinal lymphangiectasia showed that the percentage increases above normal in the catabolic rate for IgG, IgM, and albumin were similar when expressed as percentages of the corresponding intravascular pools turned over per day.

C. Functional Disorders

At times the existence of steatorrhea occurs in the presence of normal carbohydrate absorption. As the small intestinal mucosa is practically normal, apart from the existence of edema carbohydrate absorption is normal and the mechanism of steatorrhea appears to be complex. Malabsorption of fat may be accused, but exudation, "back flow," of fat from the dilated lymphatics also seems to play a role as considered by Mistilis *et al.* (1965) who observed a steatorrhea of 8.5 g/24 hr under a completely fat-free diet in a patient with intestinal lymphangiectasia. In some cases, with high steatorrhea, lesions of the exocrine pancreas secondary to lymph stasis have been observed by Mizuno *et al.* (1966) who incriminated exocrine pancreas insufficiency.

D. Specific Factors

"Intestinal lymphangiectasia" is not a disease, not even a syndrome, but only the more typical morphological anomaly resulting from lymph accumulation in the small intestinal wall due to an impaired lymph flow usually caused by anatomical lesions and possibly functional mechanisms. Therefore, it is easily conceived that many diseases with this impaired lymph flow will be found which show protein-losing enteropathies of lymphatic origin.

1. *Lymph Channels*

The lymphatic channels may be sclerotic (MacDonagh *et al.*, 1965; Mizuno *et al.*, 1966) and the thoracic duct may be the site of a localized stenosis (Mizuno *et al.*, 1966). The abdominal lymphnodes may be involved by infection, tuberculosis, lymphosarcoma, or Hodgkin's disease. Typical cases have been reported by de Ruyters *et al.* (1964) and Pequignot *et al.* (1969).

Diffuse agenesis of hypogenesis of lymphatics of the lower limbs, and of pelvic and abdominal lymphatics is relatively common. The onset is very early, sometimes at birth. Lymphedema is frequently seen in these cases as well as chylous effusions. Lymphography is very useful to assess the extension of this process, most likely malformative, as shown by Pomerantz and Waldmann (1963). The cases published have been reviewed by Waldmann (1968). This picture was also observed in the patients of Herskovic *et al.* (1967) and Marchal (1965).

2. *Compression*

Compressions of the main lymphatic channels in the mesenterium and at the origin of the thoracic duct can result in a typical picture of protein-losing enteropathy of lymphatic origin, as in the cases reported by Vesin *et al.* (1964) and Roberti (*et al.* 1965). These patients presented malignant or benign tumors. In other cases, the lymphatic flow is impaired by diseases of the mesenterium proper: mesenterial malformation as reported by Marchal (1965); retractile mesenteric panniculitis as in the case of Soergel and Hensley (1966); and mesenterial fibrosis as reported by Vesin *et al.* (1970).

3. *Cardiovascular Diseases*

Several cardiovascular affections can be accompanied by protein-losing enteropathy with leakage of lymph into the gut. The best known is constrictive pericarditis studied by Linazasoro (1962), Kaihara *et al.* (1963), Petersen and Hastrup (1963), Plauth *et al.* (1964), Cardi *et al.* (1965, Mizuno *et al.* (1966), Vesin and Cattan (1969), and Wilkinson *et al.* (1965). More seldom, some forms of cardiomyopathy, atrial septal defect, and pulmonary stenosis as reviewed by Waldmann (1968) have been complicated by lymph leakage into the gut.

4. *Lymphatic Fistulae*

Lymphatic fistulae into the small intestine are very uncommon and only two proven cases have been reported; one by Vesin *et al.* (1965), probably secondary to thoracic duct agenesis; the other by Riva *et al.* (1964), secondary to acute pancreatitis. It should be noted here that opacification of the small intestine by the dye following lymphography is not an absolute evidence of a gross fistula into the gut: As in the case of Mistilis *et al.* (1965), exploratory laparotomy failed to detect any macroscopic fistula.

In all these diseases, the lymphatic origin of the protein-losing enteropathy is unquestionable. It is less obvious in other diseases such as Whipple's disease studied by Laster *et al.* (1962) or in the peculiar case of Zollinger-Ellinson reported by Martini *et al.* (1964). It is conjectural in Crohn's enteritis or in severe malabsorption although Vesin (unpublished study) raised the possibility of a role of the lymphatics in a patient with severe malabsorption and protein leakage, in whom dilated lymphatics were found in the submucosa. At last, it should not be forgotten that lymphatic impairment is found only in a fraction of the cases of protein-losing enteropathies, as emphasized by Stewart and Stewart (1965) and others.

VII. Treatment

A. Medical

In most cases, as the lesions responsible for lymph accumulation in the small intestinal wall and subsequent leakage into the gut are not amenable to surgery (tumors, malignant diseases of abdominal lymph nodes, extensive lymphatic agenesis resulting in lymph accumulation in the greatest part of the small intestine). Medical treatment can be resorted to decrease edema: Albumin infusions and diuretics have proved useful. A more ambitious treatment aims at reducing lymph production and in consequence, lymph accumulation by practically suppressing fat from the diet as done by Jeffries *et al.* (1964) and Jarnum and Jensen (1965) or by replacing completely in the diet the long-chain triglycerides transported by the lymph by medium-chain triglycerides (MCT) transported via the portal blood as in the cases of Holt (1964) and Herskovic *et al.* (1967). This attempt has proven effective and fairly good results have been reported by these authors: Increase in serum protein level, decrease in protein leakage, reduction of diarrhea, and decrease or even disappearance of dilated intestinal lymphatics.

B. Surgical

Apart from the treatment of constrictive pericarditis which is usually followed by the disappearance of all disorders due to lymph leakage (when there is no definite impairment of lymph flow as in the cases reported by Petersen and Hastrup (1963) and by Vesin and Cattan (1969), indications for surgery are few. Extensive resection of the small intestine is usually ineffective as indicated by Waldmann *et al.* (1962), despite an occasional good result such as that reported by Jeejeeboy (1962).

Similarly, operations aiming at decreasing lymphatic pressure by lymphovenous anastomosis, as reported by Mistilis *et al.* (1965), or by peritoneovenous diversion, as reported by Vesin *et al.* (1970), have proven ineffective, despite transiently good results. Only in a patient with stenosis of the thoracic duct, observed by Mizuno *et al.* (1966), was the anastomosis between the prestenotic thoracic duct and the left vertebral vein successful. In spite of these poor results, exploratory laparotomy is frequently indicated in patients with protein-losing emteropathy of lymphatic origin to make an accurate assessment of the responsible lesions. Leakage of the dye used for lymphography into the gut requires a careful exploration of the pancreaticoduodenal area to detect a lymphatic fistula into the gut, which can be closed as in the cases reported by Wenzl *et al.* (1964) and by Vesin *et al.* (1965).

References

Bierring, F., Jarnum, S., and Swchartz, M. (1965). *In* "Physiology and Pathophysiology of Plasma Protein Metabolism" (H. Koblet, P. Vesin, H. Diggelmann, and S. Barandun, eds.), pp. 201–208. Huber, Berne.

Cardi, E., Colarizzi, P., Marzano, E., and Castello, M. A. (1965). *Riv. Clin. Pediat.* **76,** 249.

de Ruyters, A., Brassine, A., and Lisin, N. (1964). *Acta Gastro-Enterol. Belg.* **27,** 505.

Dobbins, W. O., III (1966). *Gastroenterology* **51,** 1004.

Gordon R. (1959). *Lancet* **i,** 325.

Herskovic, T. D., Winawer, S. J., Goldsmith, R., Klein, R., and Zamcheck, W. (1967). *Pediatrics* **40,** 345.

Holt, P. R. (1964). *Pediatrics* **34,** 629.

Jarnum, S., and Jensen, H. (1965). *In* "Physiology and Pathophysiology of Plasma Protein Metabolism" (H. Koblet, P. Vesin, H. Diggelmann, and S. Barandun, eds.), pp. 210–217. Huber, Berne.

Jeejeebhoy, K. N. (1962). *In* "Plasma Proteins and Gastro-Intestinal Tract in Health and Disease" (M. Schwartz and P. Vesin, eds.), pp. 161–162. Munksgaard, Copenhagen.

Jeffries, G. H., Chapman, A., and Sleisenger, M. H. (1964). *New Engl. J. Med.* **270,** 761.

Kaihara, S., Nishimura, H., Ayagi, T., Kameda, H., and Veda, H. (1963). *Jap. Heart J.* **4,** 386.

Laster, L., Waldmann, T. A., Fenster, L. F., and Singleton, J. W. (1962). *Gastroenterology* **42,** 762.

Laumonnier, R., and Marche, C. (1970). *Ann. Anat. Pathol.*, to be published.

Linazasoro, J. Ma. (1962). *In* "Plasma Proteins and Gastro-Intestinal Tract in Health and Disease" (M. Schwartz and P. Vesin, eds.), pp. 164–173. Munksgaard, Copenhagen.

MacDonagh, T. J., Gueft, B., Pyun, K., and Arias, I. M. (1965). *Gastroenterology* **48,** 642.

Marchal, C. (1965). "Les gastroenteropathies exsudatives chez l'enfant." Thomas, Nancy.

Marshall, W. H., Neyazaki, T., and Abrams, H. (1965). *New Engl. J. Med.* **273,** 1092.

Martini, G. A., Strohmeyer, G., Haug, P., and Gusek, W. (1964). *Schweiz. Med. Wochenschr.* **89,** 313.

Milhaud, G., and Guillien, P. (1966). *In* "Labeled Proteins in Tracer Studies" (L. Donato, G. Milhaud, and J. Sirchis, eds.), pp. 159–163. Euratom, Brussels.

Milhaud, G., and Vesin, P. (1961). *Nature (London)* **191,** 872.

Mistilis, S. P., and Skyring, A. P. (1966). *Amer. J. Med.* **40**, 634.
Mistilis, S. P., Skyring, A. P., and Stephen, D. D. (1965). *Lancet* **i**, 77.
Mizuno, Y., Yasuga, N., Sanada, K., Matsumoto, I., Iwasaki, M., Kayashima, Y., Kumir, K., and Kajiyama, Y. (1966). *Acta Gastro-Enterol. Belg.* **29**, 248.
Nussle, D., Kaser, H., and Barandun, S. (1962). *In* "Plasma Proteins and Gastro-Intestinal Tract in Health and Disease" (M. Schwartz, P. Vesin, eds.), pp. 97–103. Munksgaard, Copenhagen.
Pequignot, H., Guerre, J., Debray,J., Christoforov, B., Van Amerongen, P., and Cocheton, J. J. (1969). *Sem. Hop.* **45**, 1758.
Petersen, V. P., and Hastrup, J. (1963). *Acta Med. Scand.* **173**, 401.
Plauth, W. H., Jr., Waldmann, T. A., Wochner, R. D., Braunwald, N. S., and Braunwald, E. (1964). *Pediatrics* **34**, 636.
Pomerantz, M., and Waldmann, T. A. (1963). *Gastroenterology* **45**, 703.
Riva, G., Barandun, S., Koblet, H., Nussle, D., and Witschi, H. P. (1964). *Protides Biol. Fluids Proc. Colloq.* **12**, 168.
Roberti, A., Vesin, P., Milhaud, G., and Viguie, R. (1965). *Arch. Mal. App. Dig. Mal. Nutr.* **54**, 91.
Schwantz, M., and Jarnum, S. (1959). *Lancet* **i**, 327.
Soergel, K. H., and Hensley, G. T. (1966). *Gastroenterology* **51**, 529.
Stewart, R. D. H. and Stewart, J. H. (1965). *Gut* **6**, 146.
Stoelinga, G. B. A., Van Munster, E. J. J., and Sloof, J. P. (1963). *Pediatrics* **31**, 1011.
Vesin, P., and Cattan, D. (1968). *In* "Physiology and Pathophysiology of Protein Metabolism" (G. Birke, R. Norberg, and L.-O. Plantin, eds.), pp. 155–161. Macmillan (Pergamon), New York.
Vesin, P., Milhaud, G., and Cattan, R. (1964). *Protides Biol. Fluids Proc. Colloq.* **12**, 275.
Vesin, P., Roberti, A., Bismuth, V., Desprez-Curely, J. P., Desbuquois, G., and Viguie, R. (1965). *In* "Physiology and Pathophysiology of Plasma Protein Metabolism" (H. Koblet, P. Vesin, H. Diggelmann, and S. Barandun, eds.), pp. 179–185. Huber, Berne.
Vesin, P., Marche, C., Hivet, M., and Viguie, R. (1970). To be published.
Wagner, A., Plugge, H., and Neicker, H. (1965). *Deut. Arch. Klin. Med.* **210**, 22.
Waldmann, T. A. (1965). *In* "Radio-Isotope Techniques in the Study of Protein Metabolism" Tech. Report Series **45**. pp. 162–163. *Int. At. Energy Ag.*, Vienna.
Waldmann, T. A., and Strober, W. (1968). *In* "Physiology and Pathophysiology of Plasma Protein Metabolism" (G. Birke, R. Norberg, and L.-O. Plantin, eds.), pp. 171–177. Macmillan (Pergamon), New York.
Waldmann, T. A., Gordon, R. S., Jr., Dutciter, R. F., and Wertlake, P. T. (1962). *In* "Plasma Proteins and Gastro-Intestinal Tract in Health and Disease" (M. Schwartz and P. Vesin, eds.), pp. 156–169. Munksgaard, Copenhagen.
Wenzl, H., Ganzoni, A., Siegenthaler, W., Rosenmund, H., Curtius, C., Spuhler, O., and Kaiser, E. (1964). *Deut. Med. Wochenschr.* **89**, 1247.
Wilkinson, P., Pinto, B., and Senior, J. R. (1965). *New Engl. J. Med.* **273**, 1178.

VI

Acute Phase Proteins: Stress—Trauma

22

THE EFFECTS OF TRAUMA AND PARTIAL HEPATECTOMY ON THE RATES OF SYNTHESIS OF PLASMA PROTEINS BY THE LIVER

A. H. Gordon

NATIONAL INSTITUTE FOR MEDICAL RESEARCH,
MILL HILL, LONDON, N.W.7, ENGLAND

I. Introduction

As a result of injury the concentrations of many of the plasma proteins increases considerably. The nature of the stimuli which can be responsible for the increased rate of synthesis of these proteins varies widely and detailed studies have revealed some aspects of the metabolic changes which are involved. By consideration of these changes, some insight may be obtained into the mechanisms by which, under normal conditions, the rates of synthesis and of degradation of these proteins are held constant and synthesis is increased relative to degradation after tissue damage. Probably the best known of the proteins belonging to this group, which have been referred to as the acute phase (AP) proteins, are orosomucoid, the C-reactive proteins found in man and rabbit, haptoglobin, ceruloplasmin, transferrin, and fibrinogen and, in addition, in the rat, the α_2-AP globulin (Weimer and Benjamin, 1965, 1966; Heim and Ellenson, 1967), and the α_1-AP protein studied by Darcy (1963). However, this term will only be used in the following pages in referring to proteins so named by other authors, because recent work has shown that the concentrations of very many of the plasma proteins are increased by trauma to a greater or lesser degree. Thus no sharp dividing

TABLE I

PROTEINS OF HUMAN PLASMA SHOWING ALTERED CONCENTRATION AFTER TRAUMA[a]

Concentration	Proteins	Concentrations in plasma (% of preoperative values)
Increased	Fibrinogen	>200
	Haptoglobin	206
	Orosomucoid	>200
	C-Reactive protein	>200
	α_1-Antitrypsin	>200
	Slow β-globulin	173
	Inter α-globulin	189
	Complement C′3	122
	Caeruloplasmin	124
	Easily precipitable glycoprotein	140
Unchanged	More than 30 other proteins	
Decreased	Thyroxine binding prealbumin	69
	α-Lipoprotein	—
	β-Lipoprotein	77
	Transferrin	78
	Albumin	80

[a] All the patients had undergone minor surgery 8 hr or more before the second blood sample was taken. Concentrations of all proteins with the exception of C-reactive protein and orosomucoid were obtained by quantitative immunoelectrophoresis (Minchin Clarke and Freeman, 1968).

line can be drawn between reactive and nonreactive proteins. Furthermore, as shown in Table I and Fig. 1 it is now clear that the concentration of certain plasma proteins decreases as a result of those same stimuli which lead to increased concentration of many other plasma proteins. All these changes are irrespective of dilution occurring after trauma due to transfer of fluid from the extravascular compartment into the plasma (Hoffenberg, 1969).

The application of many different stimuli, including rather mild procedures such as the subcutaneous injection of talc, are capable of causing marked alterations in the relative rates of synthesis of certain plasma proteins. After partial hepatectomy, greater changes in rate of synthesis are found to occur than would be expected as a result of the preliminary operative procedure if this were carried out without removal of part of the liver. The altered rates of synthesis which follow trauma are found to depend markedly on the

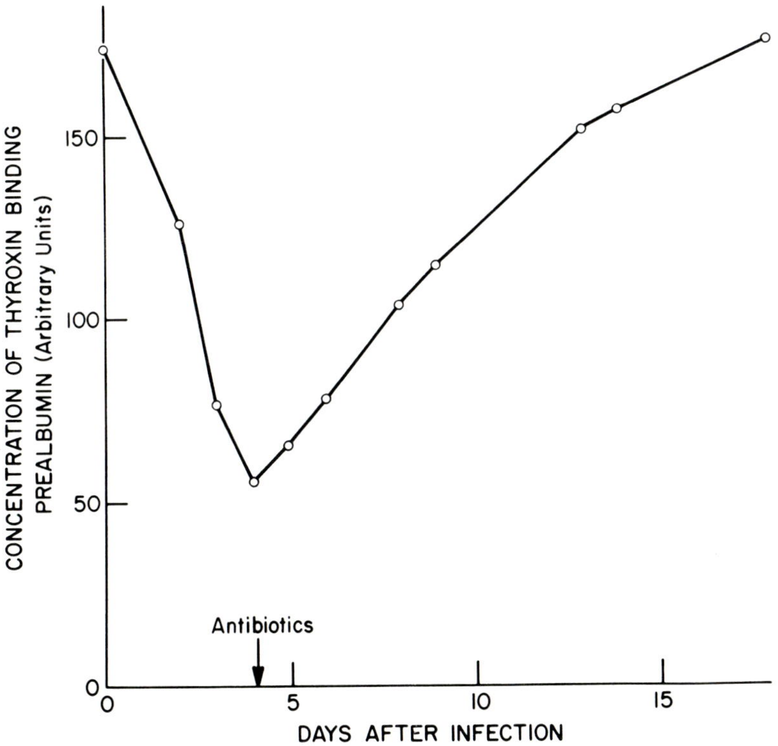

FIG. 1. Plasma concentrations of Thyroxine binding prealbumin at various times after the onset of a minor febrile illness. Estimations by quantitative immunoelectrophoresis (Minchin Clarke and Freeman, 1966).

previous history of the animal undergoing treatment. Thus a second stimulation will usually bring about a greater increase than the first (Weimer and Humelbaugh, 1967; Morgan, 1969). Stimuli of several different kinds have been shown to affect the rate of synthesis of transferrin (Morgan and Lynch, 1963). In the case of this protein, increased rates of synthesis occur not only as a result of tissue damage but also after certain nontraumatic stimuli, i.e., a reduction in the partial pressure of 0_2.

Thus an alteration in the environment to one in which an increased concentration of transferrin is physiologically advantageous acts as a stimulus for increased synthesis of this protein. Finally, it is important to note that in the unstimulated animal the concentration of many plasma proteins varies considerably according to age (Weimer *et al.*, 1967). Sometimes, as for instance in rats, in the absence of stimulation, a plasma protein (the α_2-AP globulin of Weimer) only occurs at a detectable concentration during the embryonic and neonatal stages. In this respect these plasma proteins behave similarly to the inducible liver enzymes, large changes in activity and concentrations of which also occur at birth (Greengard, 1967). That similar mechanisms control the synthesis of those liver enzymes inducible by cortisol and certain plasma proteins is also suggested by the fact that the synthesis of the α_2-AP globulin of Weimer does not take place after adrenalectomy. For a full description of the factors involved in induction of liver enzymes reference should be to Hager and Kenney (1968) and Greengard (1969).

II. Evidence Concerning Control of Rates of Synthesis Obtained from Studies *in Vivo*

A. Localization by Immunofluorescence

That liver cells are to some degree specialized for the synthesis of particular plasma proteins has been proved by means of immunofluorescent labeling (Hamashima *et al.*, 1964). Peters and Alper (1966) demonstrated by the same method that whereas haptoglobin is present in only 2.4% of the parenchymal cells of the liver of a normal dog, at 65 hr after injection of turpentine, 11.2% of the cells stained specifically for haptoglobin. Because the percentage of binucleate fluorescent cells doubled as a result of the stimulation it seems probable that the increased synthesis of haptoglobin resulted to a considerable extent from proliferation of those parenchymal cells capable of synthesizing this protein.

Increased synthesis of transferrin occurs after rats have lost blood. By means of immunofluorescence, Lane (1968) showed that after a lag period of 2 days during which much less than normal amounts of transferrin could be demonstrated in rat liver parenchymal cells the proportion of such cells containing the protein rose to between 60 and 70%. Because such a large proportion of liver cells can be caused to synthesize transferrin it seems certain that increased synthesis of more than one plasma protein must occur in certain liver cells after appropriate stimulation. In addition, however, an as yet undiscovered control mechanism capable of selecting certain cells for the preferential synthesis of particular plasma proteins may exist. Furthermore that those parenchymal cells in which the plasma proteins have been synthesized may act as a reservoir from which a rapid release into the plasma may subsequently occur seems to have been demonstrated clearly.

B. Hormonal Control

Among the numerous consequences of trauma are alterations in the plasma concentrations of several hormones. Thus it is reasonable to inquire whether these changes are in turn the main cause of the alteration in rates of synthesis of any of the plasma proteins. Unfortunately experiments involving gland ablation and replacement therapy have shed very little light on the mechanisms controlling rates of synthesis of plasma proteins. In a single instance, however, in relation to the α_2-AP globulin of rats (Heim and Ellenson, 1967; Weimer and Coggshall, 1967), it has been possible to demonstrate an absolute requirement for a hormone (cortisol) for the synthesis of a particular plasma protein. However, not only must cortisol or corticosterone be present for the synthesis of this protein to occur but also the rat must have been subjected to tissue damage. Thus for the corticosterone to be effective at least in adult rats additional, as yet unknown, substances must also be present. In very young animals and in embryos, on the other hand, as shown by Weimer *et al.* (1967), the α_2-AP globulin is a normal plasma constituent and can be demonstrated in the absence of any form of trauma. Possibly in these animals an altered balance of hormones replaces the effect of substances produced by trauma.

The specificity of the effect of corticosterone in controlling the synthesis of the α_2-AP globulin in the adult rat has also been shown by Weimer and Coggshall (1967), who noted that after stimulation of adrenalectomized rats the increase in fibrinogen concentration was similar to that found in normals. Increased plasma concentrations of haptoglobin have been found to occur after repeated injections of testosterone and parathyroid hormone as well as after cortisone (Krauss, 1968). Thus the rate of synthesis of certain plasma proteins may be partly controlled by hormones other than those which are themselves directly involved in the response to trauma.

Recently, the effect of insulin on the rate of synthesis of certain plasma proteins has been investigated by Neuhaus (1970). In these experiments, incorporation of ^{14}C-leucine into seromucoid and albumin was measured in rats rendered diabetic by means of alloxan. The response to injury was found to be less in these animals than in normals.

C. Trauma and Endotoxin

By means of the ^{14}C-urea method Koj and McFarlane (1968) have shown that the injection of endotoxin into rabbits leads to an easily measurable increase in the rate of synthesis of albumin. As will be further discussed below increased incorporation of ^{14}C-leucine into albumin has been found to follow both intraperitoneal injection of celite and partial hepatectomy (Majumdar *et al.*, 1967).

Some information concerning the nature of the processes which follow localized trauma and which ultimately lead to increased rates of synthesis of certain plasma proteins can be obtained by consideration of the time after stimulation at which maximum concentration of each protein is found to occur. Thus in rats, after laparotomy the concentration of the α_1-globulin investigated by Darcy (1966) reaches a maximum after 2–3 days.

As shown by Weimer and Humelbaugh (1967), the greatly increased concentrations of both the α_2-AP globulin and fibrinogen found at 2 days after injection of turpentine declined almost to the original levels in the course of 7 days. On restimulation in the same manner, the concentrations of both proteins rose to even higher levels and then fell at similar rates. However, when the rats were stimulated for a third time, a difference was manifested, since fibrinogen did not rise as high as after the first stimulation, whereas the response of the α_2-AP globulin was greater than previously observed. Measurements of the concentration of hemolytic complement in the same experiments showed yet another time curve. These findings appear to indicate the existence of systems capable of independent control of the rate of synthesis of each of these proteins. Furthermore, that more than one mechanism may be responsible for control of the rate of synthesis of certain plasma proteins is indicated by a consideration of the behavior of transferrin. Thus the concentration of this protein has been shown to increase following tissue damage, partial hepatectomy, oxygen depletion, and excess plasma Fe (Morgan and Brackenridge, 1962; Lane, 1968). That certain liver enzymes may also be induced by completely different control mechanisms (hormonal and co-factor) has been shown by Greengard (1967).

D. Effects of Inhibitors of Protein Synthesis

In order to ascertain whether increased synthesis of those plasma proteins which increase greatly in concentration after trauma requires previous synthesis of RNA, Neuhaus *et al.* (1966) carried out a series of experiments in which rats were given actinomycin-D at various times before and after trauma caused by injection of turpentine. Since this drug is thought to interfere with protein synthesis at the transcriptional stage by inhibiting the synthesis of mRNA, its action, if it were limited to this step, should be to reduce the rates of synthesis of all proteins, those coded for by mRNA with the shortest half-life being most rapidly affected. Presumably because the actions of actinomycin-D are much more complex, Neuhaus *et al.* (1966) observed a slight increase in incorporation of ^{14}C-glycine into seromucoid in uninjured rats at 24 hr after administration of the drug. However, they also observed that following trauma in the presence of actinomycin-D the expected increase in concentration of seromucoid which would have occurred if the drug had not been present was greatly reduced. Although this failure to increase may indeed be due at least in part to decreased synthesis of mRNA the view that increased synthesis of mRNA is the main factor allowing increased synthesis of certain plasma proteins after trauma is most unlikely to be correct. Furthermore, attempts to differentiate between the factors controlling the synthesis of albumin and of those plasma proteins the concentration of which is most elevated after trauma have led to the conclusion that the differences are quantitative rather than qualitative.

In this regard, the conditions required for the synthesis of certain liver enzymes which are subject to hormonal control are of interest. As shown by Greengard (1969), while mRNA must be formed, it is not necessarily sufficient by itself and additional factors must be present before synthesis can be increased.

E. Effects of Partial Hepatectomy

The ability of the regenerating liver to synthesize plasma proteins at altered rates while still *in vivo* cannot easily be studied because of the changes inevitably induced by operative trauma; however, Majumdar *et al.* (1967) have compared incorporation of ^{14}C-leucine into albumin and fibrinogen at various times after partial hepatectomy and after trauma caused by intraperitoneal injection of celite. Surprisingly, in view of the decreased concentration of plasma albumin known to follow trauma, increased incorporations of ^{14}C-leucine into both albumin and fibrinogen were found following this stimulus. In these experiments the ^{14}C-leucine was injected at various times after the stimulus and blood samples were taken after a further interval of 1 hr. In the case of trauma with celite, incorporation into both proteins followed a similar time curve with a maximum at 8 hr. After partial hepatectomy on the other hand, increased rates of incorporations into both proteins continued for a much longer period. The increased incorporation which occurred after partial hepatectomy was found to be adrenal independent, whereas when celite was used no increase in ^{14}C-leucine incorporation took place in the absence of the adrenals. While in experiments of this kind the relative importance of the regenerative process itself and the operative trauma cannot be distinguished, it is clear that during regeneration the rate of synthesis of plasma proteins is greatly increased and, at least in the present instance, the rate of synthesis of fibrinogen relative to that of albumin is also increased. Thus the greatly increased synthesis of cell proteins occurring during the regeneration does not prevent a considerable increase in the use of amino acids for synthesis of plasma proteins.

The evidence obtained from experiments *in vivò* as to the nature of the processes controlling the rates of synthesis of the plasma proteins is sufficient to indicate the importance of hormones as well as other substances which may perhaps be compared with the co-factors known to effect rates of synthesis of certain liver enzymes. Since no little is yet known concerning the factors which control the synthesis of the plasma proteins, it is important to emphasize that in all probability the systems are as complex and as specific as those operating in relation to the inducible liver enzymes.

III. Evidence from Experiments with the Isolated Perfused Liver

A. Livers from Normal Rats Perfused with Blood from Normal Rats

When a normal liver is perfused for more than a brief period, the possibility evidently exists that the rates of synthesis of plasma proteins may alter during the course of the experiment. In order to investigate such changes Gordon and Koj (1969) perfused livers from normal unstimulated rats for periods of 8 hr with heparinized rat blood supplemented only with glucose. ^{14}C-Lysine and ^{3}H-lysine added during the first and second 4-hr periods, respectively, allowed measurement of incorporation of lysine into albumin, fibrinogen, and α_1-AP globulin during these periods. As shown in Table II incorporation of lysine into albumin between 4 and 8 hr increased to nearly double the value found in the first period (0–4 hr). Furthermore in the second period incorporation into fibrinogen and α_1-globulin was increased above that into albumin. Since the specific activity of the intracellular lysine was not measured, the increased incorporation of this amino acid into albumin which occurred in the second period does not necessarily indicate an increased rate of synthesis of this protein during the later part of the perfusion. On the other hand, because greater increases in incorporation into the other two proteins were observed it seems certain that an alteration in the relative rates of synthesis of these proteins must have occurred. In view of the findings to be discussed below concerning inhibition of protein synthesis in soluble systems obtained from normal livers it seems probable that the increased relative rates of synthesis of fibrinogen and α_1-globulin observed during prolonged perfusion of the normal rat liver may be due to removal, perhaps by metabolic breakdown, of an inhibitory factor. Certainly these findings indicate the need for caution in the interpretation of experiments of this kind in which increased synthesis of plasma proteins by livers perfused for 6 hr or more are obtained. This behavior of the normal liver when perfused with blood from normal rats makes difficult attempts to investigate effects which might follow the use of blood from injured rats when this is used to perfuse normal livers. Evidently the interpretation of such experiments will be easier if the perfusion is not carried on for more than 4 hr.

Because of complications of this kind, it is perhaps not surprising that no convincing demonstration has yet been reported of altered rats of synthesis of plasma proteins by a normal liver when perfused with blood taken from animals which have been subjected to any form of trauma.

B. Livers from Rats after Trauma Perfused with Normal Blood

Investigations in several independent laboratories have indicated that when rat livers are perfused with heparinized rat blood, the synthesis of albumin measured by means of $^{14}CO_2$ according to the principles given by Koj and McFarlane (1968) continues at approximately half the rate which occurs *in vivo* (Hoffenberg *et al.*, 1970). Whether the rates of synthesis of other plasma proteins and in particular those plasma proteins which increase most in concentration after trauma are similarly affected after transfer of the liver to the perfusion system has not yet been established. As already mentioned, however, Gordon and Koj (1969) have shown that the relative synthesis rates of certain plasma proteins alters if livers of normal rats are perfused for more than 4 hr. Evidently therefore perfusions of livers from animals which have been treated *in vivo* to bring about alteration in rates of synthesis of plasma proteins should not be carried on for more than a rather short period, i.e., 5 hr. Using this period for perfusion of livers from protein deprived rats, Gordon (1966) found that the altered rates of synthesis occurring *in vivo* did not revert to normal on transfer of the liver to the perfusion system. Indeed, considerably increased rates of synthesis for both fibrinogen and transferrin relative to that of albumin were found when these livers were perfused with normal blood.

More recently, increased rates of synthesis of α_1-AP globulin and of fibrinogen relative to albumin have been shown to occur following subcutaneous injections of talc carried out as little as 5 hr before the commencement

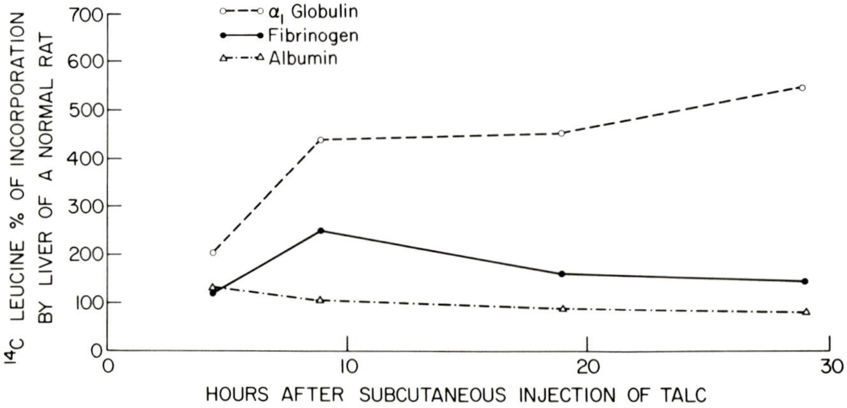

FIG. 2. Incorporation of ^{14}C-leucine into albumin, α_1-globulin, and fibrinogen after 5 hr perfusion of livers from rats at different times after subcutaneous injection of talc. Radioactivity of total amount of each protein has been expressed as percent of the average radioactivity obtained when livers from normal rats were perfused with blood from similar animals. Each point is an average of 3 values except that for at 9 hr which is an average of 2 values. Data from Gordon and Koj (1968).

of the perfusion experiment (Gordon and Koj, 1968). In these experiments, livers were transferred to the perfusion system at various times after the injection of talc. As shown in Fig. 2, whereas maximum incorporation of ^{14}C-leucine into fibrinogen occurred at 9 hr after the stimulation, increasing incorporation into α_1-globulin was found as late as at 29 hr. Thus, as early as 5 hr after subcutaneous injection of talc, sufficient of a messenger substance or substances have passed from the site of injury to the liver to alter the rates of synthesis of certain of the plasma proteins. Furthermore, a much longer period has to elapse before the messenger substance or substances have maximum effect on the rate synthesis of the α_1-globulin as compared with the rate of synthesis of fibrinogen.

Since in experiments *in vivo* the synthesis of a particular plasma protein has been found to occur only in the presence of the adrenals (Heim and Ellenson, 1965, 1967), the possibility that the rates of synthesis of other plasma proteins may be partly or wholly under adrenal or other hormonal control has to be considered. That this is true of the α_1-AP globulin of rats has recently been demonstrated by Gordon and Koj (1968), who perfused livers at 10 days after adrenalectomy. Since much greater than normal incorporation of ^{14}C-leucine into this protein was obtained, it seems necessary to conclude not only that adrenal hormones are necessary for synthesis of some plasma proteins but also that in certain circumstances they inhibit the rate of synthesis of others.

C. Perfusion after Partial Hepatectomy

Just as in experiments on incorporation of labeled amino acids into plasma proteins *in vivo*, with the perfused liver, great care is needed to distinguish the effects of partial hepatectomy itself from those due to the accompanying operative procedures. In an attempt to achieve this, Mutschler and Gordon (1966) measured incorporation of ^{14}C-leucine by livers perfused at different times after partial hepatectomy.

When the livers had been allowed to regenerate for 24 hr before perfusion, incorporation into both fibrinogen and transferrin was greatly increased over that found in laparotomized controls (Fig. 3b). Under these conditions, incorporation into albumin remained nearly the same as with a normal liver. These results indicate not only that transferrin belongs to the group of plasma proteins that are synthesized faster after trauma but that during regeneration the liver is able to synthesize these proteins even faster than occurs after major trauma.

As shown in Fig. 3a livers perfused at 12 hr after laparotomy showed maximum synthesis rates of both fibrinogen and transferrin. In respect to synthesis of fibrinogen this result is similar to that obtained after the milder stimulus of subcutaneous injection of talc used by Gordon and Koj (1968) (Fig. 2). In Fig. 3b is shown the extra incorporation of ^{14}C-leucine over and above that due to the extra hepatic trauma occurring as a result of regeneration. At 12 hr after partial hepatectomy considerably increased synthesis of fibrinogen but not of transferrin took place, whereas at a somewhat later period greatly increased synthesis of both proteins occurred as a result of regeneration. Evidently stimulation due to extra hepatic trauma at regeneration requires different periods to effect maximally the rates of synthesis of fibrinogen and transferrin.

Perhaps the simplest explanation of these findings may be that the increased rates of synthesis of both fibrinogen and transferrin occurring after laparotomy are brought about by extra nuclear RNA, the rate of synthesis of which has been found by Bucher and Swaffield (1965) to be doubled in 5 hr.

The increased rate of synthesis of extra nuclear RNA in regenerating livers may also be held to be responsible for the rapid synthesis of fibrinogen found at 12 hr after partial hepatectomy. As shown in Fig. 3b, during perfusion the rate of synthesis of this protein is considerably faster in regenerating livers compared to laparotomized controls. Because this is not true for transferrin the relative rate of synthesis of which does not reach its maximum until after 24 hr some other factor must presumably be required for the synthesis of this protein. That prior synthesis of DNA may be necessary is suggested by the longer interval after partial hepatectomy.

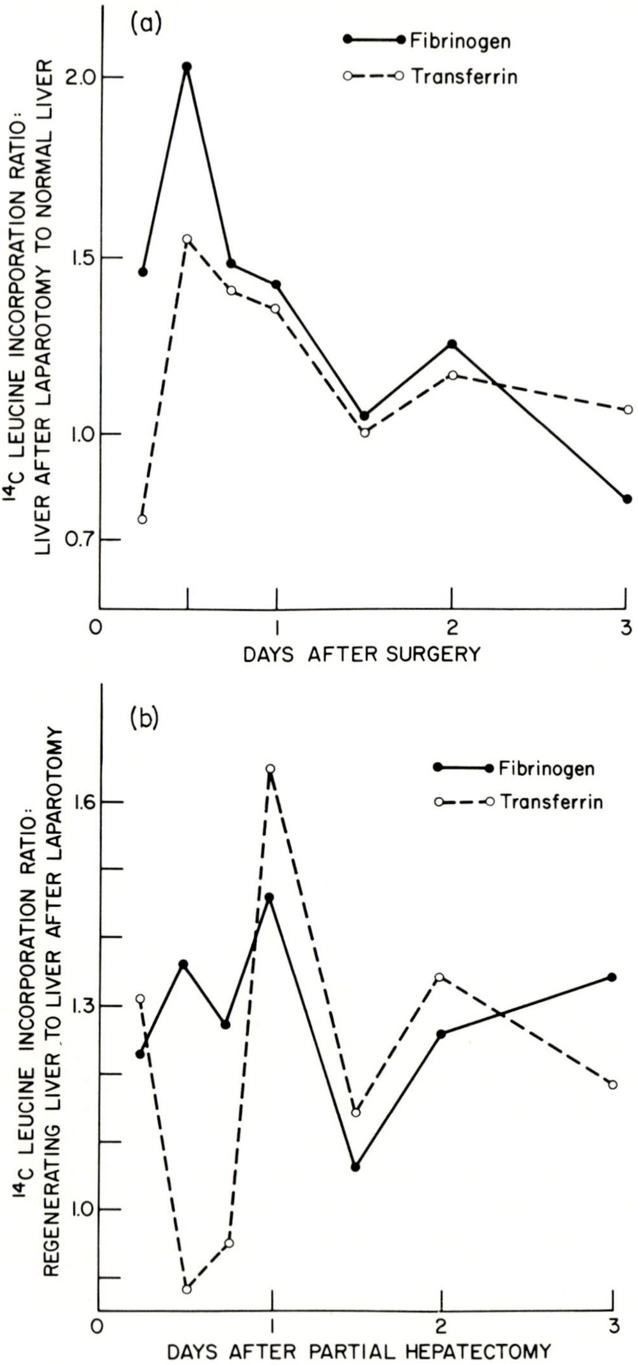

FIG. 3. (a) Incorporation of ^{14}C-leucine into fibrinogen and transferrin at the end of 6 hr of liver perfusion. The values are expressed as ratios of incorporation by livers at

In the period immediately after partial hepatectomy, the remaining lobes of the liver are subject to a number of complex changes. Unfortunately, the numerous attempts to identify which of these changes are most important in regard to initiation of regeneration have so far been unsuccessful. Since most of this work has concerned changes in synthesis of nucleic acids rather than plasma proteins, it is only indirectly relevant to the present subject. For this reason, only two experiments will be mentioned. In the first, remnants of livers removed from rats 4 hr after partial hepatectomy were perfused by Gordon and Hodgson (1966) with normal rat blood and incorporation of ^{14}C-orotic acid into RNA was measured. The conclusion was reached that the increased incorporation observed *in vivo* by remnants of livers 4 hr after partial hepatectomy must be due to humoral and/or nervous factors not present during perfusion. The rate of flow of blood through the liver remnant did not seem to be a factor. On the other hand, Majumdar *et al.* (1967), employing a technique *in vivo* which led to greatly increased flow of blood through part of the liver of adrenalectomized rats, reported increased synthesis of RNA and protein by the polyribosomes derived from livers of this kind. Thus it seems to be necessary to accept both that blood-borne factors are of importance and that pressure changes brought about by overperfusion may have very similar effects. In this respect, however, the possibility that the effect of the increased blood flow is removal of an inhibitor must be considered.

IV. Evidence from Cell-Free Systems

Comparisons of cell-free systems containing either microsomes or polyribosomes prepared from normal rat livers with similar systems from rats after trauma and also with those from regenerating livers have revealed that with both the latter faster rates of incorporation of ^{14}C-labeled amino acids always occur. That several different factors are responsible for the lower rate of incorporation given by normal liver microsomes and polyribosomes is apparent from the work of Skornik *et al.* (1967) and Von der Decken (1968), who found increased rates of incorporation of ^{14}C-leucine by both polyribosomes and cell sap prepared from normal livers after these had passed through columns of Sephadex G25. These findings indicate the

various times after laparotomy to the average obtained with livers from normal rats. Each point is an average of the results of 3 experiments. Data from Mutschler and Gordon (1966). (b) Incorporation of ^{14}C-leucine in similar experiments to those in (A), except that livers at various times after partial hepatectomy were used. The results are expressed as ratios of incorporation by the regenerating livers to those found with intact livers at the same after laparotomy. Data from Mutschler and Gordon (1966).

presence of inhibitors associated with both the polyribosomes and cell sap from normal liver which can be removed by treatment with Sephadex G25. The importance of inhibitors of this kind as factors restricting the rate of synthesis of proteins in the normal liver is clear, because in the presence of added GTP and amino acids the rates of incorporation by this system became almost as high as that of a similar system prepared from livers of rats after trauma.

Although direct proof is not yet available, evidence from studies *in vivo* suggests that the increased synthesis of protein observed in *in vitro* systems of this kind obtained from livers after injury is in fact due to more rapid synthesis of certain plasma proteins and the inducible liver enzymes sensitive to adrenal steroids. However, much more information will be needed to indicate whether the relative rates of synthesis of proteins characteristic of the intact tissue remain the same in any soluble system.

Tsukada, Moriyama, Doi, and Lieberman (1968a) have been able to demonstrate the existence of what may well be the same inhibitors by comparison of the rates of incorporation of ^{14}C-leucine by both microsomes and ribosomes from the livers of normal and traumatized rats. Thus, if DOC was used for the preparation of the microsomes, little difference was found between those from the normal animals and from those after trauma. If, on the other hand, DOC had not been used, the system from normal liver incorporated much more slowly than did that made from livers after trauma. Tsukada *et al.* (1968a) also found that microsomes prepared from regenerating livers behaved similarly to those from liver after trauma, suggesting either the absence of an inhibitor in such preparations or the presence on the ribosomes of an additional factor able to counteract the effect of the inhibitor (Tsukada *et al.*, 1968b). That the latter explanation is the more likely is suggested by the observation that the use of DOC for extraction was found to increase the incorporating power of the ribosomes from livers of rats after trauma and during regeneration, although to a lesser extent than with livers from normal animals.

Some further information about the role of the inhibitor was obtained by Tsukada *et al.* (1968a) by measurements of binding of ^{3}H-puromycin to ribosomes from livers of normal rats and those after trauma. Because puromycin inhibits protein synthesis by binding to the same ribosomal site as does amino acyl T–RNA, they were able to conclude that the binding of this latter compound is increased after trauma and partial hepatectomy, with the result that an increased rate of formation of peptide bonds is brought about.

In another study of the results of injury, Chandler and Neuhaus (1968) measured incorporation of ^{14}C-orotic acid into RNA and formation of heavier polysomes. They found that whereas ^{14}C-orotic incorporation was maximum at 8 hr after injury maximum shift of the polysome profile did not occur until after 18 hr. Possibly the newly formed polysomes remain stable during this period because of the decreased levels of ribonuclease (Lui and Neuhaus, 1968).

Interpretation of the significance of the increase in heavier polysomes is difficult, because similar changes have been observed following a number of stimuli, including treatment *in vivo* with glucocorticoids. In this instance, however, increased levels of RNA polymerase sufficient to account for a considerable part of the newly found RNA also occurred. Despite a careful attempt by Chandler and Neuhaus (1968) to detect increased RNA polymerase following injury, no such change has been found. In this respect, there is a clear difference from livers after partial hepatectomy in which the level of this enzyme is considerably increased. Since as a result of injury increased levels of adrenal steroids must be presumed to reach the liver, the absence of any change in RNA polymerase is surprising.

In this regard, the opposite effects of adrenalectomy which have been already mentioned on the ability of the liver to synthesize different clases of plasma proteins may be recalled.

As a result of both trauma and partial hepatectomy, the level of ribonuclease in liver homogenates (Arosa and de Lamirande, 1967) and in serum (Tsukada, 1969) is found to decrease. Furthermore the return to normal levels takes place much more rapidly after stress, such as intraperitoneal injection of celite (Tsukada, 1969), than after partial hepatectomy. Thus extra messenger or any other type of RNA formed in these circumstances will remain available for a longer period. Although these observations do not explain why an increased rate of synthesis of particular plasma proteins and not others occurs after injury they indicate the importance of RNA in the response to injury.

V. General Conclusions

Studies of trauma and inflammation have revealed the extraordinary complexity of the metabolic changes which follow various kinds of injury. As has already been described these include both increased and decreased rates of synthesis of a number of plasma proteins. As yet only very preliminary indications as to the nature of the factors responsible for these changed rates of synthesis have become apparent. In all probability in addition to the altered hormonal levels, factors originating at the injury site are of importance. However, experiments with the perfused liver have shown either that such factors are extremely labile or that they are effective only in conjunction with factors supplied by other organs.

Considerable evidence is now available that mechanisms exist whereby rates of synthesis of particular plasma proteins can be altered without much change in the synthesis rates of others. Perhaps the simplest hypothesis capable of explaining these observations may be that the rate of synthesis of each plasma protein is affected by the number of substances all of which reach the liver at altered concentrations after trauma. If this is true it is probable that the rate of synthesis of each plasma protein is affected at several stages as seems likely in the case of cortisol which is believed to act both at the transcriptional and translational levels.

At the moment the list of substances which when tested *in vivo* have been found to alter rates of synthesis of certain plasma proteins is remarkable mainly for its diversity. Thus Fe^{3+} after Fe^{3+} deficiency, endotoxin, ACTH at concentrations well above the physiological level (Atencio *et al.*, 1969) and the adrenal steriods are all effective.

As has been discussed above, after these and/or other competent factors have reached the liver, altered levels of ribonuclease and certain inhibitors of protein synthesis associated with microsomes, ribosomes, and cell sap can be demonstrated. Other changes, such as the greatly increased levels of ornithine decarboxylase found by Janne and Raina (1969) in the liver after trauma and not directly connected with any of the known stages of protein synthesis may yet turn out to be concerned with stimulation of the rate of synthesis of particular plasma proteins.

Thus inhibition at the ribosomal level seems likely to provide a major means of control of the rates of synthesis of a number of plasma proteins. Probably this and other controls depend on the balance between the concentrations of several substances reaching the liver after trauma rather than the presence of any single substance which determines the rates of synthesis of individual plasma proteins. It is also clear that these rates depend in part on the state of the liver at the time of stimulation. By these means, the organism is able to respond rapidly and specifically to a wide range of noxious stimuli.

VI. Summary

Increased concentrations of a number of plasma proteins are known to occur after tissue damage of various kinds. The factors controlling the rates of synthesis of these proteins have been studied by means of experiments *in vivo*, with the perfused liver and using soluble systems derived from livers of animals which have suffered tissue damage, and after partial hepatectomy. The following conclusions have been reached:

(1) Separate mechanisms capable of controlling the rates of synthesis of individual plasma proteins must exist because after tissue damage, increas-

ing or decreasing concentrations of particular proteins occur at different rates and because restimulation brings about increased or decreased maximal concentrations of the individual proteins. Furthermore, whereas the synthesis of certain plasma proteins requires the presence of the adrenals, that of others is increased after the removal of these glands.

(2) These results are due in part to the specialization of certain parenchymal cells of the liver for synthesis of particular plasma proteins and proliferation of these cells after tissue damage or partial hepatectomy.

(3) Experiments with soluble systems indicate that the presence of inhibitors in the livers of unstimulated animals restricts the rate of synthesis of certain plasma proteins and that the increased rates of synthesis which follow tissue damage are brought about by the removal of these inhibitors.

TABLE II

INCORPORATION OF ^{14}C- AND ^{3}H-LYSINE BY NORMAL RAT LIVER PERFUSED WITH NORMAL BLOOD[a]

	Average of 7 experiments	Range
Albumin	192	121–310
Fibrinogen	335	200–438
α_1-Globulin	327	124–505

Amounts of ^{14}C- or ^{3}H-lysine incorporated into each protein in the second 4-hr period expressed as a percent of the amount incorporated into the same protein in the first 4-hr period.

[a] Experimental conditions were the same as employed by Gordon and Koj (1968) except that the perfusions were carried out for 8 hr and differently labeled lysine was used in the second 4 hr.

REFERENCES

Arosa, J. D. S., and de Lamirande, G. (1967). *Can. J. Biochem.* **45**, 1021.

Atencio, A. C., Chao, P.-Y., Chen, A. Y., and Reeve, E. B. (1969). *Amer. J. Physiol.* **216**, 773.

Bucher, N. L. R., and Swaffield, M. N. (1965). *Biochim. Biophys. Acta* **108**, 551.

Chandler, A. M., and Neuhaus, O. W., (1968). *Biochim. Biophys. Acta* **166**, 186.

Darcy, D. A. (1963) *Protides Biol. Fluids Proc. Colloq.* **10**, 131.

Darcy, D. A. (1966). *Brit. J. Exp. Pathol.* **47**, 480.

Gordon, A. H. (1966). *Eur. J. Cancer* **2**, 19.
Gordon, A. H., and Hodgson, G. S. (1966). *Biochim. Biophys. Acta* **119**, 427.
Gordon, A. H., and Koj, A. (1968). *Brit. J. Exp. Pathol.* **49**, 436.
Gordon, A. H., and Koj, A. (1969). Unpublished observations.
Greengard, O. (1967). *Enzymol. Biol. Vlin.* **8**, 81.
Greengard, O. (1969). *Science* **163**, 891.
Hager, B. C., and Kenney, F. T. (1968). *J. Biol. Chem.* **243**, 3296.
Hamashima, Y., Harter, J. G., and Coons, A. H. (1964). *J. Cell Biol.* **20**, 271.
Heim, W. G., and Ellenson, S. R. (1965). *Nature (London)* **208**, 1330.
Heim, W. G., and Ellenson, S. R. (1967). *Nature (London)* **213**, 1260.
Hoffenberg, R., Gordon, A. H., and Black, E. G. (1970). To be published.
Janne, J., and Raina, A. (1969). *Biochim. Biophys. Acta* **174**, 769.
Krauss, S. (1968). *Fed. Proc. Fed. Amer. Soc. Exp. Biol.* **27**, 254.
Koj, A., and McFarlane, A. S. (1968). *Biochem. J.* **108**, 137.
Lane, R. S. (1968). *Brit. J. Haematol.* **15**, 355.
Lui, A. Y., and Neuhaus, O. W. (1968). *Biochim. Biophys. Acta* **166**, 195.
Majumdar, C., Tsukada, K., and Lieberman, I. (1967). *J. Biol. Chem.* **242**, 700.
Minchin Clarke, H. G., and Freeman, T. (1968). *Clin. Sci.* **35**, 403.
Morgan, E. H. (1969). *J. Biol. Chem.* **244**, 4193.
Morgan, E. H., and Brackenridge, C. J. (1962). *Quart. J. Exp. Physiol. Cog. Med. Sci.* **47**, 66.
Morgan, E. H., and Lynch, W. (1963). *Quart. J. Exp. Physiol. Cog. Med. Sci.* **48**, 181.
Mutschler, L. E., and Gordon, A. H. (1966). *Biochim. Biophys. Acta* **130**, 486.
Neuhaus, O. W. (1970). To be published.
Neuhaus, O. W., Balegno, H. F., and Chandler, A. M. (1966). *Amer. J. Physiol.* **211**, 151.
Peters, J. H., and Alper, C. A. (1966). *J. Clin. Invest.* **45**, 314.
Skornik, O. A., Hoagland, M. B., Pfefferkorn, L. B., and Bishop, E. A. (1967). *J. Biol. Chem.* **242**, 131.
Tsukada, K. (1969). *Biochim. Biophys. Acta* **186**, 21.
Tsukada, K., Moriyama, T., Doi, O., and Lieberman, I. (1968a). *J. Biol. Chem.* **243**, 1152.
Tsukada, K., Moriyama, T., Umeda, T., and Lieberman, I. (1968b). *J. Biol. Chem.* **243**, 1160.
Von der Decken, A. (1968). *Biochim. Biophys. Acta* **166**, 487.
Weimer, H. E., and Benjamin, D. C. (1965). *Amer. J. Physiol.* **209**, 736.
Weimer, H. E., and Benjamin, D. C. (1966). *Proc. Soc. Exp. Biol. Med.* **122**, 1112.
Weimer, H. E., and Coggshall, V. (1967). *Can. J. Physiol. Pharmacol.* **45**, 767.
Weimer, H. E., and Humelbaugh, C. (1967). *Can. J. Physiol. Pharmacol.* **45**, 241.
Weimer, H. E., Humelbaugh, C., and Roberts, D. M. (1967). *Amer. J. Physiol.* **213**, 418.

23

REGULATION OF PLASMA α_2-(ACUTE PHASE) GLOBULIN SYNTHESIS IN RAT LIVER

E. J. Sarcione

DIVISION OF MEDICINE, ROSWELL PARK MEMORIAL INSTITUTE,
NEW YORK STATE DEPARTMENT OF HEALTH, BUFFALO,
NEW YORK

Most of our current understanding of the mechanisms involved in regulation of protein synthesis has been derived from bacteria. As yet, however, it is not known to what extent the regulatory control systems of microorganisms are operative in animals. Although changes in existing levels of tissue enzyme activity and plasma proteins have provided considerable information regarding regulation of protein synthesis in mammals, interpretation of such studies is often complicated by observations that existing levels of these proteins can be altered by changes both in rates of synthesis and of degradation. The appearance of "abnormal" proteins, not normally detected in adult animals might be expected to provide a useful and somewhat different approach to our eventual understanding of regulation of protein synthesis in higher organisms. Host injury is known to produce both qualitative and quantitative changes in certain plasma proteins of man and animals. Among the qualitative changes occurring after injury is the appearance of immunologically distinct plasma proteins, such as C-reactive protein in man and rabbit and α_2-(acute phase) globulin in the rat. α_2-(Acute phase) globulin can be detected in fetal, neonatal, pregnant, tumor-bearing, and injured adult rats but cannot be detected in normal adult rats (Weimer and Benjamin, 1965). This apparently identical acute phase globulin has been studied by others and designated "abnormal" serum globulin by Lawford (1961); slow α_2-globulin, by Beaton *et al.* (1961); α_2-glycoprotein, by Bogden *et al.* (1964) and Stanislawski *et al.* (1967); and α_2-macroglobulin, by Boffa *et al.* (1965).

The mechanisms regulating the transient appearance and disappearance of this acute phase protein are not currently understood. For this reason, synthesis of plasma α_2-(acute phase) globulin detected in rats under diverse physiological and pathological conditions, but not detected in normal adult animals, was chosen as a model system to investigate the mechanisms involved in regulation of gene expression and protein synthesis in mammals.

I. Organ and Cellular Sites of Synthesis

It is known from the studies of Miller and Bale (1954) that the liver plays a dominant role in the synthesis of all plasma protein fractions except for the globulins. This organ has also been shown to synthesize several well-defined "acute phase" plasma proteins known to be elevated after injury including fibrinogen (Miller *et al.*, 1964); orosomucoid (Sarcione, 1963); C-reactive protein (Hurlimann *et al.*, 1966); haptoglobin (Krauss and Sarcione, 1964); and the α_1-globulin studied by Darcy (Weimer *et al.*, 1965). The liver has also been demonstrated to be a site of synthesis of α_2-(acute phase) globulin in injured adult rats (Sarcione and Bodgen, 1966) and tumor-bearing rats (Sarcione, 1967), using the isolated perfused intact rat liver, and by liver-cell cultures (Benjamin and Weimer, 1966).

Figure 1 shows the time course of incorporation of ^{3}H-leucine into α_2-(acute phase) globulin (left panel) and into albumin (right panel) obtained from plasma perfusing isolated rat livers *in vitro*. In these experiments, the concentration of all plasma proteins was made comparable in plasma perfusate samples obtained from both injured and normal liver perfusions. This was performed by the addition of an equal volume of nonradioactive plasma from injured adult rats to plasma perfusate samples from the normal liver–normal blood perfusion system, and the addition of an equal volume of pooled plasma from normal adult rats to perfusate samples from the injured liver–injured blood perfusions. During perfusions with livers and whole blood from injured adult rats, after an initial delay, the specific activity of immunochemically isolated plasma α_2-(acute phase) globulin increased rapidly and

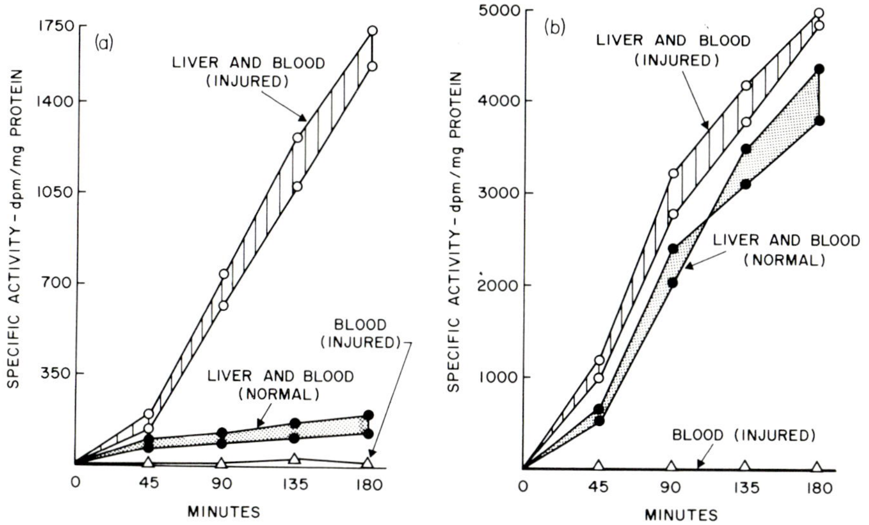

FIG. 1. Time course of incorporation of ^{3}H-leucine into plasma α_2-(acute phase) globulin (a) and albumin (b) in perfusions with livers from injured and normal adult rats, and in whole blood from injured rats circulated without a liver. Injury was produced by subcutaneous injection of 1.0 ml of turpentine, 48 hr previously. 4,5 ^{3}H-DL-leucine (100 μCi) and 1-^{14}C-glucosamine (20 μCi) were added to the perfusion blood at zero time. α_2-(Acute phase) globulin was isolated from plasma perfusates by immunochemical precipitation; the specific activity is that of the total precipitate, measured under standardized conditions. Plasma albumin was obtained by paper electrophoresis.

was linear between 45 and 180 min. On the other hand, perfusions with livers and whole blood from normal adult rats resulted in the appearance of only small and slowly increasing amounts of radioactivity in this protein. The rate of ^{3}H-leucine incorporation into comparable amounts of α_2-(acute phase) globulin isolated immunochemically under standardized conditions by perfused livers from injured rats was ten times faster than that incorporated by livers from normal adult rats.

Circulation of whole blood from injured rats for 3 hr did not result in incorporation of ^{3}H-leucine radioactivity into α_2-(acute phase) globulin indicating that the formed elements of the blood did not synthesize this protein. This information further indicates that minimal amounts of free ^{3}H-leucine were adsorbed during the immunochemical isolation procedure used.

Incorporation of ^{3}H-leucine into plasma albumin during these liver perfusions also proceeded rapidly (Fig. 1b). Throughout these liver perfusions, the specific activity of albumin obtained from plasma perfusing livers from injured rats was slightly but consistently higher than that

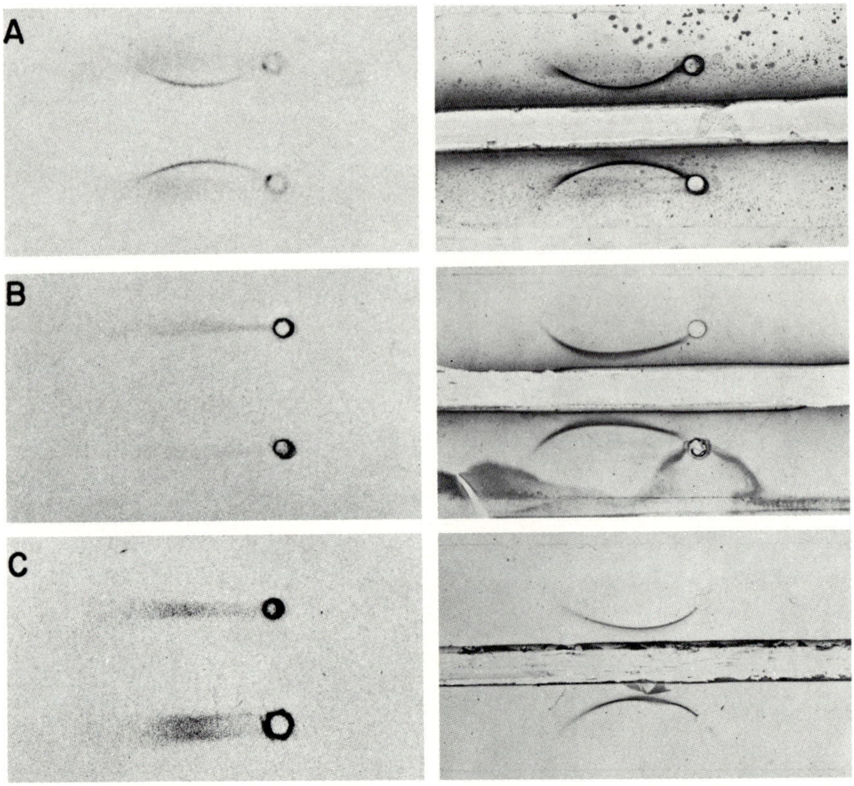

FIG. 2. Autoradiography (left panel) of immunoelectrophoretic patterns (right panel) obtained with plasma perfusing livers from injured (A) and normal (B) adult rats, and after circulation of whole blood from injured rats without the liver (C). Plasma was sampled after 3 hr. Injured rat plasma was added to (B) after sampling, to provide carrier acute phase protein. The anode was to the left and the precipitin arc was obtained with a rabbit antiserum reagent to rat α_2-(acute phase) globulin.

obtained from plasma perfusing normal livers. As would be expected, circulation of whole blood from injured rats without the liver did not result in incorporation of radioactivity into albumin.

Immunoelectrophoresis of concentrated plasma perfusates obtained after 3 hr from each of these perfusions described in Fig. 2 using monospecific rabbit antiserum reagent to rat α_2- (acute phase) globulin resulted in a single precipitin arc in the slow α_2 region (Fig. 2, right panel). Radioautography of these immunoelectrophoretic patterns (Fig. 3, left panel) showed that only perfusion of both liver and blood from injured adult rats (A, experiments 1 and 2) resulted in a positive radioautograph. Neither plasma perfusing liver and blood from normal adult rats (B, experiments 3 and 4) nor plasma from injured adult rats circulated without the liver (C, experiments 5 and 6) produced detectable labeling of these precipitin arcs. This latter finding indicates that the radioautographic labeling of α_2-(acute phase) globulin is not a result of nonspecific binding or radioactive amino acid to the precipitin arc.

To define the cellular site of synthesis within the liver, intact liver parenchymal cells from injured and normal rats, isolated by the technique described by Howard and Pesch (1968), were incubated with amino acid ^{14}C-hydrolysate. As described by these investigators, microscopic examination of suspensions of isolated liver cells showed that most of the cells were round and compact, with well-defined outline, and were almost exclusively parenchymal cells with negligible contamination with cell-free fragments. The percentage of viable cells, determined by lack of staining with trypan blue, ranged from 75 to 95%. Figure 3 shows that after incubation of parenchymal cells obtained from liver of injured rats the specific activity of immunochemically isolated α_2-(acute phase) globulin increased rapidly and progressively during 3 hr. This was in sharp contrast to normal liver parenchymal cells which incorporated small amounts of radioactivity into this globulin which did not change over time. On the other hand, both groups of liver parenchymal cell suspensions incorporated radioactive amino acids into albumin at nearly comparable rates, indicating the viability of these cells.

From these studies, it is concluded that liver parenchymal cells are a major site of synthesis and secretion of α_2-(acute phase) globulin detected in the plasma of injured and tumor-bearing animals. Furthermore, the appearance

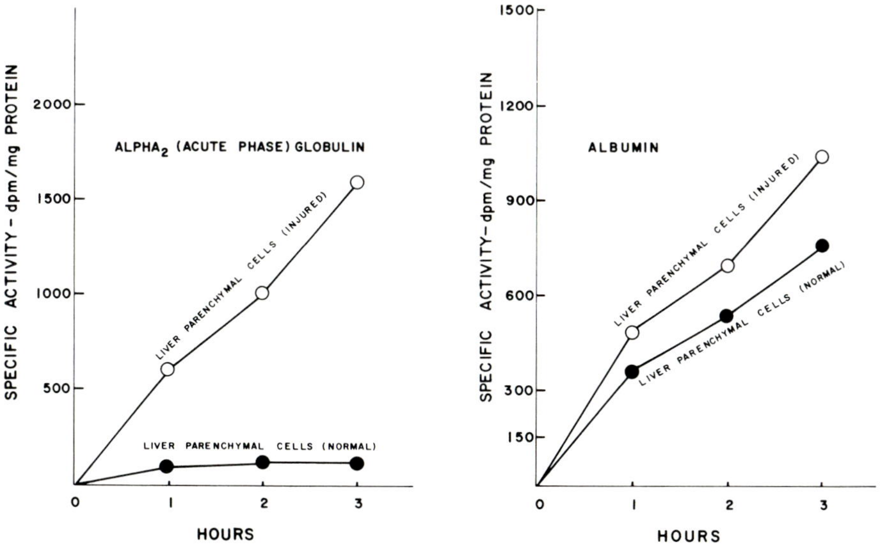

FIG. 3. Time course of incorporation of amino acid ^{14}C-hydrolysate into α_2-(acute phase) globulin (left) and albumin (right) during incubation of liver parenchymal cells isolated from injured and normal adult rats.

of this globulin in the plasma of such animals is due, at least in part, to *de novo* synthesis by the liver, rather than to release of preformed and stored globulin.

Based on the evidence that both extrahepatic tissue injury and tumors stimulate liver synthesis of this acute phase globulin, it has been postulated that this host-liver response is mediated by the release of substances from injured or necrotic tumor cells (Benjamin and Weimer, 1966; Sarcione and Bogden, 1966; Sarcione, 1967). However, the origin and identification of such postulated blood-borne stimulatory factors remain to be elucidated. Although considerable evidence indicates that adrenal glucocorticoids play a permissive role in controlling plasma levels of α_2-(acute phase) globulin, administration of these steroids to normal intact rats *in vivo* does not provoke the appearance of the globulin (Weimer and Benjamin, 1966; Heim and Ellenson, 1967; Bogden and Gray, 1968). For this reason, the adrenal glucocorticoids per se are not considered to directly stimulate liver synthesis of this acute phase protein.

II. Role of Infection

Bacterial infection and the administration of endotoxin are known to readily elicit the appearance of α_2-(acute phase) globulin in rats (Weimer and Benjamin, 1965). Bacterial contamination of tumors is also known to occur, especially during the later stages of necrosis (Patterson *et al.*, 1963; Kampschmidt and Upchurch, 1963). These observations suggest that as a result of bacterial contamination, extrahepatic injury sites and tumors release endotoxin-like products which play a primary role in stimulating hepatic synthe-

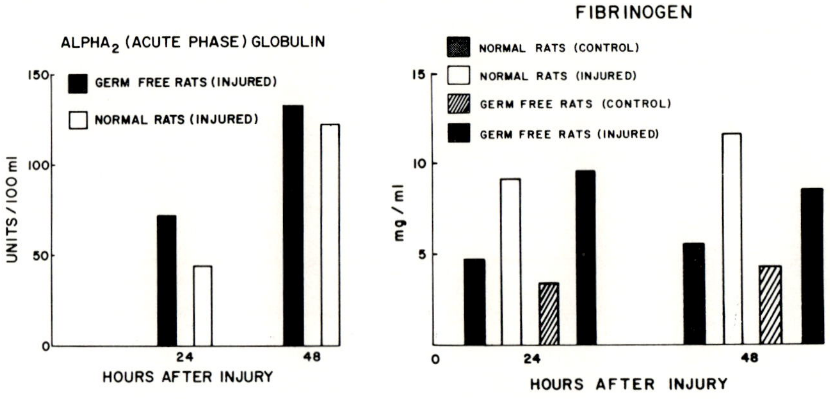

FIG. 4. Effect of injury on plasma levels of α_2-(acute phase) globulin (left) and fibrinogen (right) in germ-free and normal rats. Injury was produced by the subcutaneous injection of 1.0 ml of sterile turpentine and plasma protein measurements were made 24 and 48 hr later.

sis and secretion of this acute phase globulin. To investigate this intriguing possibility, the plasma protein response to injury was compared in normal and germ-free rats of the same strain. The germ-free rats were housed in plastic isolaters as described by Trexler and Reynolds (1967), and the feces were monitored at frequent intervals for the presence of bacterial flora as described by Wagner (1959). All of the isolated injured and control rats were found to be germ-free before, during, and after the experiment. Figure 4 shows that injury of germ-free rats produced by subcutaneous injection of sterile turpentine elevated the plasma levels of both fibrinogen and α_2-(acute phase) globulin at 24 and 48 hr to levels comparable to those found in control injured rats. α_2-(Acute phase) globulin was not detected in the plasma of either noninjured control or germ-free rats, and the fibrinogen levels were not significantly different in these same groups of rats. These findings indicate that bacterial infection of the injury site does not provide the primary stimulus either for liver synthesis and secretion of plasma α_2-(acute phase) globulin or for the elevated levels of fibrinogen found in injured rats.

III. Fetal and Neonatal Liver Synthesis

At least three mechanisms can be considered to explain regulation of α_2-(acute phase) globulin synthesis by liver parenchymal cells: (1) alteration of existing genetic information resulting in synthesis of "abnormal" protein; (2) greatly increased rate of transcription and/or translation of existing genetic information normally present in trace amounts; and (3) derepression of previously repressed gene sites coding for this protein. The detection of α_2-(acute phase) globulin in fetal, neonatal, and pregnant rats by Weimer *et al.* (1697) provides a clue that the latter mechanism may be operative. Since this globulin was detected in the plasma of both fetal and pregnant rats, however, it could not be readily determined whether it is derived from the fetus, the pregnant host, the placenta, or from two or all of these sources.

Fetal and neonatal liver synthesis of α_2-(acute phase) globulin was demonstrated by incubation of liver slices from such rats *in vitro* with ^{14}C-amino acid hydrolysate (Sarcione and Bohne, 1969). Radioactive labeling in homogenates of such slices was detected by radioautography of immunoelectrophoretic patterns obtained with appropriate rabbit monospecific antiserum reagent. Figure 5 shows that 17–19 day fetal and 6–8 day-old neonatal liver slices incorporated radioactivity into the precipitin arc corresponding to α_2-(acute phase) globulin. A comparison of the labeling intensities suggests that fetal liver synthesis of this protein is more active than neonatal liver synthesis. Under identical experimental conditions, liver slices from pregnant adult rats and from normal adult male rats did not produce detectable labeling of this acute phase globulin. The lack of detectable synthesis of acute phase globulin by normal adult liver slices correlated well

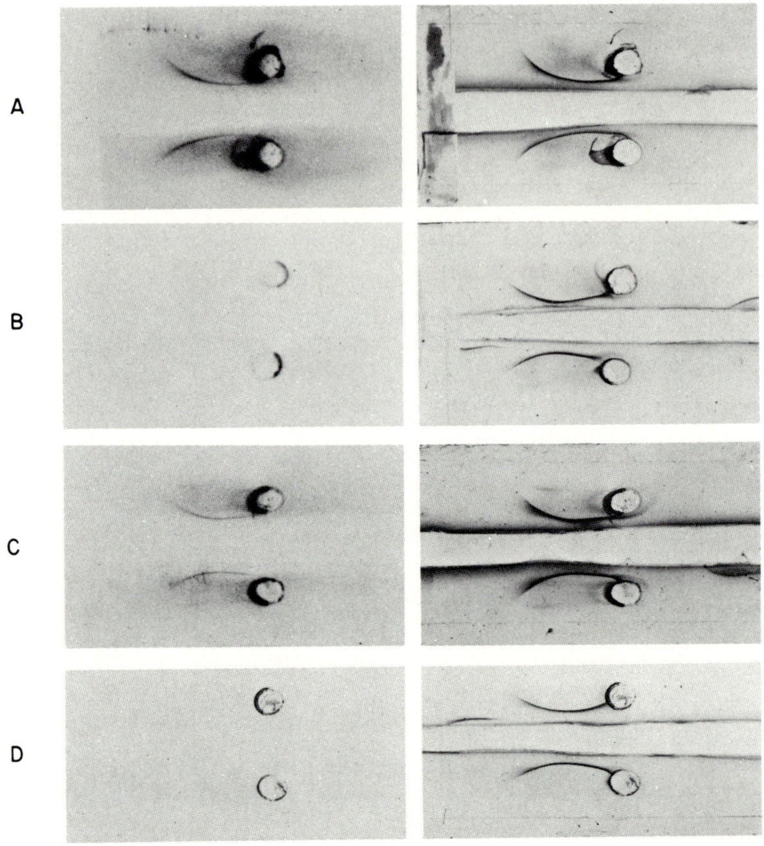

FIG. 5. Immunoelectrophoretic (right side) and autoradiographic patterns (left side) obtained by incubating liver slices with ^{14}C-amino acids *in vitro*. Serum from injured rats was added to provide carrier acute phase protein. The precipitin lines were developed using rabbit antiserum reagent to rat α_2-(acute phase) globulin. (A) Pooled 17–19 day fetal livers; (B) pregnant host liver; (C) pooled 6–8 day old neonatal livers; (D) normal adult male liver.

with its lack of detection in the sera and liver of such normal adult rats. These findings are in accord with the conclusion that in fetal and neonatal rats, the animal's own liver is the source of the α_2-(acute phase) globulin found in the plasma.

Figure 6 shows that α_2-(acute phase) globulin present in the sera of fetal, neonatal, pregnant, and rats injured by three different methods, all gave reactions of immunologic identity. Weimer *et al.* (1967) and Stanislawski *et al.*

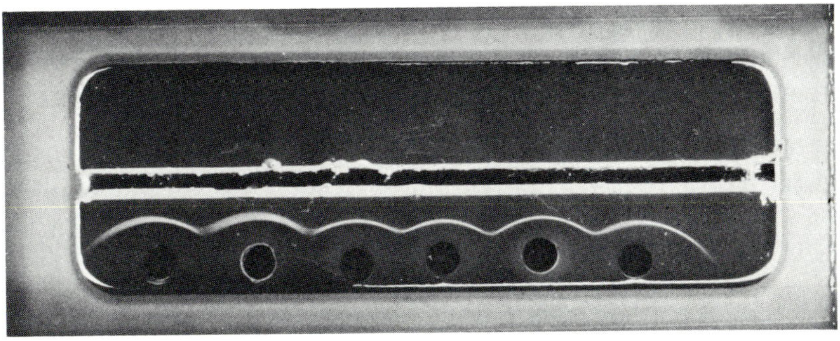

FIG. 6. Double diffusion patterns in agar demonstrating the immunologic identity of α_2-(acute phase) globulin present in different rat sera. The trough contained rabbit antiserum reagent against rat α_2-(acute phase) globulin. The wells from left to right contained pooled sera from: fetal rats 17–19 days old; adult rats 48 hr after injury produced by subcutaneous injection of 0.5 ml turpentine; pregnant rats after 17–19 days gestation; adult rats 48 hr after injury produced by abdominal surgical incision; neonatal rats 6–8 days old; adults rats 48 hr after injury produced by intraperitoneal injection of cadmium sulfate (1 mg/kg body weight), administered in 1.0 ml of sterile isotonic saline. All sera were used undiluted.

(1967) also demonstrated a similar immunologic relationship between this apparently identical acute phase globulin appearing in rats after injury and that present in fetal rats.

Taken together, these data are in accord with the conclusion that the presence of α_2-(acute phase) globulin in injured adult rats reflects hepatic synthesis of a normal fetal-specific globulin rather than an "abnormal" injury-specific protein.

IV. Relationship between Host Injury and Liver Cell Proliferation

Considerable indirect evidence suggests a possible relationship between injury-stimulated hepatic synthesis of α_2-(acute phase) globulin and the occurrence of selective liver cell proliferation: (1) appearance of this acute phase globulin in the plasma of rats undergoing rapid liver cell proliferation as found in fetal, neonatal, and partially hepatectomized rats (Weimer *et al.*, 1967; Boffa *et al.*, 1964); (2) liver synthesis of this globulin by rapidly multiplying liver cells as found in cell cultures (Benjamin and Weimer, 1966) and by fetal and neonatal rat liver slices (Sarcione and Bohne, 1969); (3) observation that production of an embryonic α-globulin by mouse liver is observed only during proliferation of normal or malignant hepatic cells (Abelev *et al.*, 1963).

To investigate whether liver cell proliferation is a prerequisite for hepatic synthesis of α_2-(acute phase) globulin, liver DNA synthesis *in vivo*, as measured by ^{14}C-thymidine incorporation into isolated DNA, was compared in normal and injured adult rats at various intervals after host injury. For these studies *in vivo*, each normal and injured rat was given a single intraperitoneal injection of 1 μCi/100 g body weight of ^{14}C-thymidine and the liver was removed exactly 2 hr later for DNA isolation. Figure 7 shows that incorporation of ^{14}C-thymidine into liver DNA of injured rats was not significantly different from that incorporated into liver of normal rats at any interval studied. The mean liver DNA levels in injured rats were also found not to differ significantly from DNA levels in normal rats.

Histologic examination of these livers from normal and injured rats at various intervals also showed no recognizable differences either in mitotic rate, enlargement of nuclei, or multinucleation. Boffa *et al.* (1964) also concluded that no correlation exists between the hepatic mitotic index of partially hepatectomized rats and the appearance of slow α_2-globulin in rats.

In this study, α_2-(acute phase) globulin was detected initially in the plasma of injured rats as early as 16 hr after injury, and reached a maximum level 48 hr after injury. It is recognized, however, that selective proliferation of a small number of minor liver cellular components might not be detected by the methods of analysis used. Both the histologic and DNA synthesis data

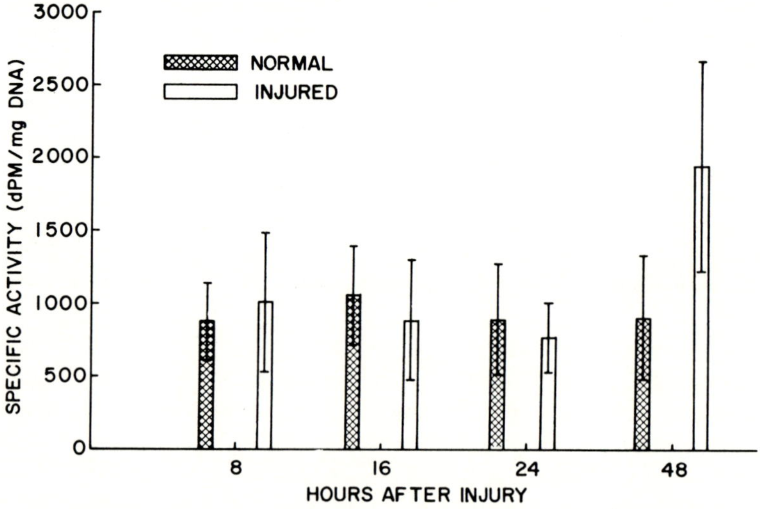

FIG. 7. Comparison of the incorporation of ^{14}C-thymidine into liver DNA *in vivo* by injured and normal adult rats at various intervals after injury. Injury was produced by subcutaneous injection of 1.0 ml of sterile turpentine. Each rat was given a single intraperitoneal injection *in vivo* of ^{14}C-thymidine at intervals after injury and the liver was removed 2 hr later for DNA isolation. Vertical lines indicate the 95% confidence limits for the means.

are consistent with the conclusion that (1) appreciable liver cell proliferation is not a prerequisite for hepatic synthesis of α_2-(acute phase) globulin, (2) differentiated adult hepatocytes present before host injury synthesize this globulin after injury.

V. Effects of Actinomycin-D *in Vivo* and *in Vitro*

In order to investigate whether a relationship exists between the appearance of plasma α_2-(acute phase) globulin and DNA-dependent RNA synthesis *in vivo*, actinomycin D was administered to injured rats at intervals after injury. Plasma levels of the acute phase plasma protein and fibrinogen were measured 24 hr after injury and compared to those of injured-control rats not receiving this antibiotic. Figure 8 shows that administration of actinomycin

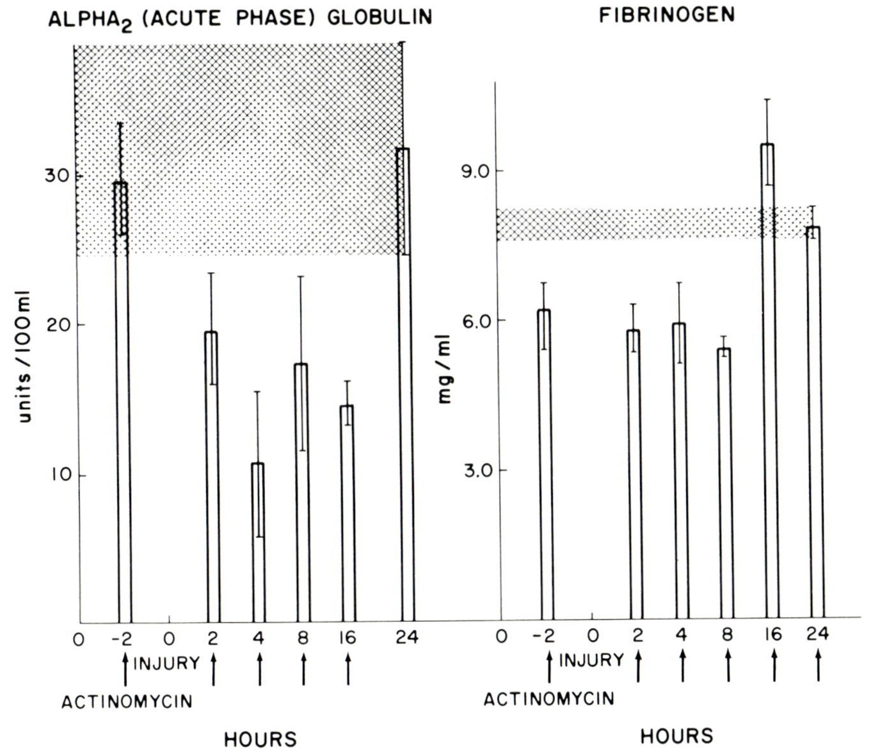

FIG. 8. Effect of actinomycin D (75 μg/100 g body weight) administered intraperitoneally at various intervals, on injury-induced increased plasma levels of α_2-(acute phase) globulin (left) and fibrinogen (right) *in vivo*. Injury was produced by subcutaneous injection of 1.0 ml of sterile turpentine at zero time and the plasma proteins chemically determined 24 hr later. Vertical lines indicate the 95% confidence limits for the means.

D at any time from 2 to 16 hr after injury, but not 2 hr before injury, consistently depressed 24 hr injury-induced α_2-(acute phase) globulin levels. In these same animals, injury-induced increased plasma levels of fibrinogen were also depressed, but some time differences were noted. Administration of actinomycin D from 2 hr before injury to 8 hr after injury depressed 24 hr fibrinogen levels, but administration of antibiotic 16 hr after injury had no effect. The observation that administration of actinomycin 16 hr after injury depressed the 24 hr response of acute phase globulin to injury, but not that of fibrinogen, suggests that this effect was produced by a differential inhibition of RNA synthesis, rather than by nonspecific generalized inhibition of all liver protein synthesis.

To study the relationship between liver synthesis of α_2-(acute phase)

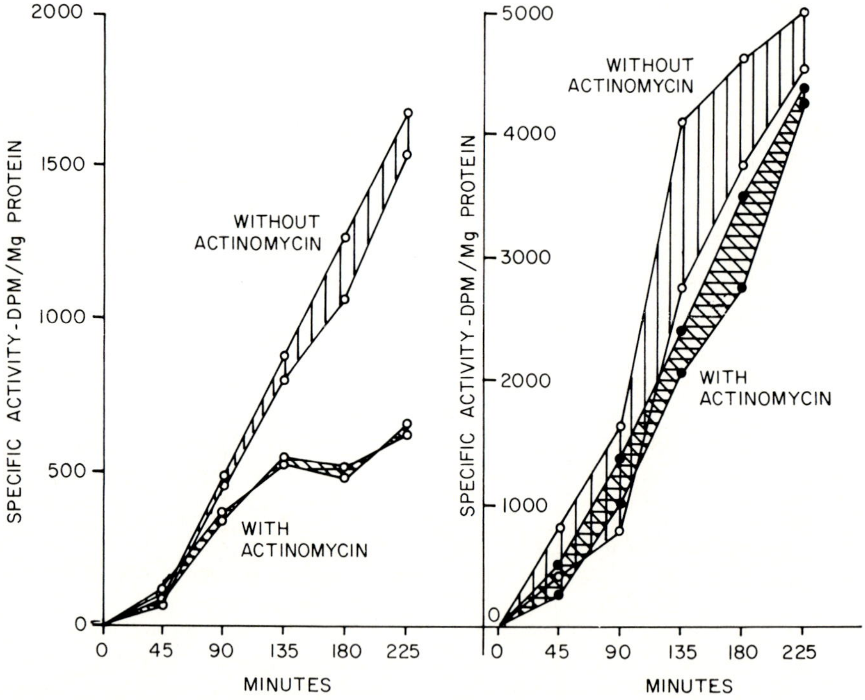

FIG. 9. Time course of incorporation of ^{3}H-leucine into plasma α_2-(acute phase) globulin (left) and into plasma albumin (right) by isolated perfused livers from injured adult rats, with and without the addition of actinomycin D to the perfusion system. The antibiotic (500 μg) was added to the blood 20 min before addition of 4,5-^{3}H-DL-leucine (100 μCi at zero minutes) and then, was infused continuously at a rate of 133 μg/hr. α_2-(Acute phase) globulin was isolated from plasma perfusates by immunochemical precipitation; the specific activity is that of the total precipitate, measured under standardized conditions. Plasma albumin was obtained by paper electrophoresis.

globulin and RNA synthesis, actinomycin D was added to the isolated perfused rat liver system *in vitro*. Figure 9 shows that the exposure of isolated perfused livers from injured adult rats to actinomycin D markedly depressed ^{3}H-leucine incorporation into α_2-(acute phase) globulin after 155 min, but was without effect on ^{3}H-leucine incorporation into plasma albumin. The lack of effect on albumin synthesis suggests that the decreased synthesis of α_2-(acute phase) globulin was due to a selective and primary effect of actinomycin D on RNA synthesis, rather than to nonspecific toxic effects on liver protein synthesis.

The results obtained with actinomycin D both *in vivo* and *in vitro* are consistent with the conclusion that liver synthesis of α_2-(acute phase) globulin is dependent on DNA-directed RNA synthesis, presumably mRNA specific for this protein. The time required for actinomycin D to reduce the synthesis rate of specific plasma proteins by the isolated perfused rat liver has been interpreted by John and Miller (1966) to measure the half-life of the specific mRNA involved; that for albumin was estimated to be 2–4 hr, and that for fibrinogen $1\frac{1}{2}$ to 2 hr. Based on this assumption, the half-life of the mRNA for hepatic synthesis of α_2-(acute phase) globulin, is estimated to be approximately 1 to $1\frac{1}{4}$ hr.

VI. Summary

Taken together, the currently available evidence supports the following conclusions. The appearance of α_2-(acute phase) globulin in the plasma of injured adult rats reflects hepatic synthesis and secretion of a fetal-specific globulin not normally detected in adult rats, rather than synthesis of an "abnormal" injury-specific protein. The mechanisms regulating liver cell synthesis involve reversible gene repression and gene activation or derepression of the genetic information coding for this fetal globulin. In fetal and neonatal rat liver, this gene site is "switched-on" to allow synthesis. In normal adult liver this gene site is "switched-off" or repressed to inhibit synthesis. In injured adult rats, this previously repressed gene site is activated or derepressed so that specific mRNA and acute phase globulin synthesis proceeds. Evidence that adult liver cells present before and after injury synthesize this acute phase globulin further suggests that fetal biosynthetic pathways can be selectively and reversibly "switched-on" and "switched-off" in such cells.

ACKNOWLEDGMENTS

Supported in part by Grant CA-11098 from the National Institutes of Health, U.S. Public Health Service.

References

Abelev, G. I., Perova, S. D., Khramkova, N. I., Postnikova, Z. A., and Irlin, I. S. (1963). *Transplantation* **1**, 174.

Beaton, G. H., Selby, A. E., Veen, M. J., and Wright, A. M. (1961). *J. Biol. Chem.* **236,** 2005.

Benjamin, D. C., and Weimer, H. E. (1966). *Nature (London)* **209,** 1032.

Boffa, G. A., Nadal, A. C., Zajdela, F., and Fine, J. M. (1964). *Nature (London)* **203,** 1182.

Boffa, G. A., Fine, J. M., Jacquot-Armand, Y., Gaudin-Harding, F., and Subsielle, H. (1965). *C.R. Soc. Biol.* **159,** 1342.

Bogden, A. E., and Gray, J. H. (1968). *Endocrinology* **82,** 1077.

Bogden, A. E., Neville, G. A., Woodward, W. E., and Gray, M. (1964). *Proc. Amer. Ass. Cancer Res.* **5,** 6.

Heim, W. G., and Ellenson, S. R. (1967). *Nature (London)* **213,** 1260.

Howard, R. B., and Pesch, L. A. (1968). *J. Biol. Chem.* **243,** 3105.

Hurlimann, J., Thorbecke, G. J., and Hochwald, G. M. (1966). *J. Exp. Med.* **123,** 365.

John, D. W., and Miller, L. L. (1966). *J. Biol. Chem.* **241,** 4817.

Kampschmidt, R. F., and Upchurch, H. F. (1963). *Cancer Res.* **23,** 756.

Krauss, S., and Sarcione, E. J. (1964). *Biochim. Biophys. Acta* **90,** 301.

Lawford, D. J. (1961. *Biochem. Pharmacol.* **7,** 109.

Miller, L. L., and Bale, W. F. (1954). *J. Exp. Med.* **99,** 125.

Miller, L. L., Hanavan, H. R., Titthasiri, N., and Chowdhury, A. (1964). *Advan. Chem. Ser.* **44,** 17.

Patterson, M. K., Jr., Maxwell, M. D., and McCoy, T. (1963). *Proc. Soc. Exp. Biol. Med.* **113,** 689.

Sarcione, E. J. (1963). *Arch. Biochem. Biophys.* **100,** 516.

Sarcione, E. J. (1967). *Cancer Res.* **27,** 2025.

Sarcione, E. J., and Bogden, A. E. (1966). *Science* **153,** 547.

Sarcione, E. J., and Bohne, M. (1969). *Proc. Soc. Exp. Biol. Med:* **131,** 1454.

Stanislawski-Birencwajg, M., Uriel, J., and Grabar, P. (1967). *Cancer Res.* **27,** 1990.

Trexler, P. C., and Reynolds, L. I. (1957). *Appl. Microbiol.* 406.

Wagner, M. (1959). *Ann. N.Y. Acad. Sci.* **78,** 89.

Weimer, H. E., and Benjamin, D. C. (1965). *Amer. J. Physiol.* **209,** 736.

Weimer, H. E., and Benjamin, D. C. (1966). *Proc. Soc. Exp. Biol. Med.* **122,** 1112.

Weimer, H. E., Benjamin, D. C., and Darcy, D. A. (1965). *Nature (London)* **208,** 1221.

Weimer, H. E., Humelbaugh, C., and Roberts, D. M. (1967). *Amer. J. Physiol.* **213,** 418.

24

ANALYSIS OF "ACUTE PHASE" PROTEIN SYNTHESIS IN THE MOUSE BY IMMUNOELECTROPHORESIS

Curtis A. Williams

THE ROCKEFELLER UNIVERSITY,
NEW YORK, NEW YORK

I. Introduction

In recent years most investigators of protein biosynthesis have been concerned with the specific identity of newly finished products. In complex systems, however, convenient and defining chemical isolations are applicable to very few proteins. Faced with this difficulty, a few biochemists turned to immunochemical methods—precipitation by specific antibody the most favored (Cohn and Torriani, 1951; Campbell and Stone, 1957; Redman, 1968). This is still the best for quantitative separation of proteins from mixture, if specific antibody is available and if great care is taken to avoid nonspecific coprecipitation.

Immunoelectrophoretic analysis (IEA), while it does not provide reliable quantitative data as customarily applied, does give critical qualitative information. In IEA, an antigen mixture is placed in a well in a layer of agar and is resolved by zone electrophoresis in a manner similar to that described by

Gordon and co-workers in 1950 (Gordon *et al.*, 1950). After separation, an antiserum is placed in a trough, lateral to the migration route, from which antibodies diffuse into the gel, encountering the corresponding antigen diffusing from its characteristic electrophoretic position. The intersecting diffusion fronts produce an arc of precipitate for each antigen–antibody reaction independently of the others. The method of Grabar and Williams (1953) made use of 13 × 18 cm gels, but the more economical method adapted by Scheidegger (1955) to microscope slides was used for the studies described here (Williams, 1970).

With IEA followed by autoradiography of the patterns (Rejnek and Bednarik, 1960) positive identification of many of the labeled products of biosynthesis is possible (Thorbecke *et al.*, 1970). In some cases estimates of relative amounts produced are obtained.

The studies discussed in this chapter have centered on the synthesis of plasma proteins by the liver in mice. The objectives were: (1) to compare qualitatively the biosynthetic activity of tissue *in vitro* and cell-free incorporating systems with *in vivo* patterns of biosynthesis; and (2) by inducing abnormal states, to discern systems for studying mechanisms of regulation. The high sensitivity and resolving power of IEA, as employed in this study, have led to several observations which we consider important. Most significant of these are: (1) the activity of cell-free systems reflect transient *in vivo* states more faithfully than do tissue cultures; (2) proteins produced for export, i.e., the plasma proteins, are produced only on membrane bound ribosomes and are enclosed by the microsomes; (3) newly made exportable proteins produced *in vivo* by the liver are resident in the cells for a few minutes only; and (4) several "acute phase" phenomena are independently regulated.

II. Tissue Culture Studies

In 1961, we (Williams and Wemyss, 1961) observed very marked qualitative and quantitative changes in mouse plasma protein patterns 4 days following staphylococcal infection. It was expected from many references in literature (Leutscher, 1947; Petermann, 1960) that we should find elevations notably among the α-globulins and eventually in the γ region. Densitometric scans of the simple gel electrophoresis patterns shown in Fig. 1 demonstrate the large quantitative changes observed. "Normal" mice in this context means young (6–8 weeks of age) specific pathogen free and inexperienced with the usual infectious stresses most animals encounter early in life. Actually the pattern after recovery from infection at 14 days has a more familiar "normal" appearance.

Hochwald *et al.* (1961) and Asofsky and Thorbecke (1961) had incubated

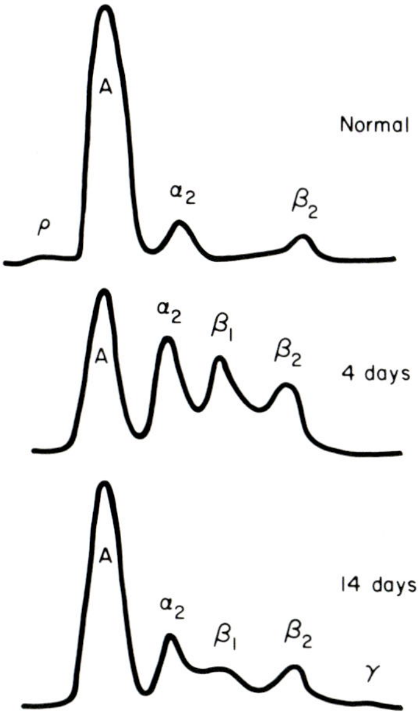

FIG. 1. Densitometric scan of agar–gel electrophoresis was carried out for 90 min at 6 V/cm in 1.5% agar buffered at pH 8.2 with 0.04 *M* barbital buffer on a microscope slide. A 2 μliter sample of fresh mouse plasma was placed in the starting well. After electrophoresis, the slides were fixed and stained in azocarmin, washed, and dried flat to the glass, Patterns shown are darkened tracings of a Joyce-Loebl Mk III scanning densitometer. Times refer to duration of an experimental staphylococcal infection.

various human and monkey tissues in a medium containing radioactive ^{14}C-amino acids in order to determine the sites of synthesis of the various plasma proteins. They had employed immunoelectrophoresis to analyze the protein secreted into the culture medium. After the slides were washed and dried, they obtained autoradiograms of the patterns. With Asofsky and Thorbecke (Williams *et al.*, 1963), we applied this technique to see what mouse liver slices, spleen minces, and lymph nodes from staphylococcus-infected mice produced in culture. The interesting question, of course, was whether these bits of tissues, removed from the animal, would continue to reflect altered biosynthetic activity.

Results obtained with liver slices indicated greatly increased radioactivity in the α- and β-globulins after 3 days of infection. This level of incorporation subsided markedly after 11 days when the animals were near recovery (Fig. 2). Cultured spleen minces labeled the immunoglobulins, but results with the 4-day infected spleens were indistinguishable from those with the control spleen. Spleens from 10-day infected mice, however, incorporated large amounts of ^{14}C-amino acids into immunoglobulin-G and β_1-C

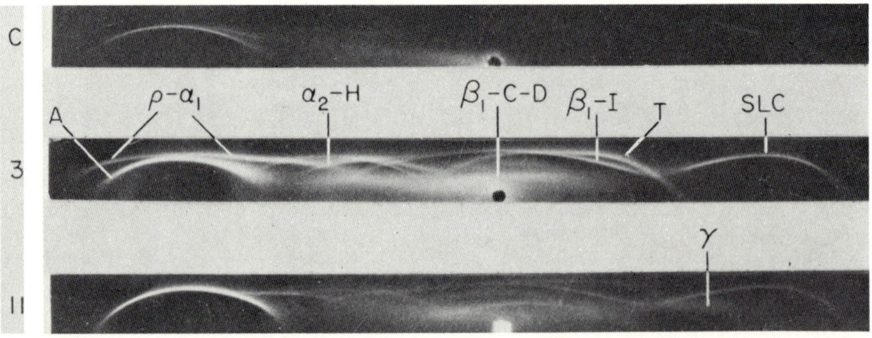

FIG. 2. ^{14}C-Labeled plasma proteins produced by liver slices in culture. Autoradiograms were made from IEA slides prepared with antigens concentrated from culture medium after 6 hr. C, normal culture; 3, 3-day staphylococcus infection; 11, 11-day staphylococcus infection. Illustration adapted from Williams *et al.* (1963).

globulin (a component of complement). The tissues in culture, therefore, appear to be making plasma proteins of defined characteristics in the relative amounts which would account for the corresponding density scans shown in Fig. 1. The outstanding exception is the strong radioactivity associated with the albumin produced by 3-day infected liver slices. These should have been less active in albumin production than the control cultures from noninfected animals.

TABLE I

PERCENT OF ^{14}C-AMINO ACID PULSE INCORPORATED INTO MOUSE PLASMA PROTEINS IN ONE HOUR[a]

Treatment of mice	Albumin	Globulins
None	0.43	0.33
Staphylococcal infection (4 days)	0.15	5.42
Killed staphylococcus (4 days)	0.18	0.33
Fasted (4 days)	0.53	0.47
Protein-free diet (6 days)	0.10	0.15

[a] Incorporation of ^{14}C-amino acids administered intravenously in a single pulse. Mice were bled from the infraocular venous plexus after a 60 min incorporation period. Incorporation is expressed as percent of administered dose. Albumin was extracted from ice-cold TCA precipitates by absolute ethanol. Infected mice received 2.7 × 10^7 viable units of bacterial culture. 10^8 killed units were given to group 3.

III. *In Vivo* Incorporation Studies

Table I compares results in *in vivo* incorporation in several abnormal states. The data are expressed in terms of percent of injected label incorporated. Note particulary that the "live" infection is required for the massive globulin response, but that killed staphylococci will depress albumin synthesis. Fasting has no effect, but a protein-free diet markedly curtails apparant plasma protein synthesis. These data indicate that the albumin and globulin responses are independently triggered phenomena.

At some time after the start of the infectious stress, we should be able to detect a sharp decrease in the relative amount of labeled amino acids incorporated into albumin and equal of greater increase of incorporation into globulins. Figure 3 further demonstrates this, showing a sharp early increase in the rate of albumin synthesis, followed by a sharp decrease after about

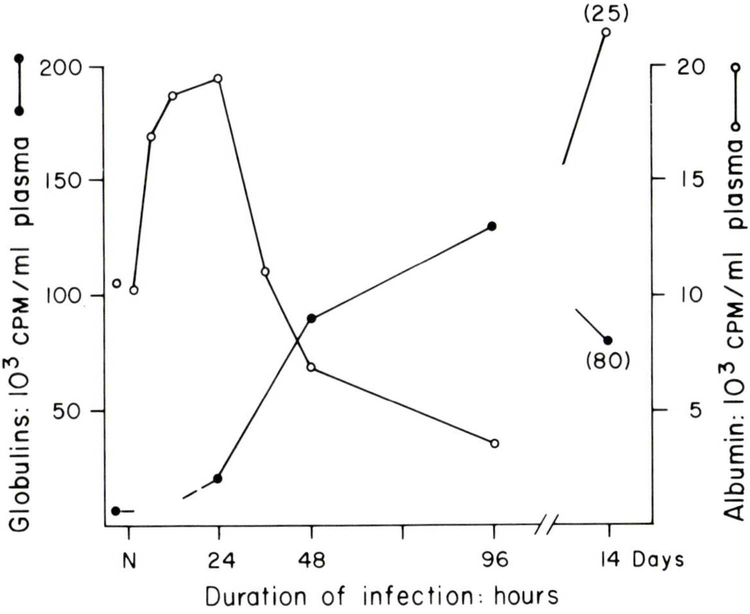

FIG. 3. Incorporation of amino acids into plasma proteins during 1 hr. Mice were infected intravenously with 2.7×10^7 viable units of *Staphylococcus aureus* (Giogio). After 2, 6, 12, 24, 36, 48, 96 hr, and 14 days, 2 μCi of ^{14}C-amino acid mixture was injected into the tail veins of three mice. One hr later, the mice were exsanguinated. Total incorporation into plasma protein was determined on 25 μliter samples. Albumin (○——○, right-hand scale) was extracted from cold TCA precipitates with absolute ethanol. The TCA–ethanol precipitate residues were considered as globulins (●——●, left-hand scale). Radioactivity was determined by liquid scintillation at 55% efficiency. N stands for non-infected control.

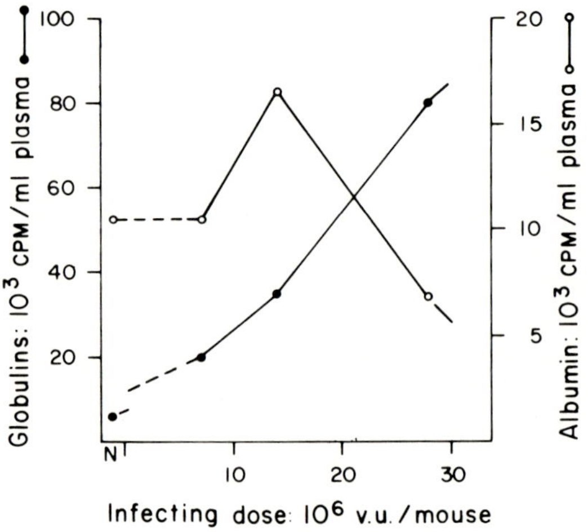

FIG. 4. Effect of dose of staphylococci on incorporation of amino acids into plasma proteins during 1 hr after 4 hr of infection. The infecting dose was varied from 8×10^6 to 2.7×10^7 viable units. The last points on the curves correspond to the 48-hr points in Fig. 3.

24 hr with the rates of incorporation approaching the nadir after about 4 days (Williams, 1965). In the case of the globulins, there is a progressive increase in rate of incorporation up to 4 days. The time of these responses is dose dependent, however, as shown in Fig. 4.

It should be mentioned at this point that, while the selection of parameters and the handling of data are not rigorous, we believe these relationships reflect real fluctuations in rates of biosynthesis of different plasma proteins. Accessible amino acid pools in the plasma and liver fluctuate, surely, but probably no more than in prolonged fasting (Henderson *et al.*, 1949; Steele *et al.*, 1950). The acutely infected animals do not eat much and they lose weight at about the same rate as starved mice. Except for any specific regulatory effects of individual amino acid pools, the effect of small fluctuations on apparent rates of protein synthesis should be largely overcome by the use of mixture of radioactive amino acids for incorporation studies. We have routinely employed a mixture of 13 purified ^{14}C-labeled precursors (" reconstituted protein hydrolysate," Schwartz Bioresearch, Inc.).

Autoradiographic patterns of *in vivo*-labeled proteins in the plasma are not very helpful but they confirm earlier studies and help to identify components into which isotope is most readily incorporated. However, when the liver microsomes were prepared after the administration of ^{14}C-amino acids for a 5-min pulse labeling, we can see what has just come of the assembly line, so to speak. The patterns of protein released from microsomes on sonication are shown in Fig. 5. The IEA pattern, *a*, is of plasma from a mouse infected

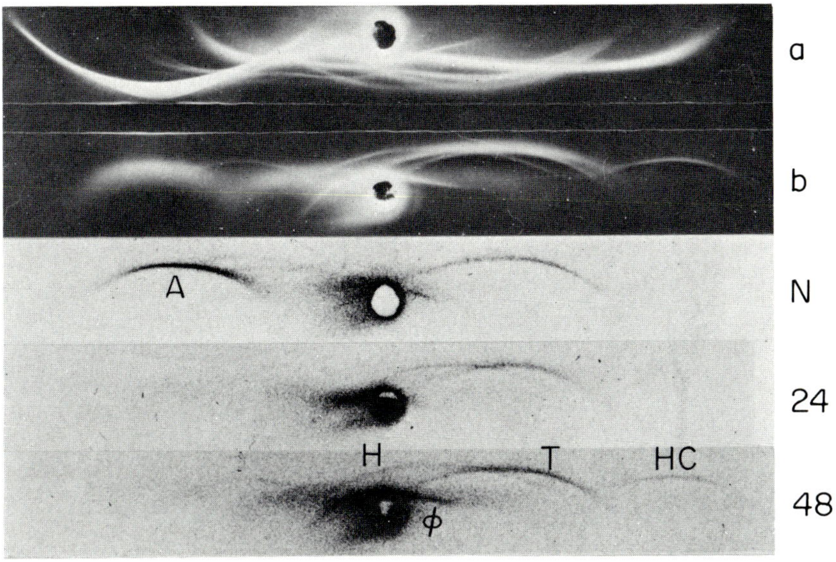

FIG. 5. ^{14}C-Labeled plasma proteins released from mouse liver microsomes by sonication. Mice were infected intravenously with staphylococci and after 24 hr and 48 hr were injected with ^{14}C-labeled amino acid (see Fig. 2). After a 5-min incorporation period, mice were exsanguinated and the livers were homogenized. The microsomal fractions, prepared by ultracentrifugation of the postmitochondrial supernatant extract, were resuspended in saline and disrupted by sonication. The suspensions were cleared by ultracentrifugation, and the released supernatant proteins were concentrated by precipitation with ammonium sulfate at 75% saturation. The immunoelectrophoretic pattern of such a preparation from 48-hr infected mice (b) is compared with the whole plasma pattern (a), both developed by rabbit antiserum against whole mouse plasma proteins. The lower three patterns are autoradiograms of slides obtained with proteins from normal (N), 24-hr infected (24), and 48-hr infected mice (48). The bottom panel is the autoradiogram of the pattern in panel b. A, albumin; T, transferrin; H, hemopexin; HC, hemopexin-cross-reacting component; ϕ, fibrinogen.

for 48 hr. The autoradiogram of the normal microsome extract pattern shows strong radioactivity in the albumin, fibrinogen, and transferrin and weaker activity in the α-1 and α-2 region. After 24 hr and 48 hr of infection, autoradiograms indicate that incorporation into albumin had been greatly reduced, transferrin remains essentially the same, fibrinogen synthesis appears progressively increased, and a β-1 component we identify as hemopexin (a hemin binding protein) appears with a structurally related component is indicated by the conjoining of the two arcs of precipitate. Compare these patterns with those of Fig. 2 obtained with the products of liver slice cultures.

No labeled plasma proteins are found in the circulation for about 10 min. After 15 min only very small amounts of identifiable labeled protein are released from microsomes. These findings are consistent with the findings of Peters (1957) that newly synthesized protein exported by the liver is rapidly passed out of the cell and Redman (1968) that most incorporation occurs in the first 5 min of a pulse label.

The distribution of labeled amino acids in fractions of the liver extract after a 5-min *in vivo* pulse is shown in Table II. It is interesting to note that

TABLE II

AMINO ACID INCORPORATION *in Vivo* DURING 5-MIN PULSE (dpm/mouse)[a]

	Normal	48-hr infection
Liver homogenate (15,000 *g*, 20 min)	17,600	30,600
Cell sap (100,000 *g*, 3 hr)	4100	8900
Sonic release fraction of microsomes	3100	7400
Voided microsomal fraction (by difference)	10,400	14,300

[a] Incorporation of ^{14}C-amino acids (2 μCi/mouse) administered intravenously in a single pulse. Livers from eight mice in each group were homogenized and fractionated between 5 and 10 min after pulse.

incorporation into cell sap proteins is also markedly increased in the infected animals. There is practically no labeled plasma protein detectable in the cell sap fraction so this must be soluble liver proteins. This could well be an unexplored aspect of the "acute phase." The increased label in the voided microsomes could be unfinished and unreleased peptides, plasma proteins not extracted or proteins of the endoplasmic reticulum made by membrane-bound ribosomes (see Williams and Ganoza, Chapter 10, this volume).

IV. Cell-Free Systems

Campbell and his colleagues demonstrated that cell-free systems prepared from rat livers would incorporate ^{14}C-amino acids into serum albumin (Campbell *et al.*, 1960) and von der Decken used autoradiography of IEA patterns for positive identification (von der Decken, 1963). With Ganoza and Lipman we have shown that many plasma proteins are labeled in a rat liver cell-free system (Ganoza *et al.*, 1965). To study the staphylococcal infection phenomena in mouse liver incorporation systems, liver microsomes and soluble (105,000 *g*) extracts from infected and control animals were prepared (Williams *et al.*, 1965). These were recombined and cross-combined in the incorporation mixture. After incubation at 37°C for 50 min, microsomes were again collected and disrupted by sonication. Released protein was dialyzed, lyophilized, and reconstituted in a minimal volume for analysis. It can be seen in Fig. 6a that the normal microsomal system produces only albumin and transferrin in easily detectable amounts. When microsomes from infected mice are combined with soluble extract of normal liver, the labeling pattern shown in Fig. 6b is obtained. There is striking similarity of these autoradiograms to those of proteins obtained from liver microsomes after *in vivo* incorporation, even down to details such as marked reduction of albumin synthesis and the production of the slow component related to hemopexin. Supernatant protein preparations from the cell-free systems revealed only faint traces of labeled albumin and transferrin, probably resulting from slight disruption of microsomes during incubation and centrifugation (Ganoza and Williams, 1969).

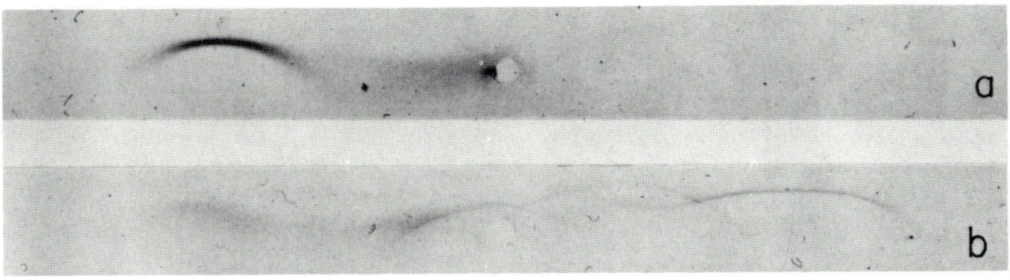

FIG. 6. Autoradiograms of IEA patterns of mouse plasma proteins made by mouse liver cell-free systems. Microsomal incorporation systems were prepared as described by Williams *et al.* (1965) from livers of normal and 4-day staph-infected mice. Proteins recovered from the microsomes following the incorporation period were dialyzed and concentrated by lyophilization. IEA patterns were developed by rabbit antiserum against whole mouse plasma. a, Autoradiogram of normal pattern; b, autoradiogram of 4-day infection pattern. Compare a with N and b with 48 on Fig. 5.

We conclude from these experiments that transitory states of plasma protein synthesis in the intact animal can be investigated with considerable confidence in organ culture, and in cell-free systems. There was more perfect correspondence between the patterns of incorporation into plasma proteins by cell-free systems and the intact mouse than there was between either of these and the tissue culture results. This difference was mainly seen in the amount of label incorporated into serum albumin, probably because this protein is the only one whose production appeared radically curtailed early in infection. Infected liver slice culture patterns show strong labeling of albumin, contrary to results obtained by other methods. This may be because the intact liver cells of the slice culture, once removed from the infected animal, partially "recover" from the albumin suppression during the prolonged incubation period.

Acknowledgments

These investigations have been supported by grants from the National Science Foundation (GB 1383 and GB 5290) and the U.S. Public Health Service (AI 07346). Parts of this chapter are adapted from a contribution to "Protein Metabolism and Biological Function" (R. Hilf and C. P. Bianchi, eds.), Rutgers University Press, New Brunswick, New Jersey (1970).

References

Asofsky, R., and Thorbecke, G. J. (1961). *J. Exp. Med.* **114**, 471.

Campbell, P. N., and Stone N. E. (1957). *Biochem. J.* **66**, 19.

Campbell, P. N., Greengard, O., and Kernot, B. A. (1960). *Biochem. J.* **74**, 107.

Cohn, M., and Torriani, A. M. (1951). *C. R. Acad. Sci.* **232**, 115.

Ganoza, M. C., and Williams, C. A. (1969). *Proc. Nat. Acad. Sci. U.S.* **63**, 1370.

Ganoza, M. C., Williams, C. A., and Lipmann, F. (1965). *Proc. Nat. Acad. Sci. U.S.* **52**, 619.

Gordon, A. H., Keil, B., Sebesta, K., Kuessl, O., and Sorm, F. (1950). *Collect. Czech. Chem. Commun.* **18**, 1.

Grabar, P., and Williams, C. A. (1953). *Biochim. Biophys. Acta* **10**, 193.

Henderson, L. M., Schurr, P. E., and Elvehjem, C. A. (1949). *J. Biol. Chem.* **177**, 815.

Hochwald, G. M., Thorbecke, G. J., and Asofsky, R. (1961). *J. Exp. Med.* **114**, 459.

Leutscher, J. A. (1947). *Physiol. Rev.* **27**, 621.

Petermann, M. L. (1960). *Plasma Proteins* **2**, 309.

Peters, T. (1957). *J. Biol. Chem.* **229**, 659.

Redman, C. M. (1968). *Biochem. Biophys. Res. Commun.* **31**, 845.

Rejnek, J., and Bednarik, T. (1960). *Clin. Chim. Acta* **5**, 250.

Scheidegger, J. J. (1955). *Int. Arch. Allergy Appl. Immunol.* **7**, 103.

Steele, B. F., Reynolds, M. S., and Baumann, C. A. (1950). *Arch. Biochem.* **25**, 124.

Thorbecke, G. J., Hochwald, G. M., and Williams, C. A. (1970). *In* "Methods in Immunology and Immunochemistry" (C. A. Williams and M. W. Chase, eds.), Vol. 3, Ch. 14E.4. Academic Press, New York.

von der Decken, A. (1963). *Biochem. J.* **88**, 385.

Williams, C. A. (1965). *Fed. Proc. Fed. Amer. Soc. Exp. Biol.* **24**(2), 506.

Williams, C. A. (1970). *In* "Methods in Immunology and Immunochemistry" (C. A. Williams and M. W. Chase, eds.), Vol. 3, Ch. 14.D.2. Academic Press, New York.

Williams, C. A., and Wemyss, C. T. (1961). *J. Exp. Med.* **114**, 311.

Williams, C. A., Thorbecke, G. J., and Asofsky, R. (1963). *J. Exp. Med.* **118**, 315.

Williams, C. A., Ganoza, M. C., and Lipmann, F. (1965). *Proc. Nat. Acad. Sci, U.S.* **53**, 622.

25

REGULATION AND METABOLISM OF THE THIRD COMPONENT OF COMPLEMENT (C3)

Chester A. Alper

DEPARTMENT OF PEDIATRICS, HARVARD MEDICAL SCHOOL;
DEPARTMENT OF MEDICINE AND BLOOD GROUPING LABORATORY,
CHILDREN'S HOSPITAL MEDICAL CENTER; DEPARTMENT OF MEDICINE,
PETER BENT BRIGHAM HOSPITAL, BOSTON, MASSACHUSETTS

I. Introduction

An understanding of the factors controlling the plasma concentration of C3,* the third component of complement, requires an examination of the

* The nomenclature for complement components and their derivatives is that recommended in the *Bull. W.H.O.* **39**, 935 (1968). In brief, the components are designated Cn, their activated forms are denoted $C\bar{n}$, and fragments are indicated by letter suffixes: Cna, Cnb, etc. Red cell intermediate complexes are denoted by E (erythrocyte), A (antibody) C142 . . ., wherein the numbers following "C" indicate the components which have interacted with EA.

molecular events which comprise complement activation *in vitro*. We shall focus on the reactions involving C3, including the elaboration of biologically active fragments, and shall also consider reactivities of the C3 molecule other than those occurring during classic complement activation. We shall attempt to relate these observations *in vitro* to alterations in C3 occurring in a variety of patients *in vivo*.

As is true of all plasma proteins, the synthesis (and catabolism) of C3 is intimately related to genetic factors and these, too, will be touched upon. Finally, because of the central role which the complement system is presumed to occupy in host resistance to injury and infection, we shall attempt to relate abnormalities affecting C3 to these important functions.

II. Complement Components and Their Interaction *in Vitro*

Intensive investigation, particularly over the past decade, has resulted in considerable clarification of our view of the complement system. Although it was recognized fairly promptly after the discovery of complement around the turn of the century (Buchner, 1889; Nuttall, 1904; Ehrlich and Morgenroth, 1899; Bordet and Gengou, 1901) that more than one substance was involved (Ferrata, 1907), its full complexity has been appreciated only recently (Nelson, 1965; Lepow, 1965; Müller-Eberhard, 1968). The currently defined components number nine, and at least three inhibitors or inactivators have been identified. All of these substances appear to be proteins, and many have been purified and characterized physicochemically. The view of complement activation that has emerged is one of sequential interaction in the order C1, C4, C2, C3, C5, C6, C7, C8, and C9 with elaboration of enzymic (esterase, protease, or peptidase) activity following activation of (at least) C1, C4 and C2, and C3. The reasonable hypothesis at present is that full activation of the system involves a limited proteolytic "cascade," similar to that which appears to occur in the coagulation system.

The prime classic initiator of complement activation is an antibody-sensitized cell (erythrocyte or bacterium) and the lysis of such cells has been utilized extensively in the study *in vitro* of complement activation. In this situation, some of the complement components (C1, C4, and C2), or large activated fragments of these components, bind to the antigen–antibody site. On the other hand, activated C3 ($C\bar{3}$) has been shown to bind to the cell surface at some distance from this site (Müller-Eberhard *et al.*, 1966). All nine complement components are required for immune cytolysis.

It is now clear that important biological functions may be elaborated at various points in complement activation short of the full sequence. For

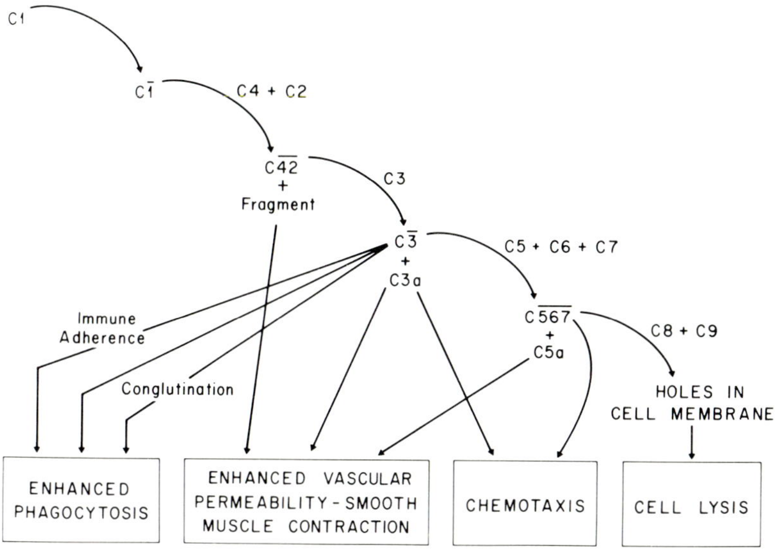

FIG. 1. Complement activation and the generation of biological activities contributing to the inflammatory response (modified from Cinader and Lepow, 1967).

example, as can be seen in Fig. 1, chemotaxis for leukocytes, anaphylatoxins, immune adherence, enhancement of phagocytosis of certain cells, and increased vascular permeability are produced with activation of only some of the components. These functions are mediated either by fragments produced by enzymic cleavage of complement components or by enzymic activities resident in the activated components themselves.

III. Alterations in Complement *in Vivo*

In attempting to relate the scheme of complement interaction outlined above to events in the living organism, several questions are immediately raised. First of all, what role does complement play in normal bodily functions such as host resistance to injury or invasion by parasitic organisms? Although on the face of the matter it would seem obvious that virtually all the complement-mediated functions already mentioned play important parts in the immune and inflammatory responses, the answer to this question is not entirely clear. Individuals with hereditary deficiency of C2 (Klemperer *et al.*, 1966) and animals with hereditary deficiency of C5 (Rosenberg and Tachibana, 1962; Cinader *et al.*, 1964) or C6 (Rother and Rother, 1961; Rother *et al.*, 1966) appear to be in excellent health and are not unduly susceptible to infection.

The second important question relates to the extent to which complement is involved in the pathophysiology and even the pathogenesis of various human ills. A first step in this direction is the measurement of hemolytic complement in the serum of individuals with a variety of disorders. If this is found to be low, measurement of individual components by immunochemical or functional assay or both may give a further clue to the abnormalities involved. If many or all of the complement components are lowered in concentration, it would seem reasonable to suppose that activation of the entire sequence, perhaps initiated by antigen–antibody interaction *in vivo*, has occurred. This subject has been recently reviewed (Schur and Austen, 1968). On the other hand, a low level of only one component suggests some other process, most reasonably decreased synthesis. Information obtained by measurements of this sort is perforce limited unless combined with other, more dynamic, analyses.

The presence of complement components (chiefly C3 and C4) on red cells in cases of hemolytic anemia and in tissues on histologic section has been taken as evidence for complement activation *in vivo*. Glomerular localization of C3 has been demonstrated in acute glomerulonephritis and in systemic lupus erythematosus (Lachmann *et al.*, 1962) as well as in experimentally induced glomerulonephritis in the rabbit (Unanue and Dixon, 1964; Vassalli and McCluskey, 1964). However, C3 may be detected on glomeruli in these diseases after serum complement has returned to normal so that its presence may indicate an immunological event in the past rather than at the time of biopsy.

Still another approach to the problem is the demonstration in fresh plasma from patients of altered forms of complement components, particularly forms known to result from full or partial complement activation. Whether or not such forms will be detectable, even though produced *in vivo*, is very much dependent upon the rapidity with which they are cleared from the plasma. Again, the presence of conversion products may result from processes other than sequential complement activation so that caution in interpretation of such findings is mandatory.

Finally, metabolic studies with isotopically labeled purified complement components can be expected to provide important information about alterations in complement *in vivo*. Such studies yield information on the rates of synthesis and catabolism and on the size of plasma and extravascular pools of specific components and thus provide an "explanation" for an increased or decreased concentration in a given subject. The relationship of this information to pathophysiological mechanisms is quite complex and remains problematical.

It is helpful to attempt to translate concepts and observations derived from the classic serological study of complement into the language of interacting protein molecules and of "turnover" studies. For example, the term "complement fixation" has been used to describe the loss of hemolytic complement activity in serum in the presence of antigen–antibody complexes *in vitro*. There is a corresponding loss in functional activity of individual components where these have been measured. Even in cases in which the antigen is particulate and hence capable of physically binding some of the complement components, there is little, if any, reduction in the concentration of individual components measured immunochemically. The reason for this is that the bulk of these proteins undergo physicochemical changes in the "fluid phase" and only a small percentage of the total number of molecules physically attach to the antigen–antibody site. If such an antigen–antibody reaction occurs within the mammalian organism, hemolytic complement falls in concentration (M'Gowan, 1908; Schwab *et al.*, 1950) but relatively little is known of the biological effects of such complement activation, nor do we have extensive knowledge of the fate of the products of the reaction.

The relation of complement activation to the fate *in vivo* of injected purified complement components labeled *in vitro* with a radioactive isotope depends intimately on the metabolic behavior of the products of activation. As pointed out earlier, if the activation–inactivation products are cleared rapidly relative to the clearance of the native protein, they will not easily be detected in the circulating blood. The major result in terms of a metabolic study in a subject with continuous complement activation will be an abnormally elevated fractional catabolic rate for the protein in question (Alper and Rosen, 1967). On the other hand, if the activation products are cleared at the same rate as the native components, fractional catabolic rates will be unaltered by complement activation *in vivo*. Theoretically, it is even possible to imagine that complement activation *in vivo* could result in a decreased fractional catabolic rate for a component with activation products cleared more slowly than the native protein.

Several other points require special attention in the interpretation of metabolic studies with labeled complement components in patients. Whereas a finding of an increased catabolic rate is consistent with complement activation *in vivo* under the conditions stipulated above, it may also result from other processes. For example, in the nephrotic syndrome of childhood, the fractional catabolic rates of many proteins are increased over and above urinary excretion (Gitlin *et al.*, 1956) for reasons that are not well understood.

Criteria for the suitability of preparations of complement components for metabolic studies are more complicated than for such proteins as albumin or immunoglobulin G. Complement proteins should not only be free of significant contamination with other proteins but also of their own conversion products. It should also be ascertained that preparations retain all the reactivities, chemical and biological, of native components. Finally, preparations should not be altered by purification in such a fashion that their behavior with respect to general metabolic processes is changed.

Treatment of data from metabolic studies also requires special consideration. For example, it is possible that clearance rates from plasma of either labeled complement components themselves or their conversion products are so rapid that a significant percentage of the administered material is removed during the "mixing period" prior to the collection of the initial plasma sample. A reasonable technique for assessing and correcting for this potential source of error is to administer albumin labeled with some tag other than that used for the complement component. In this fashion, two estimates of the plasma volume will be obtained and can be compared. In cases where the plasma volume determined from dilution of the label on the complement component is higher than can be accounted for by the error of the method, an appropriate correction can be applied to obtain the true initial complement component radioactivity (Alper and Rosen, 1967). In such a case the observed initial radioactivity should be multiplied by the ratio of complement component-determined plasma volume to that calculated by the labeled albumin.

IV. Characteristics and Reactions of C3

A. Physicochemical Characteristics

Müller-Eberhard and his colleagues (1960; Müller-Eberhard and Nilsson, 1960) first isolated a β_1-glycoprotein from human serum and showed its relation to the complement system. The protein they described, β_{1C}-globulin, has since been twice redesignated, first as C′3 and most recently as C3. It is by far the most abundant of the complement components occurring as it does at a concentration of 100–200 mg/100 ml of normal serum or plasma (Lundh, 1964; West *et al.*, 1964; Klemperer *et al.*, 1965). C3 has a sedimentation coefficient ($s_{20,w}$) of 9.5 S and its molecular weight is approximately 240,000 (Müller-Eberhard, 1968). The protein contains 2.7% carbohydrate (Müller-Eberhard *et al.*, 1960) and serine is reported to be the only amino terminal amino acid (Müller-Eberhard, 1968). Whereas the native molecule contains two reactive sulfhydryl groups, four additional sulfhydryl groups are detected after treatment with dithiothreitol (Müller-Eberhard, 1968). The structure and subunit composition are not yet established. C3 apparently binds calcium, since its electrophoretic mobility is slowed in the presence of this (Laurell *et al.*, 1956) and other divalent cations such as magnesium ions (Azen *et al.*, 1969).

B. Conversion Products, Reactions, and Biologically Active Fragments

On storage of serum and analysis in agar gel immunoelectrophoresis, C3 converts to a protein with more rapid mobility. This protein was originally termed β_{1A}-globulin (Müller-Eberhard *et al.*, 1960) and is now designated C3c. It has a sedimentation coefficient ($s_{20,w}$) of 6.7 S, has the same carbohydrate content as C3, and is not active as a hemolytic complement component (Müller-Eberhard and Nilsson, 1960). In the initial studies of this conversion product, it was shown that with some antisera to C3, a small spur could be observed, indicating that at least one antigenic determinant present on C3 was absent from C3c. Conversion of C3 to material with approximately the same electrophoretic mobility as C3c was also shown to occur on incubation of serum with antigen–antibody specific precipitates or zymosan (Müller-Eberhard and Nilsson, 1960), hydrazine or ammonium ions (Zimmer, 1960; Müller-Eberhard, 1961), proteolytic enzymes such as trypsin (Steinbuch, 1964), or $C\bar{1}$ (Peetoom *et al.*, 1964). Conversion of C3 by hydrazine and trypsin proceeds in the absence of serum, whereas the other agents require other serum proteins for the mediation of conversion (Müller-Eberhard and Nilsson, 1960; Müller-Eberhard, 1961; Müller-Eberhard *et al.*, 1967).

Conversion of fluid phase C3 is induced by cells in the state EAC142 or EAC42 (Pondman and Peetoom, 1964; Linscott and Cochrane, 1964). In addition, about 5–15% of the fluid phase C3 binds to the surface of such red cells (Müller-Eberhard *et al.*, 1966). However, in contrast to C1, C4, and C2 which bind at the antigen–antibody site, C3 appears to bind at loci on the red cell surface more or less remote from this site (Müller-Eberhard *et al.*, 1966), and the number of C3 molecules which bind is very much greater than the numbers of molecules of C1, C4, or C2. Thus, the activation of C3 apparently represents an amplification step in the complement sequence. Müller-Eberhard and his colleagues (1966) have postulated an evanescent activated form of C3, $C\bar{3}$, which is briefly capable of attaching to the red cell surface, since neither C3, C3b, nor C3c exhibit this property.

Cells in the state EAC1423 exhibit several interesting properties. As would be anticipated from earlier statements, such cells will lyse if supplied with C5–9. In addition, EAC1423 (as well as EAC143 and EAC43) will adhere to primate erythrocytes, platelets, or leukocytes, a phenomenon termed immune adherence (Nelson, 1953; Nishioka and Linscott, 1963). Moreover, EAC1423 exhibits peptidase activity for aromatic peptides such as glycyl-L-tyrosine, and since EAC142 lacks this activity, it appears that this is a function of cell-bound C3 (Cooper and Becker, 1967). There appears to be a low molecular weight substance in serum capable of enhancing this peptidase activity as well as the hemolytic and immune adherence activity of EAC1423 (Cooper, 1967). A normal serum protein which destroys the immune adherence and hemolytic activity of cell-bound C3 has been described (Nelson *et al.*, 1966) and has been termed the C3 inactivator. This protein and the conglutinogen-activating factor (KAF) of Lachmann and Müller-Eberhard (1968) are apparently the same protein (Lachmann and Müller-Eberhard, 1968; Ruddy and Austen, 1969). As conglutinogen-activating factor, it acts on EAC1423, EAC143, or EAC43 to render such cells agglutinable by a bovine serum protein, conglutinin. The latter phenomenon is known as conglutination (Lachmann and Coombs, 1965).

In studies with purified complement components, Müller-Eberhard and and his colleagues (1966) obtained evidence that $C\overline{1}$ acts on C4 and C2 in such a manner that the latter two proteins form a complex, $C\overline{42}$, which in turn acts enzymically upon C3 to activate it and produce a conversion product only slightly smaller in size (MW 220,000) than C3. On lowering the pH to 5 or below, this protein dissociates into two fragments, C3b, with a sedimentation coefficient of 9.0 S and an electrophoretic mobility similar to C3c, and C3a. The latter is a fragment with an approximate molecular weight of 9000 which possesses anaphylatoxin properties (Dias da Silva and Lepow, 1966, 1967; Cochrane and Müller-Eberhard, 1968). Anaphylatoxin derived from C3 (C3a) causes degranulation of rat peritoneal mast cells with release of histamine and also produces contraction of guinea pig ileum. Similar, if not identical, fragments may be obtained by brief treatment of purified C3 with trypsin, thrombin, and plasmin (Bokisch *et al.*, 1969). In normal serum, an inactivating protein exists which operates either by binding to (Lepow and Patrick, 1968) or proteolytically degrading (Bokisch *et al.*, 1969) C3a.

Although it had been known for many decades that the venom of the cobra *Naja naja* destroyed hemolytic complement (Flexner and Noguchi, 1903; Ritz, 1912), it is only recently that this phenomenon has been shown to occur *in vivo* and to result specifically from inactivation and depletion of C3 (Klein and Wellensiek, 1965; Nelson, 1966). The active principle in the venom, designated CoF, is a 7 S protein of beta mobility which has no effect on isolated C3 (Müller-Eberhard *et al.*, 1966). However, CoF interacts in the presence of Mg^{2+} with a 5 S β-mobility pseudoglobulin (proinactivator) from normal human serum to form a 9 S bimolecular complex capable of converting C3 to C3b and C3a.

In addition to the conversion products already mentioned, several others have been found in serum and have been produced from purified C3 under various conditions. Lundh (1965a) showed that in agarose electrophoresis the major C3 conversion product on treatment of serum or purified C3 with hydrazine migrated in a position cathodal to C3 at pH 8.6. He called this conversion product β_{1AH}-globulin. It has subsequently been observed that a conversion product of identical mobility is also seen in agarose electrophoresis on aging of serum (particularly in the cold), on treatment of serum with $C\bar{1}$ or with zymosan (Alper and Johnson, 1969), or on storage of purified C3 in the cold. A storage conversion product in serum with an electrophoretic mobility anodal to C3c was called α_{2D} by West and co-workers (1965). They reported that with certain antisera to C3 prepared in goats, complete immunological nonidentity could be shown between this conversion product and either C3 or C3c (West *et al.*, 1966). Lundh and Laurell (1967) also observed an α_2 mobility conversion product of C3 on storage of serum and concluded that it was probably an aggregation product since its molecular weight, as judged by gel filtration on Sephadex G200, was over 200,000 and hence greater than that of C3c. By contrast, Pollěy and Müller-Eberhard (1966) found the molecular weight of α_{2D} to be about 75,000, and they therefore assumed that it represented a further fragment in C3 degradation. An α_2 fragment can be produced from C3b by prolonged treatment with trypsin, and this has been named C3d. Its molecular weight is reported to be 27,000 (Bokisch *et al.*, 1969). The situation with respect to the degradation of C3 on storage either in the pure state or in serum is even more complicated than already outlined. Lundh and Laurell (1967) have described an interbeta conversion product with an electrophoretic mobility in agarose between C3 and C3c. A conversion product with α_1 mobility is also visible on acrylamide or agarose gel electrophoresis (Alper and Rosen, unpublished observations) as can be seen in Fig. 2. To render the picture even more complex, the various conversion products differ in mobility reflecting the genetic polymorphism of C3 described below (Alper and Propp, 1967; Azen and Smithies, 1968).

C. Genetic Polymorphism

There are at present nine electrophoretically distinct forms of C3 which are inherited as autosomal codominant traits (Alper and Propp, 1968; Azen and Smithies, 1968; Alper and Propp, 1970). The most common gene in the populations studied to date is designated $C3^S$. In Caucasians, this gene has a frequency of approximately 0.77, and a gene with an electrophoretically faster product, $C3^F$, occurs with a frequency of about 0.22. Of 80 Orientals studied, no example of $C3^F$ was found, whereas this gene has a frequency in American Negroes of approximately 0.07. Other variants are rare in all populations thus far investigated except for the Bantus of Rwanda in whom variants rare in other populations have a combined frequency of almost 0.04 (Wieme, 1968). In heterozygous individuals, both C3 variants occur in serum at approximately equal concentrations, as a general rule. An exception will be discussed below. All of the variants known at present possess C3 functional activity in a hemolytic assay.

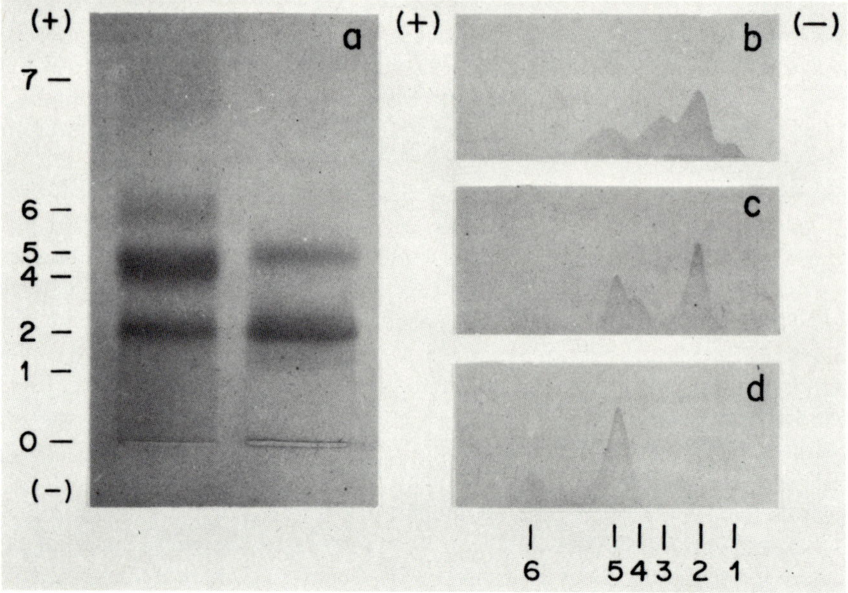

FIG. 2. Conversion products of C3 as analyzed by immunofixation (a) and antigen–antibody crossed electrophoresis (b–d) in agarose gel using antibody specific for C3. The origin is indicated by (0). The left-hand sample in (a) was purified C3 incubated for 5 days at 37°C, whereas the right-hand sample was slightly aged whole serum. The electrophoretic positions of the various bands are shown at the left and represent (1) γ_1 conversion product, (2) C3, (4) C3b, (5) C3c, (6) α_2 D (or C3d), and (7) α_1 conversion product. The numbers at the bottom of (b–d) correspond to those in (a). Position (3) is that of the interbeta conversion product.

D. Site of Synthesis

Attempts to define the site of synthesis of C3 by *in vitro* incubation of tissue slices with radioactive amino acids followed by radioimmunoelectrophoresis led to the conclusion that this protein was synthesized throughout the reticuloendothelial system and that the macrophage was most active in this synthesis (Hochwald *et al.*, 1961; Asofsky and Thorbecke, 1961; Thorbecke *et al.*, 1965). The findings in a case of orthotopic liver transplantation cast considerable doubt on the relevance of these conclusions to the synthesis of C3 *in vivo* (Alper *et al.*, 1969a). It was observed that the C3 genetic type of the recipient in this case changed completely to that of the donor and remained so until the patient died of sepsis 45 days after operation. These observations constitute strong evidence that the liver is the primary, if not sole, site of synthesis of C3 *in vivo*.

Using the genetic markers on C3, it was shown that the human fetus synthesizes C3 and that transplacental passage is either very minimal or does not occur at all (Propp and Alper, 1968; Azen and Smithies, 1968). It has been demonstrated that C3 is detectable in the serum of the fetus quite early in gestation (Adinolfi and Gardner, 1967; Gitlin and Biasucci, 1969).

V. Abnormalities of C3 Concentration

A. Acquired Abnormalities

C3 concentration in serum, determined immunochemically, has been observed to be lowered in systemic lupus erythematosus (Seligmann and Hanau, 1958; West *et al.*, 1964; Lundh, 1965a; Klemperer *et al.*, 1965), acute glomerulonephritis (West *et al.*, 1964; Klemperer *et al.*, 1965), progressive glomerulonephritis (Gotoff *et al.*, 1965), subacute bacterial endocarditis with focal or embolic glomerulonephritis (Williams and Kunkel, 1962), in some patients with severe liver disease (West *et al.*, 1964), and in a few patients with acquired hemolytic anemia (Petz *et al.*, 1968).

The serum concentration of C3 may, on the other hand, be increased following tissue necrosis or inflammation as in acute myocardial infarction or pneumonia (Lundh, 1965b). In contrast to other acute phase reactants such as haptoglobin or α_1 antitrypsin, the increase in C3 concentration is modest and occurs more sluggishly.

B. Genetic Abnormalities

A kindred has been described (Alper *et al.*, 1969b) in which some members in each of three generations had half-normal concentrations of C3. From the C3 types of affected individuals in this kindred, it was concluded that they had inherited an allele of the C3 structural system which failed to produce a detectable product—a so-called silent allele. Although hemolytic complement was slightly low in affected individuals, they were entirely free of symptoms.

In another kindred, three individuals, one in each of three generations, were found to be heterozygous for C3 structural alleles. On prolonged electrophoresis in agarose gel and on antigen–antibody crossed electrophoresis, sera from these subjects was unusual in that, although they appeared to be C3 FS, the concentration of C3 F was visibly less than that of C3 S (Alper and Rosen, 1969). It seemed probable that in this family a hypomorphic form of C3 F, designated C3 f, was inherited as an allele of the other variants. The concentration of C3 in the serum of these persons was, as might be expected, within the normal range.

A complex disorder affecting the serum concentration of C3 has also been briefly described (Alper *et al.*, 1968, 1970). While rigorous proof that this disorder is hereditary is lacking in the single case so far uncovered, an inherited defect of some normal serum protein seems highly likely. The affected individual was a 24-year-old man with Klinefelter's syndrome who gave a life-long history of infections with pyogenic organisms. *In vitro* tests of complement-mediated functions in his serum disclosed gross defects in bactericidal activity for smooth gram-negative bacilli, in chemotaxis for normal leukocytes, in hemolytic complement, and in the enhancement of phagocytosis of pneumococci by normal peripheral blood leukocytes. Measurements of serum complement components yielded normal values with the notable exception of C3, which was about 30 mg/100 ml by an immunochemical method. However, of this immunoreactive material, about three-fourths was in the form of conversion product which had the electrophoretic and size characteristics of C3b. This pattern was found in all fresh plasma samples taken from the patient over a 2-year period, whether or not he had infection.

It was observed that most complement-mediated functions in his serum could be improved by the addition of a heat-labile 5–6 S β-pseudoglobulin. Enhancement of phagocytosis required the addition of both this protein and C3. Although the nature of the restoring factor is unclear, it is probably identical to that co-factor shown to be necessary for maximal enhancement of pneumococcal phagocytosis in the presence of C1, C4, C2, C3, and a dialyzable substance (Johnston *et al.*, 1969).

An additional finding was that the patient's red cells reacted with an anti-C3 Coombs reagent (Abramson *et al.*, 1968) without manifestations of hemolytic anemia. It seems probable that this C3 coating is the result of the transient generation of $\overline{C3}$ during the continuous attack on C3 in this patient. It was further observed (Abramson *et al.*, 1969) that on incubation in normal serum the patient's red cells lost their reactivity with anti-C3. No such effect was seen when the cells were incubated in the patient's serum. These observations suggested that he was lacking a normal serum constituent. On purification from normal serum, this proved to have the characteristics of the C3 inactivator or KAF. In addition, evidence was obtained that the C3 proinactivator was absent from the patient's serum.

VI. Studies of Alterations of C3 *in Vivo*

A. Presence of Conversion Products

In examining C3 by immunoelectrophoresis of fresh EDTA plasma from patients with systemic lupus erythematosus, Morse and co-workers (1962) found significant C3c in one of nine samples. In a later study (Alper and Rosen, 1967) utilizing antigen–antibody crossed electrophoresis (Laurell,

1965) as a more sensitive means of detecting low concentrations of conversion products, none were detected in samples from patients with systemic lupus erythematosus. The difference in results may have reflected differences in the patients studied.

Immunoelectrophoretic studies by West and co-workers (1967) of plasma and serum from patients with acute glomerulonephritis and hypocomplementemic persistent (chronic) glomerulonephritis (West *et al.*, 1965), a disease possibly the same as progressive glomerulonephritis (Gotoff *et al.*, 1965), indicated the presence of circulating α_{2D} in all patients who had reduced concentrations of C3. They took this finding as evidence for breakdown of C3 *in vivo* in patients with these two disorders. They failed to find other conversion products, such as C3b or C3c.

Soothill (1967) noted irregularities of the C3 arc in immunoelectrophoresis of fresh plasma from patients with a variety of forms of glomerulonephritis, irrespective of C3 concentration, duration of disease, or other variables. These findings are difficult to interpret.

Using antigen–antibody crossed electrophoresis, Alper and Rosen (1967) detected material with the electrophoretic mobility of C3b or C3c in the fresh EDTA plasma from three of eight patients with acute glomerulonephritis. Total C3 concentration, including conversion products, was markedly lowered in these plasma samples (between 7 and 18 mg/100 ml), and the conversion product amounted to 10–35% of the total C3. Samples were obtained from these three patients within the first 48 hr from the onset of symptoms, whereas most of the remaining five patients were studied later. No conversion products were detected in the fresh plasma of patients with progressive glomerulonephritis or of a single patient with subacute glomerulonephritis, despite C3 concentrations as low, in some of the former patients, as those seen in patients with acute glomerulonephritis. In no instance was any α_2 conversion product detected, although the antiserum used reacted with such material in aged serum.

In a rabbit model for serum sickness (Schwab *et al.*, 1950), studies with antigen–antibody crossed electrophoresis using antiserum to rabbit C3 revealed striking and complex changes in this protein during antigen–antibody interaction *in vivo* as well as *in vitro* (Propp and Alper, 1969). Immediately following the addition of specific precipitates to fresh rabbit serum at 37°C, a fall in the concentration of the peak corresponding to native C3 was observed with the concomitant appearance of at least seven conversion products, most of which were anodal to C3. Qualitatively similar changes were seen in rabbit plasma following intravenous injection of antigen (bovine serum albumin) into immune animals. It was also observed that the prompt fall in concentration of native C3 paralleled the fall in hemolytic complement, whereas, as expected, there was a lag in the fall of total immunochemically

TABLE I

METABOLISM OF ISOTOPICALLY LABELED C3 IN CONTROL SUBJECTS[a]

Control subjects	C3 concentration (mg/100ml)	Fractional catabolic rate (% plasma pool/hr)	Synthesis rate (mg/kg/hr)	Extravascular pool/plasma ratio	Reference
10 Normals	148.4 ± 35.7	2.35 ± 1.05	1.38 ± 0.51	0.90 ± 0.41	Alper and Rosen (1967)
12 Normals	134.0 ± 33.9[b]	2.12 ± 0.56	1.17 ± 0.26	0.52 ± 0.39	Petz *et al.* (1968)
2 Normals, 1 Anephric, 8 stable renal allograft	117.5 ± 39.2[c]	1.44 ± 0.55[d]	1.50 ± 1.67[c]	0.80 ± 1.27	Carpenter *et al.* (1969)

[a] Values are given as the mean ±2 SD.
[b] Based on determinations in 13 normal subjects.
[c] Based on determinations in four control subjects.
[d] Fractional catabolic rates "corrected for denatured protein" (see text).

detected C3 owing to the presence of circulating conversion products. By 24 hr, the C3 had returned to normal concentration, and circulating conversion products were no longer detectable.

B. Metabolic Studies with Labeled Purified C3

The results of metabolic studies with labeled purified C3 in normal subjects or individuals thought not to have abnormalities of metabolism of C3 are given in Table I. Two types of preparations have been used and, although the earlier material (Müller-Eberhard *et al.*, 1960) is only 80–85% pure as compared with 95% or better for the newer material (Nilsson and Müller-Eberhard, 1965), their metabolic behavior is essentially the same (Alper and Rosen, 1967). Trace labeling with ^{125}I or ^{131}I may be achieved with either the iodine monochloride technique (McFarlane, 1958) or the chloramine T method (McConahey and Dixon, 1966) without alteration in the functional properties or reactivities of C3 (Müller-Eberhard *et al.*, 1966; Alper *et al.*, 1966) if, in the iodine monochloride technique, the iodination is carried out at one atom or less of iodine per molecule of C3.

The values presented in Table I were obtained in standard fashion from semilogarithmic plots of protein-bound radioactivity and all studies used the Matthews (1957) method of analysis. In some instances, the method of Nosslin (Donato *et al.*, 1966) was also applied and yielded similar results. It will be noted that the results in all three laboratories are in good agreement except for the fractional catabolic rates, wherein those obtained by Carpenter *et al.* (1969) are significantly lower than those obtained in the two earlier studies. The lower catabolic rates are apparently the result of subtraction of the initial exponential from curves that required three exponentials instead of two for their resolution. The justification for this procedure resides in the belief of these authors that only two exponentials should be necessary to describe the plasma disappearance curve of undenatured protein, and that therefore the initial exponential must entirely reflect denatured labeled material. They have presented no evidence that the label from such material can be quantitatively accounted for in the earlier urine samples.

Studies with purified labeled C3 conversion products (C3b, C3c, or both) have indicated that the latter are cleared at a rate 5–16 times that of C3 (Alper and Rosen, 1967; Petz *et al.*, 1968). For the reasons outlined earlier, this means that the finding of an increased fractional catabolic rate for C3 in patients is at least compatible with complement activation and conversion of C3 *in vivo*.

Some conclusions can be drawn from the results of metabolic studies with purified labeled C3 in patients summarized in Table II. Of the three patients with acute glomerulonephritis studied, fractional catabolic rates were mildly

TABLE II

METABOLISM OF ISOTOPICALLY LABELED C3 IN PATIENTS

Diagnosis[a]	Number	C3 concentration (mg/100 ml)		Fractional catabolic rate (% plasma pool/hr)		Synthesis rate (mg/kg/hr)		Reference[b]
		Mean	Range	Mean	Range	Mean	Range	
AGN	3	22	6–47	8.7	3.6–15.2	0.8	0.1–1.5	(1)
SLE	4	98	57–150	3.7	2.8–5.4	1.7	1.2–2.0	(1)
PGN	4	18	7–30	3.2	1.7–5.2	0.3	0.1–0.6	(1)
NS	2	153	134–171	5.8	2.8–8.7	2.9	1.8–3.9	(1)
AHA	2	172	125–220	2.0	1.5–2.5	1.8	1.5–2.1	(1)
	7	100	58–149	3.2	2.0–4.9	1.7	1.3–2.8	(2)
PNH	2	113	75–150	2.4	1.8–3.0	1.8	1.1–2.5	(1)
	3	129	110–152	2.5	1.8–2.9	1.8	1.1–2.2	(2)
HANE	2	145	135–155	3.3	3.0–3.6	1.6	1.5–1.8	(1)
	4	115	91–137	2.9	2.6–3.2	1.5	1.1–1.8	(3)
SGN	1	77	—	6.6	—	2.5	—	(1)
CGN	1	144	—	2.6	—	1.1	—	(3)
	3	"low"	—	—	5.8–15.0	—	0.3 to 'high"	(4)
RAR	3	130	—	2.3	2.1–2.5	2.0	—	(3)

[a] AGN, acute glomerulonephritis; SLE, systemic lupus erythematosus; PGN, progressive glomerulonephritis; NS, nephrotic syndrome of childhood; AHA, acquired hemolytic anemia; PNH, paroxysmal nocturnal hemoglobinuria; HANE, hereditary angioneurotic edema; SGN, subacute glomerulonephritis; CGN, chronic glomerulonephritis; RAR, renal allograft rejection.

[b] (1) Alper and Rosen (1967); (2) Petz *et al.* (1968); (3) Carpenter *et al.* (1969); (4) Kohler *et al.* (1969).

to markedly increased. The patients with the highest fractional catabolic rates were those studied earliest during the course of their illness. The C3 synthesis rate was markedly low in one patient of this group, at the lower limit of normal in the second patient, whereas the third had a normal synthesis rate. These findings are consistent with complement activation, at least during the early phase of acute glomerulonephritis. The lowered synthesis rate of C3 is consistent with the finding that, after the first few days of illness, purified C3 alone added to the serum of patients with acute glomerulonephritis produces a normalization of serum hemolytic complement (Klemperer *et al.*, 1965), and with the observation that although the concentration of C3 does not return to normal until 4–6 weeks after the onset of clinical acute glomerulonephritis (West *et al.*, 1964), the concentration of C4 returns to normal within the first week or so (Kohler and Müller-Eberhard, 1967).

In progressive glomerulonephritis, a form of the nephrotic syndrome of childhood and early adolescence in which the concentration of C3 is regularly lowered (Gotoff *et al.*, 1965), depression of C3 synthesis was observed in each of four studies (Alper and Rosen, 1967) and, despite mild to moderate increase in the fractional catabolic rate also seen in two of the four studies, was clearly the chief factor explaining the low serum C3 concentrations in these patients.

The results in subacute and chronic glomerulonephritis and idiopathic nephrotic syndrome were mixed. As can be seen, some patients showed moderate to marked increases in fractional catabolic rates in association with lowered serum C3 concentrations, and both high and low C3 synthesis rates were observed (Alper and Rosen, 1967; Carpenter *et al.*, 1969; Kohler *et al.*, 1969). The moderate increases in both the fractional catabolic and synthesis rates of C3 in one patient with the nephrotic syndrome are similar to the findings of Gitlin *et al.* (1956) for albumin, IgG, and transferrin in some patients with this disorder. Three of four patients with systemic lupus erythematosus had elevated fractional catabolic rates for C3 (Alper and Rosen, 1967), a finding consistent with complement activation *in vivo* in some of these patients.

Of nine patients with acquired hemolytic anemia who have been studied (Alper and Rosen, 1967; Petz *et al.*, 1968), five had no increase in the fractional catabolic rate for C3, whereas in four patients with warm antibodies and positive C3 Coombs antiglobulin tests, mild to moderate increases in the catabolic rate were observed (Petz *et al.*, 1968) in association, in most of these cases, with a depression of the serum C3 concentration. C3 synthesis rates were normal or elevated in patients with acquired hemolytic anemia. In neither of the two series of studies could labeled C3 be detected on the patients' red cells after injection, despite positive antiglobulin tests. Five studies in patients with paroxysmal nocturnal hemoglobinuria failed to disclose important abnormalities of catabolism, although an increase in the C3 synthesis rate was observed in three of the studies (Alper and Rosen, 1967; Petz *et al.*, 1968).

Hereditary angioneurotic edema is a disorder associated with a lack of serum C1 esterase inhibitor activity (Donaldson and Evans, 1964). During attacks, free C1 esterase activity is demonstrable, there is a marked diminution of C4 and C2 as measured by functional assays, and conversion of C4 *in vivo* can be detected (Donaldson and Rosen, 1964). No important reduction in the concentration of C3 is seen and conversion products of C3 are not found in fresh plasma. As might be expected from these observations, fractional catabolic rates of C3 in patients with this disorder are normal or at most only slightly elevated (Alper and Rosen, 1967; Carpenter *et al.*, 1969). On the other hand, as one might also expect, fractional catabolic rates for C4 are markedly increased (Carpenter *et al.*, 1969).

The metabolic behavior of purified C3 FF was very similar to that of C3 SS in various subjects of different C3 type and was normal in all subjects tested, including one who was C3 fS. This suggests, but does not prove, that C3 f is a variant with a diminished synthesis rate (Alper and Rosen, 1969).

In the patient with the complex disorder characterized by increased susceptibility to infection, decreased concentration of C3 in the serum, and the continual presence of C3b, metabolic studies with labeled purified C3 were performed on two occasions. In the first of these, the fractional catabolic rate was found to be markedly elevated at about 10% of the plasma pool per hour, whereas the C3 synthesis rate was normal. Furthermore, on preparative agarose gel electrophoresis about 40% of the administered radioactivity had the electrophoretic mobility of C3b 2 hr after injection in contrast with the radioactivity in the 10-min sample from the patient and both the 10-min and 2-hr samples from normal controls, all of which was found in the C3 area exclusively. Thus, there was prompt conversion of the labeled C3 to C3b in the patient (Alper *et al.*, 1968). These results suggested that the patient was missing a factor necessary for the stability of C3 *in vivo*. Further evidence for this hypothesis was obtained following an infusion of 500 ml of normal plasma. Over a 24-hr period, C3 rose in concentration to 60 or 70 mg/100 ml, and C3b rapidly fell in concentration so that by 8 hr it was no longer detectable. Complement-mediated functions in the patient's serum improved gradually but dramatically, and his C3 Coombs test became negative. This state of affairs persisted for the following 16 days, after which all measurements reverted to the original pattern.

A second metabolic study was performed during the patient's normalization period, and it was observed that the fractional catabolic rate of C3 had fallen to about 5% of the plasma pool per hour. The C3 synthesis rate was again normal and similar to that of the first study.

VII. Conclusions

While investigation of the factors controlling the serum concentration of C3 is still in its early phases, much has already been learned. These studies have been greatly facilitated by the happy circumstance that this protein occurs normally in substantial concentration in serum so that it can be purified and antiserum can easily be prepared against it. Such antisera have been fruitfully used to examine tissue binding in various disease states, to measure the concentration in serum from normal subjects and patients, and to define the presence of circulating conversion products in certain disease states. The presence of genetically controlled structural polymorphism in C3 has permitted some definition of genetic factors in the control of the serum concentration of this protein. Finally, the ability to trace label the purified protein without alteration in its reactivities has provided a powerful means of studying the metabolic behavior of C3 and to relate such behavior to the reactions of C3 during complement activation in *in vitro* systems.

References

Abramson, N., Alper, C. A., Rosen, F. S., and Jandl, J. H. (1968). *Proc. 12th Congr. Int. Soc. Hematol., New York* p. 122 (FF–1). (Abstr.)

Abramson, N., Alper, C. A., Rosen, F. S., and Jandl, J. H. (1969). *Clin. Res.* **17**, 318. (Abstr.)

Adinolfi, M., and Gardner, B. (1967). *Acta Paediat. Scand.* **56**, 450.

Alper, C. A., and Johnson, A. M. (1969). *Vox Sang.* **17**, 445.

Alper, C. A., and Propp, R. P. (1967). *Clin. Res.* **15**, 291. (Abstr.)

Alper, C. A., and Propp, R. P. (1968). *J. Clin. Invest.* **47**, 2181.

Alper, C. A., and Propp, R. P. (1970). In preparation.

Alper, C. A., and Rosen, F. S. (1967). *J. Clin. Invest.* **46**, 2021.

Alper, C. A., and Rosen, F. S. (1969). *J. Clin. Invest.* **48**, 2a. (Abstr.)

Alper, C. A., Levin, A. S., and Rosen, F. S. (1966). *Science* **153**, 180.

Alper, C. A., Abramson, N., Johnston, R. B., Jr., McCall, C. E., Jandl, J. H., and Rosen, F. S. (1968). *J. Clin. Invest.* **47**, 1a. (Abstr.)

Alper, C. A., Johnson, A. M., Birtch, A. G., and Moore, F. D. (1969a). *Science* **163**, 286.

Alper, C. A., Propp, R. P., Klemperer, M. R., and Rosen, F. S. (1969b). *J. Clin. Invest.* **48**, 553.

Alper, C. A., Abramson, N., Johnston, R. B., Jr., Jandl, J. H., and Rosen, F. S. (1970). *New Engl. J. Med.* **282**, 349.

Asofsky, R., and Thorbecke, G. J. (1961). *J. Exp. Med.* **114**, 471.

Azen, E. A., and Smithies, O. (1968). *Science* **162**, 905.

Azen, E. A., Smithies, O., and Hiller, O. (1969). *Biochem. Genet.* **3**, 215.

Bokisch, V. A., Müller-Eberhard, H. J., and Cochrane, C. G. (1969). *J. Exp. Med.* **129**, 1109.

Bordet, J., and Gengou, O. (1901). *Ann. Inst. Pasteur Paris* **15**, 129.

Buchner, H. (1889). *Zentralbl. Bakteriol. Parasitenk.* **5**, 817.

Carpenter, C. B., Ruddy, S., Shehadeh, I. H., Müller-Eberhard, H. J., Merrill, J. P., and Austen, K. F. (1969). *J. Clin. Invest.* **48**, 1495.

Cinader, B., and Lepow, I. H. (1967). *In* "Antibodies to Biologically Active Molecules" (B. Cinader, ed.), Vol. 1, p. 15. Macmillan (Pergamon), New York.

Cinader, B., Dubiski, S., and Wardlaw, A. C. (1964). *J. Exp. Med.* **120**, 897.

Cochrane, C. G., and Müller-Eberhard, H. J. (1968). *J. Exp. Med.* **127**, 371.

Cooper, N. R. (1967). *J. Immunol.* **98**, 132.

Cooper, N. R., and Becker, E. L. (1967). *J. Immunol.* **98**, 119.

Dias da Silva, W., and Lepow, I. H. (1966). *J. Immunol.* **95**, 1080.

Dias da Silva, W., and Lepow, I. H. (1967). *J. Exp. Med.* **125**, 921.

Donaldson, V. H., and Evans, R. R. (1964). *Am. J. Med.* **35**, 37.

Donaldson, V. H., and Rosen, F. S. (1964). *J. Clin. Invest.* **43,** 2204.
Donato, L., Matthews, C. M. E., Nosslin, B., Segre, G., and Vitek, F. (1966). *J. Nucl. Biol. Med.* **9,** 3.
Ehrlich, P., and Morgenroth, J. (1899). *Berlin. Klin. Wochenschr.* **36,** 6.
Ferrata, A. (1907). *Berlin. Klin. Wochenschr.* **44,** 366.
Flexner, S., and Noguchi, H. (1903). *J. Exp. Med.* **6,** 277.
Gitlin, D., and Biasucci, A. (1969). *J. Clin. Invest.* **48,** 1433.
Gitlin, D., Janeway, C. A., and Farr, L. E. (1956). *J. Clin. Invest.* **35,** 44.
Gotoff, S. P., Fellers, F. X., Vawter, G. F., Janeway, C. A., and Rosen, F. S. (1965). *New Engl. J. Med.* **273,** 524.
Hochwald, G. M., Thorbecke, G. J., and Asofsky, R. (1961). *J. Exp. Med.* **114,** 459.
Johnston, R. B., Jr., Klemperer, M. R., Alper, C. A., and Rosen, F. S. (1969). *J. Exp. Med.* **129,** 1275.
Klein, P. G., and Wellensiek, H. J. (1965). *Immunology* **8,** 590.
Klemperer, M. R., Gotoff, S. P., Alper, C. A., Levin, A. S., and Rosen, F. S. (1965). *Pediatrics* **35,** 765.
Klemperer, M. R., Woodworth, H. C., Rosen, F. S., and Austen, K. F. (1966). *J. Clin. Invest.* **45,** 880.
Kohler, P. F., and Müller-Eberhard, H. J. (1967). *Clin. Res.* **15,** 296. (Abstr.)
Kohler, P. F., Hutt, M. P., and Riley, C. (1969). *J. Clin. Invest.* **48,** 45a. (Abstr.)
Lachmann, P. J., and Coombs, R. R. A. (1965). *Complement Ciba Found. Symp.,* **1964** pp. 242–280.
Lachmann, P. J., and Müller-Eberhard, H. J. (1968). *J. Immunol.* **100,** 691.
Lachmann, P. J., Müller-Eberhard, H. J., Kunkel, H. G., and Paronetto, F. (1962). *J. Exp. Med.* **115,** 63.
Laurell, C.-B. (1965). *Anal. Biochem.* **10,** 358.
Laurell, C.-B., Laurell, S., and Skoog, N. (1956). *Clin. Chem.* **2,** 99.
Lepow, I. H. (1965). *In* "Immunological Diseases" (M. Samter and H. L. Alexander, eds.), pp. 188–210. Little, Brown, Boston, Massachusetts.
Lepow, I. H., Dias da Silva, W., and Patrick, R. A. (1969). *In* "Cellular and Humoral Mechanisms in Anaphylaxis and Allergy" (H. Z. Movat, ed.), S. Karger, New York, p. 247.
Linscott, W. D., and Cochrane, C. G. (1964). *J. Immunol.* **93,** 972.
Lundh, B. (1964). *Scand. J. Clin. Lab. Invest.* **16,** 108.
Lundh, B. (1965a). *Acta Univ. Lund. Sect.* **2, 25,** 3.
Lundh, B. (1965b). *Acta Pathol. Microbiol. Scand.* **63,** 266.
Lundh, B., and Laurell, C.-B. (1967). *Immunology* **13,** 649.
McConahey, P. J., and Dixon, F. J. (1966). *Int. Arch. Allergy Appl. Immunol.* **29,** 185.
McFarlane, A. S. (1958). *Nature* (*London*) **182,** 53.
M'Gowan, J. P. (1908). *J. Pathol. Bacteriol.* **12,** 519.
Matthews, C. M. E. (1957). *Phys. Med. Biol.* **2,** 36.
Morse, J. H., Müller-Eberhard, H. J., and Kunkel, H. G. (1962). *Bull. N.Y. Acad. Med.* **38,** 641.
Müller-Eberhard, H. J. (1961). *Acta Soc. Med. Upsal.* **66,** 152.
Müller-Eberhard, H. J. (1968). *Advan. Immunol.* **8,** 1.
Müller-Eberhard, H. J., and Nilsson, U. (1960). *J. Exp. Med.* **111,** 217.
Müller-Eberhard, H. J., Nilsson, U., and Aronsson, T. (1960). *J. Exp. Med.* **111,** 201.
Müller-Eberhard, H. J., Nilsson, U. R., Dalmasso, A. P., Polley, M. J., and Calcott, M. A. (1966). *Arch. Pathol.* **82,** 205.
Müller-Eberhard, H. J., Polley, M. J., and Calcott, M. A. (1967). *J. Exp. Med.* **125,** 359.
Nelson, R. A., Jr. (1953). *Science* **118,** 733.

Nelson, R. A., Jr. (1965). *In* "The Inflammatory Process" (B. W. Zweifach, R. T. McCluskey, and L. H. Grant, eds.), pp. 819–872. Academic Press, New York.
Nelson, R. A., Jr. (1966). *Surv. Ophthalmol.* **11**, 498.
Nelson, R. A., Jr., Jensen, J., Gigli, I., and Tamura, N. (1966). *Immunochemistry* **3**, 111.
Nilsson, U. R., and Müller-Eberhard, H. J. (1965). *J. Exp. Med.* **122**, 277.
Nishioka, K., and Linscott, W. D. (1963). *J. Exp. Med.* **118**, 767.
Nuttall, G. (1904). "Blood Immunity and Blood Relationship." Cambridge Univ. Press, Cambridge, England.
Peetoom, F., van der Hart, M., and Pondman, K. W. (1964). *Vox Sang.* **9**, 85.
Petz, L. D., Fink, D. J., Letsky, E. A., Fudenberg, H. H., and Müller-Eberhard, H. J. (1968). *J. Clin. Invest.* **47**, 2469.
Polley, M. J., and Müller-Eberhard, H. J. (1966). *In* "Progress in Hematology" (E. B. Brown, and C. V. Moore, eds.), pp. 2–25. Grune and Stratton, New York.
Pondman, K. W., and Peetoom, F. (1964). *Immunochemistry* **1**, 65.
Propp, R. P., and Alper, C. A. (1968). *Science* **162**, 672.
Propp, R. P., and Alper, C. A. (1969). *Immunology* **17**, 693.
Ritz, H. (1912). *Z. Immunitaetsforsch. Exp. Ther.* **13**, 62.
Rosenberg, L. T., and Tachibana, D. K., (1962). *J. Immunol.* **89**, 861.
Rother, U., and Rother, K. (1961). *Z. Immunitaetsforsch. Exp. Ther.* **121**, 224.
Rother, K., Rother, U., Müller-Eberhard, H. J., and Nilsson, U. R. (1966). *J. Exp. Med.* **124**, 773.
Ruddy, S., and Austen, K. F. (1969). *J. Immunol.* **102**, 533.
Schur, P. H., and Austen, K. F. (1968). *Annu. Rev. Med.* **19**, 1.
Schwab, L., Moll, F. C., Hall, T., Brean, H., Kirk, M., Hawn, C. V. Z., and Janeway, C. A. (1950). *J. Exp. Med.* **91**, 505.
Seligmann, M., and Hanau, C. (1958). *Rev. Hematol.* **13**, 239.
Soothill, J. F. (1967). *Clin. Exp. Immunol.* **2**, 83.
Steinbuch, M. (1964). *Vox Sang.* **9**, 96.
Thorbecke, G. J., Hochwald, G. M., van Furth, R., Müller-Eberhard, H. J., and Jacobson, E. B. (1965). *Complement Ciba Found. Symp.* (*1964*) pp. 99–119.
Unanue, E., and Dixon, F. J. (1964). *J. Exp. Med.* **119**, 965.
Vassalli, P., and McCluskey, R. T. (1964). *Amer. J. Pathol.* **45**, 653.
West, C. D., Northway, J. D., and Davis, N. C. (1964). *J. Clin. Invest.* **43**, 1507.
West, C. D., McAdams, A. J., McConville, J. M., Davis, N. C., and Holland, N. H. (1965). *J. Pediat.* **67**, 1089.
West, C. D., Davis, N. C., Forristal, J., Herbert, J., and Spitzer, R. (1966). *J. Immunol.* **96**, 650.
West, C. D., Winter, S., Forristal, J., McConville, J. M., and Davis, N. C. (1967). *J. Clin. Invest.* **46**, 539.
Wieme, R. J. (1968). *Nature* (*London*) **220**, 176.
Williams, R. C., Jr., and Kunkel, H. G. (1962). *J. Clin. Invest.* **41**, 666.
Zimmer, M. J. (1960). *Bull. Soc. Fr. Dermatol. Syphiligr.* **67**, 616.

26

REGULATION OF PROTEIN METABOLISM IN BURNS

Gunnar Birke

KING GUSTAF V RESEARCH INSTITUTE,
STOCKHOLM, SWEDEN

I. Introduction

The body's responses to trauma, cancer, and coronary thrombosis, as manifested by changes in the protein pattern, are relatively uniform. There is an initial rise in α_1-, but notably in α_2-globulins, transferrin, and fibrinogen (Fig. 1). The increase in α-globulins is brought about mainly by glucoproteins which, like fibrinogen, are synthesized in the liver. Fibrinogen and α_2-globulin increase rapidly and reach a peak within 4 days (Birke *et al.*, 1965). An interesting fact is that α_2-macroglobulin is the protein that shows the quickest

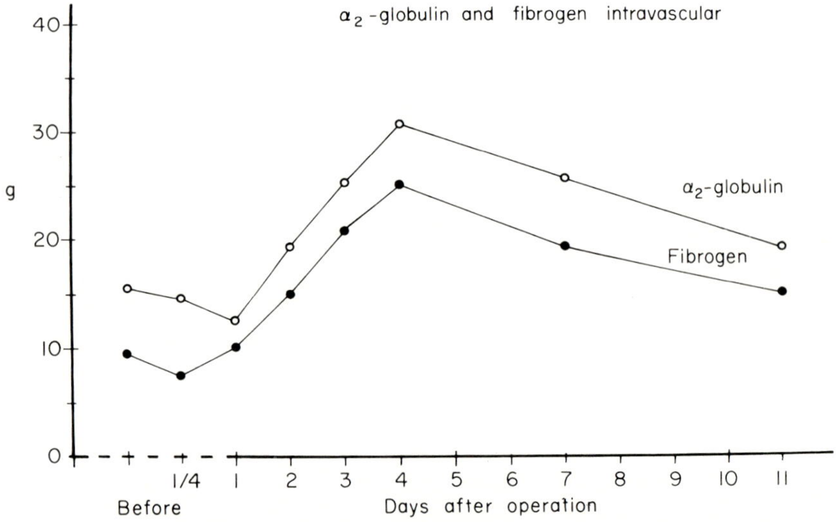

FIG. 1. α_2-Globulin and fibrinogen concentration intravascularly after operation

and perhaps most marked initial increase. Thereafter the albumin concentration decreases, while the immunoglobulins show an undefinable pattern, if there is no antigen stimulation in the form of infection. The amounts of protein lost from the circulation after large burns are well correlated with the size of the burn. Albumin concentration is elevated in the skin and in the adjacent tissue suggesting increased leakage and increased extravascular distribution in these areas (Birke *et al.*, 1968 ; Norberg *et. al.* 1970).

II. Methods

A. Treatment

The patients received antishock treatment based on plasma, blood, albumin, electrolytes, water, antiseptsis therapy, and skin grafting (Birke, 1960; Birke *et al.*, 1964; Liljedahl *et al.*, 1963). The methods of management are continually changing in view of the clinical research and the results obtained at the burns unit (Birke *et al.*, 1968).

B. Analytical Methods

Total serum protein concentration was determined by the Biuret method. Albumin was analyzed conventionally by paper electrophoresis and the other serum proteins by standard immunologic methods. Human albumin obtained after purification of serum from healthy donors was labeled with ^{131}I by

McFarlane's method (1958). 8–15 μCi of ^{131}I-labeled albumin was injected intravenously.

Plasma samples were taken at 9, 20, and 30 min, 1, 2, and 4 hr, and daily after injection of the dose. The radioactivity was measured in these samples and in 24 hr urine samples. The measurements were continued until the radioactivity was equal to the background activity or until the patient was discharged from the hospital. The retained dose was calculated by subtracting the cumulative urinary excretion from the administered dose.

In the interpretation of the results obtained it should be borne in mind that the measurements are made on the labeled albumin, which is injected at a certain time during the course of the burn. Obviously, injections made at other times can be expected to result in a different pattern, as the patient is not in a steady state. Beside the changes in the kinetics of distribution there occurs continuous treatment by infusion of both fluid and albumin. Therefore, in each individual case, the discussion is applicable mainly to the albumin situation at the time of injection.

Whole-body counting was performed with the patient lying supine on a stretcher in a room with walls consisting of a 15 cm steel and a 3 mm lead layer. One scintillation detector with a cylindrical sodium–iodide crystal, 12.5 cm in diameter and 7.5 cm in height, moved above and another one below the stretcher. The detector pulses were counted with a conventional pulse height analyzer and scaler. As a rule, all pulses over 0.1 MeV were counted when ^{131}I was used. With this setting no counts from ^{125}I are obtained. The patient's background activity was measured before injection of the ^{131}I proteins. The efficiency of measurement of ^{131}I was about 1%.

III. Results

A. Immunoglobulin Levels

The data on serum protein levels from the severely burned patients are shown in Fig. 2 (Birke *et al.*, 1969). As no significant variations were noted in the plasma volume, the recorded changes in plasma concentration represent the actual intravascular content of the respective immunoglobulins. All three analyzed immunoglobulins decrease and reach minimum values within 5 days; however, Fig. 2 shows data only for IgM and IgG. The decrease in protein is due to losses, partly extravascularly and partly through the burned skin. But because of the greatly increased synthesis, the concentration quickly reaches supernormal levels with maximum for IgM on day 10, for IgA within 14 days, but for IgG not until the third week. It is interesting to note that the immunoglobulins do not increase in the very severely burned patients, who later die of sepsis, as is illustrated by one of the cases (Fig. 2). This is

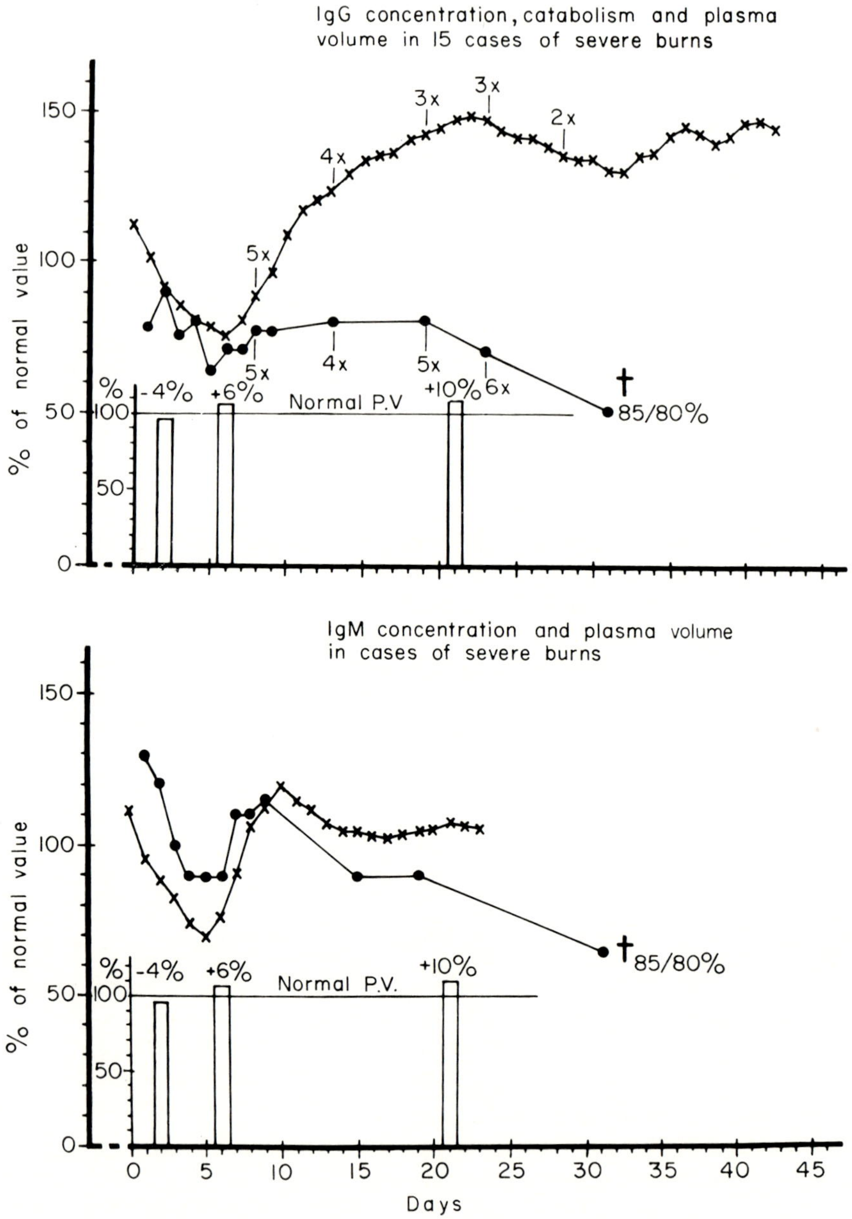

FIG. 2. IgG and IgM concentration, catabolism, and plasma volume in 15 cases of severe burns. The figures 2x–6x denote catabolism 2–6 times the normal level.

partly accounted for by the fact that the catabolic rate increases to six times the normal; but the main explanation is that, for reasons not yet made clear, the synthesizing capacity cannot increase adequately in these cases.

Since the problem of infection is so important to the preservation of life in severe burns, the metabolism of γ-globulin must undergo close scrutiny. Initially in patients with severe burns, that is patients with more than 60% of the body surface involved, 90% of the intravenously injected γ-globulin leaves the blood stream within the first 48 hr. The elimination was most rapid in the first 24-hr period but still rapid 4 days after injury. The distribution was considerably increased extravascularly, particularly in edema beneath the burned skin area, and elevations in γ-globulin levels in blisters occurred within 12–16 hr remaining high for 4–5 days. If γ-globulin is injected into the burned area, 50% of the dose remains at the site of injection after 24 hr although this value has varied and slower absorption rates than have been noted. Specific activity of γ-globulin plasma in blisters may exceed that in the plasma. γ-Globulin activity was high in all lymph nodes and elevated in all tissues in which active sepsis was noted. The activity in the spleen was not found to be elevated. The main factor in the stimulation of the IgG synthesis in cases of trauma is most certainly strong antigenic stimulation but the question why the synthesizing capacity fails in some cases of severe trauma remains to be answered. However, the amount of γ-globulin degraded can be remarkably high in cases of acutely burned patients, exceeding 19 g in one 24-hr period and it is unlikely that synthesis could keep pace with this excessive degradation for any period of time (Liljedahl *et al.*, 1963).

B. Albumin and IgG Metabolism

The results of the metabolism for albumin and IgG in studies on 30 burn cases are summarized in Table I (Davies *et al.*, 1969). This, as well as the other studies reported here, was made both in the conventional way with ^{131}I- or ^{125}I-labeled proteins and with whole-body counting, as above, for the studies with ^{131}I-labeled proteins (Birke *et al.*, 1968).

The protein concentration was measured by immunological methods. Dilution of the labeled protein in serum with newly synthesized protein and the changes in circulating protein pools made possible the rough calculation of maximum and minimum limits for the rate of synthesis. The plasma volume was also measured in the conventional way. The absolute catabolic rate for both labeled proteins is very high, the extravascular distribution is increased, most markedly in the most extensive burns (Fig. 1 and Table I). Great losses have occurred through the skin, and the synthesis of albumin is reduced.

TABLE I

PROTEIN METABOLISM IN SEVERE BURNS[a,b]

	Albumin		IgG	
	22°	32°	22°	32°
Catabolism	20–40 g/day	20 g/day	4–6 g/day	3 g/day
EV/IV	5.0	2.5	4	1.5
Losses	10 g/day	5 g/day	—	—
Synthesis	3–4 g/day	6–8 g/day	6 g/day	3 g/day

[a] Mean values.

[b] Treated in room temp. (22°); treated in dry warm air (32°).

The studies of albumin and γ-globulin metabolism have shown that while the catabolic rate is raised and losses through the skin occur, it is the extravascular distribution that plays the predominant role in altering the protein level. At a burn of 30%, albumin catabolism accounts for about 20 g and the losses through the skin amount to about 10 g. However, the extravascular distribution implies that about 100 g have passed from the intravascular to the extravascular compartment.

From the viewpoint of intravascular protein, the studies of the initial phase of burns have given the following results: After an initial reduction, the albumin concentration gradually returns to normal within 5–6 weeks, when the three negative factors (increased catabolism, increased extravascular distribution, and the great protein loss through the skin) have been eliminated. IgG shows initially the same pattern but reaches the supernormal values up to a maximum within 3 weeks, when its synthesis, unlike that of albumin, is quickly and markedly increased. The factors that cause the increase of IgG synthesis have already been discussed; it is remarkable that at severe bacterial invasion of the body, the synthesis cannot be increased to more than four or five times the normal, although in some pathological conditions it can be as high as 60 times the normal.

Albumin Synthesis

It is remarkable that the albumin synthesis shows subnormal values for several weeks, although albumin concentration intravascularly is clearly below the normal for the same period of time. The plasma volume is not reduced, which means that the total amount of circulating albumin is clearly reduced. Would this defect in the regulation of synthesis be due to an insufficient supply of essential amino acids? The fact that these patients lose very little in weight during this phase and receive 4000 calories with protein in abundance argue against this possibility. It is also evident that the lysosomes are not concerned in the initiation of albumin synthesis (Koj and Allison, 1969). A marked extravascular distribution of albumin has occurred and in these conditions about 90 g are present intravascularly and about 270 g extravascularly as against the normal of 120 g intravascularly and 180 extravascularly. As has been shown (Rothschild *et al.*, 1969), albumin synthesis seems to be regulated by the albumin compartment in the liver. If this is the responsible mechanism, we may presume that it would be generally valid for trauma. It must be considered, however, that although this could be the explanation of the low values for intravascular albumin recorded in cases of trauma, it is not a very expedient mechanism. The colloid osmotic effect, among other factors, would mean a greater need of albumin intravascularly in these cases, so as to influence the plasma volume and to improve the vital transport capacity of albumin. It would therefore be of great importance to find a way of stimulating albumin synthesis in cases of trauma.

Albumin synthesis can be improved by anabolic steroids if the supply of essential amino acids is adequate; but it is evident from the literature that this would hardly be a feasible approach in cases of trauma. We have shown that human growth hormone can promote synthesis in cases of trauma, but the degree of stimulation was not of an order of magnitude sufficient to make this the solution of the protein problem in association with trauma (Birke *et al.*, 1961). The synthesis can be only partly influenced in traumatic cases, but we know that it is a question of very great changes and that under optimal conditions the synthesis can be increased threefold. Thus efforts must be directed toward the other three factors that act negatively in these cases (namely, the increased catabolism, the increased extravascular distribution, and the great protein losses through the skin) and toward finding out what factors give rise to these pathological disturbances. If the distribution can be influenced, it may be presumed that the synthesis would improve secondarily in accordance with the views presented earlier.

The losses through the skin are, of course, due to severe and extensive vascular damage. The extravascular distribution has been attributed to increased capillary permeability, possibly caused by vasoactive substances. For a long time efforts have been made to find substances that would influence permeability in a favorable direction, but unfortunately without success.

C. Warm Air Treatment

On the basis of our earlier observations on the markedly raised basal metabolic rate and the increased evaporation of water, we have initiated treatment with warm dry air as a means of relieving the body of the energy production required for the evaporation (Barr *et al.*, 1968). This treatment resulted in a significant reduction of the basal metabolic rate and a return to normal of essential metabolic processes. Warm dry air causes a highly significant reduction of the basal metabolic rate ($p < 0.001$), as the body is relieved of the energy production necessary to evaporate the great amount of water lost through the skin, particularly through the burned area. This may account for the observed changes in protein distribution and catabolism as this form of therapy is continued.

The catabolic rate is greatly reduced, the distribution tends to be normal, and the losses through the skin decrease for both albumin and γ-globulin. The albumin synthesis, which is subnormal, tends to return to normal, and for IgG it is, as an average, four times the normal (Table I). The factors that influence the catabolism have not been made fully clear, but according to these studies, there is a good correlation between the basal metabolic rate and albumin catabolism ($p < 0.001$), which thus seems to be one of the factors that regulate albumin catabolism.

IV. Extravascular–Intravascular Protein Distribution and the Lymph: Effects of Warm Air Treatment

The extravascular–intravascular distribution for albumin and γ-globulin is changed in a normal direction. This result is very probably due to two factors: that (1) the warm dry air treatment improves the circulation and, in particular, (2) the patients are in better condition. This would promote the return flow through the lymphatic system. The correlation between oxygen uptake and cardiac output is highly significant (Birke *et al.*, 1958, 1959). Therefore, if the oxygen uptake is reduced, the cardiac output is correspondingly normalized, thus promoting a better circulation (Fig. 3). These improvements lead to a significantly increased albumin concentration intravascularly.

When the distribution is normalized, the synthesis also improves, which may indicate that the synthesis regulating factor discussed in the foregoing would be operative in severe burns.

V. Regulation of Synthesis of Acute Phase Proteins

Recent studies (Koj and Allison, 1969) have shed some light on the factors that initiate synthesis of acute phase proteins in cases of trauma. It has been found that drugs which affect the permeability of the lysosomes can initiate the synthesis of glucoproteins, fibrinogen, and ceruloplasmin but

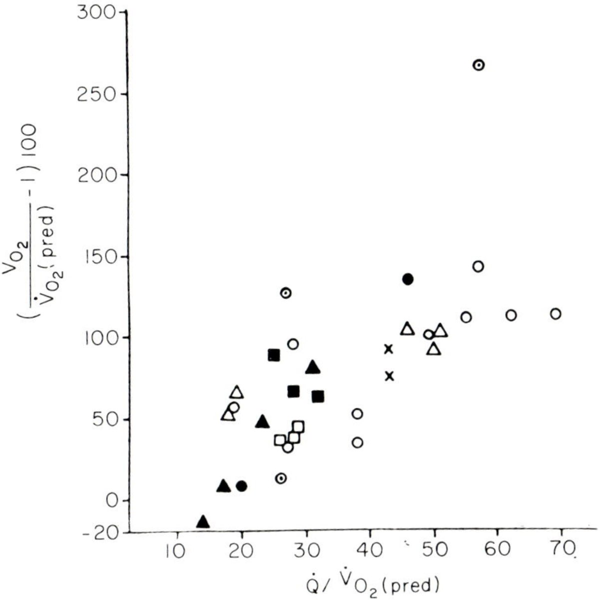

FIG. 3. Percentage increase in oxygen uptake over predicted basal value, 100 [V_{O_2}/V_{O_2} (pred) − 1], in relation to cardiac output (liter/min), Q/V_{O_2} (pred). The relation between the two functions is statistically significant.

not that of albumin. Koj and Allison postulated that disruption of lysosomes might cause alterations in the synthesis of acute phase proteins by at least more main methods. Local tissue damage might cause the release of lysosomal enzymes which would be transported to the liver where protein synthesis would be stimulated. The other possibility given major consideration was a specific effect of lysosomal enzymes on the liver cell acting to release inhibitors to the synthesis of new RNA and even DNA. The administration of excess vitamin A, known to decrease the stability of lysosomes, is followed by a rapid increase in the concentration of the acute-phase proteins. Albumin synthesis was not stimulated. The stimulation of synthesis of acute phase proteins is not dependent on the integrity of the central nervous system. In paraplegics, treated surgically for various reasons, measurements of postoperative fibrinogen and α_2-globulin gave the same results as in other forms of trauma (Davies *et al.*, 1969). This indicates that the factor or factors which initiate protein synthesis in cases of trauma are not mediated via the nervous system or the hormones regulated by it.

Electronmicroscopy has added evidence to the specific role of lysosomes in the stimulation of these serum proteins 1969) (Fig. 4). The increase of these enzymes were studied in ischemia of the liver, both experimentally induced

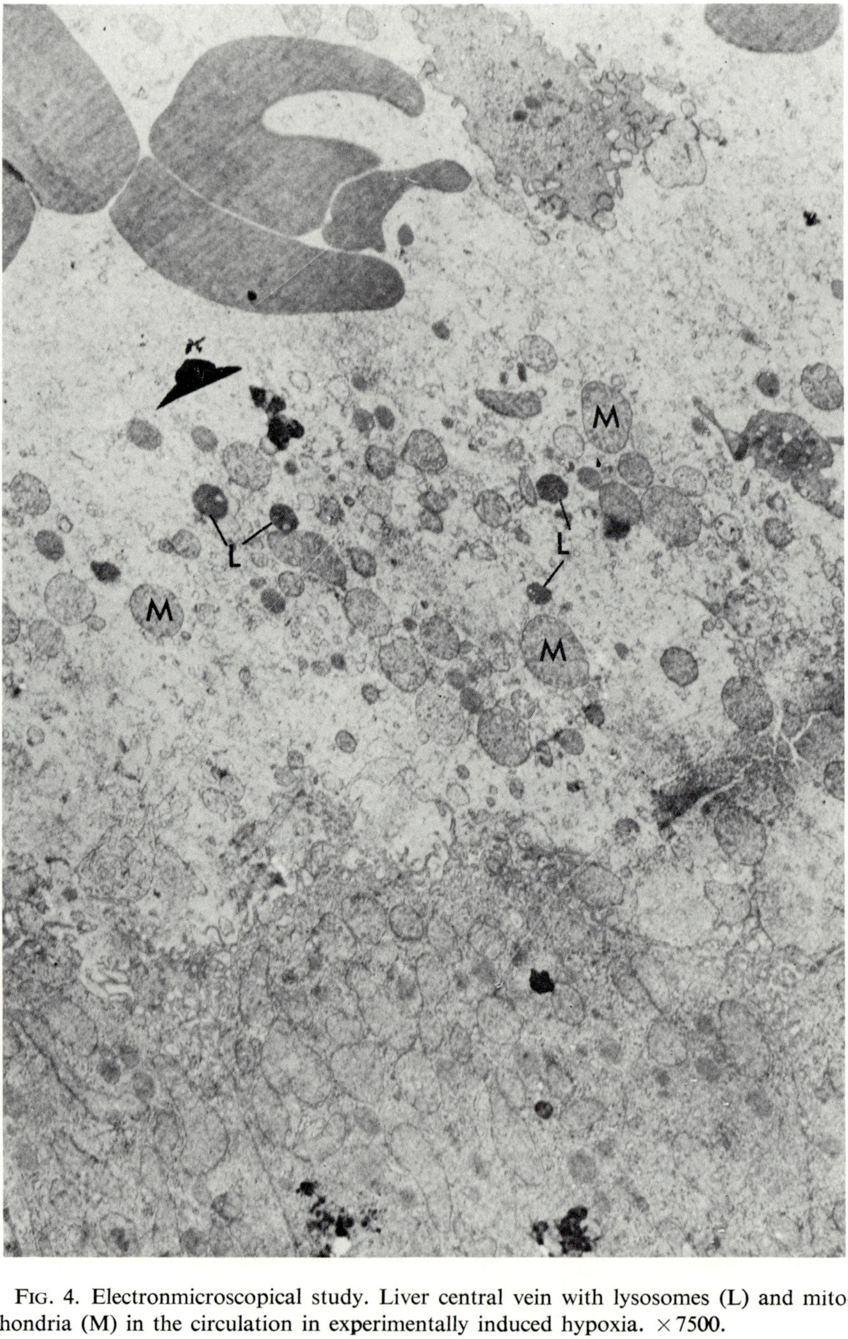

FIG. 4. Electronmicroscopical study. Liver central vein with lysosomes (L) and mitochondria (M) in the circulation in experimentally induced hypoxia. ×7500.

and in man in connection with surgery. Figure 4 shows experimentally induced hypoxia. In the central veins, there is a profuse mount of mitochondrial and lysosomal fragments, which indicate that at least in this form of trauma the lysosomes would be able to act on the liver.

VI. Summary

Severe burns, dramatically alter the distribution of serum proteins with a marked increase in the amount that is located in the extravascular pool. These shifts in protein distribution may influence albumin synthesis in particular. The improvement in the circulation and the decrease in the loss of water through the skin by warm air treatment of these severe burns prevent some of the shifts in protein distribution and returns metabolism toward normal. Each type of trauma should be investigated because it is probable that the interplay of nutrition, hormones, trauma induced lysosomal disruption, and altered protein distribution may have different interrelated effects.

References

Barr, P.-O., Birke, G., Liljedahl, S.-O., and Plantin, L.-O. (1968). *Lancet* **i**, 164.
Birke, G. (1960). *Acta Chir. Scand.* **118**, 353.
Birke, G., Liljedahl, S.-O., and Linderholm, H. (1958–1959). *Acta Chir. Scand,* **116**, 370.
Birke, G., Gemzell, C. A., Liljedahl, S.-O., and Plantin, L.-O. (1961). *Acta Chir. Scand.* **122**, 1.
Birke, G., Liljedahl, S.-O., Backdahl, M., and Alylen, B. (1964). *Acta Chir. Scand. Suppl.* **337**.
Birke, G., Bottiger, L. E., Engstedt, L., Liljedahl, S.-O., Zetterstrom, B., and Asen, P. (1965). *Acta Chir. Scand.* **130**, 399.
Birke, G., Liljedahl, S.-O., Plantin, L.-O., and Reizenstein, P. (1968). *Acta Chir. Scand.* **134**, 27.
Birke, G., Liljedahl, S.-O., and Norberg, R. (1969). *Scand. J. Plast. Reconstr. Surg.* **3**, 39.
Davies, J., Birke, G., and Liljedahl, S.-O. (1969). *Injury* **1**, 43.
Davies, J., Lilhedahl, S.-O., and Reizenstein, P. (1969). Unpublished observations.
Koj, A., and Allison, A. C. (1969). *Folia Biol.* (*Warsaw*) **17**, 1.
Liljedahl, S.-O., Olhagen, B., Plantin, L.-O., and Birke, G. (1963). *Acta Chir. Scand. Suppl.* **309**.
McFarlane, A. S. (1958). *Nature* (*London*) **182**, 53.
Norberg, R., Birke, G., Hedfors, E., and Plantin, L.-O. (1970). (This volume).
Rothschild, M. A., Oratz, M., and Schreiber, S. S. (1969). *In* "Physiology and Patho physiology of Plasma Protein Metabolism" (G. Binhe, R. Norberg, and L. O. Plantin, eds.), Pergamon Press, London.

27

METABOLISM OF α_2-MACROGLOBULIN

Regulation and Distribution

Renee Norberg, G. Birke, Eva Hedfors, and L.-O. Plantin

KING GUSTAF V RESEARCH INSTITUTE, STOCKHOLM, SWEDEN

I. Introduction

Although α_2-macroglobulin is one of the major serum protein components —it comprises about one-third of the α_2-proteins— fairly little information about its function and physiologic importance has been available.

α_2-Macroglobulin was first isolated and characterized, mainly according to its carbohydrate content by Brown *et al.* (1954). Later, Schultze *et al.* (1955) as well as Müller-Eberhard *et al.* (1956) presented further physicochemical characteristics. Its molecular weight is 820,000 with a sedimentation coefficient of 18–19 S. The total carbohydrate content is 8.4%. α_2-Macroglobulin seems to be involved in a variety of enzyme mechanisms. It combines with and inhibits the proteolytic activity of, for instance, trypsin and plasmin, but the α_2-macroglobulin–enzyme complex remains its esterolytic activity and the complex is not susceptible to various trypsin inhibitors, e.g., soybeans. This property of the complex has been utilized by Ganrot (1966) in the elaboration of a method for quantitative determination of α_2-macroglobulin. α_2-Macroglobulin also binds thrombin, inactivating the thrombin clotting activity (Lanchantin *et al.*, 1966). The ability of α_2-macroglobulin to act as a carrier protein, e.g., for various hormones, has been studied, but the results are contradictory (Adham *et al.*, 1968).

α_2-Macroglobulin comprises about two-thirds of the serum macroglobulin fraction. Relevant serum values reported in the literature for male controls vary from 180 to 284 mg/100 ml. The corresponding values for females are usually reported to be 30% higher, but there are also studies which cannot verify the difference between men and women (Adham *et al.*, 1968; Housley, 1968; Ganrot and Schersten, 1967; James *et al.*, 1966).

α_2-Macroglobulin has been demonstrated in secretions and exudates, but in relatively lower amounts than in serum.

There are only a limited number of studies concerning the serum values of α_2-macroglobulin in different diseases and the results are rather conflicting, as in rheumatoid arthritis in which elevated, normal, or depressed levels are reported (Adham *et al.*, 1968; Housley, 1968; James *et al.*, 1966).

As to the metabolism, only incomplete data are available (Reuge *et al.*, 1966; Kluthe *et al.*, 1967). The half-life was found to be about 10 days in controls.

Therefore, it seemed worthwhile to analyze the serum α_2-macroglobulin values in some groups of diseases and to perform metabolic studies in selected cases.

II. Methods

The *quantitative estimation* of α_2-macroglobulin was made by the method of Mancini *et al.* (1965). α_2-Macroglobulin, prepared from pooled normal sera by preparative ultracentrifugation, gelfiltration on Sephadex G200, and zone electrophoresis in polyvinyl chloride was used as reference antigen and in metabolic studies (Norberg, 1964). The *labeling* of the protein and the *performance of the metabolic studies* have been described earlier (Ahlinder *et al.*, 1968).

III. Serum Levels in Disease

The results of the quantitative analyses in burned patients and in patients with rheumatoid arthritis (RA), systemic lupus erythematosus (SLE), and hypogammaglobulinemia are shown in Figs. 1 and 2.

A. Normal

The values in the patient groups are given as percentages of the mean normal value, which for men equalled 240 mg/100 ml and for women 295 mg/100 ml. The standard variation was $\pm 21\%$ of the normal mean for both men and women. As the normal values are higher in childhood and decrease with age (Ganrot and Schersten, 1967), mainly patients more than 20 years old were accepted.

B. Burns

We studied 12 severely burned patients between 20 and 82 years old with one exception, a boy of 15 (in Fig. 1A indicated by ▲-▲-▲-▲). The burned areas varied from 20 to 60% of the body surface. In nearly all the cases, the α_2-macroglobulin serum values were elevated in the first few days after the injury, then decreased successively to normal within the first week, and remained at that level during the rest of the course. The immunoglobulin values, on the other hand, were over normal on the first day, but fell to about 50% of the mean normal level after 5–6 days. After the first week, the immunoglobulins successively increased to values significantly higher than the normal ones (Birke *et al.*, 1969).

The cause of the initial increase of the α_2-macroglobulin is unknown. It cannot be explained by a relative increase of high molecular weight serum proteins for the plasma volume on day 1, was about 96% of the predicted volume, and the other serum macroglobulin (IgM) was not correspondingly increased (Birke *et al.*, 1969).

A possible explanation of this early increase in α_2-macroglobulin might be a very rapid release or synthesis of this protein. It is not clear whether the rapid fall of the serum values is due to losses from the burned area or to increased catabolism. It might be argued that the elevated α_2-macroglobulin values observed would represent an increased need of proteins with ability to act as carriers and to combine with and inactivate proteolytic enzymes, which might have been released by the burns.

IV. Systemic Lupus, Rheumatoid Arthritis, and Hypogammaglobulinemia

The rest of the clinical material investigated comprised 61 patients with rheumatoid arthritis, 14 with systemic lupus erythematosus, and 11 with hypogammaglobulinemia (IgG values <400 mg/100 ml). As seen from Fig. 2 most of the values for RA patients are within the normal range. The mean value of the group was not elevated. Three of the five patients in this group, with values exceeding the mean + 2 SD, had renal amyloidosis with high protein excretion. The diagnosis was verified by biopsy.

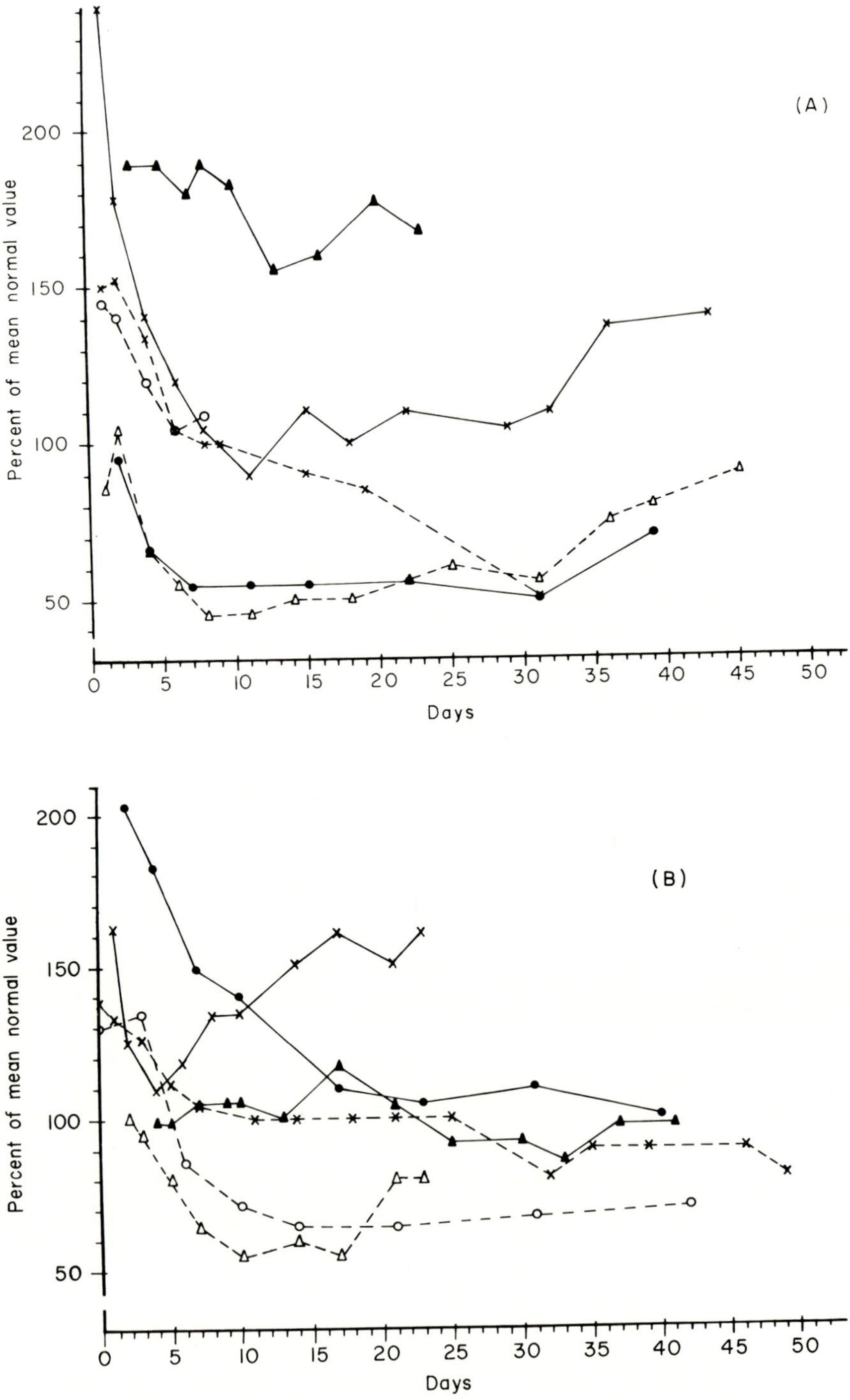

(A)
Percent of mean normal value
200
150
100
50
0
5
10
15
20
25
30
35
40
45
50
Days
(B)
Percent of mean normal value
200
150
100
50
0
5
10
15
20
25
30
35
40
45
50
Days

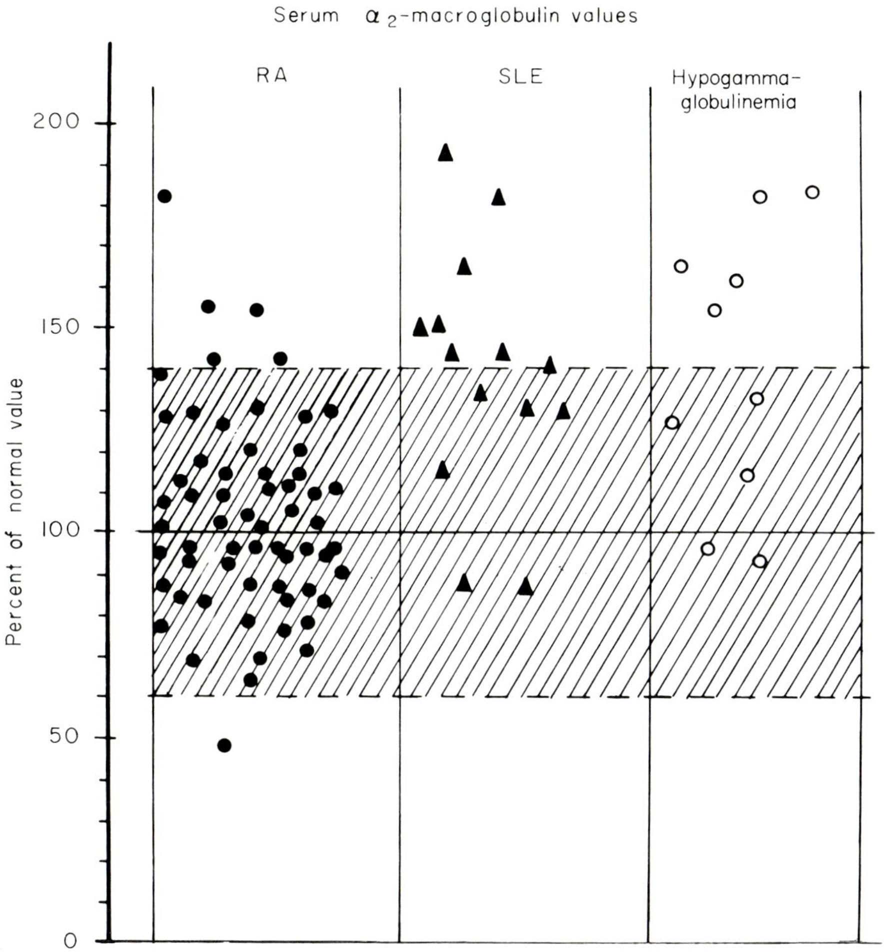

FIG. 2. Serum α_2-macroglobulin values as percentages of mean value for controls in rheumatoid arthritis (RA), systemic lupus erythematosus (SLE), and hypogammaglobulinemia (IgG < 400 mg/100 ml).

In systemic lupus erythematosus most α_2-macroglobulin values were elevated. All these patients had an active disease and most of them had signs of renal damage. The highest values were found in two patients with nephrotic syndrome.

Most patients with hypogammaglobulinemia had increased α_2-macroglobulin values. The group did not include any patients with obvious pathologic

---

FIG. 1. (A) Serum α_2-macroglobulin values as percentages of mean value for controls in six burned patients. Burned areas 40–60% of body surface. (B) Serum α_2-macroglobulin values as percentages of mean value for controls in six burned patients. Burned areas 20–35% of body surface.

protein loss from the intestine or the kidneys. The cause of the elevated values is not clear. There was, however, no correlation with concomitant acute infections.

V. Metabolic Studies

A. Isolation

The preparation of a native, metabolically homogeneous α_2-macroglobulin is very difficult. α_2-Macroglobulin is a very labile protein and during the preparation it will very easily split into fragments with lower sedimentation coefficients, usually 15 S, but occasionally lower.

These fragments have the same mobility in polyvinyl chloride electrophoresis as the native protein, they are precipitated by a monospecific anti-α_2-macroglobulin serum indicating impaired antigenic determinants (Fig. 3). Gelfiltration on Sephadex G200 and on Sepharose gives two peaks, however, but so close to each other that they cannot be separated without recycling procedure, with the continuous risk of molecule splitting (Fig. 4).

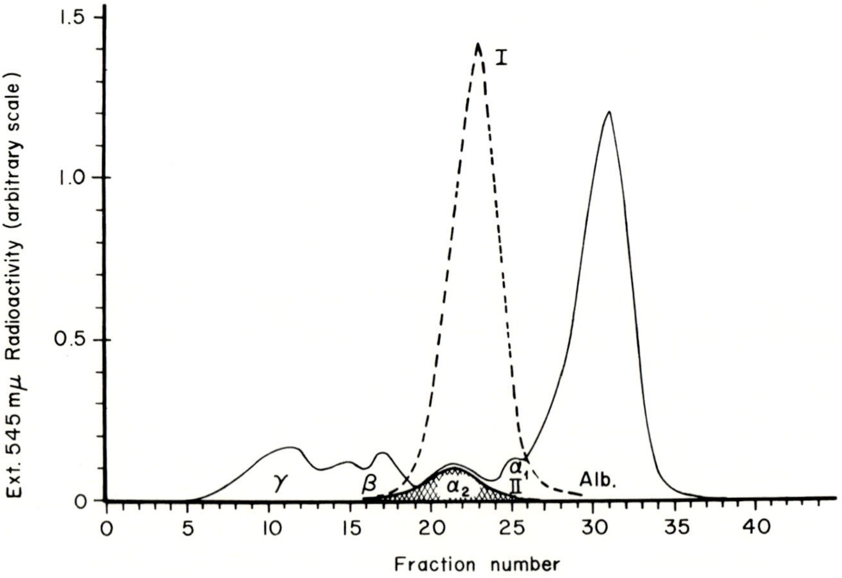

FIG. 3. Zone electrophoresis in polyvinyl chloride of a control serum, containing ^{125}I-α_2 macroglobulin with 40% of 15 S fragments, before (I) and after (II) precipitation with anti-α_2-macroglobulin serum and removal of the precipitate.

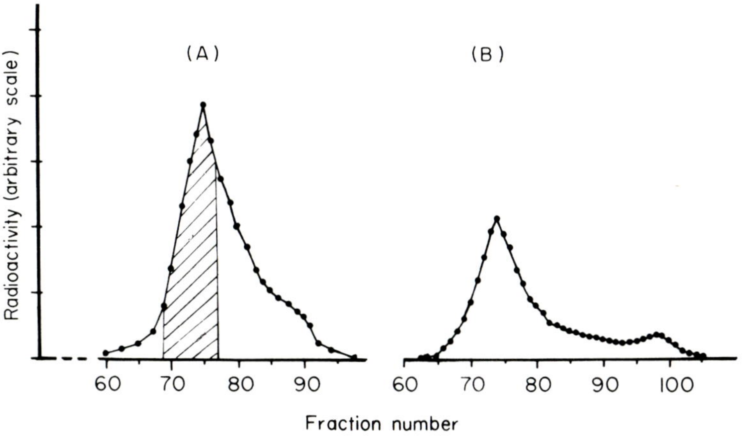

FIG. 4. Radioactivity curve after gel filtration on Sephadex G200 of a ^{125}I-α_2-macroglobulin. The fractions corresponding to the dotted areas (A) were collected, ultrafiltered, and submitted to refiltration on the same Sephadex G200 column (B). The curves illustrate the continuous risk of molecule splitting at gel filtration.

The 15 S fragments are withdrawn from the circulation at a very high rate. Given to a volunteer we recovered after 10 min about 50% of predicted activity and 82% of given amount was collected in the urine within 3 days.

B. CATABOLISM

α_2-Macroglobulin metabolism was studied in ten cases, six with serum α_2-macroglobulin values within the normal ranges and four with elevated serum values. All ^{125}I-α_2-macroglobulin preparations used contained 15–20% of 15 S material. The catabolic rate was calculated by the U/P ratio. The excretion of radioactivity was very high during the first few days, probably depending upon the excretion of α_2-macroglobulin fragments, as previously mentioned. The results of the metabolic studies are presented in Table I and Fig. 5.

The mean fractional catabolic rate was 8.2% of the intravascular pool per day (range 7.0–9.5) both in cases with normal and elevated serum values. This is contradictory to the results of Kluthe *et al.* (1967), who found a longer half-life in patients with elevated α_2-macroglobulin values than in controls.

Virtually all the α_2-macroglobulin administered seemed to be distributed intravascularly. In the case, illustrated by Fig. 5, there is throughout the study, a constant extravascular pool comprising 6–8% of the given dose.

The metabolic data obtained support the theory of an increased synthetic

TABLE I

METABOLISM OF α_2-MACROGLOBULIN IN TEN CASES, SIX WITH NORMAL AND FOUR WITH INCREASED SERUM VALUES

Case No.	Sex	Age (yr)	α_2-M (g/100 ml)	Plasma vol (liter)	I.V. pool (g)	U/P %/24 hr	U/P g/24 hr
1	♀	48	0.30	3.0	9.00	7.9	0.71
2	♂	28	0.21	3.5	7.18	7.8	0.56
3	♀	74	0.35	3.0	10.35	9.5	0.98
4	♂	44	0.24	2.7	6.48	8.2	0.53
5	♂	28	0.26	2.9	7.40	8.0	0.59
6	♂	26	0.22	3.1	6.67	8.0	0.53
Mean			0.26	3.0	7.80	8.2	0.65
7	♂	22	0.48	2.7	12.96	9.1	1.18
8	♂	63	0.40	2.9	11.46	7.0	0.80
9	♀	60	0.68	2.7	18.36	7.8	1.43
10	♂	54	0.44	3.1	13.64	9.0	1.23
Mean			0.50	2.9	14.11	8.2	1.16

rate or release causing the elevated serum α_2-macroglobulin values observed in some diseases. The number of metabolic studies, however, is too small to allow definite conclusions.

VI. Summary

The serum α_2-macroglobulin levels were determined in controls and in patients with severe burns, rheumatoid arthritis, disseminated lupus erythematosus, and hypogammaglobulinemia. In burned patients the values were elevated in the first few days after the injury. They decreased to normal within the first week and remained at that level during the rest of the course. The values of patients with rheumatoid arthritis did not differ from those of controls. In patients with systemic lupus the α_2-macroglobulin values were mostly elevated, especially in patients with concomitant renal disease. Also in hypogammaglobulinemia the serum values were found to be increased. The cause of the elevated serum α_2-macroglobulin levels is unknown, but some aspects have been discussed.

The difficulties in preparing a pure and undenatured α_2-macroglobulin were pointed out. The properties of the preparation used for metabolic

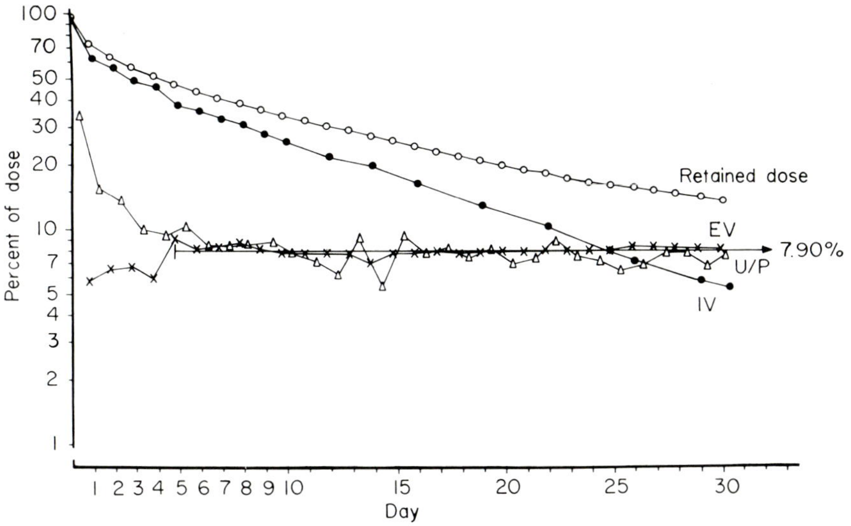

FIG. 5. Catabolism of α_2-macroglobulin in a normal subject.

studies were analyzed with regard to their influence upon the metabolic data. The metabolism of α_2-macroglobulin was studied in six patients with normal serum levels and in four with elevated ones. The catabolic rate in percent of the intravascular pool was the same in all the cases investigated. Virtually all α_2-macroglobulin was intravascularly distributed.

ACKNOWLEDGMENTS

Supported by grants from Konung Gustaf V:s 80-arsfond and Magn. Bergvalls stiftelse.

REFERENCES

Adham, N., Wilding, P., Mehl, J., and Haverback, B. J. (1968). *J. Lab. Clin. Med.* **71**, 271

Ahlinder, S., Birke, G., Norberg, R., Olhagen, B., Plantin, L.-O., and Reizenstein, P. (1968). *Acta Med. Scand.* **184**, 25.

Birke, G., Liljedahl, S. O., and Norberg, R. (1969). *Scand. J. Plast. Reconstr. Surg.* **3**, 201.

Brown, R. K., Baker, W. H., Peterkofsky, A., and Kauffman, D. L. (1954). *J. Amer. Chem. Soc.* **76**, 4244.

Ganrot, P. O. (1966). *Clin. Chim. Acta* **14**, 493.

Ganrot, P. O., and Schersten, B. (1967). *Clin. Chim. Acta* **15**, 113.

Housley, J. (1968). *J. Clin. Pathol.* **21**, 27.

James, K., Johnson, G., and Fudenberg, H. H. (1966). *Clin. Chim. Acta* **14**, 207.

Kluthe, R., Hagemann, U., and Kleine, N. (1967). *Vox Sang.* **12**, 308.

Lanchantin, G. F., Plesset, M. L., Friedmann, J. A., and Hart, D. W. (1966). *Proc. Soc. Exp. Biol. Med.* **121**, 444.

Mancini, G., Carbonara, A. O., and Heremans, J. F. (1965). *Immunochemistry* **2**, 235.

Müller-Eberhard, H. J., Kunkel, H. G., and Franklin, E. (1956). *Proc. Soc. Exp. Biol. Med.* **93**, 146.

Norberg, R. (1964). *Clin. Chim. Acta* **9**, 89.

Reuge, C., Blatrix, C., Brevet, J. P., and Steinbuch, M. (1966). "Labelled Proteins in Tracer Studies," (L. Donato, ed.), p. 143. Euratom, Brussels.

Schultze, H. E., Gallner, I., Heide, K., Schonenberger, M., and Schwick, G. (1955). *Z. Naturforsch. B.* **10**, 463.

VII

Fibrinogen and Prothrombin

28

REGULATION OF THE METABOLISM OF PROTEINS INVOLVED IN BLOOD COAGULATION

*Erwin Regoeczi**

BIOPHYSICS DIVISION, NATIONAL INSTITUTE FOR MEDICAL RESEARCH, LONDON N.W.7, ENGLAND

In the field of blood coagulation an effort appears to be made to amalgamate two different faces into a single Janus head: whereas blood must remain fluid under physiological conditions, on the one hand, it must also speedily clot whenever the circumstances necessitate it, on the other. It may be quite justified, therefore, to simultaneously pose the following questions: How does it happen that blood does not coagulate in the normal circulation, and what initiates clotting when an appropriate pathological stimulus is present?

These questions will have to remain unanswered until considerably more data are available, but meantime it is understood that most of the unique features of the hemostatic mechanism are connected with the closely interrelated functioning of over a dozen specific plasma proteins. These can purposefully be divided into three categories, depending on whether they promote coagulation (factors I, II, V, VII–XIII), fibrinolysis (plasminogen, intrinsic proactivators), or whether they neutralize activated coagulant and fibrinolytic enzymes (antithrombin II and III, α_2-macroglobulin). Except for the group of protease inhibitors, the other hemostatic proteins occur in the plasma as inactive precursors, from which, in some known instances, the active substance emerges as the result of partial proteolysis, the classic examples being fibrinogen, prothrombin, and plasminogen.

Turning to the regulation of this complex system, the first thing to be pointed out is that the presently available information is insufficient to

*Present address: Department of Pathology, McMaster University, Hamilton, Ontario, Canada.

permit final conclusions. This should not be too surprising, considering that while metabolic studies cannot even be thought of without pure and undenatured tracers at hand, the isolation and physicochemical characterization of many of the hemostatic proteins has not been achieved until the past few years owing to formidable difficulties. It is a natural consequence of this particular situation that our current attitude to all but a few of these proteins is still more or less an empirical one: We measure "activities" and "units" instead of physical quantities, and we express changes as percent deviations from a pool of normal sera, not knowing in many instances precisely whether they reflect altered synthesis, or consumption, or both. Therefore, to raise the question of the metabolic regulation of the hemostatic system may appear somewhat premature at the present stage.

The only two proteins studied so far in any detail are fibrinogen (factor I) and prothrombin (factor II), certain aspects of their metabolic behavior being the subject of the following chapters.

Some idea of the speed by which other factors disappear from the plasma was obtained by measuring the biological half-life of the activity of preparations infused into patients with the corresponding deficiency. According to such studies, the activity of proconvertin (factor VII; MW 63,000) falls to half in the plasma in 1–2 hr, that of the antihemophilic globulin (factor VIII; MW >200,000) in 10 hr, that of proaccelerin (factor V; MW 291,000) in 12–36 hr, and that of the Christmas factor (factor IX; MW 50,000) in 20–40 hr. In relation to the others, the longest half-life of 48–72 hr was obtained in the case of the Stuart-Prower factor (factor X; MW 86,000). Whereas data of this kind are useful as a guide for treatment schedules, they are of restricted value for assessing the true turnover rate of the relevant proteins, since a great deal of the initial losses must inevitably be due to diffusion into an extravascular compartment of unknown size, rather than to genuine catabolism.

In case the foregoing considerations give a misleadingly negative impression about our present state of knowledge concerning the regulation of hemostatic proteins, it should be mentioned that even the relatively small amount of metabolic information available has produced some new ideas. Thus, experiments with labeled fibrinogen and prothrombin have shown that both these proteins are normally catabolized directly, that is, without a preliminary conversion to fibrin and thrombin, respectively. Clearly, the popular, but never fully proven theory, according to which the fluidity of the blood is maintained by a subtle balance between continuous coagulation and fibrinolysis, will have to be critically reevaluated.

If, as it appears, fibrinogen and prothrombin are directly catabolized, then the same may be true for the other components of the hemostatic chain. The fact that minute fractions of these may behave differently and be dealt with by unknown mechanisms escaping detection by the current techniques should not be allowed to prejudice seriously the view, that an immense number of highly specific molecules are being continuously synthesized and destroyed unused by the body in order that a few will always be available for any eventuality. Assuming that the catabolic rates of the hemostatic proteins are indeed as heterogenous as the scattered half-lives of infused activities suggests, then it becomes fascinating to ask why this should be so within a functionally closely related group of proteins? Finding the correct answer to this question would certainly contribute a great deal to our general understanding of the specificity of plasma protein catabolism.

29

EFFECTS OF TYPHOID ENDOTOXIN ON FIBRINOGEN AND PROTHROMBIN METABOLISM IN CALVES

Y. Takeda

DEPARTMENT OF MEDICINE,
UNIVERSITY OF COLORADO SCHOOL OF MEDICINE,
DENVER, COLORADO

I. Introduction

Infectious disease is occasionally complicated by multiple thrombosis, but its mechanisms are poorly understood despite the vast amount of work in this field (Chien *et al.*, 1964, 1965; Good and Thomas, 1953; Hardaway, 1962; McKay and Shapiro, 1958). The previous studies employed the main approach of following the response of several plasma proteins to the injections of endotoxins in animals. Although these studies have provided some valuable information, full interpretation of the data so obtained was often difficult or impossible, since the observed behavior and response of plasma proteins are a complex function of several factors as illustrated in Fig. 1. In the present study the response of exogenously administered ^{125}I-fibrinogen and ^{125}I-prothrombin, as well as that of native fibrinogen and prothrombin, to a single injection of typhoid endotoxin was observed in order that further interpretation and insight may be obtained. This chapter describes studies of fibrinogen and prothrombin metabolism in calves first under control conditions, second under heparinized conditions, and finally under conditions in which a single injection of typhoid endotoxin was made. In addition the effects of heparin on fibrinogen and prothrombin responses to a single injection of typhoid endotoxin will be described.

II. Methods

A. Fibrinogen

96% pure bovine fibrinogen was prepared by the method previously reported (Takeda, 1966) and was labeled with ^{125}I in a ratio of 0.5 or less atoms iodine per molecule of fibrinogen, using the iodine monochloride method (Atencio *et al.*, 1965b; McFarlane, 1958). The general methods are similar to those reported elsewhere (Takeda, 1966). Typhoid endotoxin* of 3 ml was intravenously injected after a 24-hr sample was obtained and the experiments lasted for 4–6 days. The plasma fibrinogen concentration was measured by the isotope dilution method of Atencio *et al.* (1965b). Heparin was injected every 2–6 hr in the amount of 7000 units intravenously and equal amounts subcutaneously during the experiment, and after the endotoxin injection, every 2 hr. The Lee-White clotting time was maintained at greater than 1 hr during the study. The data obtained in the steady-state experiments were analyzed as described elsewhere (Atencio *et al.*, 1965a).

* Typhoid vaccine, USP Eli Lilly and Company, Indianapolis, Indiana, contains 1000 million killed typhoid bacilli (*S. typhosa*) in each milliliter.

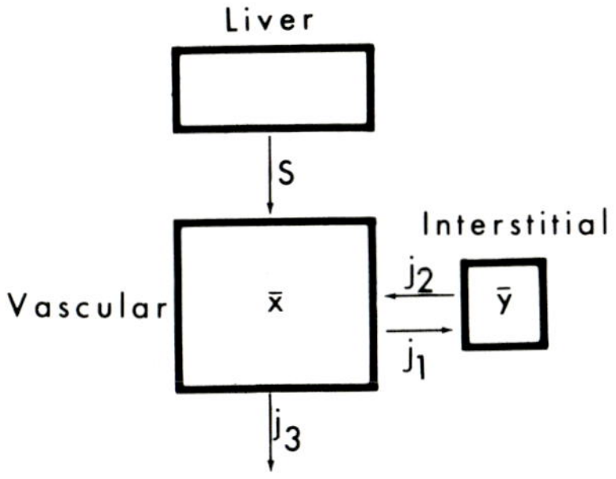

FIG. 1. Control system of plasma fibrinogen and prothrombin. S is the rate of synthesis in mg/day/kg, $\bar{x}$ and $\bar{y}$ are plasma and interstitial fibrinogen or prothrombin in mg/kg, j_1 is the fractional transcapillary transfer rate of $\bar{x}$/day, j_2 is the fractional return rate of $\bar{y}$/day, and j_3 is the fractional catabolic (or deposition or both) rate of $\bar{x}$. The addition rate of fibrinogen or prothrombin to plasma is then $(S + j_2\bar{y})$ mg/day/kg and the removal rate of plasma fibrinogen or prothrombin is $(j_1 + j_3)\bar{x}$ mg/day/kg. In a steady state $S = j_3\bar{x}$, $j_2\bar{y} = j_1\bar{x}$, and $S + j_2\bar{y} = (j_1 + j_3)\bar{x}$.

B. PROTHROMBIN

Bovine prothrombin of high specific activity was prepared by the method of Seegers (1962). The specific activity ranged from 20,000 to 39,000 units/mg tyrosine. Prothrombin preparations above 30,000 units/mg tyrosine were then labeled with ^{125}I in a ratio of 0.25 or less atoms iodine per molecule of prothrombin (MW 68,900) (Seegers, 1962), and free ^{125}I in a preparation was removed by passing it through a Sephadex G25 column. The ^{125}I-prothrombin thus obtained was first tested with respect to its thrombin generation ability in comparison with that of the unlabeled prothrombin, using the two-stage method of Ware and Seegers (1949). No appreciable difference was found between the labeled and unlabeled prothrombins. Then, the behavior of labeled prothrombin was compared with that of unlabeled prothrombin on Sephadex G200 and DEAE cellulose columns, showing no appreciable differences between the two. The labeled prothrombin was further studied with respect to its *in vivo* behavior in a calf and compared with that of the "screened" labeled prothrombin. No significant difference was observed. Studies were then made of the metabolism and distribution of prothrombin, using the ^{125}I-prothrombin, by the general methods described above. The plasma prothrombin concentration was measured by a radial immunodiffusion method as shown in Fig. 2. The antiprothrombin was produced in rabbits by injecting intramuscularly a few mg prothrombin in aluminum hydroxide gel. The produced antiprothrombin was absorbed with $BaSO_4$ absorbed bovine serum using the method of Bloch and Bunim (1959). The absorbed antiprothrombin formed a single precipitin arc against

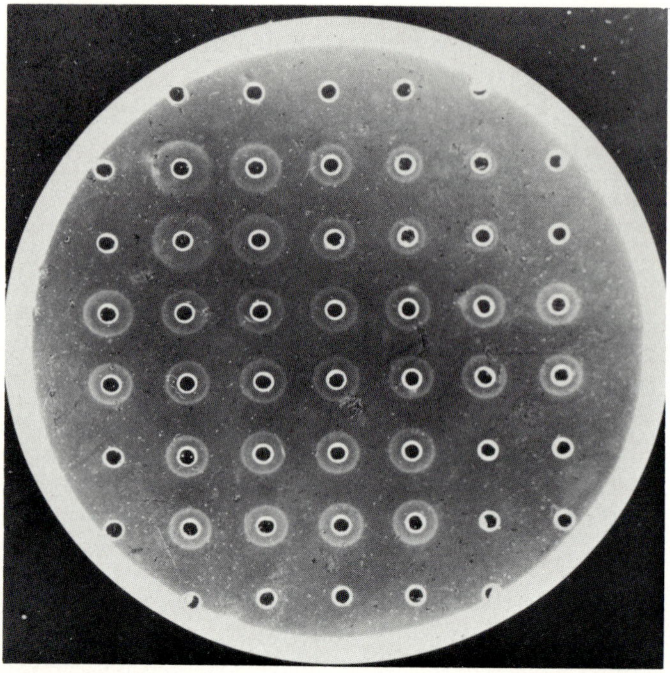

FIG. 2. A radial immunodiffusion method for the measurement of plasma prothrombin concentration. The top two rows are the precipitin rings against standards of known varying concentrations in duplicate. The rest are those against various normal bovine plasmas. The antiprothrombin is mixed in a volume ratio of 0.5–10 ml 0.1% agarose solution in pH 8.6 barbital buffer.

bovine plasma and purified bovine prothrombin; no arcs were formed against $BaSO_4$ absorbed bovine serum, bovine albumin, and fibrinogen on an agarose plate. The tracer data in the steady state experiments were analyzed by the previously published methods (Atencio *et al.*, 1965a).

III. Results

A. CONTROL STUDIES ON FIBRINOGEN AND PROTHROMBIN METABOLISM

1. *Fibrinogen*

Five healthy calves were studied under control conditions using ^{125}I-fibrinogen. The results are summarized in Tables I, II, and III. The hematocrit and plasma fibrinogen concentration stayed relatively constant during the study, and it may be assumed that the animals were in the steady state during the study. The half-lives of plasma ^{125}I-fibrinogen averaged 3.1

TABLE I

PLASMA FIBRINOGEN DATA

Calf no.	Plasma volume (ml/kg)	Plasma fibrinogen concentration (mg %)	Plasma fibrinogen ($\bar{x}$, mg/kg)[a]	Hematocrit (%)
66	50.0	610	305	35.3
67	54.3	750	407	40.7
68	46.5	670	311	41.5
69	44.1	890	392	39.0
70	49.5	710	351	39.0
Mean	48.9	728	353	39.1
SD	±3.8	±105	±46	±2.2

[a] $\bar{x}$ is the amount of fibrinogen in plasma in mg/kg.

days and the entire plasma curve was closely fitted by a 2-exponential function by the graphical analysis (Matthews, 1957). The catabolic rate ($j_3\,\bar{x}$) averaged 102 mg/day/kg, which is equal to the synthetic rate of fibrinogen (S) since the animals were in the steady state. The interstitial fibrinogen ($\bar{y}$) averaged 92 mg/kg and constitutes 26% of the plasma fibrinogen ($\bar{x}$).

TABLE II

PLASMA ^{125}I-FIBRINOGEN DATA[a]

Calf no.	$x = C_1 e^{-at} + C_2 e^{-bt}$				$T_{1/2}$ days
	C_1	a	C_2	b	
66	0.75	0.231	0.25	4.05	3.0
67	0.78	0.231	0.22	3.95	3.0
68	0.79	0.231	0.21	3.50	3.0
69	0.79	0.210	0.21	4.50	3.3
70	0.72	0.216	0.28	3.90	3.2
Mean	0.76	0.223	0.24	3.98	3.1
SD					±0.1

[a] x is the plasma fibrinogen radioactivity curve and is expressed by a 2-exponential equation; $T_{1/2}$ is the half-life of x in days.

TABLE III

FIBRINOGEN FLUXES AND INTERSTITIAL FIBRINOGEN[a]

Calf no.	$j_1\bar{x}$ (mg/day/kg)	$j_3\bar{x}$ (mg/day/kg)	$\bar{y}$ (mg/kg)
66	268	92.4	88
67	309	118.4	98
68	202	84.0	62
69	330	105.1	92
70	335	103.2	118
Mean	289	101.6	92
SD	±50	±13.1	±20

[a] $j_1\bar{x}$ is the amount of plasma fibrinogen filtered through the capillaries in mg/day/kg; $j_3\bar{x}$ is the amount of plasma fibrinogen catabolized in mg/day/kg and equals the amount synthesized; $\bar{y}$ is the amount of interstitial fibrinogen in mg/kg.

2. *Prothrombin*

Nine healthy calves were studied, using ^{125}I-prothrombin, under control conditions. Preliminary reports have appeared elsewhere (Takeda, 1968). Additional studies showed the following average results: plasma volume, 49.9 ml/kg; plasma prothrombin concentration, 18.7 mg %; hematocrit, 42%; plasma prothrombin ($\bar{x}$), 9.3 mg/kg; the half-life of plasma ^{125}I-prothrombin, 1.77 days; the transcapillary transfer rate of prothrombin ($j_1\bar{x}$), 25.3 mg/day/kg; the catabolic rate of prothrombin ($j_3\bar{x}$), 7.53 mg/day/kg; and the interstitial prothrombin ($\bar{y}$), 8.24 mg/kg. The catabolic rate of prothrombin ($j_3\bar{x}$) equals the synthetic rate (S) since the animals were in the steady state. The interstitial prothrombin ($\bar{y}$) constituted 93% of the plasma prothrombin ($\bar{x}$) on the average. The details of the study will be published elsewhere.

B. EFFECTS OF HEPARIN ON PROTHROMBIN METABOLISM

Seven healthy calves were studied with respect to the metabolism and distribution of prothrombin under heparinized conditions, using ^{125}I-prothrombin as a tracer. The preliminary results have been reported elsewhere (Takeda, 1969). The average results were as follows: plasma volume, 49.2 ml/kg; plasma prothrombin concentration, 19.3 mg %; hematocrit, 39.3%; plasma prothrombin ($\bar{x}$), 9.5 mg/kg; the half-life of plasma ^{125}I-prothrombin, 1.86 days; the transcapillary transfer rate of prothrombin ($j_1\bar{x}$), 25.1 mg/

day/kg; the catabolic rate of prothrombin ($j_3 \bar{x}$), 7.32 mg/day/kg; and the interstitial prothrombin ($\bar{y}$), 8.45 mg/kg. The comparison of these results with those under the control condition showed no significant difference. The details of the study will be published elsewhere.

C. EFFECTS OF A SINGLE INJECTION OF TYPHOID ENDOTOXIN ON LABELED AND NATIVE FIBRINOGENS AND PROTHROMBINS

1. ^{125}*I-Fibrinogen*

Three healthy calves were studied. They were intravenously injected with about 50–100 μCi of ^{125}I-fibrinogen and its *in vivo* behavior was followed

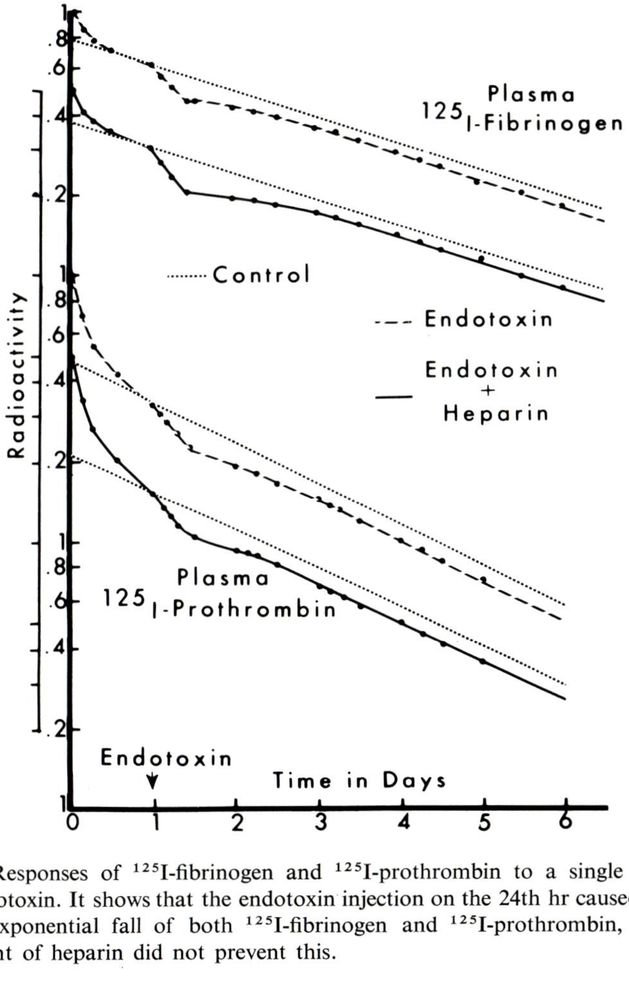

FIG. 3. Responses of ^{125}I-fibrinogen and ^{125}I-prothrombin to a single injection of typhoid endotoxin. It shows that the endotoxin injection on the 24th hr caused the precipitous and exponential fall of both ^{125}I-fibrinogen and ^{125}I-prothrombin, and that a large amount of heparin did not prevent this.

for 4–6 days. At the twenty-fourth hr following ^{125}I-fibrinogen injection, 3 ml of typhoid endotoxin was intravenously injected and the response of the ^{125}I-fibrinogen as well as that of native fibrinogen and prothrombin were observed. One of the typical experiments is shown in Fig. 3. It shows that upon injection of typhoid endotoxin the levels of plasma ^{125}I-fibrinogen precipitously fell and continued to fall exponentially with a half-life of 0.91 days for a period of about 11 hr. Thereafter, the rate of decline slowed for a period of about 1.5 days and then returned to the control level with a half-life of 3.0 days. The magnitude of the fall of plasma ^{125}I-fibrinogen level during the 11-hr period was 12% of the level which would have existed had there not been the endotoxin injection. In all three calves studied the general response pattern was similar to this, although some quantitative differences were observed. The duration in which ^{125}I-fibrinogen fell exponentially and precipitously following the endotoxin injection was 11, 10, and 8 hr and the half-lives during these periods were 0.91, 0.7, and 1.5 days, respectively. The magnitude of the fall of plasma ^{125}I-fibrinogen levels was 12, 20, and 16% and averaged 16%.

2. *Native Fibrinogen*

The response of native fibrinogen to the endotoxin injection was observed in eight calves. The results are shown in Fig. 4. It shows that the endotoxin injection caused a fall of native plasma fibrinogen concentration in all the calves studied during the 2-hr period following the endotoxin injection. The magnitude of the fall was 5, 3, 9, 6, 3, 6, 4, and 5% of each control level, and averaged 5%. Thereafter, the plasma fibrinogen concentration increased rapidly over the control level and kept increasing during the experimental period.

3. *^{125}I-Prothrombin*

The response of ^{125}I-prothrombin to the endotoxin injection was observed in five calves. One of the typical experiments is shown in Fig. 3. It is seen that upon injection of typhoid endotoxin at the twenty-fourth hour the level of plasma ^{125}I-prothrombin fell rapidly and exponentially with a half-life of 0.91 days for the 11-hr period following the injection of typhoid endotoxin. Thereafter, the decline slowed for approximately 1.5 days and then returned to the control level with a half-life of 1.9 days (Section III, A, 2). The magnitude of the fall during the 11-hr period following the endotoxin injection was 20% of the plasma ^{125}I-prothrombin level which would have existed had there not been the endotoxin injection. All five calves responded similarly. The half-lives following the typhoid endotoxin injection were 1.3, 1.1, 1.1, 0.91, and 1.0 days and averaged 1.1 days. The duration in which these half-lives prevailed was 8, 10, 9, 11, and 12 hr. and 10 hr on the average

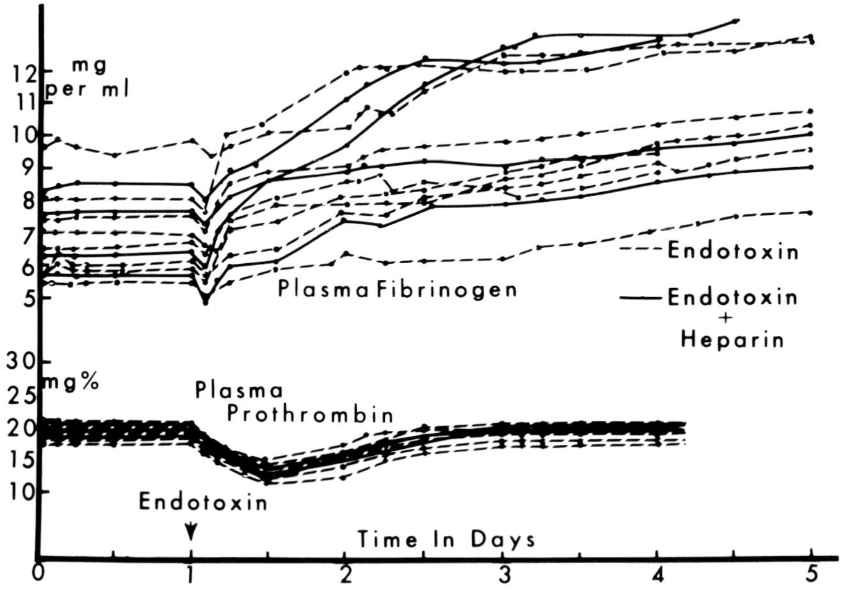

FIG. 4. Responses of native plasma fibrinogen and prothrombin to a single injection of typhoid endotoxin. It shows a decrease of plasma fibrinogen concentration during the 2 hr following a typhoid endotoxin injection and its increase thereafter, a decrease of plasma prothrombin concentration during the approximately 12 hr following the injection and its increase thereafter, and its return to the control level on the third day. It also shows that heparin did not appreciably alter these responses.

following the injection. The magnitude of the fall of the plasma ^{125}I-prothrombin level during these periods was 20, 17, 20, 20, and 23% of the plasma level which would have existed under the control condition and averaged 20%.

4. *Native Prothrombin*

The response of native prothrombin to a single injection of typhoid endotoxin was studied in eight calves. The data are shown in Fig. 4. It is seen that the endotoxin injection at the twenty-fourth hour caused a fall of native plasma prothrombin concentration during the period of 10–12 hr following the injection, and thereafter the plasma prothrombin concentration increased rapidly for 1.5 days and returned to the control level. The magnitude of the fall of native plasma prothrombin concentrations was 22, 24, 23, 26, 22, 20, 26, and 25% of each control level during the 12-hr period following the endotoxin injection, and 21% on the average during the 10-hr period.

D. Effects of Heparin on the Responses of Labeled and Native Fibrinogens and Prothrombins to a Single Injection of Typhoid Endotoxin

1. ^{125}*I-Fibrinogen and* ^{125}*I-Prothrombin*

Two calves were used for ^{125}I-fibrinogen study and two for ^{125}I-prothrombin study. The animals were heparinized at zero time as described in Section II, A, and injected with ^{125}I-fibrinogen or ^{125}I-prothrombin. On the twenty-fourth hour 3 ml of typhoid endotoxin was intravenously injected, and the heparin injection was increased to every 2 hr for the next 12-hr period and thereafter reduced to every 3–6 hr during the rest of the experimental period. One of the experiments is shown in Fig. 3. It is seen that heparin did not appreciably alter responses of ^{125}I-fibrinogen as well as those of ^{125}I-prothrombin.

2. *Native Fibrinogen and Prothrombin*

Responses of native fibrinogen and prothrombin to a single injection of typhoid endotoxin were observed in four calves each; the data are shown in Fig. 4. It is seen that heparin did not have appreciable effects on the fibrinogen and prothrombin responses to the endotoxin injection.

IV. Discussion

A. Prothrombin Degradation under Normal Conditions

Thrombin is an activation product of prothrombin, at least *in vitro*, and the claims by Reich and Sternberger (1956), Sternberger (1952), and Sternberger and Maltaner (1951) that thrombin can be recovered from plasma

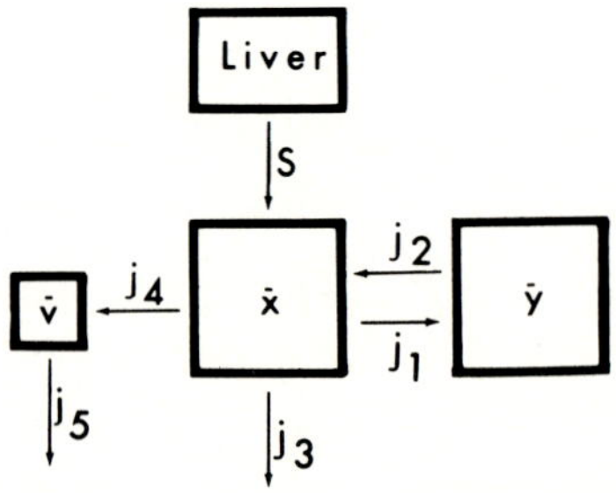

FIG. 5. A model of the metabolism and distribution of native prothrombin. $\bar{x}$ and $\bar{y}$ are the plasma and interstitial prothrombins, respectively. $\bar{v}$ is the thrombin in its own space and S is the rate of prothrombin synthesis. j_1 and j_2 are the fractional rates of prothrombin exchange per day between the plasma and interstitial fluids. j_4 is the fractional rate of thrombin formation per day and j_5 is the fractional breakdown rate of thrombin per day; j_3 is the fractional rate of direct breakdown of prothrombin per day.

were disputed by Godal *et al.* (1961). Thus, it is not known whether or not thrombin is generated *in vivo* under normal conditions, although it seems to be true that sufficient amount of thrombin must be generated at least when a thrombus is formed. The present studies do not answer these questions, but provide some further insights. Fig. 5 shows a model of native prothrombin metabolism and distribution, where j_4 is the fractional rate of thrombin formation per day, j_5 is the fractional rate of thrombin degradation per day and $\bar{v}$ is the amount of thrombin present in its own space. This system can be described by the following sets of differential equations:

$$\frac{d\bar{x}}{dt} = S + j_2\bar{y} - (j_1 + j_3 + j_4)\bar{x} \quad (1)$$

$$\frac{d\bar{y}}{dt} = j_1\bar{x} - j_2\bar{y} \quad (2)$$

$$\frac{d\bar{v}}{dt} = j_4\bar{x} - j_5\bar{v} \quad (3)$$

Since the animals were in a steady state, $d\bar{x}/dt = 0$, $d\bar{y}/dt = 0$, and $d\bar{v}/dt = 0$ (Reeve and Roberts, 1959). Therefore, $j_1\bar{x} = j_2\bar{y}$, $S = (j_3 + j_4)\bar{x}$ and from the latter and also as reported previously (Atencio *et al.*, 1965a)

$$j_3 + j_4 = \frac{S}{\bar{x}} = (C_1/a + C_2/b)^{-1} \quad (4)$$

where C_1, a, C_2, and b are the parameters of a function describing the behavior of plasma ^{125}I-prothrombin. If j_4 were significantly large, its abolition would cause an increase in $\bar{x}$ at least temporarily or a decrease in S or both and also must alter the values of C_1, a, C_2, and b or at least some of them to balance the decrease in the left side of Eq. (4). Heparin is known to retard or inhibit, depending on the dose, the generation of plasma thromboplastin and thus retard or inhibit the formation of thrombin from prothrombin (Biggs *et al.*, 1953; Engelberg, 1963). Thus, the present findings that heparin did not cause appreciable alterations of prothrombin metabolism and distribution (Section III, B) indicate that j_4 was very small or nonexistent and support the concept that a major fraction of prothrombin is directly degraded without the formation of thrombin under normal conditions.

B. Labeled and Native Fibrinogens during the Typhoid Endotoxin Injection

1. *Rate of Fibrinogen Synthesis* (*S*)

The data were obtained in the unsteady state, and it was not possible to analyze them exactly. Therefore, an approximate interpretation is given below. The studies showed some differences in the response of ^{125}I-labeled and native fibrinogens to a single injection of typhoid endotoxin. Namely, ^{125}I-fibrinogen levels decreased to the extent of 16%, on the average, of control levels during an average of 9.6 hr following the endotoxin injection, whereas native fibrinogen concentration decreased by 5% of control levels during the 2-hr period following the endotoxin injection, and thereafter it kept increasing over the control values (Section III, C, 1, and 2). The percent decrease (d) of ^{125}I-fibrinogen over the control level during the 2-hr period following the endotoxin injection is calculated from $d = 1 - e^{-0.672t}$, where 0.672 is obtained from log 2 divided by the average half-life of ^{125}I-fibrinogen during the 9.6-hr period which is 1.03 days, and t is the time in days. The calculated d was 5.4% which is very close to the average decrease of 5% for native fibrinogen, indicating that during the 2-hr period following the endotoxin injection there was no significant addition of native fibrinogen to plasma over the normal amount of fibrinogen added. The amount of fibrinogen added to plasma is a sum of the rate of fibrinogen synthesis (S) and the amount of interstitial fibrinogen returned to plasma ($j_2 \bar{y}$) (Fig. 1). Therefore, the rate of fibrinogen synthesis was at most at the control level during the 2-hr period. During the next 7.6 hr (9.6–2 hr), it is evident that fibrinogen over the normal amount ($S + j_2 \bar{y}$) was added to plasma, since despite the increased rate of removal of plasma ^{125}I-fibrinogen (shorter half-lives) the native plasma fibrinogen concentration increased (Section III, C, 2). How much of this increase is due to the increased addition of interstitial fibrinogen ($j_2 \bar{y}$) and how much to the increased rate of synthesis (S) cannot be answered, but the interstitial fibrinogen ($\bar{y}$) constitutes only 26% of plasma fibrinogen ($\bar{x}$) (Section III, A, 1), so that the increase in plasma fibrinogen concentration can be accounted for at most up to 26% of the increase by the increased amount of return of interstitial fibrinogen ($j_2 \bar{y}$). This is the theoretical maximum with complete emptying of interstitial fibrinogen into plasma; realistically it would be much less than this, since an approximately equal amount of plasma fibrinogen must enter the interstitial space. Most calves during this period showed more than a 26% increase in the plasma fibrinogen concentration (Fig. 4). Therefore, it is evident that some increase in the synthetic rate of fibrinogen (S) has occurred. During the next 1.5 days in which slower removal rate (longer half-life) than that under the control condition prevailed (Fig. 3), it may be reasonably assumed that the effects of endotoxin on the removal of native as well as labeled fibrinogens have ceased, based on the abrupt change of half-lives from 0.91 days to about 5 days, and that the longer half-life during the 1.5-day period is due to the increased return of ^{125}I-fibrinogen from the interstitial space with normal or near normal removal rate of plasma ^{125}I-fibrinogen. Then, the rate of addition of fibrinogen to plasma ($S + j_2 \bar{y}$) must be increased, since the plasma fibrinogen concentration still kept increasing. As discussed above, up to 26% of the increase in the plasma fibrinogen concentration can be accounted for by the increased amount of return of interstitial fibrinogen ($j_2 \bar{y}$). Since most of the calves showed an increase greater than 26% of control levels, the synthetic rate of fibrinogen (S) was still increased. In the period following this, which is 3 days after the start of the experiment, the half-life returned to the control level of 3.0 days, indicating the end of the redistribution phase which preceded it and the return of the removal rate of plasma fibrinogen ($j_1 + j_3$) to the control level. During this period the rate of addition of fibrinogen to plasma ($S + j_2 \bar{y}$) must be still increased, since the plasma fibrinogen concentration was still increasing slowly. Since the amount of plasma fibrinogen transferred into the interstial space ($j_1 \bar{x}$) can be assumed approximately equal to the amount of return of interstial fibrinogen to plasma ($j_2 \bar{y}$) during this period, the synthetic rate of fibrinogen (S) must be increased.

2. *Removal Rate of Plasma Fibrinogen* $[(j_1 + j_3)\bar{x}]$

The amount of plasma fibrinogen removed is a sum of the removal by transcapillary filtration ($j_1 \bar{x}$), and catabolic or deposition processes (or both) of plasma fibrinogen ($j_3 \bar{x}$). As shown above, the removal rate of plasma ^{125}I-fibrinogen ($j_1 + j_3$) averaged 16% above the control level during the 9.6-hr period following the endotoxin injection. How much of this is due to the increased catabolic (j_3) (or irreversible deposition or both) or transcapillary transfer rate (j_1) cannot be determined, but both must have occurred (Chien *et al.*, 1965; Good and Thomas, 1953). During the next 1.5 days, the removal rate ($j_1 + j_3$) may be assumed to have returned to the near normal or normal level as discussed above. During the period following this, the normal half-life suggests the return of the removal rate ($j_1 + j_3$) to the control level, but the actual amount removed $[(j_1 + j_3)\bar{x}]$ was still increased, since the plasma fibrinogen concentration was increased over the control level.

C. Labeled and Native Prothrombins during the Typhoid Endotoxin Injection

1. *Rate of Prothrombin Synthesis* (S)

The studies showed very similar responses by both ^{125}I-prothrombin and native prothrombin to a single injection of typhoid injection. Both decreased about 20% of respective control level during the 10-hr period following the endotoxin injection (Section III, C, 3, and 4); for an additional 2 hr, native prothrombin concentration kept decreasing while the removal rate of ^{125}I-prothrombin started slowing down. This indicates that during the 12-hr period following the typhoid injection the amount of prothrombin added to plasma remained at the control level. Since the amount of addition is a sum of the rate of prothrombin synthesis (S) and addition of the interstitial prothrombin to plasma ($j_2 \bar{y}$), the rate of prothrombin synthesis (S) was at most at the control level during this period. The longer half-life of plasma ^{125}I-prothrombin during the next 1.5 days (Fig. 3) is assumed to be due to the increased amount of return of interstitial ^{125}I-prothrombin to plasma with normal or near normal rate of removal ($j_1 + j_3$) of plasma ^{125}I-prothrombin as discussed for fibrinogen in Section IV, A, 1, but the actual amount of plasma prothrombin removed $[(j_1 + j_3)\bar{x}]$ would be less than the control value, since plasma prothrombin concentration was less than the normal value during this period. Then, the return of native plasma prothrombin concentration during this period toward the control level (Fig. 4) indicates a greater amount of prothrombin addition ($S + j_2 \bar{y}$) over the removal rate during this period. Analysis by mass balance showed that the synthetic rate of prothrombin during this period appeared somewhat increased over the control value and that the return to the initial level of plasma prothrombin concentration was due to the lesser removal rate of prothrombin $[(j_1 + j_3)\bar{x}]$ than the control value with the normal or somewhat increased rate of prothrombin synthesis (S). During the period following this which is 3 days after the start of ^{125}I-prothrombin injection, the half-life of plasma ^{125}I-prothrombin returned to the control level of 2 days, and the native plasma prothrombin concentration returned to the control level (Fig. 4) and remained at this level thereafter. This indicates reaching a normal steady state and the return to the control level of the transcapillary transfer rate (j_1), catabolic rate (j_3), and the rate of interstitial prothrombin return (j_2), and that the addition rate of prothrombin to plasma ($S + j_2 \bar{y}$) is now equal to the removal rate of plasma prothrombin $[(j_1 + j_3)\bar{x}]$. Since the transcapillary filtration rate ($j_1\bar{x}$) is equal to the return rate of interstitial prothrombin ($j_2 \bar{y}$), the catabolic rate ($j_3 \bar{x}$) equals the synthetic rate (S). Since the catabolic rate (j_3) and plasma prothrombin ($\bar{x}$) are at the control levels, the synthetic rate (S) is now at the control level.

2. *Removal Rate of Plasma Prothrombin* $[(j_1 + j_3)\bar{x}]$

During the 10–12-hr period following the endotoxin injection, both plasma ^{125}I-prothrombin and native plasma prothrombin concentration decreased by about 20% of respective control level. How much of this is due to the increased consumption (catabolic) processes (j_3) and how much to the increased rate of transcapillary filtration (j_1) cannot be determined, but both must have occurred (Chien *et al.*, 1964; Good and Thomas, 1953). During the next 1.5 days, the decline rate of plasma ^{125}I-prothrombin slowed. This may be assumed as for fibrinogen as being due to the increased return rate of interstitial ^{125}I-prothrombin (j_2), with normal or near normal rate of removal of plasma ^{125}I-prothrombin ($j_1 + j_3$). During the period following this, which is 3 days after the start of the experiment, the half-life of plasma ^{125}I-prothrombin and native plasma prothrombin concentration returned to the control level. This suggests, as discussed in Section IV, B, 1, the return to the control levels of the transcapillary transfer rate (j_1), catabolic rate (j_3), and the return rate of interstitial prothrombin (j_2).

D. Effects of Heparin on Fibrinogen and Prothrombin Responses to a Single Injection of Typhoid Endotoxin

The studies showed that large amounts of heparin did not appreciably alter responses of both labeled and native fibrinogens and prothrombins (Figs. 3 and 4). This suggests main mechanisms other than coagulation for the observed response of both fibrinogen and prothrombin to the injection of typhoid endotoxin.

V. Conclusions

Using ^{125}I-fibrinogen and ^{125}I-prothrombin, studies were first made of the metabolism and distribution of fibrinogen and prothrombin in the steady state, and also of the effects of heparin on prothrombin metabolism. Heparin did not appreciably alter the metabolism and distribution of prothrombin; this supports the concept that a major fraction of prothrombin under normal conditions is directly catabolized without the formation of thrombin.

Following a single injection of typhoid endotoxin, responses of both labeled and native fibrinogens and prothrombins were observed. The plasma fibrinogen concentration decreased by about 5% of the control level during the 2-hr period following the endotoxin injection; thereafter it rapidly increased and remained elevated during the experimental period of 4–6 days. On the other hand, the plasma prothrombin concentration decreased by about 20% of the control level during the 12-hr period following the endotoxin injection, and thereafter started increasing and kept increasing for the next 1.5 days and reached the control level on the third day from the start of the experiment. Thereafter it remained at this level. Interpretation of these data with the tracer data led to the following conclusions: The rate of fibrinogen synthesis increased rapidly at 2 hr following the endotoxin injection and remained greater than the control value during the rest of the experimental period. On the other hand, the rate of prothrombin synthesis did not appear to have altered much during the entire experimental period and the return to the initial level of plasma prothrombin concentration during the period of 1.5–3 days was due to the decreased removal rate of plasma prothrombin with the normal or somewhat increased rate of prothrombin synthesis.

Large amounts of heparin did not appreciably alter responses described above of fibrinogen and prothrombin to the injection of endotoxin. This suggests that the main mechanisms for the observed changes of fibrinogen and prothrombin metabolism are other than coagulation processes.

Acknowledgments

This work was jointly supported by Research Grants HE-11686 and HE-02262 from the National Heart Institute and a research grant from the Colorado Heart Association.

My sincere thanks are due to Ming Sun, Barbara Johnson, and Elizabeth DeNiro for their technical assistance, and also to Ellen Burney for her secretarial work.

Y. Takeda is a recipient of a Career Development Award from the National Institute of Health.

References

Atencio, A. C., Bailey, H. R., and Reeve, E. B. (1965a). *J. Lab. Clin. Med.* **66**, 1.
Atencio, A. C., Burdick, D. C., and Reeve, E. B. (1965b). *J. Lab. Clin. Med.* **66**, 137.
Biggs, R., Douglas, A. S., and MacFarlane, R. G. (1953). *J. Physiol.* (*London*) **122**, 554.
Bloch, K. J., and Bunim, J. J. (1959). *J. Amer. Med. Ass.* **169**, 79.
Chien, S., Sinclair, D. G., Dellenback, R. J., Chang, C., Peric, B., Usami, S., and Gregersen, M. I. (1964). *Amer. J. Physiol.* **207**, 518.
Chien, S., Dellenback, R. J., and Usami, S. (1965). *Proc. Soc. Exp. Biol. Med.* **118**, 1187.
Engelberg, H. (1963). "Heparin." Thomas, Springfield, Illinois.
Godal, H. C., Hasselback, R., and Hjort, P. F. (1961). *Scandinav. J. Clin. Lab. Invest.* **13**, 532.
Good, R. A., and Thomas, L. (1953). *J. Exper. Med.* **97**, 871.
Hardaway, R. M. (1962). *Ann. Surg.* **155**, 325.
McFarlane, A. S. (1958). *Nature* (*London*) **182**, 53.
McKay, D. G., and Shapiro, S. S. (1958). *J. Exp. Med.* **107**, 353.
Matthews, C. M. E. (1957). *Phys. Med. Biol.* **2**, 36.
Reeve, E. B., and Roberts, J. E. (1959). *J. Gen. Physiol.* **43**, 415.
Reich, T., and Sternberger, L. A. (1956). *Surg. Gynecol. Obstet.* **102**, 463.
Seegers, W. H. (1962). "Prothrombin." Harvard Univ. Press, Cambridge, Massachusetts
Sternberger, L. A. (1952). *J. Amer. Med. Ass.* **150**, 1591.
Sternberger, L. A., and Maltaner, F. (1951), *Science* **114**, 414.
Takeda, Y. (1966). *J. Clin. Invest.* **45**, 103.
Takeda, Y. (1968). *J. Lab. Clin. Med.* **72**, 1023.
Takeda, Y. (1969). *Fed. Proc. Fed. Amer. Soc. Exp. Biol.* **28**, 386.
Ware, A. G., and Seegers, W. H. (1949). *Amer. J. Clin. Pathol.* **19**, 471.

30

ABNORMAL FIBRINOGEN METABOLISM

*Erwin Regoeczi**

BIOPHYSICS DIVISION,
NATIONAL INSTITUTE FOR MEDICAL RESEARCH,
LONDON N.W.7, ENGLAND

I. Introduction

Metabolism of plasma proteins may become abnormal either as the result of an off-balance in the existing physiological regulatory mechanisms or by the appearance of additional factors which are absent under normal conditions. In either case, to qualify a change as "abnormal," some idea of that which may be regarded as "normal" is required. In accordance with this principle, the concept of normal fibrinogen metabolism used as a background for preparing this chapter is compiled in Table I. Since a detailed

*Present address: Department of Pathology, McMaster University, Hamilton, Ontario, Canada.

TABLE I

Concept of Normal Fibrinogen Metabolism

1. Catabolism
   (a) First-order kinetic process, i.e.,
       mass of fibrinogen catabolized is proportional to the concentration of this protein in the plasma, the fraction of the pool degraded being constant
   (b) Intracellular
   (c) The catabolic process itself is uncontrolled
   (d) Catabolic rate is determined by the endocytic rate of fibrinogen
2. Distribution between pools
   Largely proportional to the concentration of fibrinogen in the plasma
3. Synthesis
   (a) Regulation unknown
   (b) Labile (unlike catabolism)
   (c) Primarily responsible for changes in pool size
   (d) The synthesis process does not determine or influence the catabolic process

explanation of this concept would be beyond the scope of this chapter, comments will be restricted to its essential features.

The amount of fibrinogen in the body ("pool") is determined by the dynamic balance of two continuous processes, i.e., (1) the degradation of fibrinogen in the tissues ("catabolism"), and (2) the production, that is the delivery into the circulation of new molecules by the liver ("synthesis").

There is sufficient experimental evidence indicating that fibrinogen catabolism is a first-order kinetic process, and these were reviewed recently by Regoeczi (1970). Kinetic considerations lead Reeve *et al.* (1967) to conclude that the catabolic process itself is uncontrolled. Tracer studies also suggest that before catabolism actually takes place, fibrinogen is probably taken up by cells (Regoeczi, 1967). If this is so, then the factor determining the rate of degradation is most likely to be the rate of transfer of fibrinogen from the extracellular fluid into the cytoplasm of the catabolic cells.

Distribution of fibrinogen in the body is a function of the capillary transfer rate and of the return flow, both rates obeying the same kinetic order as does the catabolic rate (Atencio *et al.*, 1965).

Whereas the fractional catabolic rate of fibrinogen appears to be rather steady, the synthesis rate is subject to considerable "spontaneous" fluctuations. The implications of this metabolic pattern were recently approached mathematically by Atencio *et al.* (1969a). In the light of numerous isotopic studies, it is now clear that the majority of factors known to affect fibrinogen metabolism do so by modifying the synthesis rate, but how, still remains a mystery.

Changes in fibrinogen metabolism cannot be correctly understood without keeping in mind that the synthesis process has no bearing on the catabolic process.

II. Forms of Abnormal Fibrinogen Metabolism

Like in the case of other plasma proteins too, it is essential to distinguish between *absolute* catabolic and synthesis rates of fibrinogen on one hand and *fractional* catabolic and synthesis rates on the other. From the metabolic concept given for fibrinogen in Table I it follows that of these four rates the diagnostically important ones for this protein are the fractional catabolic rate and the absolute synthesis rate. Changes in the absolute catabolic rate of fibrinogen alone are inconclusive for the characterization of any disease, considering that increases in this rate together with a normal fractional catabolic rate mean stimulated synthesis and not a pathological catabolic process. Reference to fractional synthesis rates of fibrinogen only may turn out to be similarly misleading, particularly in situations in which enlarged pools are being turned over at more or less normal fractional rates, as, for example, in the nephrotic syndrome.

In spite of a rapidly increasing number of studies on fibrinogen metabolism in the recent literature, surprisingly little effort seems to have been made so far toward the classification of the abnormal metabolic conditions

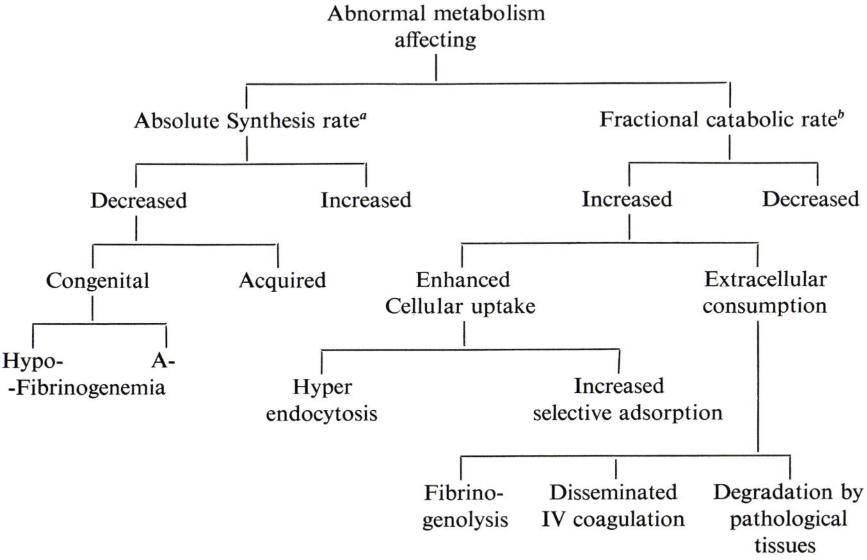

FIG. 1. Forms of abnormal fibrinogen metabolism. *a*, mg fibrinogen synthesized/kg/day; *b*, % of the intravascular (IV) pool broken down per day.

according to pathophysiological aspects. To fill this gap, a system which is based on the above considerations is presented in Fig. 1, and its elucidation will be the aim of the subsequent sections.

III. Changes in the Absolute Synthesis Rate of Fibrinogen

A. Decreased Synthesis

Intervals during which fibrinogen synthesis falls to subnormal (or perhaps more precisely, subaverage) values play an important role in the normal regulatory mechanism by which episodes of stimulated synthesis are compensated for (Atencio *et al.*, 1969a). Whether acquired conditions with sustained depression of fibrinogen synthesis also exist is less well known.

Studies in a group of patients with advanced cirrhosis of the liver using the ^{14}C-carbonate technique showed markedly reduced synthesis rates (12–17 mg/kg/day) in some cases but not in others (Jones, 1969). Considering that thyroid and other developmental hormones [by the virtue of their capacity to increase RNA transport from the nucleus to cytoplasm, to accelerate proliferation of microsomal membranes, and to induce redistribution of the ribosomes (cf. Tata, 1969)] profoundly affect protein synthesis, low fibrinogen synthesis rates would not be surprising at all as a symptom accompanying certain endocrine disorders.

Ham and Curtis (1938) studied the concentration of fibrinogen in the plasma of patients with pernicious anemia. They found that remission did not result in a normalization of the low fibrinogen levels unless the diet contained at least 50 g protein per day. Nevertheless, nutritional deficiency in pernicious anemia did not prevent the development of hyperfibrinogenemia following intercurrent infections. These findings encourage one to think that further studies of fibrinogen metabolism in this disease using more advanced techniques could be rewarding.

Since low fibrinogen synthesis rates per se have no rebound effect on the fractional catabolic rate (Gitlin and Borges, 1953; Werder, 1963), the leading symptom in such conditions will be hypofibrinogenemia of varying severity. Congenital hypofibrinogenemia with plasma fibrinogen levels ranging between 0.6 and 1.6 mg/ml has been observed among members of several families (Hasselback *et al.*, 1963; Phillips *et al.*, 1963). However, if they represent clear cases of deficiency, synthesis awaits future confirmation by metabolic balance studies. That fibrinogen synthesis may completely be absent congenitally is supported by reports on antibody production to fibrinogen in afibrinogenemic patients who had been given blood transfusions or fibrinogen infusions (Brönnimann, 1954; de Vries *et al.*, 1961).

B. Increased Synthesis

This is the most commonly observed form of altered fibrinogen metabolism appearing with or without concomitant changes in the fractional catabolic rate. Enhanced production of fibrinogen is not linked in any obligatory way with changes in the fractional catabolic rate, and therefore these should be regarded as an additional, independent symptom. Conditions in which the fractional catabolic rate is reduced are probably very rare (cf. below), whereby hyperfibrinogenemias can be taken with little exception as a reflection of stimulated synthesis.

There are numerous clinical conditions in which fibrinogen synthesis is increased, e.g., acute and chronic bacterial infections (Ham and Curtis, 1938; McFarlane *et al.*, 1964), postoperative states (Izak *et al.*, 1967; Hickman, 1969), cancer (McFarlane *et al.*, 1964), burns (Davies *et al.*, 1966), myocardial infarction (Hill *et al.*, 1966), hyperthyroidism (Hoak *et al.*, 1969), nephrotic syndrome (Takeda and Chen, 1967), essential hypertension (Takeda 1966a), rheumatoid arthritis (Takeda, 1967), and pregnancy (Regoeczi and Hobbs, 1969).

As exemplified by the data in Table II, hyperfibrinogenemia can experimentally be induced by a variety of means. Some of the methods included in this table—which by no means is intended to be comprehensive—probably work on identical pathological principles, for example, by causing tissue damage. However, it is also obvious that the same stimulatory effect can be achieved by techniques which seem to have nothing in common with each other. Attempts to isolate and identify a hypothetical humoral factor (cf. Sarcione *et al.*, 1965) which may be mobilized by all these experimental procedures have so far been unsuccessful.

Another problem requiring elucidation in this connection is whether fibrinogen synthesis can be stimulated independently of the other acute phase proteins. Adrenalectomy abolishes the response to turpentine injections of the α_2-acute phase globulin but not that of fibrinogen (Weimer and Coggshall, 1967). Similarly, this globulin does not respond to endotoxin injections following adrenalectomy (Heim and Ellenson, 1965), whereas fibrinogen does to some extent (Atencio *et al.*, 1969b).

The real merit in the controversy about the somewhat mysterious effects of cooked pig stomach (Foster and Whipple, 1921a) seems to be that it has focused early attention on the relationship between nutritional state and fibrinogen synthesis. Now it would appear that while high protein intake may moderately and transitorily stimulate fibrinogen production in the dog (Vars, 1930), the human synthesis process is essentially independent of dietary measures (Ham and Curtis, 1938).

TABLE II

SOME AGENTS AND TECHNIQUES USED EXPERIMENTALLY TO STIMULATE FIBRINOGEN SYNTHESIS

Agent or technique	Method	Species	Studied by
Thrombin	Plasma conc.	Dog, rat, rabbit	Monkhouse and Milojevic (1960)
Thromboplastin	Plasma conc.		Monkhouse and Milojevic (1960)
Thromboplastin	^{131}I-fibrinogen	Dog	Adelson (1965)
Free fatty acids (e.g., palmitate)	^{14}C-glycine (liver-biopsy)	Human	Pilgeram and Pickart (1968)
Lymphosarcoma homogenate	Plasma conc.	Rat	Wycoff *et al.* (1959)
Heterologous proteins			
Peptone	Plasma conc.	Dog	Foster and Whipple (1921c)
Fibrinogen	Plasma conc.	Rabbit	Kropatkin and Izak (1968)
Homologous proteins			
Serum ip	^{14}C-arginine	Rat	Winkert *et al.* (1964)
Fibrinogen	^{131}I-fibrinogen	Rabbit	Regoeczi (1970)
Large volume of			
Locke's soln.	Plasma conc.	Dog	Foster and Whipple (1921b)
Saline	Plasma conc.		Monkhouse and Milojevic (1960)
Endotoxins	^{131}I-fibrinogen	Rabbit	Regoeczi *et al.* (1963)
	^{14}C-carbonate	Rabbit	Koj and McFarlane (1968)
Tissue injuries			
Turpentine-abscess	Plasma conc.	Dog	Foster and Whipple (1921c)
	^{14}C- and ^{3}H-amino acids	Rabbit	Koj (1968)
Surgical trauma	Plasma conc.	Rat	Chanutin *et al.* (1938)
X ray	Plasma conc.	Dog	Foster and Whipple (1921c)
Talc ip	^{14}C-leucine	Rat	Gordon and Koj (1968)
Hormones			
ACTH, β^{1-23} corticotropin	^{131}I-fibrinogen	Rabbit	Atencio *et al.* (1969b)
l-thyroxine, T_3	^{75}Se-methionine	Human	Hoak *et al.* (1969)
Adrenaline, isopropylnor-A	Plasma conc.	Rat	Henriques *et al.* (1956)
Cooked pig stomach	Plasma conc.	Dog	Foster and Whipple (1921a)

IV. Changes in the Fractional Catabolic Rate of Fibrinogen

A. DECREASED CATABOLISM

Comparative measurements by Atencio and Reeve (1965) in groups of young and older rabbits provided data which suggest that the catabolic rate of fibrinogen (both fractional and absolute) moderately decreases with aging.

Pathological conditions with reduced catabolism of fibrinogen seem to be rare. Hart (1964) described observations with labeled fibrinogen in a case of myxedema, and it is likely that the markedly prolonged survival of fibrinogen in the plasma of this patient ($t_{\frac{1}{2}}$ = 194 hr) reflected a true reduction in the catabolic rate. However, it is desirable that this finding will be further furnished by proper turnover studies since, as rightly pointed out by Takeda (1966b), "half-lives alone are insufficient to define fibrinogen metabolism."

There is indirect evidence suggesting that synthesis of the clotting factors is diminished in hypothyroidism (Simone *et al.*, 1965), and that plasma fibrinogen concentration yet remains about normal would also suggest concomitant changes in the fractional catabolic rate.

B. Increased Catabolism

Because of its diagnostic significance it must be emphasized that increases in the fractional catabolic rate, implying a reduced average life-span for fibrinogen molecules, are a genuine symptom and not merely a reflection of the functional state of fibrinogen synthesis in the liver. It is hardly surprising therefore that comparably accelerated catabolic rates may occur in association with normal fibrinogen levels as well as with hypo- or hyperfibrinogenemias.

Increased fractional catabolic rates can arise as a result of difuse or circumscript consumption of fibrinogen in extracellular sites or of an enhanced uptake by catabolic cells.

1. *Extracellular Fibrinogen Consumption*

a. *Fibrinogenolysis and Disseminated Intravascular Coagulation.* Although neither of these processes represent increased fibrinogen catabolism in a true metabolic sense, this review would not be complete without at least briefly mentioning them.

Recognition of fulminant and acute cases requires no isotopic assays and is usually based on standard hematological tests, for which the reader is referred to the relevant literature (e.g., Johnson and Merskey, 1966; Lasch *et al.*, 1966; Merskey, 1968; Merskey and Johnson, 1966; Merskey *et al.*, 1967; Verstraete and Vermylen, 1968; Verstraete *et al.*, 1963).

By contrast, administration of labeled fibrinogen may prove valuable in identifying mild forms (Baker *et al.*, 1968a), where changes in clotting factors and in the components of the fibrinolytic system are often inconclusive. Recent investigations, showing that fibrinogen is capable of forming complexes with fibrin monomer (Shainoff and Page, 1962; Sasaki *et al.*, 1966) as well as with its own proteolytic derivatives (Lipinski *et al.*, 1967), lend support to the assumption that fibrinogen consumption during disseminated intravascular coagulation and fibrinogenolysis is of a complex nature: In addition to the fraction lost by the action of enzymes, a further fraction is utilized for the removal of the altered protein from the circulation in the form of complexes. Obviously, if the clearance of the proteolytic products of fibrinogen involves such a mechanism, this must amplify the diagnostic sensitivity of turnover studies in these conditions.

Thrombotic events confined to one or a few vessels seem to have little or no effect on the fractional catabolic rate of fibrinogen (Flute, 1969), the coagulated protein usually amounting to only a small proportion of the pool. Exceptions are, however, clearly possible (Amris and Amris, 1964).

b. *Degradation of Fibrinogen by Pathological Tissues.* Local processes involving inflammation and tissue damage may result not only in enhanced synthesis of fibrinogen but also in increased rates of degradation. Wounds, tumors, sterile and septic abscesses, as well as sites of burns are all capable of accumulating and degrading fibrinogen. By this means complicated isotopic patterns arise which are often difficult to interpret quantitatively. This is illustrated in Fig. 2 by the behavior of ^{131}I-fibrinogen in a rabbit bearing a turpentine abscess. Before injecting turpentine, the catabolic rate equaled 33% of the total body fibrinogen or 46% of the intravascular pool, the pool ratio (K_{13}/K_{31}) being 0.29. The absolute catabolic rate was 190 mg/day (= 52 mg/kg).

The first changes induced by turpentine were retention of iodide in the body and a sharp rise in plasma fibrinogen concentration. Some 2 days later, labeled fibrinogen began to escape from the intravascular compartment at an accelerated rate. This was accompanied by a slowing-down of the slope of the total body protein-bound activity curve. Obviously, activity was now accumulating in an additional extravascular "pool," and this process was still going on at the time when the experiment was terminated. Deposition of radioactive fibrinogen in tissues surrounding abscesses was demonstrated via direct measurements by Mutschler (1964), and the data obtained by Davies *et al.* (1966) in burned patients suggest that discrepancies between total body and plasma radioactivity slopes, similar to the one in Fig. 2, may persist for considerable periods of time.

The intriguing question as to how much fibrinogen is involved in local processes cannot be answered satisfactorily because of the uncertainty of the extent to which local inflammations affect the distribution and catabolism of fibrinogen in other regions of the body. If we assume—and admittedly, without any good reason—that such overall effects were absent, then the data in Fig. 2 would imply that during the time between hours 120 and 190 of the experiment, fibrinogen was moving into the inflamed area at an approximate rate of 674 mg/day (= 185 mg/kg). This is about 3.5 times more than the quantity degraded in the normal catabolic process.

The above assessment can be grossly erroneous if the underlying

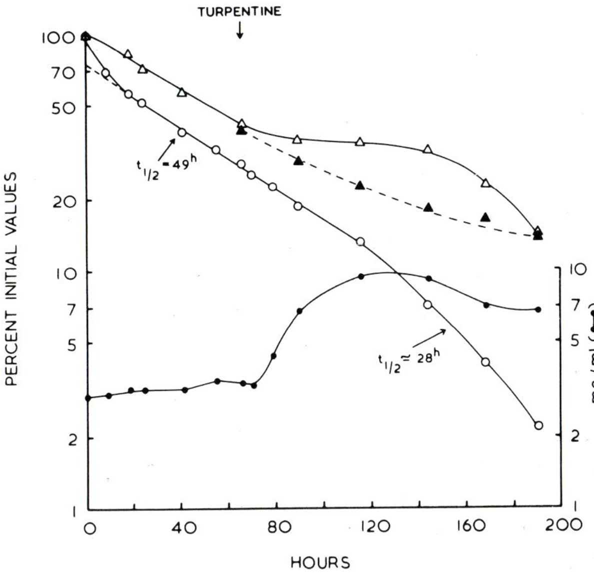

FIG. 2. Effect of subcutaneously injected turpentine (0.5 ml/kg) on the turnover of ^{131}I-fibrinogen in a rabbit. The curves denote total body γ-radiation (△), clottable intravascular activity (○), plasma fibrinogen concentration (●), and total body protein-bound activity (▲). The latter curve was calculated from measurements of the nonprotein activity in the plasma, assuming that the body iodide space remained constant. For explanations see text.

assumption is incorrect, but the fact remains that injured tissues turn over fibrinogen more slowly than normal ones. This may partly be explained by the fact that in pathological tissues most of the fibrinogen is present as fibrin (McCutcheon, 1961). In addition to the histological evidence, this conclusion is also supported by studies with labeled fibrinogen in rats bearing transplantable tumors or turpentine abscesses: The fraction of the dose accumulating locally could significantly be reduced or increased by the respective use of anticoagulant and antifibrinolytic agents (Shaeffer, 1963, 1964a, b; Mutschler, 1964).

2. *Increased Cellular Uptake of Fibrinogen*

As pointed out in Section I, fibrinogen catabolism is most likely to be limited by the rate of transfer of fibrinogen from the extracellular into the intracellular compartment. The simplest way of visualizing this is to assume that fibrinogen becomes attached to cell membranes by selective adsorption and is transported into the cells by endocytosis. The possibility that these fundamental biological functions may play a significant role in fibrinogen degradation has not yet been duly considered in the literature, and therefore the aim of the subsequent paragraphs should be to discuss some experimental observations relevant to this concept.

a. *Endocytosis: Iodine-labeled Polyvinylpyrrolidone (I-PVP) as an Indicator.* The term "endocytosis" is used here in the framework of a nomenclature introduced by de Duve (1963) to replace a multitude of expressions (e.g., cytopempsis, pinocytosis, micropinocytosis, phagocytosis, ultraphagocytosis, phagotrophy, and colloidopexy), all referring to the ability of living cells to ingest droplets from the surrounding medium with the aid of specific membrane movements. Current aspects of endocytosis have been reviewed by several authors (Holter, 1965; Jacques, 1966; de Duve and Wattiaux, 1966; Jacques, 1969).

Although endocytosis can only be seen in tissue cultures (Lewis, 1931), there is sufficient indirect evidence to postulate that the same mechanism also operates in the intact mammalian body. Cellular uptake of homologous plasma proteins has so far been the subject of a relatively small number of studies (Freeman *et al.*, 1958; Holtzer and Holtzer, 1960; Gordon and Jacques, 1966).

If it is assumed that fibrinogen is transported into the catabolic cells by endocytosis, then a possible experimental approach to study this process *in vivo* would be to use an independent indicator which is taken up at rates comparable to the catabolic rate of fibrinogen. Such a substance is PVP with the clear advantage that once inside a cell, it will stay there for

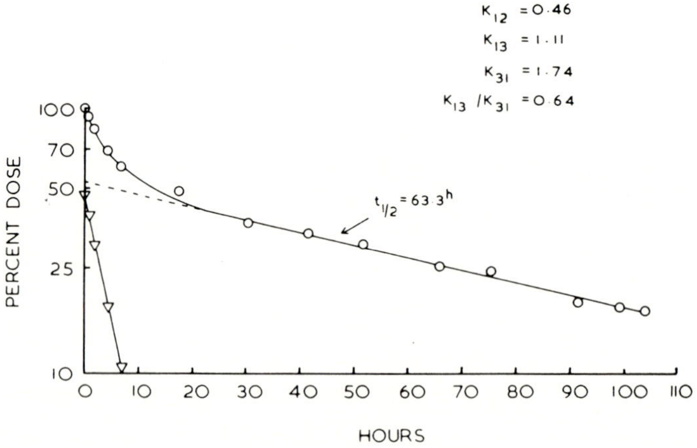

FIG. 3. Behavior of ^{125}I-PVP (2 mg/kg, approximate MW 50,000) in a rabbit. Clearance rate (K_{12}), capillary transfer rate (K_{13}), the return flow (K_{31}) and the partition between intra- and extravascular spaces (K_{13}/K_{31}) were calculated by compartmental analysis (Matthews, 1957). Excretion rate (not shown); 0.04; endocytic rate, $0.46 - 0.04 = 0.42$/day.

considerable periods of time (Ravin *et al.*, 1952). In contrast to other colloidal materials in current use (carbon particles, colloidal chromic phosphate, heat-aggregated albumin, lipid emulsions, etc.), trace doses of I-PVP are cleared slowly, whereby the activity of the reticuloendothelial cells can conveniently be followed up for days or even weeks from a single injection. This is illustrated in Fig. 3 by showing the clearance of ^{125}I-PVP from the plasma of a normal rabbit. It will be seen that with respect to distribution and disappearance rates the labeled material behaved in the recipient very similarly to an iodinated plasma protein; in fact, its endocytic rate fell within the normal range of the fractional catabolic rate of fibrinogen in this species.

b. *Interrelationship between Fibrinogen and PVP in Vivo.* The recognition of this unique feature of PVP fractions with average molecular weights around 50,000 stimulated further studies aimed to find out whether the endocytosis of PVP can be influenced through fibrinogen and vice versa.

The first experiments were disappointing because, as it turned out, neither was fibrinogen turnover affected even by large (300–400 mg) doses of PVP on one hand, nor was PVP endocytized more readily following the injection of 30–50 mg undenatured fibrinogen into rabbits on the other.

The picture was quite different, however, when the fibrinogen was

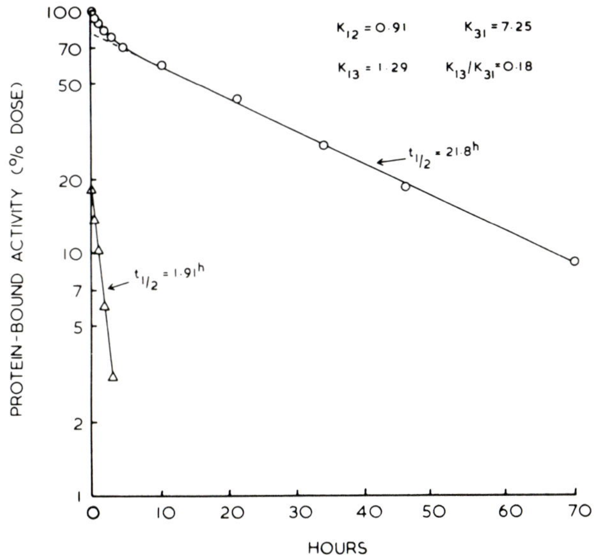

FIG. 4. Behavior of trypsin-digested homologous fibrinogen in a rabbit. Digestion was for 16 min at 25°C, stopped by adding an excess of soybean trypsin inhibitor, and subsequently the mixture was fractionated on a column of Sephadex G150. For abbreviations see Fig. 3.

modified enzymically so that its catabolic rate became somewhat faster. This was done by exposing fibrinogen for short periods of time to the action of trypsin under conditions as described by Mihályi and Godfrey (1963). According to these authors, in the first phase of the reaction trypsin splits 12 bonds in fibrinogen without appreciably degrading the molecule but making it incoagulable by thrombin; some 80 bonds are attacked during the second phase thus giving rise to the formation of large fragments with an approximate molecular weight of 95,000.

Such digests were prepared in this laboratory from purified homologous fibrinogen and studied after labeling in rabbits. (One experiment is shown in Fig. 4.) From the data in this figure it is evident that trypsin-digested fibrinogen (TDF) distributed in the body in much the same way as does native fibrinogen but its catabolic rate was significantly faster. The catabolic rate of TDF was found to be variable within certain limits depending on the length of the tryptic digestion.

As shown in Fig. 5, TDF profoundly affected the endocytic rate of I-PVP. Its promoting effect, by which PVP clearance has increased 1.68 times, manifested itself within a few hours of injection and lasted for over 2.5 days. The influence of TDF on endocytosis of PVP was less pronounced when the enzymic digestion was allowed to continue for prolonged intervals of time, and these digestion products were cleared from the circulation with half-lives sometimes as short as 3 hr. Obviously, the whole effect on endocytosis critically depends on the size and hence the clearance rate of TDF.

In similar studies with TDF rabbits were preinjected with trace doses of labeled fibrinogen instead of I-PVP. One of these experiments is shown in Fig. 7 from which it is evident that after injecting the large fragments of

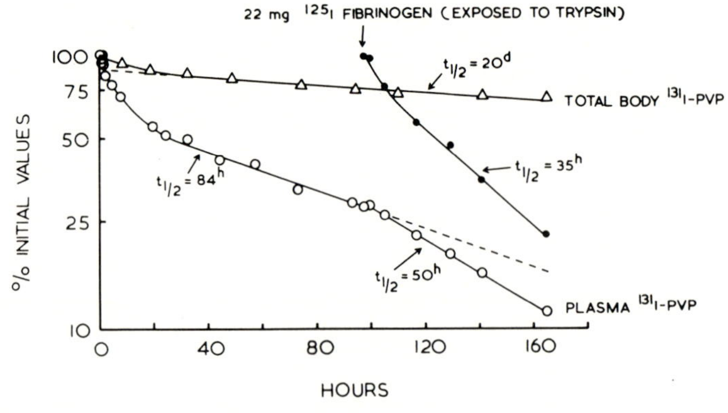

FIG. 5. Effect of TDF on the endocytic rate of I-PVP in a rabbit. TDF was prepared by reacting fibrinogen with trypsin at a weight ratio of 155:1 for 60 sec. Further treatment as described in Fig. 4.

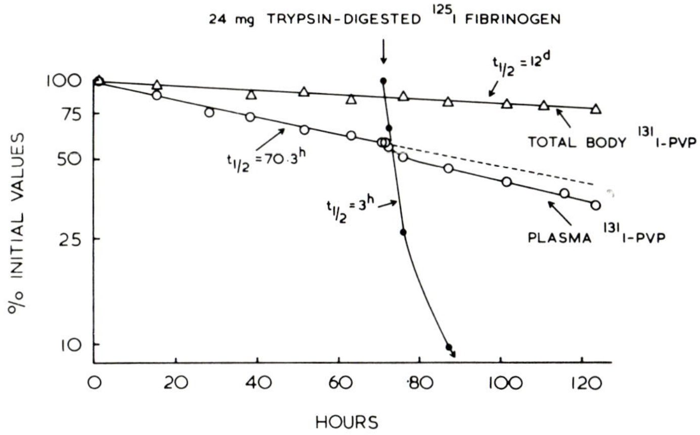

FIG. 6. Effect of TDF, obtained by incubating fibrinogen with trypsin for 30 min, on the endocytosis of I-PVP. Note the difference between this experiment and the one shown in Fig. 5.

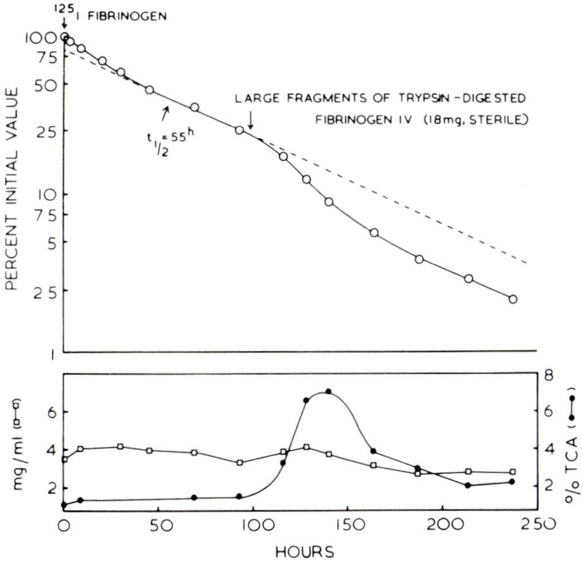

FIG. 7. Effect of TDF on fibrinogen catabolism in a rabbit, the curves denoting intravascular clottable activity (○), proportion of nonprotein activity in the plasma (●), and plasma fibrinogen concentration (□). For explanations see text.

TDF the fractional catabolic rate of fibrinogen soon began to rise and it remained elevated for about 2 days. That we were really dealing with accelerated catabolism and not merely with redistribution is also indicated by the appearance of a "bulge" in plasma nonprotein activities at the same time. Analogous to the situation with PVP, the effect of exhaustively digested fibrinogen on fibrinogen catabolism was relatively small (Fig. 8).

These observations indicate that certain derivatives of fibrinogen arising during the early stages of the proteolysis are able both to enhance the endocytosis of PVP and to increase the fractional catabolic rate of fibrinogen. How this is achieved is not quite clear. Considering that cellular uptake of circulating colloids is a biphasic process (Jancsó, 1955), two explanations are possible. First, TDF may promote selective adsorption of fibrinogen on the membranes of the catabolic cells, for example, by forming soluble complexes with fibrinogen. Indeed, studies in the analytical ultracentrifuge suggest that the two proteins can interact at ionic conditions close to those in the plasma. Fletcher *et al.* (1952) reported that under certain conditions PVP may form precipitates with fibrinogen.

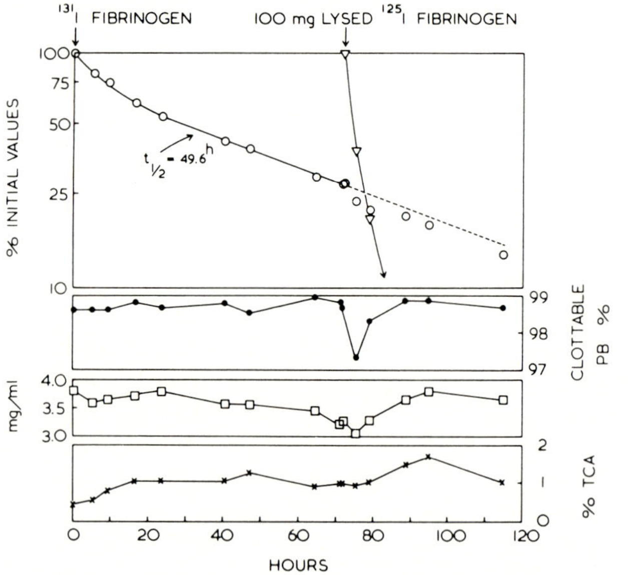

FIG. 8. Effect of plasmin-digested homologous fibrinogen on the catabolism of fibrinogen in a rabbit. The curves are: intravascular clottable activity (○); clottability of the circulating protein-bound activity (●); plasma fibrinogen concentration (□); and proportion of nonprotein activity in the plasma (×). Degradation was induced by activating the contaminant plasminogen with urokinase and self-degradation was allowed to reach completion. Fragments E and D were separated from the mixture by gel filtration on Sephadex G150.

However, analysis of the data obtained *in vivo* does not fully support this theory. In the experiment shown in Fig. 7, the rabbit was catabolizing a daily average of 125 mg fibrinogen before receiving 18 mg TDF compared with 246 mg on the first day thereafter, indicating that fibrinogen consumption was considerably more than one would have expected even if all the injected TDF had complexed with plasma fibrinogen. Thus it seems unlikely that changes in the selective adsorption of fibrinogen alone would adequately explain the effect of TDF on fibrinogen catabolism.

An alternative possibility to be considered is that TDF increases the endocytic activity directly at the cellular level by influencing the engulfment rate of the membrane of cells which are involved both in fibrinogen catabolism and uptake of PVP. Factors maintaining and regulating endocytosis in the intact mammalian body are still largely unknown, but studies in ameba and cell cultures suggest that proteins as well as amino acids, depending on their electrical charge, can induce endocytosis (Chapman-Andresen, 1962, 1964). According to the interpretation by Jacques (1969), induction actually means the stimulation of a continuously existing submicroscopic process to the degree of detectability (spikepinocytosis), and the general shape of the TDF-stimulated I-PVP and ^{125}I-fibrinogen curves in Figs. 5 and 7 are much in the favor of this concept. Future studies will have to show whether the stimulatory effect of TDF is a generalized one or if it is confined to certain types of cells.

Although TDF is only a laboratory product with no immediate relevance to the pathophysiology of fibrinogen metabolism, nevertheless it provides a useful tool for demonstrating that hyperendocytosis can affect fibrinogen metabolism.

c. *Increased Selective Adsorption.* This last form of abnormal fibrinogen catabolism to be discussed here is classified in Fig. 2 beside hyperendocytosis as the second subgroup of the hypercatabolic conditions due to increased cellular uptake of fibrinogen. During hyperendocytosis, fibrinogen molecules and their affinity to the catabolic cells are assumed to be normal, the only difference envisaged being a change in the cell membrane activity. In other words, hyperendocytosis is a functional state of the catabolic cells. By contrast, increased selective adsorption primarily reflects changes in the physicochemical properties of circulating fibrinogen with corresponding alterations in its affinity to the catabolic cells, the endocytic activity of the latter remaining either normal or changing in a secondary manner only.

It is remarkable that the biological survival of fibrinogen will be shortened irrespective of whether the molecule is rendered smaller or larger, more hydrophilic or hydrophobic, and it is tempting to assume in this connection that in the mammalian blood there exists a very sensitive screening system engaged in preferential fixation of altered homologous proteins to catabolic cells.

Increased selective adsorption of fibrinogen is likely to play a far more important role in various conditions associated with abnormal fibrinogen catabolism than it is generally realized. To a large extent this is due to the pronounced tendency of fibrinogen to form complexes with its own derivatives as well as with other substances and plasma proteins. Of the various forms of interaction between fibrinogen and other molecules the present discussion will only be restricted to cryoprofibrin.

The origin of cryoprofibrin can briefly be outlined as follows. Aggregation of fibrinogen under physiological circumstances is prevented by the repulsive negative charge of the fibrinopeptides (Laki, 1968), which is eliminated by the action of thrombin. The reaction scheme for the fibrinogen–fibrin conversion consists of three reversible steps (Scheraga and Laskowski, 1957): First, thrombin cleaves the fibrinopeptides, whereby tyrosyl, lysyl, and α-amino groups become unmasked to serve as donors for the hydrogen bonds in the polymerization of fibrin monomers (Laki and Mihályi, 1949; Donnelly *et al.*, 1955), the acceptors probably being imidazole groups of histidyl residues (Shulman and Ferry, 1950; Mihályi, 1954). In the second step, intermediate polymers form, and these are united into a coherent, visible fibrin during the third step. Cryoprofibrin is a normal product of the second step and it is a combination of fibrinogen with fibrin monomer. Donnelly *et al.* (1955), assuming that—unlike the donor of the hydrogen bond—the acceptor is not masked in native fibrinogen, predicted the existence of "copolymers" on the basis of a competition between acceptor fibrinogen and acceptor fibrin monomer for the donor monomer. Cryoprofibrin was isolated and studied by Shainoff and Page (1962), Sasaki *et al.* (1966), and Cohen *et al.* (1966).

Cryoprofibrin may be the only product of a clotting process in the presence of excess fibrinogen and low thrombic activity, hence its crucial importance for the recognition of low-grade intravascular coagulation. The characteristic feature of these polymers is a reduced solubility compared with native fibrinogen, this providing the common denominator for a number of seemingly unrelated techniques designed to demonstrate their presence, such as exposure to the cold, precipitation with heparin (Smith and von Korff, 1957), β-naphtol or protamine (Lipinski and Worowski, 1968), and withdrawal of solvent (Morrison, 1946). A simple technique to prepare fibrinogen free of cryoprofibrin for turnover studies was recently described by Regoeczi and Stannard (1969).

The question as to whether low solubility fibrinogen fractions are a normal constituent of plasma is very difficult to answer because in this particular field not even extra care provides guarantee against artifacts. Nevertheless,

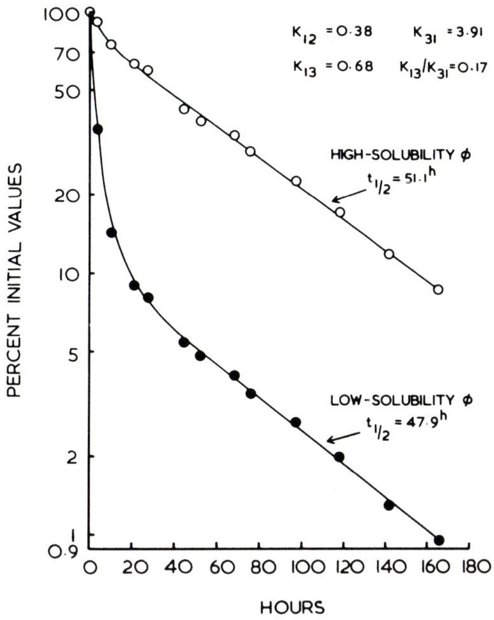

FIG. 9. Catabolism of homologous complex-free ^{125}I-fibrinogen (high-solubility fibrinogen) and complex-rich ^{131}I-fibrinogen (low-solubility fibrinogen) in a normal rabbit, both curves denoting intravascular clottable protein-bound activities. The rate constants (cf. Fig. 3) refer to the ^{125}I-labeled preparation. For explanations see text.

there is no doubt that after certain treatments, like endotoxin (Shainoff and Page, 1960), thrombin infusion (Lipinski and Worowski, 1968), or subcutaneous turpentine injection (Regoeczi, unpublished), as well as in a number of clinical conditions (Ritzmann *et al.*, 1963), significant quantities of fibrinogen complexes can be detected in blood samples.

The metabolic behavior of a low-solubility fibrinogen preparation is illustrated in Fig. 9. This was isolated from the deprothrombinized plasma of a rabbit 48 hr after subcutaneously injecting 0.5 ml turpentine/kg with ammonium sulphate at 0.19 saturation, whereas the simultaneously injected high-solubility fraction was obtained by increasing the degree of saturation to 0.238. When finally purified, 96.6% of the low-solubility and 97.7% of the high-solubility-labeled material was clottable with thrombin.* The difference between their behavior *in vivo* was striking (Fig. 9), and it may be regarded as an example of increased selective adsorption with respect to the low-solubility preparation since the high-solubility ^{125}I-fibrinogen failed to

* To avoid confusion it should be pointed out that the fractions referred to here as being of high or low solubility are not identical with those described by Mosesson and Sherry (1966) and Mosesson *et al.* (1967) using similar terminology.

indicate any signs of hyperendocytosis. It is noteworthy that some 10–15% of the low-solubility material was catabolized at the same rate as normal fibrinogen, suggesting either that the preparation contained a corresponding amount of uncomplexed fibrinogen or that the fibrinogen component of the complexes interchange to some extent with plasma fibrinogen. Although this experiment makes it clear that fibrinogen complexes are subject to increased selective adsorption and thus to a relatively fast catabolism, it would probably be a mistake to elaborate an overall catabolic rate for fibrinogen copolymers, their average size and composition being greatly dependent on the kinetics of the first (enzymic) phase of the fibrinogen–fibrin transition.

The low-solubility fibrinogen fraction shown in Fig. 9 was highly clottable with thrombin, but this is not always the case. In contrast, the occurrence of cryofibrinogens with coagulabilities as low as 35% have been reported (Mosesson *et al.*, 1968), and this poses the questions whether (1) fibrinogen in such instances forms complexes with a globulin rather than with its own derivatives, and (2) if so, which of the participating proteins is primarily responsible for the abnormal association taking place?

At present, neither of these questions can be answered satisfactorily. Morrison *et al.* (1948) as well as Mosesson *et al.* (1968) found that the so called "cold insoluble globulin" present in cryofibrinogen preparations had the electrophoretic mobility of a β_1-globulin and it sedimented with an approximate Svedberg value of 11. Edsall *et al.* (1955) calculated that this "globulin" was about of the length of fibrinogen but its mass was approximately twice that. Careful immunoelectrophoretic studies were performed on a purified cryofibrinogen preparation using four different antisera by Mosesson *et al.* (1968). With unabsorbed rabbit serum against human fibrinogen, the preparation yielded two precipitation lines, one of which disappeared when the same antiserum was made specific to fibrinogen by adsorption with normal human serum. This would imply that the batch of cryofibrinogen under investigation contained a protein component present in the normal plasma and not identical with fibrinogen. However, immunoelectrophoresis of the preparation against goat or horse antiserum to human serum failed to produce any precipitation line.

It is conceivable that plasma fibrinogen of subjects with idiopathic cryofibrinogenemia differs from the normal fibrinogen molecule so that in some way it is predisposed to pathological associations. This possibility was investigated by Baker *et al.* (1968b) in a patient suffering from severe cryofibrinogenemia over 4 years without any detectable underlying disease. His fibrinogen was isolated within 24 hr of an episode of transient venous occlusion, labeled with ^{131}I and reinjected together with ^{125}I-fibrinogen from a noncryofibrinogenemic donor. The curves obtained (Fig. 10) showed that, despite a continuous heparin therapy (450 units/kg/day) throughout the observation period, both labeled fibrinogens were catabolized at markedly

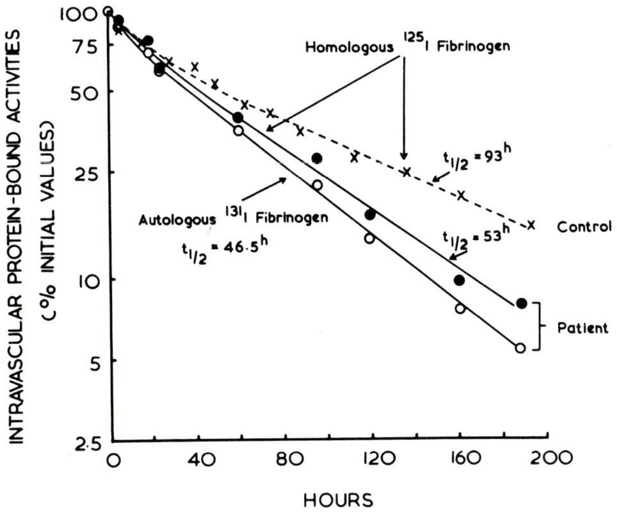

FIG. 10. Catabolism of autologous ^{131}I-fibrinogen and homologous ^{125}I-fibrinogen in a cryofibrinogenemic patient under continuous therapy with heparin. The dotted curve shows the behavior of the same ^{125}I-fibrinogen preparation in a control subject.

elevated rates, the autologous protein perhaps even more so than the homologous one. Whereas this discrepancy remains to be explained, the shortened survival of the homologous fibrinogen in this patient strongly suggests that cryofibrinogenemia is caused by secondary alterations to the fibrinogen molecule in a pathological milieu rather than by a structurally different fibrinogen.

REFERENCES

Adelson, E. (1965). *Fed. Proc. Fed. Amer. Soc. Exp. Biol.* **24**, 810.
Amris, A., and Amris, C. J. (1964). *Thromb. Diath. Haemorrh.* **11**, 404.
Atencio, A. C., and Reeve, E. B. (1965). *J. Lab. Clin. Med.* **66**, 20.
Atencio, A. C., Bailey, H. R., and Reeve, E. B. (1965). *J. Lab. Clin. Med.* **66**, 1.
Atencio, A. C., Joiner, K., and Reeve, E. B. (1969a). *Amer. J. Physiol.* **216**, 764.
Atencio, A. C., Chao, P.-Y., Chen, A. Y., and Reeve, E. B. (1969b). *Amer. J. Physiol.* **216**, 773.
Baker, L. R. I., Rubenberg, M. L., Dacie, J. V., and Brain, M. C. (1968a). *Brit. J. Haematol.* **14**, 617.
Baker, L. R. I., Brain, M. C., and Regoeczi, E. (1968b). Unpublished observations.
Brönnimann, R. (1954). *Acta Haematol.* **11**, 40.
Chanutin, A., Hortenstine, J. C., Cole, W. S., and Ludewig, S. (1938). *J. Biol. Chem.* **123**, 247.
Chapman-Andresen, C. (1962). *C. R. Trav. Lab. Carlsberg* **33**, 73.
Chapman-Andresen, C. (1964). *Methods Cell Physiol.* **1**, 277.
Cohen, C., Slayter, H., Goldstein, L., Kucera, J., and Hall, C. (1966). *J. Mol. Biol.* **22**, 358.

Davies, J. W. L., Ricketts, C. R., and Bull, J. P. (1966). *Clin. Sci.* **30**, 305.
de Duve, C. (1963). *Lysosomes, Ciba Found. Symp.* p. 126.
de Duve, C., and Wattiaux, R. (1966). *Annu. Rev. Physiol.* **28**, 435.
de Vries, A., Rosenberg, T., Kochwa, S., and Boss, J. H. (1961). *Amer. J. Med.* **30**, 486.
Donnelly, T. H., Laskowski, M., Jr., Notley, N., and Scheraga, H. A. (1955). *Arch. Biochem. Biophys.* **56**, 369.
Edsall, J. T., Gilbert, G. A., and Scheraga, H. A. (1955). *J. Amer. Chem. Soc.* **77**, 157.
Fletcher, F., Martin, L. E., and Ratcliffe, A. H. (1952). *Nature (London)* **170**, 319.
Flute, P. T. (1969). Personal communication.
Foster, D. P., and Whipple, G. H. (1921a). *Amer. J. Physiol.* **58**, 379.
Foster, D. P., and Whipple, G. H. (1921b). *Amer. J. Physiol.* **58**, 393.
Foster, D. P., and Whipple, G. H. (1921c). *Amer. J. Physiol.* **58**, 407.
Freeman, T., Gordon, A. H., and Humphrey, J. H. (1958). *Brit. J. Exp. Pathol.* **39**, 459.
Gitlin, D., and Borges, W. H. (1953). *Blood* **8**, 679.
Gordon, A. H., and Jacques, P. (1966). *In* "Labelled Proteins in Tracer Studies" (L. Donato, G. Milhaud, and J. Sirchis, eds.), pp. 127–132. Euratom, Brussels.
Gordon, A. H., and Koj, A. (1968). *Brit. J. Exp. Pathol.* **49**, 436.
Ham, T. H., and Curtis, F. C. (1938). *Medicine (Baltimore)* **17**, 413.
Hart, H. C. (1964). *Thromb. Diath. Haemorrh. Suppl.* **17**, 121.
Hasselback, R., Marion, R. B., and Thomas, J. W. (1963). *Can. Med. Ass. J.* **88**, 19.
Heim, W. G., and Ellenson, S. R. (1965). *Nature (London)* **208**, 1330.
Henriques, S. B., Henriques, O. B., and Levy, A. (1956). *Brit. J. Pharmacol.* **11**, 99.
Hickman, J. A. (1969). Personal communication.
Hill, R. G., Cox, G. E., and Wexler, B. C. (1966). *Fed. Proc. Fed. Amer. Soc. Exp. Biol.* **25**, 445.
Hoak, J. C., Wilson, W. R., Warner, E. D., Theilen, E. O., Fry, G. L., and Benoit, F. L. (1969). *J. Clin. Invest.* **48**, 768.
Holter, H. (1965). *Symp. Soc. Gen. Microbiol.* **15**, 89.
Holtzer, M., and Holtzer, S. (1960). *C. R. Trav. Lab. Carlsberg*, **31**, 373.
Izak, G., Galewski, K., and Eyal, Z. (1967). *Thromb. Diath. Haemorrh.* **18**, 544.
Jacques, P. (1966). *Rev. Quest. Sci.* **27**, 99.
Jacques, P. (1969). *In* "Lysosomes in Biology and Pathology" (J. T. Dingle and H. B. Fell, eds.), Vol. 2, North-Holland Publ., Amsterdam.
Jancsó, N. (1955). "Die Speicherung," p. 189. Akadémiai Kiadó, Budapest.
Johnson, A. J., and Merskey, C. (1966). *Thromb. Diath. Haemorrh. Suppl.* **20**, 161.
Jones, E. A. (1969). M.D. Thesis, p. 228. Univ. of Birmingham, Birmingham, England.
Koj, A. (1968). *Biochim. Biophys. Acta* **165**, 97.
Koj, A., and McFarlane, A. S. (1968). *Biochem. J.* **108**, 137.
Kropatkin, M. L., and Izak, G. (1968). *Thromb. Diath. Haemorrh.* **19**, 547.
Laki, K. (1968). *In* "Fibrinogen" (K. Laki, ed.), p. 6. Dekker, New York.
Laki, K., and Mihályi, E. (1949). *Nature (London)* **163**, 66.
Lasch, H. G., Róka, L., and Heene, D. (1966). *Thromb. Diath. Haemorrh. Suppl.* **20**, 97.
Lewis, W. H. (1931). *Johns Hopkins Hosp. Bull.* **49**, 17.
Lipinski, B., and Worowski, K. (1968). *Thromb. Diath. Haemorrh.* **20**, 44.
Lipinski, B., Wegrzynowicz, Z., Budzynski, A. Z., Kopec, M., Latallo, Z. S., and Kowalski, E. (1967). *Thromb. Diath. Haemorrh.* **17**, 65.
McCutcheon, M. (1961). *In* "Pathology" (W. A. D. Anderson, ed.), 4th Ed., p. 31. Mosby, St. Louis, Missouri.
McFarlane, A. S., Todd, D., and Cromwell, S. (1964). *Clin. Sci.* **26**, 415.
Matthews, C. M. E. (1957). *Phys. Med. Biol.* **2**, 36.
Merskey, C. (1968). *Brit. J. Haematol.* **15**, 523.

Merskey, C., and Johnson, A. J. (1966). *Thromb. Diath. Haemorrh. Suppl.* **21**, 555.
Merskey, C., Johnson, A. J., Kleiner, G. J., and Wohl, H. (1967). *Brit. J. Haematol.* **13**, 528.
Mihályi, E. (1954). *J. Biol. Chem.* **209**, 733.
Mihályi, E., and Godfrey, J. E. (1963). *Biochim. Biophys. Acta* **67**, 73.
Monkhouse, F. C., and Milojevic, S. (1960). *Amer. J. Physiol.* **199**, 1165.
Morrison, I. R. (1946). *Amer. J. Med. Sci.* **211**, 325.
Morrison, P. R., Edsall, J. T., and Miller, S. G. (1948). *J. Amer. Chem. Soc.* **70**, 3103.
Mosesson, M. W., and Sherry, S. (1966). *Biochemistry* **5**, 2829.
Mosesson, M. W., Alkjaersig, N., Sweet, B., and Sherry, S. (1967). *Biochemistry* **6**, 3279.
Mosesson, M. W., Colman, R. W., and Sherry, S. (1968). *New Engl. J. Med.* **278**, 815.
Mutschler, L. E. (1964). *Proc. Soc. Exp. Biol. Med.* **115**, 1024.
Phillips, L. L., Skrodelis, V., and Wolff, J. A. (1963). *Acta Haematol.* **30**, 244.
Pilgeram, L. O., and Pickart, L. R. (1968). *J. Atheroscler. Res.* **8**, 155.
Ravin, H. A., Seligman, A. M., and Fine, J. (1952). *New Engl. J. Med.* **247**, 921.
Reeve, E. B., Takeda, Y., and Atencio, A. C. (1967). *Protides Biol. Fluids Proc. Colloq.* **14**, 283.
Regoeczi, E. (1967). *Thromb. Diath. Haemorrh.* **18**, 276.
Regoeczi, E. (1970). *Clin. Sci.* **38**, 111.
Regoeczi, E., and Hobbs, K. R. (1969). *Scand. J. Haematol.* **6**, 175.
Regoeczi, E., and Stannard, B. A. (1969). *Biochim. Biophys. Acta* **181**, 287.
Regoeczi, E., Henley, G. E., Holloway, R. C., and McFarlane, A. S. (1963). *Brit. J. Exp. Pathol.* **44**, 397.
Ritzmann, S. E., Hamby, C., Cooper, R., and Levin, W. C. (1963). *Texas Rep. Biol. Med.* **21**, 262.
Sarcione, E. J., Bohne, M., and Krauss, S. (1965). *Fed. Proc. Fed. Amer. Soc. Exp. Biol.* **24**, 230.
Sasaki, T., Page, I. H., and Shainoff, J. R. (1966). *Science* **152**, 1069.
Scheraga, H. A., and Laskowski, M., Jr. (1957). *Advan. Protein Chem.* **12**, 1.
Shaeffer, J. R. (1963). Rep. No. UR-630, pp. 1–41. U.S. At. Energy Comm. Univ. of Rochester, Rochester, New York.
Shaeffer, J. R. (1964a). *Amer. J. Physiol.* **206**, 573.
Shaeffer, J. R. (1964b). *Amer. J. Physiol.* **206**, 1251.
Shainoff, J. R., and Page, I. H. (1960). *Circulation Res.* **8**, 1013.
Shainoff, J. R., and Page, I. H. (1962). *J. Exp. Med.* **116**, 687.
Shulman, S., and Ferry, J. D. (1950). *J. Phys. Chem.* **54**, 66.
Simone, J. V., Abildgaard, C. F., and Schulman, I. (1965). *New Engl. J. Med.* **273**, 1057.
Smith, R. T., and von Korff, R. W. (1957). *J. Clin. Invest.* **36**, 596.
Takeda, Y. (1966a). *Fed. Proc. Fed. Amer. Soc. Exp. Biol.* **25**, 445.
Takeda, Y. (1966b). *J. Clin. Invest.* **45**, 103.
Takeda, Y. (1967). *J. Lab. Clin. Med.* **69**, 624.
Takeda, Y., and Chen, A. Y. (1967). *J. Lab. Clin. Med.* **70**, 678.
Tata, J. R. (1969). *Gen. Comp. Endocrinol. Suppl.* **2**, 385.
Vars, H. M. (1930). *Amer. J. Physiol.* **93**, 554.
Verstraete, M., and Vermylen, J. (1968). *Thromb. Diath. Haemorrh.* **20**, 444.
Verstraete, M., Amery, A., and Vermylen, C. (1963). *Acta Clin. Belg.* **18**, 480.
Weimer, H. E., and Coggshall, V. (1967). *Can. J. Physiol. Pharmacol.* **45**, 767.
Werder, E. (1963). *Helv. Paediat. Acta* **18**, 208.
Winkert, J., Ball, R., Barone, A., Winkert, E., and Kleiner, I. S. (1964). *Fed. Proc. Fed. Amer. Soc. Exp. Biol.* **23**, 355.
Wycoff, H. D., Agee, J., and Parsons, J. (1959). *Proc. Soc. Exp. Biol. Med.* **100**, 859.

VIII

Lipoproteins

31

THE METABOLISM OF PLASMA LIPOPROTEINS

Terry Langer, Warren Strober, and Robert I. Levy*

THE MOLECULAR DISEASE BRANCH, NATIONAL HEART AND LUNG INSTITUTE, BETHESDA, MARYLAND; THE METABOLISM BRANCH, NATIONAL CANCER INSTITUTE, BETHESDA, MARYLAND; THE MOLECULAR DISEASE BRANCH, NATIONAL HEART AND LUNG INSTITUTE, BETHESDA, MARYLAND

*Present address: Department of Medicine, Columbia University College of Physicians and Surgeons, Presbyterian Hospital, New York, New York.

I. Introduction

The plasma lipoproteins comprise a unique group of macromolecules which serve to solubilize and transport cholesterol, glycerides, and phospholipids through the plasma from sites of lipid absorption and synthesis to sites of storage and catabolism. In the past decade, it has become apparent that lipoproteins are fundamentally involved in the pathogenesis of atherosclerosis and this has stimulated investigation into their structure and metabolism. Recently, the introduction of a new classification of hyperlipidemic states based on lipoprotein patterns instead of on the presence of increased concentrations of cholesterol and/or glycerides (Fredrickson *et al.*, 1967) has led to the differentiation of five clinically and metabolically distinct lipid transport disorders and has provided new insights into the function of the plasma lipoproteins.

Four lipoproteins can be identified in plasma by paper electrophoresis: chylomicrons; very low density lipoprotein (VLDL, pre-β- or α_2-lipoprotein); low density lipoprotein (LDL, β-lipoprotein); and high density lipoprotein HDL, α-lipoprotein). In this chapter we shall review the factors controlling the synthesis and catabolism of these plasma lipoproteins and the derangements of lipoprotein regulatory mechanisms responsible for pathologic elevations or depressions of plasma lipoprotein levels. A summary of the important physical properties of the lipoproteins is shown in Table I.

TABLE I

PHYSICAL CHARACTERISTICS OF PLASMA LIPOPROTEINS

Lipoprotein	S_f[a]	Density (g/ml)	Mobility[b]	Molecular weight	Size (Å)
Chylomicrons	>400	<0.95	Origin	10^3–$10^4 \times 10^6$	750–10,000
VLDL	20–400	0.95–1.006	Pre-β	5–10×10^6	300–800
LDL	0–20	1.006–1.063	β	2.1–2.6×10^6	215–220
	0–12	1.019–1.063			
HDL	—	1.063–1.21	α	200,000	75–100

[a] Lipoprotein flotation rate in Svedberg units (10^{-13} cm/sec/dyne/g) in a sodium chloride solution of density 1.063 g/ml (26°C).

[b] Paper electrophoresis.

II. Turnover Methodology

Lipoproteins may be labeled biosynthetically with ^{35}S- or ^{14}C-amino acids and ^{14}C- and ^{3}H-fatty acids, cholesterol, phospholipids, or glycerol.

However, the use of lipoproteins labeled in this manner for metabolic studies is complicated by significant reincorporation of the label resulting from the rapid turnover of lipoproteins. The use of labeled lipids as markers for intact lipoproteins has the additional disadvantage arising from the considerable exchange of lipid moieties from one lipoprotein to another *in vitro* and *in vivo*. *In vitro* labeling of the protein component of lipoproteins with radioiodine provides a nonreutilizable, nonexchangeable label which will reflect the metabolism of the intact lipoprotein. Of the techniques available for labeling peptides, the iodine monochloride method (McFarlane, 1958) and the electrolytic method (Rosa *et al.*, 1964) appear to be associated with minimal alteration of the native protein.

Special precautions must be taken during the isolation and labeling of lipoproteins to be used in metabolic studies. Since the lipoproteins are easily denatured, especially when purified and concentrated, one must be careful to avoid extremes of temperature (freezing irreversibly denatures lipoproteins), foaming, and prolonged contact with foreign surfaces. The ease with which lipoproteins may be denatured requires that the ratio of iodine atoms bound to lipoprotein molecules be kept below 1.0 and that the disodium salt of ethylenediaminetetraacetic acid (EDTA) be present at all times to act as an antioxidant. In addition, since the double bonds of unsaturated fatty acids are susceptible to iodination, one must provide labeling conditions which minimize this reaction, especially when one is attempting to label lipoproteins of low protein content, i.e., chylomicrons and VLDL.

It appears that most of the lipid-bound radioiodine is associated with the phospholipids of the lipoproteins (Langer, unpublished observations). Since phospholipids have a long plasma survival, a significant amount of radioactivity bound to lipoprotein lipid could alter the terminal portion of the plasma decay curve of the lipoprotein and lead to erroneous determinations of the fractional catabolic rate.

In order to be certain that the manipulations involved in the isolation and radioiodination of the lipoproteins have not resulted in significant denaturation, appropriate screening experiments must be performed in animals. The presence of significant amounts of denatured protein may also be ascertained by making careful urine collections and determining the ratio of the daily urinary radioactivity to the mean daily plasma radioactivity. If the U/P ratio thus determined does not remain constant throughout the study, one must suspect heterogeneous or contaminated preparations, the presence of rapidly degradable protein or incomplete urine collections. The U/P ratio should also provide an independent estimate of the fractional catabolic rate and must therefore approximate the value of this parameter calculated from the plasma decay curve.

III. Chylomicrons

A. Characterization

Chylomicrons may be defined as lipoproteins of S_f* >400 which arise from the gastrointestinal tract and serve to transport glycerides of dietary (exogenous) origin from the intestine into the lymph. They are the largest of the lipoproteins, ranging in size from approximately 750–10,000 Å with a molecular weight of 10^3–10^4 million. On paper and agarose gel electrophoresis they remain at the origin, while on starch and cellulose acetate they have α_2 mobility and migrate with VLDL. Chylomicrons consist predominantly of glycerides (80–95 %) with smaller amounts of phospholipid (3–15 %), cholesterol (2–12 %), and protein (0.5–2.5 %) (Zilversmit, 1965; Fredrickson *et al.*, 1967; Lossow *et al.*, 1969). The nature of this small protein component remains uncertain. Even after repeated washings by ultracentrifugation, several plasma proteins, including albumin and γ-globulin, can be identified immunologically, presumably adsorbed to the surface of the chylomicrons (Dole and Hamlin, 1962). Certain peptides characteristic of other lipoproteins may, however, be intrinsic components of the chylomicron since the peptides associated with HDL and LDL can be consistently isolated from delipidated chylomicrons (Fredrickson *et al.*, 1967). Furthermore, chylomicrons are not synthesized at all in patients with abetalipoproteinemia, who are presumably unable to synthesize the apoprotein of LDL (Levy *et al.*, 1966a), thus suggesting that LDL peptides are an essential component of chylomicrons.

Chylomicrons transport exogenous lipid from the sites of lipid absorption in the small intestine into the circulation via the thoracic duct. They are absent in the plasma of normal subjects after an overnight fast.

B. Synthesis

Chylomicrons are produced in the intestinal mucosa in response to the ingestion of dietary fat. In man, chylomicrons are normally absent in plasma after an overnight (12–16 hr) fast, even in subjects given diets consisting nearly entirely of fats. The *de novo* synthesis of chylomicron protein in the intestine of animals has been suggested by the studies of several investigators (Rodbell *et al.*, 1959; Isselbacher and Budz, 1963; Hatch *et al.*, 1966). The synthesis of chylomicrons in the rat intestine is inhibited by ethionine (Hyams *et al.*, 1966). Upon entering the blood, chylomicrons probably acquire additional protein at the expense of circulating lipoproteins of

* S_f refers to the rate of lipoprotein flotation in Svedberg units in a sodium chloride solution of density 1.063 g/ml (26°C). At this density, chylomicrons, VLDL, and LDL float, while HDL does not.

intestinal or hepatic origin either by exchange or adsorption. In the absence of certain gastrointestinal disorders which lead to delayed absorption of dietary fat, fasting chylomicronemia is usually a result of impaired chylomicron catabolism.

C. Catabolism

The uncertain nature of chylomicron protein and the preponderance of lipid in this macromolecule have precluded rigorous investigation of the metabolism of chylomicron proteins. When chylomicrons containing labeled glycerides are infused into humans, they disappear with a half-life of 5–15 min (Nestel, 1964). This catabolic pathway probably involves a hydrolytic reaction mediated by a group of lipolytic enzymes. Lipolytic activity is normally absent in fasting plasma but the intravenous administration of small amounts of heparin results in the appearance of these lipolytic enzymes in the plasma and this activity is often denoted postheparin lipolytic activity (PHLA).

The site of chylomicron catabolism is uncertain, but it appears that in several species, after an overnight fast, the major site of chylomicron removal may be the liver, while in animals allowed access to food, most of the chylomicrons may be removed by adipose tissue (Havel, 1965). Recent studies have suggested that in the absence of heparin, the liver plays a very minor role in the uptake of chylomicrons (Felts and Mayes, 1965; Mayes and Felts, 1967).

D. Regulation

1. *Abetalipoproteinemia*

The production of chylomicrons is dependent on the amount of dietary fat ingested and an intact absorptive mechanism. Inability to synthesize chylomicrons is a characteristic feature of subjects with abetalipoproteinemia; the intestinal mucosa is laden with lipids but in the absence of the LDL apoprotein, this lipid apparently cannot be transferred into the lymph in the form of chylomicrons (Farquhar and Ways, 1966).

2. *Type I Hyperlipoproteinemia* (*Hyperchylomicronemia*)

This disorder is characterized by fasting chylomicronemia and may be familial or acquired. In familial Type I hyperlipoproteinemia, there is a deficiency in lipolytic activity, specifically involving triglyceride lipase, one of the postheparin lipolytic enzymes (Greten *et al.*, 1969). Chylomicrons accumulate in the plasma and the glyceride level may exceed 10,000 mg/100 ml, often with accompanying hepatosplenomegaly, abdominal pain, lipemia retinalis, and eruptive xanthomata (Fredrickson *et al.*, 1967).

Severe insulinopenic diabetes mellitus (Bagdade *et al.*, 1967) and rarely myxedema (Porte *et al.*, 1966) may also be associated with hyperchylomicronemia and a low PHLA. Plasma inhibitors of PHLA have been postulated to underlie the chylomicronemia occasionally observed in acute pancreatitis (Greenberger *et al.*, 1966) and alcoholism (Kessler *et al.*, 1963).

Recently the Type I pattern has been reported in association with heparin-binding globulins in several patients with systemic lupus erythematosus, dysglobulinemias, and lymphoma (Glueck *et al.*, 1969a). If heparin or a heparin-like substance plays a functional role in the *in vivo* activation or stimulation of lipolytic activity, then globulins which bind heparin *in vitro* may prevent the elaboration of the lipolytic activity in the tissues or plasma. The administration of a pharmacologic dose of heparin to these patients will induce lipolysis and clearing of the chylomicronemia.

3. *Type V Hyperlipoproteinemia*

In Type V hyperlipoproteinemia, chylomicronemia is accompanied by elevated levels of VLDL and a low PHLA is sometimes found. The administration of progestational agents to some female patients with familial Type V hyperlipoproteinemia has been shown to increase PHLA and lower plasma triglyceride levels (Glueck *et al.*, 1969b), while patients with familial Type I are refractory to pharmacologic therapy.

IV. Very Low Density Lipoprotein (VLDL, Pre-β-Lipoprotein)

A. Characterization

VLDL comprises a family of macromolecules with a broad S_f range of 20–400, isolated between density 1.006 and 0.95 g/ml. They are roughly spherical particles with a diameter of 300–800 Å and a molecular weight of $5\text{–}10 \times 10^6$ (Lindgren and Nichols, 1960). As the density of VLDL increases, the relative protein content increases and the lipid content decreases (Gustafson *et al.*, 1965). The protein content ranges from approximately 6% in the lowest density fraction to 13% in the highest density fraction (Lossow *et al.*, 1969). The predominant lipid of VLDL is glyceride which comprises approximately 50–70% of the molecule by weight, with smaller quantities of phospholipid (15–25%) and cholesterol (10–25%). VLDL migrates with pre-β or α_2 mobility on most media but with decreasing density and increasing size they may mimic chylomicrons in composition and electrophoretic behavior. The apoproteins of LDL and HDL have been identified in VLDL (Levy *et al.*, 1966b; Gustafson *et al.*, 1965) and recent studies have confirmed the presence of at least three additional peptides in VLDL (Brown *et al.*, 1969; Shore and Shore, 1969). VLDL function primarily to transport glycerides of endogenous origin out of the liver, and to a lesser extent, the gastrointestinal tract. The concentration of VLDL in normal, fasting plasma is approximately 20–400 mg/100 ml.

B. Synthesis

Abundant evidence is available to demonstrate the synthesis of VLDL in the liver (Marsh, 1958; Marsh and Whereat, 1959; Radding *et al.*, 1958; Haft *et al.*, 1962; Ruderman *et al.*, 1968). In addition, the gastrointestinal tract may contribute smaller amounts of VLDL to the circulation (Rodbell *et al.*, 1959; Hatch *et al.*, 1966; Roheim *et al.*, 1966; Windmueller and Levy, 1968; Ockner *et al.*, 1969). Subjects with abetalipoproteinemia fail to synthesize VLDL, presumably because of the inability to synthesize LDL apoprotein, a constituent protein of VLDL. In the absence of a lipoprotein to transport them into the plasma, triglycerides accumulate in hepatic cells just as they do in the intestinal mucosa (Levy *et al.*, 1966a). In the rat, the synthesis or release of VLDL from the liver is inhibited by a diet containing 1% orotic acid; plasma VLDL (and LDL) become undetectable while intestinal synthesis and release of VLDL continue (Windmueller and Levy, 1967, 1968). This effect can be reversed by the addition of 0.25% adenine sulfate to the diet. Other substances which suppress protein synthesis, such as puromycin and actinomycin, will also decrease VLDL production (Marsh, 1963; Jones *et al.*, 1967; Faloona *et al.*, 1968).

The synthesis of VLDL can be stimulated by a variety of metabolic agents including the endogenous release of free fatty acids from adipose tissue (Steinberg and Vaughan, 1965), exogenous medium chain length fatty acids reaching the liver directly via the portal vein (Furman *et al.*, 1965), and carbohydrate feeding (Fredrickson *et al.*, 1967).

C. Catabolism

The sites and mechanisms for the removal of VLDL from the circulation remain controversial. It is not clear whether the pathways for the catabolism of exogenous, chylomicron-glycerides and endogenous, VLDL-glycerides are identical. VLDL can be metabolized by the isolated perfused rat heart (Scheuer and Olson, 1967), adipose tissue (Markscheid and Shafrir, 1965), and skeletal muscle (Olivecrona and Belfrage, 1965). The role of the liver in the removal of endogenous glyceride may be of greater importance in the fasting state than in the fed state when adipose tissue may be the predominant site of glyceride uptake (Havel *et al.*, 1962). Other studies have concluded that the liver is relatively unimportant in the catabolism of VLDL (Friedberg and Estes, 1964). The role of postheparin lipolytic enzymes in the catabolism of VLDL is also uncertain.

The metabolism of the protein moiety of VLDL (S_f 10–400) labeled *in vitro* with ^{131}I has been studied by Gitlin *et al.* (1958) in a small number of normal and nephrotic subjects. The striking findings were the very rapid disappearance of radioactivity in VLDL of S_f 10–400 and the appearance of label in LDL of S_f 0–10. No conversion to lipoproteins of higher density was noted. This latter observation is surprising in view of the presence of the apoproteins of HDL in VLDL. The conversion of VLDL protein to LDL protein has also been noted in recent studies of LDL turnover (Walton *et al.*, 1965). Gitlin did not estimate the half-life of survival or the fractional catablic rate of VLDL, but within 8 hr following injection of labeled VLDL to normal subjects, only 30% of the label remained in the VLDL, while approximately 40% had been converted to LDL. If VLDL is predominantly intravascular, as one would expect from its large size, then the half-life of plasma survival of about 6–12 hr, demonstrated by Gitlin *et al.*, would correspond to a fractional clearance rate for intact VLDL of approximately 100–250% of the intravenous pool per day. The labeled LDL derived from VLDL catabolism was metabolized with a half-life and fractional catabolic rate similar to that of LDL labeled *in vitro*, suggesting that a least a portion of circulating LDL is derived from the breakdown of VLDL. Interconversions between VLDL and LDL can also be induced by the administration of heparin *in vivo* and *in vitro* (Shore and Shore, 1962; Nichols *et al.*, 1968; Lossow *et al.*, 1963).

D. Regulation

The plasma concentration of VLDL is ultimately determined by the balance of synthesis and catabolism. The inflow of free fatty acids, carbohydrates, and other precursors of hepatic triglyceride synthesis appears to stimulate hepatic VLDL production. The synthetic and catabolic rates of VLDL have been estimated by the measurement of VLDL–triglyceride turnover using glycerol (Farquhar *et al.*, 1965; Reaven *et al.*, 1965), fatty acid labeled VLDL (Frank, 1969; Eaton *et al.*, 1969), or by continuous infusion studies with labeled fatty acids (Havel, 1969; Ryan and Schwartz, 1965). Conflicting results have appeared for the magnitude of VLDL–glyceride synthesis and the fractional turnover of VLDL, presumably because of the rapid exchange of glycerides between VLDL and higher density lipoproteins, recycling of labeled fatty acids, partially denatured or altered VLDL preparations and incorrect assumptions regarding the relationship of the initial portion of the VLDL turnover curves to the fractional catabolic rate of VLDL. It is apparent that additional investigation of VLDL turnover using peptide-labeled VLDL will circumvent some of these difficulties.

1. *Type IV Hyperlipoproteinemia*

This disorder is characterized by fasting hypertriglyceridemia due to increased concentrations of VLDL. It may be familial or secondary to many disease states (Fredrickson *et al.*, 1967). Type IV individuals often demonstrate a marked rise in VLDL when an excessive amount of carbohydrate is provided in the diet. Frank (1969) and Havel (1969) have suggested that the persistence of abnormal VLDL concentrations in Type IV hyperlipoproteinemia results primarily from a defect in the removal of VLDL from the plasma, although Havel's data suggested that the Type IV patients also tended to have somewhat higher synthetic rates than normal. In contrast to these findings, Reaven *et al.* (1965) have concluded that increased production of VLDL is the primary mechanism responsible for the elevated level of VLDL in Type IV hyperlipoproteinemia and have suggested that hyperglycemia and hyperinsulinism are the stimuli for excessive VLDL synthesis (Reaven *et al.*, 1967). Although a large proportion of subjects with familial Type IV hyperlipoproteinemia have abnormal glucose tolerance, Levy and Glueck (1969) have shown that many of these individuals have normal or subnormal insulin secretion.

Using a multicompartmental model for VLDL–triglyceride metabolism, Frank (1969) also showed that the rise in plasma VLDL in two normal subjects fed a high carbohydrate diet apparently resulted from the stimulation of VLDL synthesis. Two Type IV subjects also demonstrated increased VLDL synthesis in response to a high carbohydrate intake and one of these subjects had a reduction in the fractional turnover rate of VLDL as well. Excessive VLDL synthesis may be responsible for the Type IV pattern often seen in patients with diabetic ketoacidosis, glycogen storage disease, and pheochromocytoma where increased lipolysis results in the release of excessive amounts of free fatty acids into the circulation.

The role of PHLA in the regulation of VLDL catabolism is conjectural. A depression of lipolytic activity has been suggested to be responsible for the modest hypertriglyceridemia seen in normal subjects receiving estrogen–progestin preparations (Hazzard *et al.*, 1969) and in uremia (Bagdade *et al.*, 1968). However, patients with familial Type I hyperlipoproteinemia, who have markedly diminished PHLA and absent triglyceride lipase activity, do not accumulate VLDL on a normal diet and display only modest elevations of VLDL on fat-free, high carbohydrate diets, while patients with familial Type IV hyperlipoproteinemia and elevated VLDL levels all have normal PHLA.

2. *Nephrotic Syndrome*

The nephrotic syndrome is frequently accompanied by hyperlipoproteinemia with elevations of VLDL and/or LDL (Baxter *et al.*, 1960). Studies of

VLDL turnover in nephrotic children by Gitlin *et al.* (1958), using protein-labeled VLDL, have demonstrated a decrease in the rate of conversion of VLDL to LDL. Other studies have suggested an increased synthesis of VLDL from the liver (Marsh and Whereat, 1959; Marsh, 1963; Radding and Steinberg, 1960). The defective conversion of VLDL to LDL may be related to the finding of decreased PHLA in nephrotics (Nikkilä and Grasbeck, 1954), while the stimulation of hepatic lipoprotein synthesis may be part of a generalized increase in the hepatic synthesis of plasma proteins in nephrosis (Marsh and Drabkin, 1960).

E. Summary

VLDL concentrations are usually maintained by the balance between the endogenous production of VLDL, in response to the demand for transport of glycerides of hepatic and gastrointestinal origin, and the clearance of VLDL and uptake of glycerides by adipose and other tissues. If there is an excessive flow of glyceride precursors to the liver and lipoprotein synthesis is unimpaired, then VLDL concentrations will increase. If VLDL synthesis is normal or increased and catabolism is defective, then elevated plasma VLDL levels will be manifest. At present, techniques for estimating VLDL turnover by measuring the removal of labeled VLDL-glycerides may be subject to errors in interpretation due to the instability of VLDL isolated *in vitro* and the lability of VLDL lipids. Havel (1969) has recently described a technique for the direct estimation of VLDL production by determination of the rate of *in vivo* incorporation of labeled free fatty acids into VLDL. This procedure, however, is still subject to the same reservations concerning the pathways of VLDL catabolism as any technique which utilizes VLDL labeled in the lipid moiety. In view of the importance of the lipoprotein apoproteins in the structure and metabolism of lipoproteins and the recent finding of additional peptides in VLDL, it would appear that peptide-labeled VLDL preparations will provide a significant amount of vital information on the catabolism and fate of VLDL and its protein constituents.

V. Low Density Lipoprotein (LDL, β-Lipoprotein)

A. Characterization

LDL can be isolated by preparative ultracentrifugation between densities 1.006 and 1.063 g/ml, corresponding to S_f 0–20, and has beta mobility on most electrophoretic media. Normally, most of the LDL fall in the S_f range of 0–12 (density 1.019–1.063) with a peak S_f of 6–7 (Lindgren *et al.*, 1969). The molecule is apparently a spherical particle of about 215–220 Å in diameter with a molecular weight of approximately 2.1–2.6 × 10^6 (Adams and Schumaker, 1969; Lindgren *et al.*, 1969). Cholesterol and cholesterol esters account for 40–45% of LDL by weight with the remainder consisting of approximately 20–25% protein, 25% phospholipid, and 10% glyceride (Bragdon *et al.*, 1956). In a normal subject, one-half to two-thirds of the plasma cholesterol circulates bound to LDL. The concentration of LDL in normal fasting plasma is approximately 200–400 mg/100 ml.

B. Synthesis

As previously mentioned, there is considerable evidence for the synthesis of immunochemically identifiable LDL apoprotein in the intestinal mucosa and liver. LDL apoprotein probably gains entry into the plasma as a component of VLDL. Intact LDL is then presumably released as a catabolic byproduct of VLDL metabolism. It is uncertain whether LDL is synthesized independent of VLDL by these organs. Particles approximately the size of VLDL have been identified in electron micrographs of hepatic cells from livers actively synthesizing lipoproteins and *in vitro* hepatic perfusion studies have demonstrated the *de novo* production of VLDL but not LDL (Ruderman *et al.*, 1968). Moreover, in orotic acid treated rats, hepatic production of VLDL and LDL are both markedly suppressed when the synthesis of the LDL apoprotein is impaired (Windmueller and Levy, 1967, 1968) and subjects with abetalipoproteinemia have no detectable VLDL or LDL in the plasma. The inability to move glycerides from the tissues into the plasma in the absence of LDL apoprotein has led to the concept that this protein is of great importance in the transport of endogenous and exogenous glyceride out of the cells.

Langer *et al.* (1969a) have reported a synthetic rate of approximately 15 mg/kg/day for the peptide moiety of LDL in normal subjects of both sexes, on low cholesterol, high polyunsaturated fat diets, corresponding to about 60 mg/kg/day of intact LDL. These data are similar to values for LDL synthesis in males reported by Walton *et al.* (1963, 1965) and Scott and Winterbourn (1967) but these investigators found synthetic rates to be about one-third higher in females. LDL synthesis does not appear to be stimulated by cholesterol feeding or restriction, or by the administration of cholestyramine which increases cholesterol turnover and stimulates cholesterol synthesis (Langer *et al.*, 1969b).

C. Catabolism

Studies of the turnover of LDL have revealed a biological $t_{1/2}$ of 2–4 days in normal subjects given LDL labeled with ^{35}S (Volwiler *et al.*, 1955) and ^{125}I or ^{131}I (Gitlin *et al.*, 1958; Walton *et al.*, 1963, 1965). Langer *et al.*

(1969a) have recently studied the turnover of autologous S_f 0–12 LDL isolated by preparative ultracentrifugation and radioiodinated with ^{125}I by the McFarlane method (McFarlane, 1958). In normal subjects on a low cholesterol diet supplemented with polyunsaturated fats, the half-life of LDL was 3.1 ± 0.5 days (mean $\pm$ SD) and the fractional catabolic rate, determined from analysis of the plasma decay curve according to Nosslin (1964) and from the urinary excretion of iodine (Campbell *et al.*, 1956) was $46.7 \pm 9.7\%$ of the intravenous LDL pool per day. The distribution of LDL could be simulated by a two-compartment model with about 70–80% of the protein restricted to the intravascular compartment. The labeled LDL was not converted to lipoproteins of other densities and no evidence for the reutilization of LDL apoprotein was obtained. The data of Langer *et al.* (1969a) suggest that in normal subjects the concentration of LDL is inversely related to the fractional catabolic rate. Since the synthetic rate of LDL is identical in normal subjects with LDL concentrations ranging from 160 to 300 mg %, it appears that the fractional catabolic rate is the factor which determines the plasma level of this lipoprotein. The rapid turnover of LDL suggests that it may be performing a transport function similar to other proteins such as haptoglobin and transferrin, which also have short half-lives and high fractional catabolic rates relative to other large plasma proteins.

The site and mechanism of LDL catabolism are unknown. There is no evidence that lipolytic enzymes are involved in the catabolism of LDL but the reticuloendothelial system may be active in removing LDL from the circulation. Reticuloendothelial cells have been reported to take up lipoproteins, probably by pinocytosis (Stein and Stein, 1967). Since cholesterol metabolism may be fundamentally related to lipoprotein turnover, it is conceivable that the major site of lipoprotein catabolism may be the liver, where synthesis as well as catabolism of cholesterol are known to occur. The proximity of this relationship would permit changes in cholesterol turnover to modulate lipoprotein metabolism.

D. Regulation

1. *Type II Hyperlipoproteinemia*

This disorder is characterized by an increase in plasma LDL concentration resulting in hypercholesterolemia with normal or only moderately elevated glyceride concentrations. Type II hyperlipoproteinemia is frequently familial and is apparently transmitted as an autosomal dominant trait. The heterozygote often develops tendon xanthomatosis and signs of coronary atherosclerosis before age 50 (Fredrickson *et al.*, 1967). Langer *et al.* (1969a) found that the synthetic rate of LDL apoprotein in patients with familial Type II hyperlipoproteinemia was normal (about 15 mg/kg/day); no abnormalities in the distribution of LDL between the plasma and the extravascular compartment were detected. The fractional catabolic rate, however, was markedly reduced to 21.9 ± 1.4% of the intravenous pool per day compared to the normal rate of 46.7 ± 9.7% and the half-life of survival was prolonged to 4.9 ± 0.2 days compared to the normal $t_{1/2}$ of 3.1 ± 0.5 days. This decrease in fractional catabolism could completely account for the elevation of the plasma pool of LDL. In contrast to these findings, Walton *et al.* (1963) and Scott and Hurley (1969) reported that the fractional catabolic rate of LDL was normal in several patients wuth hypercholesterolemia and a predominance of LDL of S_f 0–20, suggesting that an increase in the synthetic rate of LDL was responsible for the disorder. These patients differ from the familial Type II hyperlipoproteinemics studied by Langer *et al.* in that they all had substantially increased levels of VLDL (S_f 20–400) and all responded to clofibrate therapy, while the subjects in the study by Langer *et al.* did not have elevations of VLDL and were unresponsive to treatment with this agent.

A clear-cut reduction of plasma LDL concentration is seen in patients with familial Type II hyperlipoproteinemia following treatment with cholestyramine (Langer *et al.,* 1969b). This agent is an anion exchange resin which binds bile salts in the gastrointestinal tract, preventing their reabsorption and resulting in the stimulation of cholesterol synthesis and an increase in the rate of conversion of cholesterol to bile acids. Langer *et al.* (1969b) demonstrated that the cholestyramine-induced reduction in the plasma pool of LDL apparently results from a marked increase in the fractional catabolic rate of LDL and not from reduced synthesis or altered distribution of LDL. Similarly, the reduction of LDL concentration in normal individuals and familial Type II heterozygotes that follows the restriction of dietary cholesterol intake and the substitution of polyunsaturated for saturated dietary fats appears to result from a significant increase in fractional catabolism of LDL (Langer *et al.*, 1969b). Walton *et al.* (1963) and Scott and Hurley (1969) observed a reduction in the synthesis of LDL and no change in fractional catabolic rate when their subjects with elevated LDL were treated with clofibrate.

2. *Type III Hyperlipoproteinemia*

This is an uncommon disorder characterized by the accumulation of an abnormal lipoprotein in the plasma which displays the electrophoretic mobility of normal LDL but has the lighter density and the ultracentrifugal properties of VLDL. This "floating β-lipoprotein" has a lipid composition which differs from LDL and VLDL; it is richer in cholesterol than normal VLDL and is laden with more glyceride than normal LDL. Subjects with Type III hyperlipoproteinemia usually present marked hypercholesterolemia and hypertriglyceridemia, tendon, tuberous, and planar xanthomata as well as accelerated atherosclerotic vascular disease. Although Walton *et al.* (1963) found that two subjects with Type III synthesized excessive amounts of LDL, the labeled lipoprotein preparations were heterogeneous, containing lipoproteins of S_f 20–400 (VLDL) as well as S_f 0–20 (LDL). This heterogeneity would probably be relatively insignificant in normal subjects and in patients with familial Type II, where the concentration of VLDL is quite small compared to LDL, but would lead to serious error in the interpretation of data from patients with Type III, since their plasma contains large amounts of lipoproteins of S_f 12–20 and 20–100.

The origin of the abnormal lipoprotein remains obscure. It might arise from abnormal catabolism of VLDL, or directly from the liver as an LDL molecule with some structural abnormality, perhaps in the protein moiety. If the former occurs, then it is conceivable that this lipoprotein persists in the plasma because of difficulty in the catabolism of an abnormal metabolite of VLDL or of a normal intermediary product of VLDL breakdown which usually has only a fleeting existence in plasma. In the latter case, one could postulate either overproduction of this abnormal LDL or a defect in its removal from plasma.

3. *Nephrotic Syndrome*

The elevated LDL of nephrosis may result from an increase in the hepatic synthesis of lipoproteins, especially VLDL, the immediate precursor of LDL. Gitlin *et al.* (1958) found normal or slightly increased LDL fractional catabolic rates in nephrotics, implying excessive synthesis as a cause of the elevated LDL concentrations, although only two of the six nephrotic children studied had elevated LDL levels.

4. *Dysglobulinemias*

The presence of excessive amounts of abnormal plasma proteins in patients with various dysglobulinemic states, such as multiple myeloma, cryofibrinogenemia, and Waldenström's macroglobulinemia may be associated with elevations of LDL or VLDL. The hyperlipoproteinemia is occasionally complicated by xanthomatosis. It has been suggested that these abnormal globulins form complexes with lipoproteins and retard their normal catabolism.

5. *Thyroid Disease*

Thyroid hormone is intimately involved in the metabolism of cholesterol and lipoproteins. The hyperlipoproteinemia of myxedema is usually manifest by an elevation of plasma LDL. Walton *et al.* (1965) found the synthetic rate of LDL in four subjects with hypothyroidism and hyperlipoproteinemia to be normal, but the fractional catabolic rate of LDL was reduced to 16.6% of the intravenous pool per day compared with the normal rate of 33.7% of the intravenous pool per day. This corresponded to an increase in the mean $T_{1/2}$ from 2.4 days in normals to 6.0 days in the hypothyroid individuals. Conversely, two hyperthyroid subjects had shortened half-lives and increased fractional catabolic rates with normal synthetic rates for LDL.

E. Summary

LDL concentrations in normal individuals appear to be regulated primarily by the rate of LDL catabolism. Pathologic elevations of LDL may result from a defect in the catabolism of LDL or from excessive LDL synthesis. Familial Type II hyperlipoproteinemia represents a unique disorder in which elevated levels of a plasma protein appear to result from defective catabolism rather than increased synthesis. Since the catabolic rate of LDL is not fixed but can be increased by appropriate dietary or pharmacologic perturbation, it is possible that this catabolic disturbance is secondary to a resetting of the degradative rate of the enzymes responsible for the catabolism of LDL. The factors which influence the catabolic rate of LDL at the cellular level await further investigation but it appears likely that the rate of hepatic cholesterol turnover may be important in regulating the catabolism of LDL.

The function of LDL remains obscure. The inability to move glycerides from tissue into plasma in the absence of LDL apoprotein has led to the concept that this peptide is of fundamental importance in the transport of endogenous and exogenous lipid out of cells. While this role may be fulfilled when LDL apoprotein is an intrinsic component of VLDL, the actual function of intact LDL may be to rapidly dispose of residual cholesterol and phospholipids which remain after the glycerides of VLDL are removed.

VI. High Density Lipoprotein (HDL, α-Lipoprotein)

A. Characterization

HDL is the smallest of the lipoproteins with a molecular weight of about 200,000 and a diameter of 75–100 Å. This lipoprotein can be isolated between density 1.063 and 1.21 g/ml in the ultracentrifuge and migrates electrophoretically with the α-globulins. Recent evidence indicates some heterogeneity in HDL protein (Shore and Shore, 1968). This lipoprotein class is often divided into two subclasses: HDL_2 (density 1.085–1.12); and HDL_3 (1.12–1.21). With increasing density, the relative amount of protein and the ratio of esterified to free cholesterol increase, while the total cholesterol decreases (Glomset *et al.*, 1966). Average values for HDL composition are: 45–55% protein; 30% phospholipid; 18% cholesterol; and less than 5%

glyceride (Lindgren and Nichols, 1960). HDL levels are normally higher in females (250–650 mg/100 ml) than in males (125–425 mg/100 ml).

The function of HDL is uncertain. It appears to be intimately associated with a plasma enzyme, lecithin-cholesterol acyl transferase (LCAT) which apparently catalyzes the esterification of cholesterol by the transfer of an acyl group from lecithin to free cholesterol (Glomset, 1968). It has been suggested that the apoprotein of HDL is acquired by VLDL and chylomicrons when they enter the plasma, thereby facilitating the action of LCAT. Cholesterol ester is then presumably transferred to HDL or remains associated with LDL as VLDL is catabolized. HDL may then facilitate the transport of cholesterol ester to the liver or other sites of cholesterol catabolism. Some evidence in support of the role of HDL in the disposal of cholesterol ester is provided by subjects with Tangier disease who have a marked deficiency of normal HDL and accumulate cholesterol esters in the cells of the reticuloendothelial system, especially the tonsils, liver, and spleen (Fredrickson *et al.*, 1967). An alternate explanation for the deposition of cholesterol esters in Tangier subjects may be that the absence of normal HDL apoproteins leads to the production of an abnormal, unstable VLDL which is more susceptible to removal by the reticuloendothelial system. The VLDL of subjects with Tangier disease is abnormal to the extent that the absence of HDL apoproteins results in alteration of the electrophoretic mobility of VLDL yielding a lipoprotein which displays broad beta electrophoretic mobility, thus partially mimicking the pattern seen in Type III hyperlipoproteinemia.

B. Synthesis

The primary site of HDL synthesis appears to be the liver (Radding *et al.*, 1958; Radding and Steinberg, 1960; Marsh, 1963; Windmueller and Levy, 1967), although some synthesis of HDL may occur in hepatectomized, eviscerated dogs (Roheim *et al.*, 1966). In addition there is some evidence that HDL peptide is synthesized in the intestine, at least as part of the VLDL of intestinal origin (Windmueller and Levy, 1968). When hepatic LDL synthesis is blocked with orotic acid, VLDL and LDL levels fall rapidly and are usually undetectable within 7–10 days, while HDL concentrations are reduced only 40–50%. The peptides of HDL can also be detected in delipidated VLDL (Levy *et al.*, 1966b; Gustafson *et al.*, 1966) and the administration of heparin will acutely lower VLDL and increase HDL concentrations *in vivo* and *in vitro* (Shore and Shore, 1962; Nichols *et al.*, 1968; Lossow *et al.*, 1963; LaRosa *et al.*, 1969). The magnitude of the contribution of HDL apoproteins derived from VLDL breakdown to the pool of circulating HDL is not known. Gitlin *et al.* (1958) did not observe the conversion of VLDL to HDL using peptide-labeled VLDL and it may be that only acute heparin-induced intravascular lipolysis can release enough HDL from VLDL to alter plasma concentrations of HDL.

C. Catabolism

The half-life of radioiodinated HDL is 3–5 days in normal subjects (Gitlin *et al.*, 1958; Scanu and Hughes, 1962; Furman *et al.*, 1964) and no interconversion to other lipoproteins occurs. The fractional catabolic rate and distribution of HDL can be approximated from the data of Gitlin *et al.* and Scanu and Hughes by analysis of the plasma decay curves of labeled HDL. This procedure yields a fractional catabolic rate of approximately 25–35% of the intravenous pool per day with about 60% of the protein restricted to the vascular compartment. The synthetic rate of HDL apoprotein is then calculated to be approximately 10–20 mg/kg/day, strikingly similar to the synthetic rate of LDL apoprotein.

Furman *et al.* (1964) noted that in subjects with hyperchylomicronemia, there was apparently some association of labeled HDL with chylomicrons, especially newly synthesized chylomicrons rather than those already present in plasma. Moreover, they also noted that the half-life of HDL decreased from 3.3–3.8 days to 1.7–1.9 days in patients with hyperchylomicronemia and very low plasma HDL levels. Intermediate values for the half-life of HDL (2.5–2.7 days) were found in subjects with hyperglyceridemia, who displayed elevated VLDL and moderately depressed HDL concentrations. These data suggest that an increased fractional catabolic rate of HDL might be responsible for the reduced levels of HDL in hyperglyceridemic individuals (Types I, IV, and V of Fredrickson *et al.*, 1967).

Scanu and Hughes (1962) and Furman *et al.* (1964) observed that delipidated HDL apoprotein labeled with ^{131}I appeared to recombine with lipids *in vivo* and then acquired the ultracentrifugal properties and half-life characteristic of intact HDL.

The site of HDL catabolism is not known although the liver and other reticuloendothelial organs remain the leading candidates.

D. Regulation

The synthesis and catabolism of HDL, unlike LDL and VLDL, are apparently not markedly influenced by perturbations affecting cholesterol, carbohydrate, or free fatty acid metabolism. Markedly low levels of HDL are occasionally seen in patients with acute hepatitis or terminal hepatic failure and in patients with familial LCAT deficiency (Norum and Gjone, 1967). Familial elevations of HDL have not been described. The concentration of HDL is higher in women than in men of comparable age

(Fredrickson *et al.*, 1967) and this may be related to hormonal influences since estrogens or estrogen–progestin combinations may markedly elevate HDL concentrations by mechanisms which are unknown (Eder, 1965).

VII. Concluding Remarks

In this chapter we have discussed the areas of lipoprotein metabolism and turnover with special emphasis on the regulation of plasma lipoprotein levels. Table II summarizes the available data on the metabolism of lipoproteins. It is apparent that as a group, lipoproteins are rapidly catabolized and extremely active metabolically. Further research into the metabolism of the lipoprotein apoproteins and lipids will yield information which may further define the interactions of the lipoprotein classes and the mechanisms by which the metabolism of lipids and lipoproteins are interrelated.

TABLE II

TURNOVER OF PLASMA LIPOPROTEINS

Lipoprotein	Concentration (mg/100 ml) (fasting plasma)	$t_{1/2}$	Fractional catabolic rate (% IV pool/day)	Apoprotein synthetic rate (mg/kg/day)	% IV
Chylomicrons	None	5–15 min[a]	—	—	—
VLDL	20–400	6–12 hr[b]	100–250[b]	—	—
LDL	200–400	2–4 days	35–50	12–18	75
HDL	125–425 (males) 250–650 (females)	3–5 days	25–35[b]	10–20[b]	60[b]

[a] $t_{1/2}$ of glyceride turnover.
[b] Estimated from the literature (see text).

REFERENCES

Adams, G. H., and Schumaker, V. N. (1969). *Anal. Biochem.* **29**, 117.
Bagdade, J. D., Porte, D., Jr., and Bierman, E. L. (1967). *New Engl. J. Med.* **276**, 427.
Bagdade, J. D., Porte, D., Jr., and Bierman, E. L. (1968). *New Engl. J. Med.* **279**, 181.
Baxter, J. H., Goodman, H. C., and Havel, R. J. (1960). *J. Clin. Invest.* **39**, 455.
Bragdon, J. H., Havel, R. J., and Boyle, E. (1956). *J. Lab. Clin. Med.* **48**, 36.
Brown, W. V., Levy, R. I., and Fredrickson, D. S. (1969). *J. Biol. Chem.* **244**, 5687.
Campbell, R. M., Cuthbertson, D. P., Matthews, C. M. E., and McFarlane, A. S. (1956). *Int. J. Appl. Radiat. Isotop.* **1**, 66.
Dole, V. P., and Hamlin, J. T. (1962). *Physiol. Rev.* **42**, 674.
Eaton, R. P., Berman, M., and Steinberg, D. (1969). *J. Clin. Invest.* **48**, 1560.

Eder, H. A. (1965). *In* "The Pharmacological Basis of Therapeutics" (L. S. Goodman and A. Gilman, eds.), p. 754. Macmillan, New York.
Faloona, G. R., Stewart, B. N., and Fried, M. (1968). *Biochemistry* **7**, 720.
Farquhar, J. W., and Ways, P. (1966). *In* "The Metabolic Basis of Inherited Disease" (J. B. Stanbury, J. B. Wyngaarden, and D. S. Fredrickson, eds.), 2nd Ed., p. 509. McGraw-Hill, New York.
Farquhar, J. W., Gross, R. C., Wagner, R. M., and Reaven, G. M. (1965). *J. Lipid Res.* **6**, 119.
Felts, J. M., and Mayes, P. A. (1965). *Nature (London)* **206**, 195.
Frank, A. (1969). *In* "Proceedings of the 1968 Deuel Conference on Lipids" (G. Cowgill, D. L. Estrich, and P. D. Wood, eds.), p. 95. U.S. Gov. Printing Office, Washington, D.C.
Fredrickson, D. S., Levy, R. I., and Lees, R. S. (1967). *New Engl. J. Med.* **276**, 32.
Friedberg, S. J., and Estes, E. H., Jr. (1964). *J. Clin. Invest.* **34**, 129.
Furman, R. H., Sanbar, S. S., Alaupovic, P., Bradford, R. H., and Howard, R. P. (1964). *J. Lab. Clin. Med.* **63**, 193.
Furman, R. H., Howard, R. P., Brusso, O. J., and Alaupovic, P. (1965). *J. Lab. Clin. Med.* **66**, 912.
Gitlin, D., Cornwell, E. T., Nakasato, D., Oncley, J. L., Hughes, W. L., Jr., and Janeway, C. A. (1958). *J. Clin. Invest.* **37**, 172.
Glomset, J. A. (1968). *J. Lipid Res.* **9**, 155.
Glomset, J. A., Janssen, E. T., Kennedy, R., and Dobbins, J. (1966). *J. Lipid Res.* **7**, 638.,
Glueck, C. J., Levy, R. I., Glueck, H. I., Gralnick, H. R., Greten, H., and Fredrickson. D. S. (1969a). *Amer. J. Med.* **47**, 318.
Glueck, C. J., Levy, R. I., Brown, W. V., Greten, H., and Fredrickson, D. S. (1969b) *Lancet* **i**, 1290.
Greenberger, N. J., Hatch, F. T., Drummey, G. D., and Isselbacher, K. J. (1966). *Medicine (Baltimore)* **46**, 161.
Greten, H., Levy, R. I., and Fredrickson, D. S. (1969). *J. Lipid Res.* **10**, 326.
Gustafson, A., Alaupovic, P., and Furman, R. H. (1965). *Biochemistry* **4**, 596.
Gustafson, A., Alaupovic, P., and Furman, R. H. (1966). *Biochemistry* **5**, 632.
Haft, D. E., Roheim, P. S., White, A., and Eder, H. A. (1962). *J. Clin. Invest.* **41**, 842.
Hatch, F. T., Aso, Y., Hagopian, L. M., and Rubinstein, J. J. (1966). *J. Biol. Chem.* **241**, 1655.
Havel, R. J. (1965). *In* "Handbook of Physiology, Sect. V: Adipose Tissue" (A. E. Renold, and G. F. Cahill, Jr., eds.), p. 449. Waverley Press, Baltimore, Maryland.
Havel, R. J. (1969). *In* "Proceedings of the 1968 Deuel Conference on Lipids" (G. Cowgill, D. L. Estrich, and P. D. Wood, eds.), p. 117. U.S. Govt. Printing Office, Washington, D.C.
Havel, R. J., Felts, J. M., and Van Duyne, M. (1962). *J. Lipid Res.* **3**, 297.
Hazzard, W. R., Spiger, M. J., Bagdade, J. D., and Bierman, E. L. (1969). *New Engl. J. Med.* **280**, 471.
Hyams, D. E., Sabesin, S. M., Greenberger, N. J., and Isselbacher, K. J. (1966). *Biochim. Biophys. Acta* **125**, 166.
Isselbacher, K. J., and Budz, D. H. (1963). *Nature (London)* **200**, 364.
Jones, A. L., Ruderman, N. B., and Herrera, M. G. (1967). *J. Lipid Res.* **8**, 429.
Kessler, J. I., Kniffen, J. C., and Janowitz, H. D. (1963). *New Engl. J. Med.* **269**, 943.
Langer, T., Strober, W., and Levy, R. I. (1969a). *J. Clin. Invest.* **48**, 49a.
Langer, T., Levy, R. I., and Fredrickson, D. S. (1969b). *Circulation* **50**, Suppl. III, 14.
LaRosa, J. C., Brown, W. V., Levy, R. I., and Fredrickson, D. S. (1969). *Circulation* **50,** Suppl. III, 15.

Levy, R. I., and Glueck, C. J. (1969). *Arch. Intern. Med.* **123**, 220.
Levy, R. I., Fredrickson, D. S., and Laster, L. (1966a). *J. Clin. Invest.* **45**, 531.
Levy, R. I., Lees, R. S., and Fredrickson, D. S. (1966b). *J. Clin. Invest.* **45**, 63.
Lindgren, F. T., and Nichols, A. V. (1960). *Plasma Proteins* **2**, 1.
Lindgren, F. T., Jensen, L. C., Wills, R. D., and Freeman, N. K. (1969). *Lipids* **4**, 337.
Lossow, W. J., Naidoo, S. S., and Chaikoff, I. L. (1963). *J. Lipid Res.* **4**, 417.
Lossow, W. J., Lindgren, F. T., Murchio, J. G., Stevens, G. P., and Jensen, L. C. (1969). *J. Lipid Res.* **10**, 68.
McFarlane, A. S. (1958). *Nature (London)* **182**, 53.
Markscheid, L., and Shafrir, E. (1965). *J. Lipid Res.* **6**, 247.
Marsh, J. B. (1958). *Fed. Proc. Fed Amer. Soc. Exp. Biol.* **17**, 270.
Marsh, J. B. (1963). *J. Biol. Chem.* **238**, 1752.
Marsh, J. B., and Drabkin, D. L. (1960). *Metab. Clin. Exp.* **9**, 946.
Marsh, J. B., and Whereat, A. F. (1959). *J. Biol. Chem.* **234**, 3196.
Mayes, P. A., and Felts, J. M. (1967). *Biochem. J.* **105**, 180.
Nestel, P. J. (1964). *J. Clin. Invest.* **43**, 943.
Nichols, A. V., Strisower, E. H., Lindgren, F. T., Adamson, G. L., and Coggiola, E. L. (1968). *Clin. Chim. Acta* **20**, 277.
Nikkilä, E. A., and Grasbeck, R. (1954). *Acta Med. Scand.* **150**, 39.
Norum, K. R., and Gjone, E. (1967). *Biochim. Biophys. Acta* **144**, 690.
Nosslin, B. (1964). *In* "Metabolism of Human Gamma Globulin" (S. B. Andersen, ed.), p. 34. Blackwell, Oxford.
Ockner, P. K., Hughes, F. B., and Isselbacher, K. J. (1969). *J. Clin. Invest.* **48**, 2079.
Olivecrona, T., and Belfrage, P. (1965). *Biochim. Biophys. Acta* **98**, 81.
Porte, D., Jr., O'Hara, D. D., and Williams, R. H. (1966). *Metab. Clin. Exp.* **15**, 107.
Radding, C. M., and Steinberg, D. (1960). *J. Clin. Invest.* **39**, 1560.
Radding, C. M., Bragdon, J. H., and Steinberg, D. (1958). *Biochim. Biophys. Acta* **30**, 443.
Reaven, G. M., Hill, D. B., Gross, R. C., and Farquhar, J. W. (1965). *J. Clin. Invest.* **44**, 1826.
Reaven, G. M., Lerner, R. L., Stern, M. P., and Farquhar, J. W. (1967). *J. Clin. Invest.* **46**, 1756.
Rodbell, M., Fredrickson, D. S., and Ono, K. (1959). *J. Biol. Chem.* **234**, 567.
Roheim, P. S., Gidez, L. I., and Eder, H. A. (1966). *J. Clin. Invest.* **45**, 291.
Rosa, U., Scassellati, G. A., Pennisi, F., Riccioni, N., Giagnoni, P., and Giardani, R. (1964). *Biochim. Biophys. Acta* **86**, 519.
Ruderman, N. B., Richards, K. C., Valles, D. E., Bourges, V., and Jones, A. L. (1968). *J. Lipid Res.* **9**, 613.
Ryan, W. G., and Schwartz, T. B. (1965). *Metab. Clin. Exp.* **14**, 1243.
Scanu, A., and Hughes, W. L. (1962). *J. Clin. Invest.* **41**, 1681.
Scheuer, J., and Olson, R. E. (1967). *Amer. J. Physiol.* **212**, 301.
Scott, P. J., and Hurley, P. J. (1969). *J. Atheroscler. Res.* **9**, 25.
Scott, P. J., and Winterbourn, C. C. (1967). *J. Atheroscter. Res.* **7**, 207.
Shore, B., and Shore, V. (1962). *J. Atheroscler. Res.* **2**, 104.
Shore, B., and Shore, V. (1968). *Biochemistry* **7**, 2773.
Shore, B., and Shore, V. (1969). *Biochemistry* **8**, 4510.
Stein, O., and Stein, Y. (1967). *Lab. Invest.* **17**, 436.
Steinberg, D., and Vaughan, M. (1965). *In* "Handbook of Physiology. Sect. V: Adipose Tissue" (A. E. Renold and G. F. Cahill, Jr., eds.), p. 335. Waverley Press, Baltimore, Maryland.
Volwiler, W., Goldsworthy, P. D., MacMartin, M. P., Wood, P. A., Mackay, I. R., and Fremont-Smith, K. (1955). *J. Clin. Invest.* **34**, 1126.

Walton, K. W., Scott, P. J., Verrier Jones, J., Fletcher, R. F., and Whitehead, T. (1963). *J. Atheroscler. Res.* **3**, 396.
Walton, K. W., Scott, P. J., Dykes, P. W., and Davies, J. W. L. (1965). *Clin. Sci.* **29**, 217.
Windmueller, H. G., and Levy, R. I. (1967). *J. Biol. Chem.* **242**, 2246.
Windmueller, H. G., and Levy, R. I. (1968). *J. Biol. Chem.* **243**, 4878.
Zilversmit, D. B. (1965). *J. Clin. Invest.* **44**, 1610.

IX

Hormone Transport

32

REGULATION OF HORMONES BY BINDING PROTEINS

With Special Reference to the Transport of Thyroxine in the Circulation

Kenneth Sterling

PROTEIN RESEARCH LABORATORY, BRONX VETERANS ADMINISTRATION HOSPITAL, BRONX, NEW YORK, AND DEPARTMENT OF PATHOLOGY, COLUMBIA UNIVERSITY COLLEGE OF PHYSICIANS AND SURGEONS, NEW YORK, NEW YORK

I. Introduction

In considering hormone transport in the circulation, it seems reasonable to emphasize the existence of two distinct biological classes. The protein and peptide hormones are water soluble and evidently exist in the plasma without demonstrable interaction with other serum proteins. These hormones have a rapid metabolic turnover with a half-time in minutes. Therefore, it is not surprising to observe rather striking fluctuations in concentration in the blood within a few minutes in the case of all these peptide hormones, including parathyroid, thyrocalcitonin, growth hormone, FSH, LH, prolactin, ACTH, MSH, TSH, vasopressin, insulin, and glucagon (Table I).

TABLE I

SOLUBLE AND HYDROPHOBIC HORMONES

Soluble: Protein and peptide hormones	Hydrophobic: Small-molecule target hormones
Parathyroid	Cortisol
Thyrocalcitonin	Progesterone
Growth hormone	Estradiol
FSH, LH, prolactin	Testosterone
ACTH, MSH	Thyroxine
TSH	
Vasopressin	
Insulin	
Glucagon	

In contrast, the small-molecule "target" hormones, cortisol, progesterone, estradiol, testosterone, and thyroxine, are relatively hydrophobic and exist in aqueous solution in the blood by virtue of their firm binding by one or more serum protein carriers (Table I). The biologic half-times vary with the magnitude of the "free" or nonprotein-bound moiety. Thus cortisol, which is about 95% protein bound and 5% unbound, has approximately a 1-hr half-time, and thyroxine, more than 99.95% protein bound and less than 0.05% unbound, has a biologic half-time of turnover approximating 1 week. It is not unexpected, therefore, that the concentration of thyroxine ordinarily remains constant for long periods in health and in disease, although it may vary somewhat within a day or two in the postoperative period after major surgery. Cortisol can fluctuate markedly within hours and shows pronounced diurnal variation. Cortisol and thyroxine may be considered the extreme examples within the category of the protein-bound hormones of small molecular size; however, even the fastest on the list, cortisol, is sluggish compared to the turnover of the peptide hormones. It should be noted that aldosterone, which is appreciably less firmly bound than cortisol, has, as expected, a faster turnover than cortisol.

II. Significance of Thyroxine and Triiodothyronine

The thyroid hormones, thyroxine (T4) and triiodothyronine (T3), differ strikingly with regard to the onset and duration of their various physiologic and biochemical effects. This is exemplified by the time sequences of the

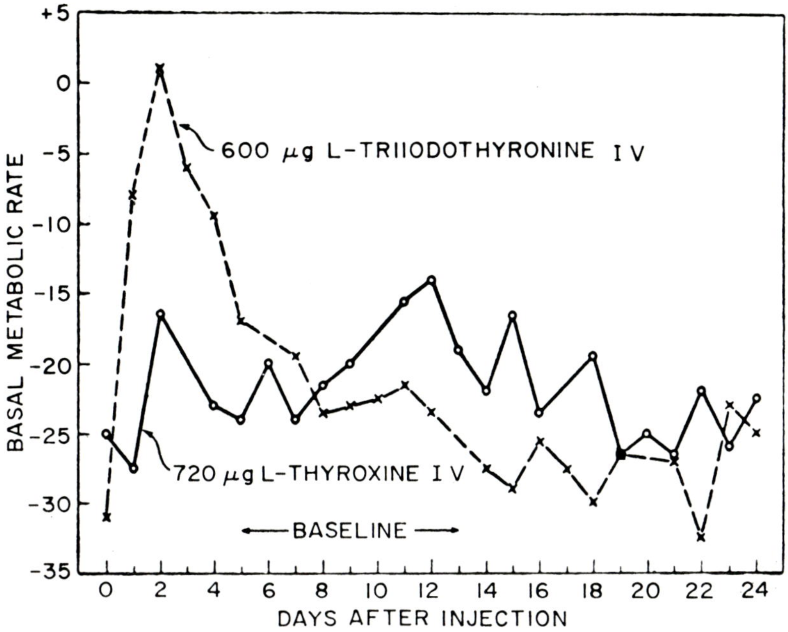

FIG. 1. Basal metabolic rates after injection of equimolar amounts of T3 and T4, into an untreated myxedematous patient. In this series of eight myxedematous patients, the peak calorigenic effect was (mean $\pm$ SD) 2.2 ± 0.1 days after intravenous injection of T3 and 9.4 ± 0.9 days after intravenous injection of T4 (from Blackburn *et al.*, 1954, p. 821).

calorigenic responses to equimolar amounts of T4 and T3 after single large intravenous doses were administered to myxedematous patients (Fig. 1). In the plot of basal metabolic rate (BMR) against time (in days), it is evident that T3 attained peak calorigenic action in 2 days while T4 required approximately 12 days in this study; in this series, (Blackburn *et al.*, 1954) carried out at the Mayo Clinic, a mean of 9 days was observed with no difficulties.

The exponential disappearance curves (Sterling *et al.*, 1954) of serum radioactivity after intravenous administration of ^{131}I-labeled T4 and T3 are illustrated in Fig. 2. The pronounced difference in the rates of disappearance may be largely attributed to the firmer binding of T4 by the serum proteins. Thyroxine has an affinity several times as great as that of T3 (Robbins and Rall, 1955, 1957, 1960; Ingbar and Freinkel, 1960a,b; Deiss *et al.*, 1953; Larson and Albright, 1955).

Thyroxine is the most abundant form of circulating hormone; T3 has been demonstrated in minute amounts. The T3 concentration in normal serum approximates one-thirtieth that of T4 (Sterling *et al.*, 1969). The simplified

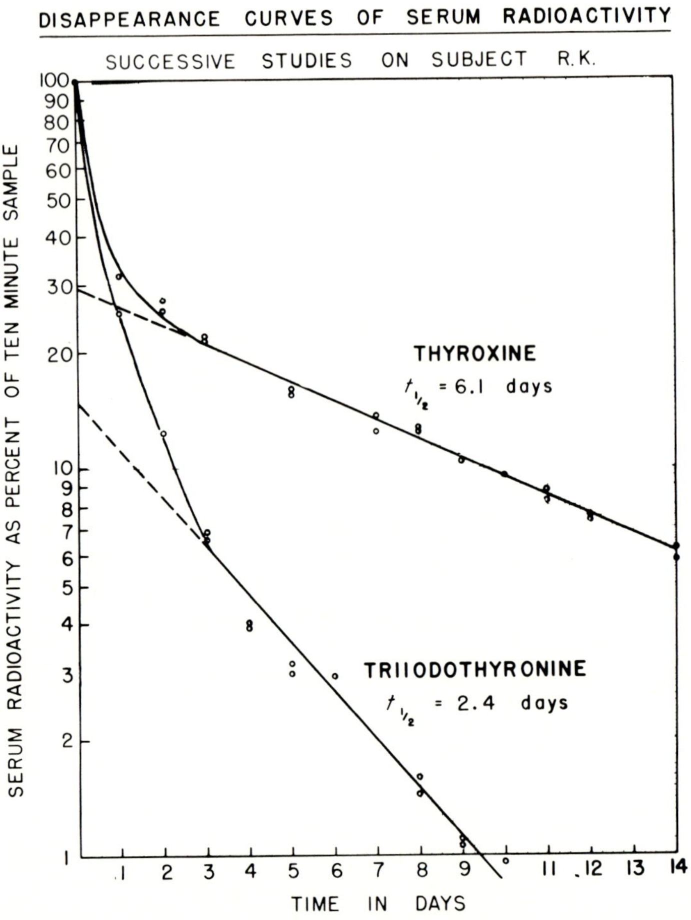

FIG. 2. Comparison of serum disappearance curves of T4 and T3. Successive studies of same subject. Semilogarithmic plots of serum radioactivity after intravenous injection of the labeled tracers (from Sterling *et al.*, 1954, p. 1033).

method recently developed is sufficiently sensitive for determinations on hypothyroid as well as normal and thyrotoxic sera. The values obtained have been in reasonable agreement with a double isotope derivative assay.

The normal T3 concentration in human serum approximates 0.2 μg/100 ml; the mean $\pm$ SD of 31 normal sera was 220 $\pm$ 27 nanograms (ng)/100 ml. Appropriate elevations were observed in sera from 40 patients with thyrotoxicosis (752 $\pm$ 282 ng/100 ml), and diminutions were found in sera from 10 hypothyroid patients (98 $\pm$ 48 ng/100 ml).

In rare instances thyrotoxicosis may be due to elevated serum T3 with normal thyroxine (T4) concentration. The incidence of this condition remains to be determined.

In approximately half the cases with low serum T4 after ^{131}I therapy, the eumetabolic state may be maintained by normal or elevated T3 concentration.

From these data and kinetic studies indicating a rapid turnover it may be inferred that T3 rather than T4 may be the more important hormone in health and in disease.

The extrathyroidal conversion of thyroxine (T4) to triiodothyronine (T3) has been demonstrated in athyreotic and hypothyroid human subjects on long-term oral maintenance therapy (Braverman, Ingbar, and Sterling, 1970).

This monodeiodination was not peculiar to the oral route of administration but was indeed a function of extrathyroidal metabolism, as shown by similar results after daily intravenous administration of stable and radioactive T4 over the course of one week.

The conversion of T4 to T3 has also been demonstrated in euthyroid volunteers employing thyroxine labeled with ^{14}C in Ring A and in the alanine side chain. The subjects received intravenous injections of purified T4-^{14}C, 1.4 μCi (8.1 μg T4), daily for 10 consecutive days. Evidence for the conversion of T4 to T3 was provided by the finding of the ^{14}C label in the T3 fraction isolated from sera, with maximal specific activity values after 72 hours (Sterling, Brenner, and Newman, 1970).

The absence of artefactual conversion of T4 to T3 was assured by the addition of ^{3}H-labeled T4 tracer to sera prior to separation of the T3 fraction. The lack of significant T3-^{3}H ruled out *in vitro* conversion.

Serial determinations of increasing specific activity of T3-^{14}C, formed during the studies, permitted preliminary computations of the rate and fraction of conversion of T4 to T3. It appears from the earliest findings of work still in progress that appreciable conversion may require 3 days and that one-third or more of the circulating T3 may arise from T4 by extrathyroidal conversion. Indeed, current investigation raises once again the question as to whether T4 itself has a primary action or exerts its effect only after transformation to T3 in the peripheral tissues.

T3 is much less firmly bound than T4 by the thyroxine binding α-globulin (TBG), the major protein carrier of thyroid hormones (Sterling, 1954, 1958, 1964a,b, 1968; Larson and Albright, 1955; Deiss *et al.*, 1953; Ross and Tapley, 1966). It is much less firmly bound by albumin (Sterling and Tabachnick, 1961b; Sterling *et al.*, 1962; Sterling, 1964b) and there is no detectable binding of T3 by the thyroxine-binding prealbumin (TBPA) (Ingbar, 1958, 1960, 1963).

Since T4 is more firmly bound than T3, it is not surprising that the diffusible (dialyzable or ultrafiltrable) fraction of this hormone is less than that of T3. The role of the free thyroxine fraction is discussed in detail below. Suffice it to say that the normal free T4 fraction approximates 0.04%

whereas the normal free T3 fraction is roughly 10 times greater (about 0.4%) (Ingbar *et al.*, 1965). The metabolism of T3 is, consequently, less influenced by variations in TBG, a subject currently undergoing intensive study. The remainder of the chapter will deal principally with the subject of T4 transport, kinetics, and protein binding.

III. Thyroxine Kinetics

If ^{131}I-labeled T4 is injected intravenously, the serum radioactivity falls rapidly for 2 or 3 days owing to distribution of the injected tracer throughout the body's extrathyroidal pool (Fig. 3). The rapid fall is followed by an exponential component (i.e., a linear disappearance on semilogarithmic plot); the slope of the slow component reflects the fractional rate of turnover of T4 in the steady state (Sterling, 1958; Sterling and Chodos, 1956). The

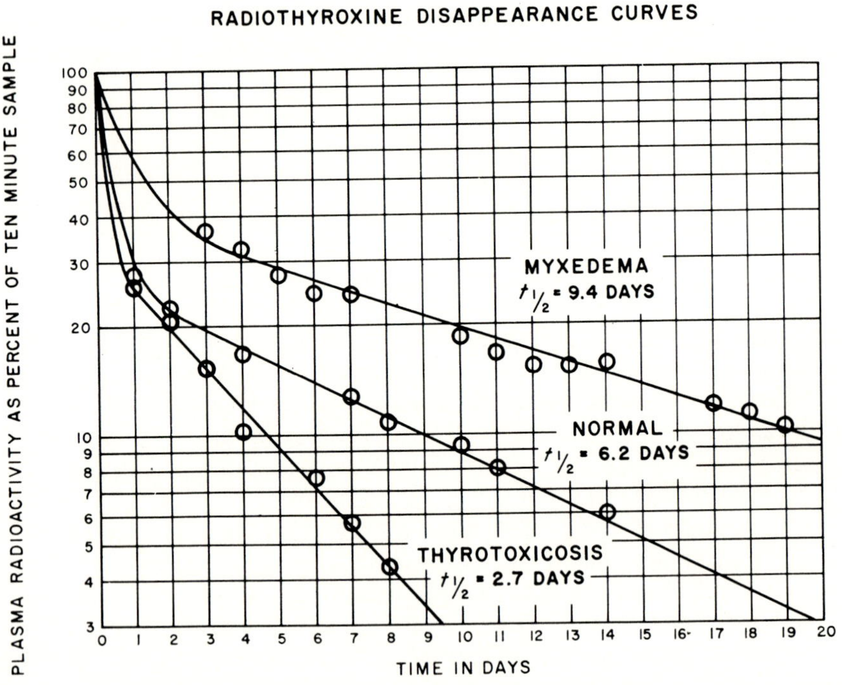

FIG. 3. Comparison of serum radiothyroxine disappearance curves in thyroid disease. Semilogarithmic plots of serum radioactivity after intravenous injection of ^{131}I-labeled T4. The slow turnover in myxedema and the accelerated turnover in thyrotoxicosis are clearly evident. Nonendocrine hypermetabolism, such as acute leukemia, also showed accelerated turnover (from Sterling and Chodos, 1956, p. 807).

disappearance is accelerated in thyrotoxicosis and is slower than normal in myxedema (Fig. 3) (Sterling and Chodos, 1956; Sterling, 1958). From the biologic half-time, obtained graphically, the turnover rate can be computed. The extrathyroidal pool is estimated by the isotope-dilution formula.

According to distribution studies (Oppenheimer *et al.*, 1967) using ^{125}I-labeled albumin and ^{131}I-labeled T4, in normal subjects after 6 hr about one-third of the labeled T4 remains in the plasma, one-third has been taken up in the liver, and one-third is in the extravascular spaces.

Nevertheless, the simplified model of a single compartment to represent the extrathyroidal pool of T4 does not appear to introduce appreciable error under most circumstances (Oppenheimer *et al.*, 1967). Studies in human subjects using ^{125}I-labeled T4 together with ^{131}I-labeled iodide have suggested a close relation between deiodination rate and hormone removal rate (Anbar *et al.*, 1965).

Figure 4 illustrates the values for the extrathyroidal organic iodine (EOI) pool, which is normally approximately 500 μg of T4 iodine (Sterling, 1964a). This is multiplied by the turnover of about 10%/day, and the product represents 50–65 μg of T4 iodine degraded or removed daily (corresponding to

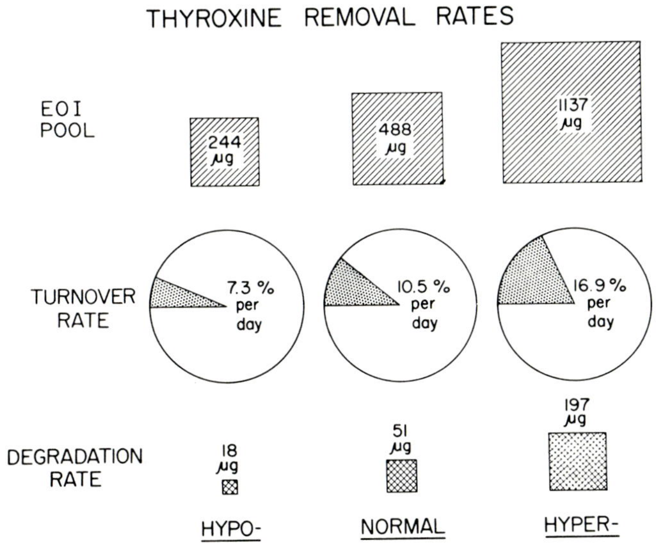

FIG. 4. Kinetics of T4 in the extrathyroidal body pool. The EOI pool refers to the "extrathyroidal organic iodine pool," signifying the body's total store of T4 outside the thyroid gland. The product of the EOI pool (in μg) and the turnover rate (in % per day) gives an estimated daily degradation or removal rate in μg of organic iodine. The normal mean values in various series have been from 50 to 65 μg of organic iodine, equivalent to 75–100 μg of T4, as the daily removal or degradation rate (from Sterling, 1964a, p. 590).

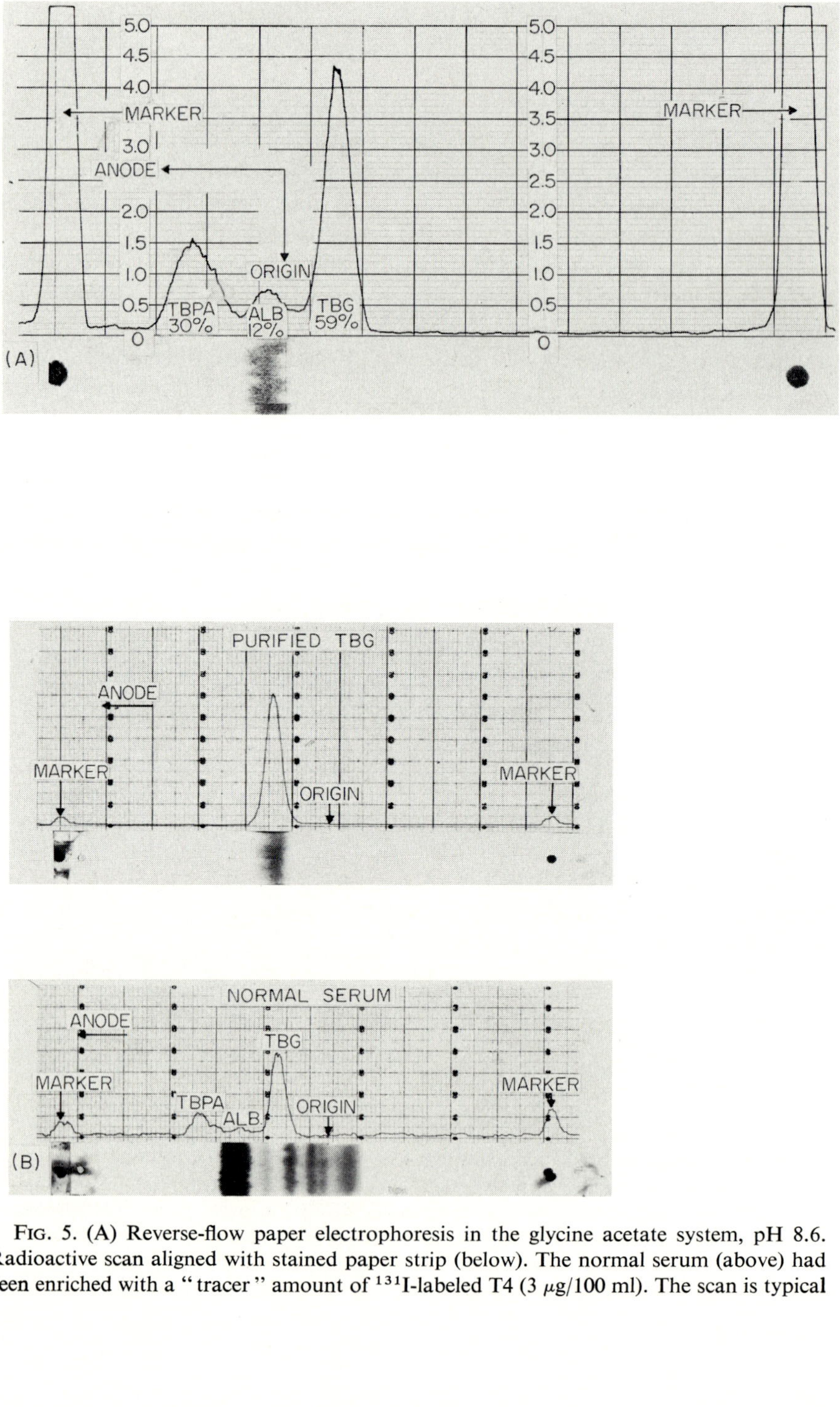

FIG. 5. (A) Reverse-flow paper electrophoresis in the glycine acetate system, pH 8.6. Radioactive scan aligned with stained paper strip (below). The normal serum (above) had been enriched with a "tracer" amount of ^{131}I-labeled T4 (3 μg/100 ml). The scan is typical

approximately 75–100 μg of T4. In thyrotoxicosis with an enlarged pool and a faster rate of turnover, the T4 removed per day is four times the normal (Fig. 4) or greater.

IV. Serum Protein Binding of Thyroxine

In contrast to the rapid removal of other amino acids from the circulation, the slow disappearance of T4 may be attributed to its firm binding by the serum proteins. Three principal carriers of the hormone have been observed, most accurately separated by reverse-flow paper electrophoresis in a glycine acetate system (Fig. 5). The binding affinity for T4 is greatest for the T4-binding alpha globulin (TBG), second for T4-binding prealbumin (TBPA), and weakest for albumin. When "tracer" amounts of labeled T4 (3 μg/100 ml or less) are added to normal human serum, approximately 60% is bound to TBG, 30% to TBPA, and 10% to albumin, with a glycine acetate system (Sterling and Tabachnick, 1961a; Oppenheimer *et al.*, 1963; Inada and Sterling, 1967b). The total binding capacities of these proteins, determined at saturation concentrations of T4, are in the reverse order. The maximal binding capacity (Oppenheimer *et al.*, 1963; Inada and Sterling, 1967b) of TBG is 21 μg of T4/100 ml of serum and that of TBPA is 260 μg/100 ml.

Although most studies employing tracer amounts of T4 have shown as much as one-third bound to TBPA, the actual portion of hormone bound by this carrier *in vivo* appears to be much less, probably not over 15% or only slightly greater than the amount bound by albumin. This has been shown by immunoprecipitation of TBPA by a rabbit antihuman prealbumin antiserum (Woeber and Ingbar, 1968) in which the amount of labeled hormone precipitated was usually less than 15%.

Additional confirmation has been provided by electrophoresis at pH 7.4 in which the pH throughout the system has been rigorously controlled and verified (Lutz and Gregerman, 1969).

Prealbumin has been isolated and crystallized; its molecular weight

---

of sera with "tracer" as opposed to "loading" or saturating concentrations of the hormone (from Inada and Sterling, 1967b, p. 1443). (B) Electrophoretic separation and radioactive scans of purified TBG and normal human serum. The scan of the purified TBG is above, with the bromphenol blue stained paper electrophoretic strip immediately beneath the scan; the stained protein band with mobility between that of α1- and α2-globulins is readily apparent. Below is the scan and stained paper strip of a simultaneously run normal human serum for comparison. The markers applied to the paper strips after electrophoresis are indicated, as well as the position of the origin, and the direction of the anode (to the left). In these studies the procedure was conventional (rather than reverse flow) paper electrophoresis in the glycine acetate system, pH 8.6.

approximates 73,000 and it is a glycoprotein (Oppenheimer *et al.*, 1965; Purdy *et al.*, 1965).

V. Molecular Characteristics of Thyroxine-Binding α-Globulin (TBG)

Thyroxine-binding α-globulin (TBG) is the major carrier of the hormone in human serum, yet its concentration evidently does not exceed 1 mg/100 ml. This protein has recently been isolated in pure form (Sterling, Hamada, Inada, and co-workers, to be published).

The TBG was isolated from Cohn Fraction IV-5,6 labeled with ^{125}I-thyroxine. The steps were (1) chromatography on CM cellulose, (2) gel filtration on Sephadex G-200, (3) chromatography on DEAE Sephadex, (4) a novel procedure of "double-gel" electrophoresis, and (5) preparative polyacrylamide gel electrophoresis. The first step, an overflow phenomenon, carried out at pH 5 allows the carboxymethyl cellulose column to retain about 97% of the protein of the starting material, while more than 70% of the TBG runs through with the initial buffer and is recovered.

Gel filtration with Sephadex G-200 achieves some further separation, as shown in Fig. 6. After this, a gradient elution on a DEAE Sephadex column with rising salt concentration from 0.13 to 0.21 *M* chloride produces further purification (Fig. 7). TBG was eluted mainly from 0.15 to 0.18 *M* chloride at pH 8. A very appreciable amount of inactive protein was retained on the column and could be eluted by alkaline solution of high ionic strength.

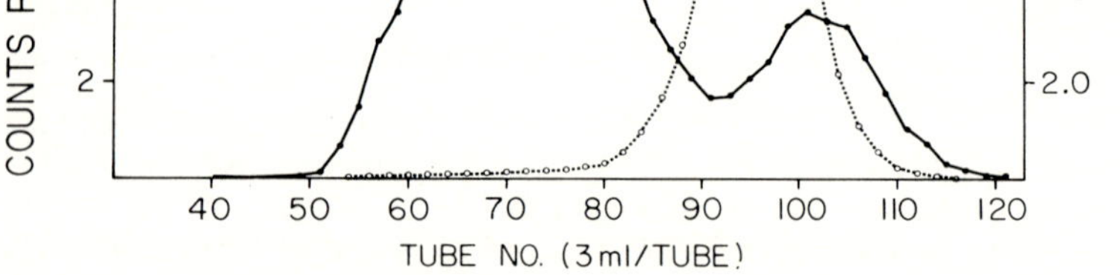

FIG. 6. Gel filtration with Sephadex G-200. The large initial protein peakwith maximum about tube 70 contains various globulins including the γ-globulins. The smaller protein peak with maximum around tube 100 is composed principally of human serum albumin. The peak radioactivity with maximum cpm usually about tube 96 or 97 contains virtually all the TBG as well as appreciable TBPA and albumin.

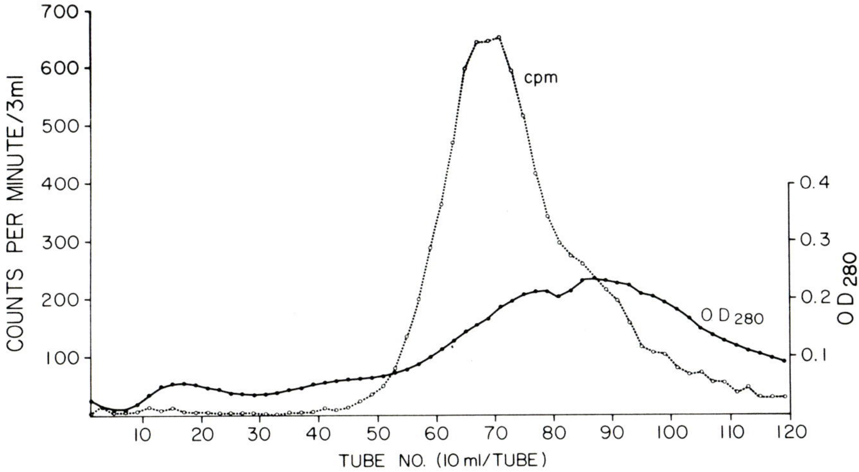

FIG. 7. DEAE Sephadex (A-50) chromatography. The gradient elution with rising salt concentration from 0.13 to 0.21 *M* chloride is illustrated. Most of the applied protein was retained on the column. Alkaline solution of high ionic strength could be employed subsequently for elution and recovery of virtually all of this protein material which contained insignificant radioactivity or T4 binding capacity.

Thereafter, fractionation is carried out by two polyacrylamide column electrophoreses, including a novel "double-gel" procedure which utilizes the negative charge TBG possesses at pH 4.5 and below. In this procedure an upper gel at pH 4.0 is placed above an ordinary pH 8.6 resolving gel. With an upper starting buffer of pH 4.5, TBG has sufficient negative charge to enter the upper gel whereas almost all other proteins are excluded and are recovered from the upper electrode buffer. The acidic buffer is then replaced by an ordinary pH 7.4 buffer, and electrophoresis yields principally TBG with minimal contamination. An additional conventional preparative polyacrylamide electrophoresis run yields a highly purified TBG preparation.

Figure 8 shows disc electrophoretic runs on polyacrylamide gel of some of the stages in purification; moving left to right the stages are whole serum, Fraction IV-5,6, DEAE Sephadex, the first electrophoresis (the so-called "double gel"), and finally the conventional polyacrylamide electrophoresis. The final preparation showed a single band.

Figure 9 shows the immunoelectrophoretic arc of purified TBG in comparison with arcs of whole serum. A single arc was observed in the best TBG preparations.

The final material was homogeneous by sedimentation analysis and diffusion.

The sedimentation constant was found to be $s_{20,w} = 3.00 \times 10^{-13}$ sec;

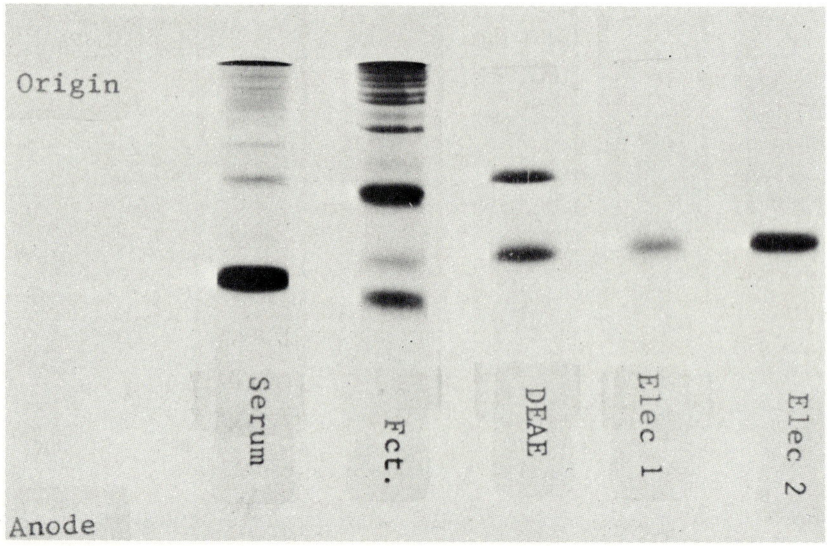

FIG. 8. Disc electrophoresis in polyacrylamide gel. Whole serum is illustrated at the left. Fraction IV-5,6, the starting material used in the present procedure, is designated as "Fct." and has different bands, as shown in this simultaneous run. Although human serum albumin should not be present in appreciable amounts, it is exceptional to obtain Fraction IV-5,6 or IV-4 without a visible albumin band. "DEAE" refers to the material after the stage of DEAE Sephadex column chromatography. Two major bands are visible. The faster (more anodal) band is not TBG but an inactive glycoprotein with alpha mobility. "Elec 1" refers to the preparation after "double gel" electrophoresis. For the first time, TBG is visible on staining with amido black. Although only a single band is clearly evident in this illustration, several additional bands could be seen with difficulty when the tube was carefully viewed against a strong light source. "Elec 2" shows the protein stain of purified TBG; only a single band could be seen by the unaided eye and scanning by Chromoscan has confirmed purity as great as 98%. Amounts of the order of 2% were occasionally discovered slightly anodal to the major band, but also containing radioactivity. Slower or faster bands in "Elec 1" were devoid of radioactivity.

diffusion constant $D_{20,w} = 8.05 \times 10^{-7}$ cm^2/sec frictional ratio $f/fo = 1.24$. Molecular weight calculated from sedimentation equilibrium was 36,650. The association constant for thyroxine was approximately 10^{10} by fluorescence quenching, and by competitive binding studies employing TBG and thyroxine binding prealbumin (TBPA). A summary of the molecular characteristics is given in Table II. Amino acid composition differed from previous reports, but shows some resemblance to the preparation of Seal and Doe (1964). Carbohydrate analyses were similar to other glycoproteins, but additional studies are pending.

The molecular weight mentioned above was somewhat greater than half those reported previously (Giorgio and Tabachnick, 1968; Seal and Doe,

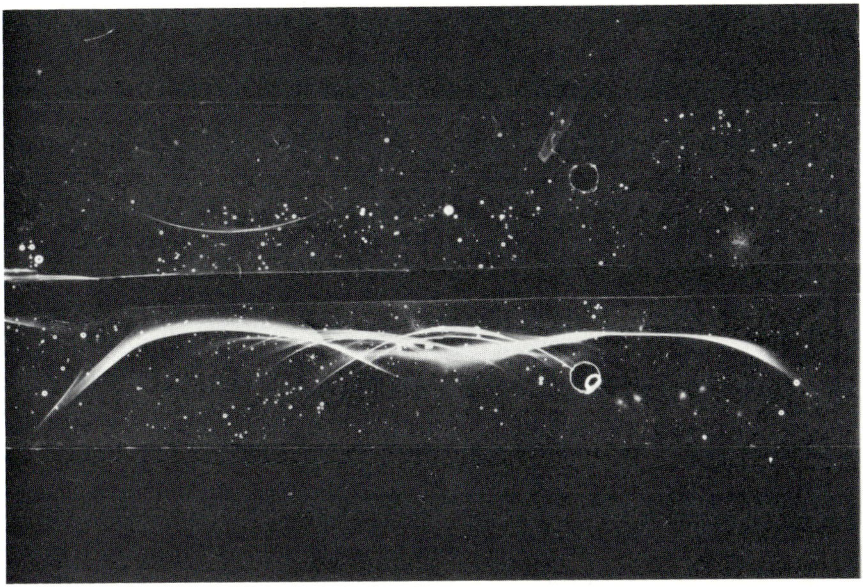

FIG. 9. Immunoelectrophoresis of purified TBG (above) in comparison with whole serum (below). The trough contained a potent antiserum against human serum.

1964). It raised a question whether this purified TBG could be a subunit of native TBG or not. It was shown that the purified TBG emerged exactly in the same position as that of serum TBG on Sephadex G-200 column chromatography. Therefore, it was apparent that the purified TBG was not a subunit of natural TBG.

The specific binding capacity of purified TBG preparations exceeded 2300 times that of the starting material, or more than 5300 times that of serum. Moreover, constant binding activity could be demonstrated through rechromatographed protein peaks, corroborating the other evidence of homogeneity.

Studies by reverse-flow paper electrophoresis suggested interaction between TBG and albumin (Fig. 10). For reference the radioactive scan of normal serum is shown with TBG in its usual position between α_1- and α_2-globulin. A partially purified TBG showed a slightly but significantly faster mobility. When this was diluted 1:3 with buffer, the mobility became greater than that of albumin. At this dilution with protein concentration of 56 μg/ml the maximal binding capacity was 25 μg T4/100 ml—slightly above that of normal serum. The same diluted preparation with an added equimolar amount of albumin showed the same mobility as that of TBG in serum. Further, thyroxine binding affinity estimated by competition with pure

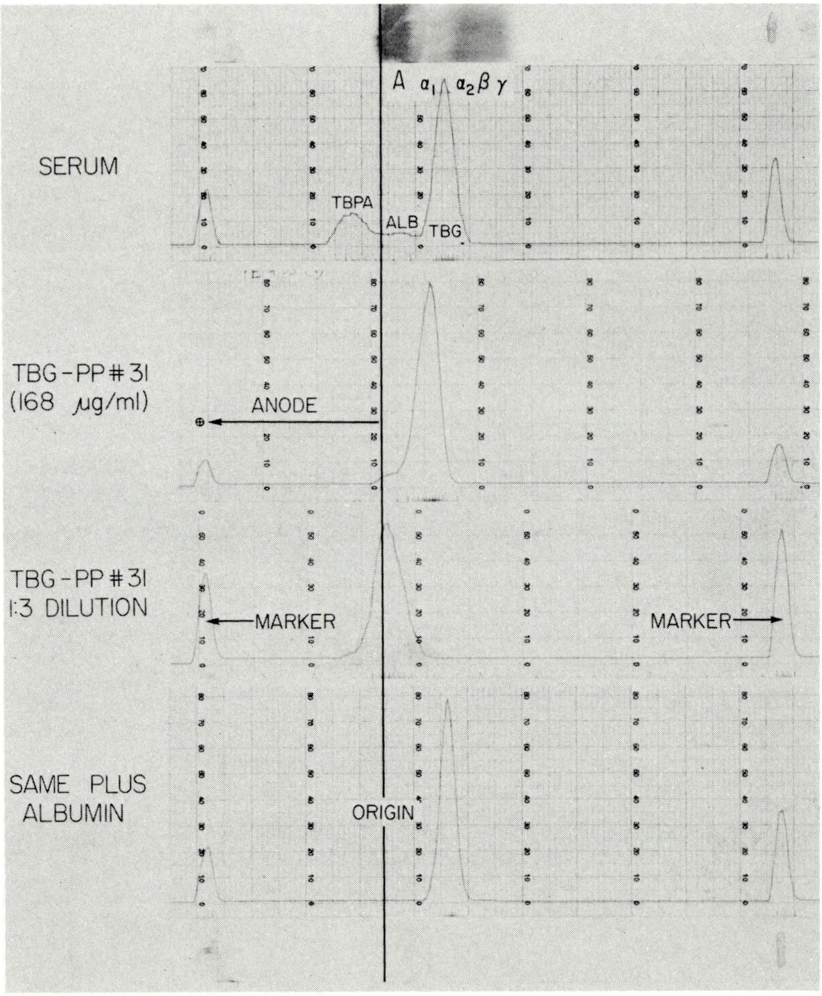

FIG. 10. The paper strip at the top, stained with bromphenol blue, is human serum subjected to reverse flow electrophoresis in the glycine acetate system and its radioactive scan is depicted immediately below. Below these are scans of a preparation of TBG illustrating the slightly faster mobility exemplified by PP 31 (protein concentration approximately 168 μg/ml). The same preparation diluted 1:3 exhibited even greater mobility with a peak slightly anodal to that of serum albumin. With the addition of human serum albumin to this diluted TBG, its mobility in the scan became identical with that of TBG in serum. The direction of the anode is indicated and radioactive protein markers are evident at each end of all strips, including the bottom strip in which no stained protein is evident except that of the markers applied to the dried paper strips after electrophoresis and before radioactive scanning.

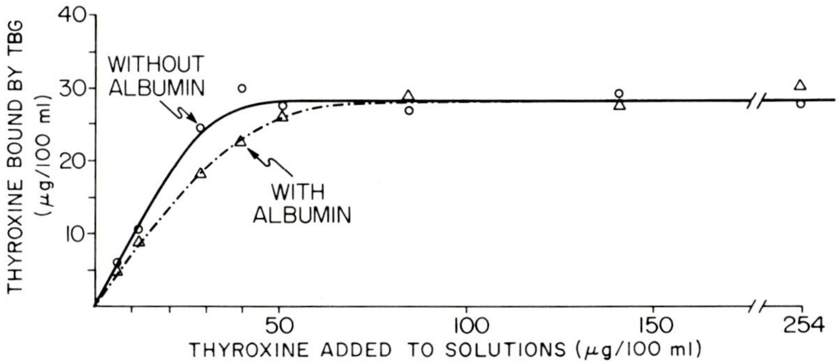

FIG. 11. TBG binding in the presence and absence of albumin.

TBPA indicated that TBG without albumin had 2–3 times higher affinity for thyroxine than that with added albumin, although maximal binding capacities remained the same (Fig. 11). It was inferred that the T4 binding affinity of TBG, and presumably its molecular structure, may be affected by protein–protein interactions.

VI. Thyroxine-Binding Proteins and Free Thyroxine

Dowling, Ingbar, and others (Engbring and Engstrom, 1959; Dowling *et al.*, 1956a, b, c, 1960; Robbins and Nelson, 1958; Nicoloff *et al.*, 1962) have reported that in pregnancy or during estrogen administration, the amount of TBG (or less likely, its binding affinity) increases, with a consequent rise of protein-bound iodine (PBI) concentration, despite maintenance of euthyroid status.

This rise is associated with a slower fractional turnover rate, as illustrated

TABLE II

MOLECULAR CHARACTERISTICS OF TBG

$s_{20,w}$	3.00×10^{-13} sec
$D_{20,w}$	8.05×10^{-7} cm^2/sec
M_{equil}	36,650
Stokes radius, a	2.64×10^{-7} cm
Frictional ratio, f/fo	1.24
Binding constants	$n = 1$, $k = 10^{10}$
Absorbance 280/260 mμ	1.24
Isoelectric point	pH 3.8

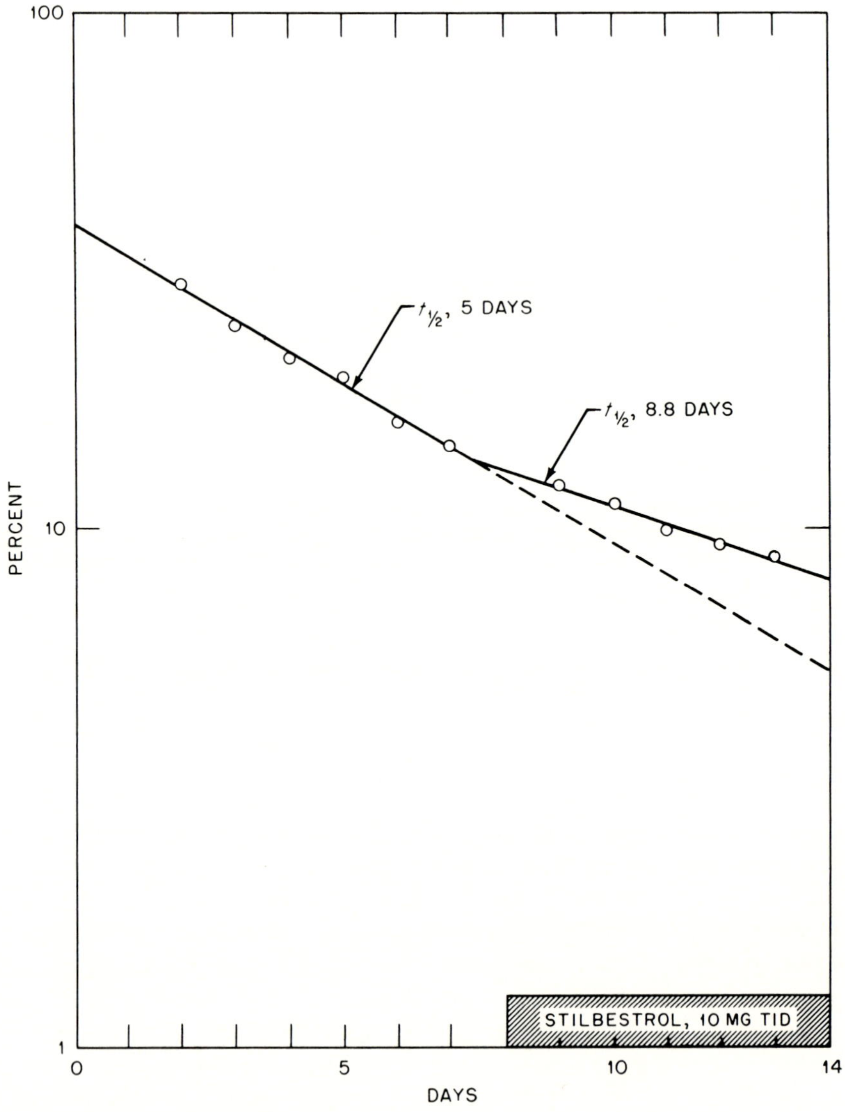

FIG. 12. Effect of estrogen upon serum radiothyroxine disappearance curve. Semilogarithmic plot of serum radioactivity after intravenous injection of ^{131}I-labeled T4. The slowing of the rate of disappearance soon after diethylstilbesterol therapy is clearly evident (from Engbring and Engstrom, 1959, p. 787).

in Fig. 12. In the radiothyroxine disappearance curve of Fig. 12, the administration of large doses of estrogen resulted in an abrupt change of slope (Engbring and Engstrom, 1959). Because of the increased PBI and pool, however, the absolute removal rate remained normal. The new equilibrium established after estrogen-induced rise in TBG does not require pituitary–thyroid regulation since identical results were found in hypothyroid patients on adequate replacement therapy (Engbring and Engstrom, 1959).

A phenomenon similar to the high TBG of estrogen administration has been demonstrated in the genetic abnormality of excessive TBG with PBI about twice normal (Table III). The initial instances observed in a father

TABLE III

GENETIC ABNORMALITIES OF TBG[a]

	Normal	Excess TBG	No TBG
PBI (μg/100 ml)	4.0–8.0	12–16	2.0
BEI (μg/100 ml)[b]	3.3–6.3	12–13	—
BMR	Normal	Normal	Normal
Cholesterol	Normal	Normal	Normal
^{131}I uptake	Normal	Normal	Normal
$T_{1/2}$ of T4 (days)	6.6	12.2	3.0
k (%/day)[c]	10.5	5.7	23.1
EOI pool (μg)	548	1099	264
Removal (μg/day)	57	63	61
In vitro uptake	Normal	Low	High

[a] "Excess TBG" data are from Beierwaltes and Robbins (1959). The "No TBG" data are from Ingbar (1961). "*In vitro* uptake" refers to T3 red cell uptake or T3 resin uptake.

[b] Butanol extractable iodine.

[c] Fractional turnover rate.

and daughter in Michigan (Beierwaltes and Robbins, 1959) were obviously euthyroid with normal BMR, serum cholesterol, and thyroidal uptake of radioiodine. The PBI values of about twice normal were associated with consequent doubling of the EOI pool, but the half-time of disappearance was also twice normal, that is, about 5% rather than 10% as the daily fractional turnover. The product of the smaller fractional turnover rate and the larger pool gave a normal quantitative removal rate. Of interest was the discovery by Dr. William Beierwaltes of a large number of relatives in the United Kingdom who were euthyroid and who had this disorder; a family member, an ornithologist, helped track down these rare birds. The converse situation of genetically absent TBG with consequently low PBI results in a smaller pool turning over more rapidly than normal and again a normal absolute removal rate and euthyroid status (Ingbar, 1961).

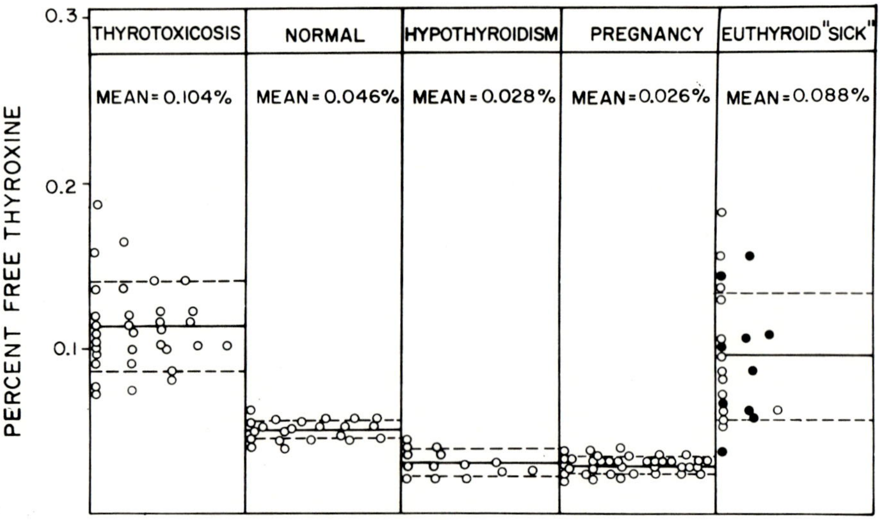

FIG. 13. Free T4 in serum as percentage of total T4 in health and disease. (—) mean values; (- - -) standard deviations; (●) patients with liver disease (from Sterling and Brenner, 1966, p. 157).

smaller pool turning over more rapidly than normal and again a normal absolute removal rate and euthyroid status (Ingbar, 1961).

Equilibrium dialysis (Sterling and Hegedus, 1962; Ingbar *et al.*, 1965; Sterling and Brenner, 1966) as well as ultrafiltration have been used to measure the unbound or diffusible fraction of hormone with such techniques as magnesium precipitation to separate the T4 in dialysates from radioactive iodide. Typical findings of the percentage free T4 are illustrated in Fig. 13. The free T4 fraction in normal serum was less than one two-thousandthof the total with elevation in thyrotoxicosis and diminution in hypothyroidism. In pregnancy the free fraction was diminished, a finding ascribed to TBG elevation. Variable elevation was observed in euthyroid "sick" patients. The product of free T4 fraction and PBI value gave an estimate of absolute free T4 iodine concentration which revealed values more than four times the normal in thyrotoxicosis and less than one-seventh normal in hypothyroidism (Fig. 14). In pregnancy the values approximated normal since they were the product of a low percentage of free T4, but a high PBI.

It seems reasonable to infer that the binding proteins, particularly TBG, determine the free T4 fraction. Support is afforded by studies in thyroid disease, liver disease, and nonthyroidal illness (mainly neoplastic) done in collaboration with Drs. Mitsuo Inada and Diego Bellabarba (Inada and Sterling, 1967a,b,c; Bellabarba *et al.*, 1968). The free thyroxine fraction bore an inverse relation to the maximal binding capacity of TBG with a highly significant correlation. The findings in 23 cases of thyrotoxicosis and 16 cases of hypothyroidism are illustrated (Fig. 15); the normal areas are cross-hatched.

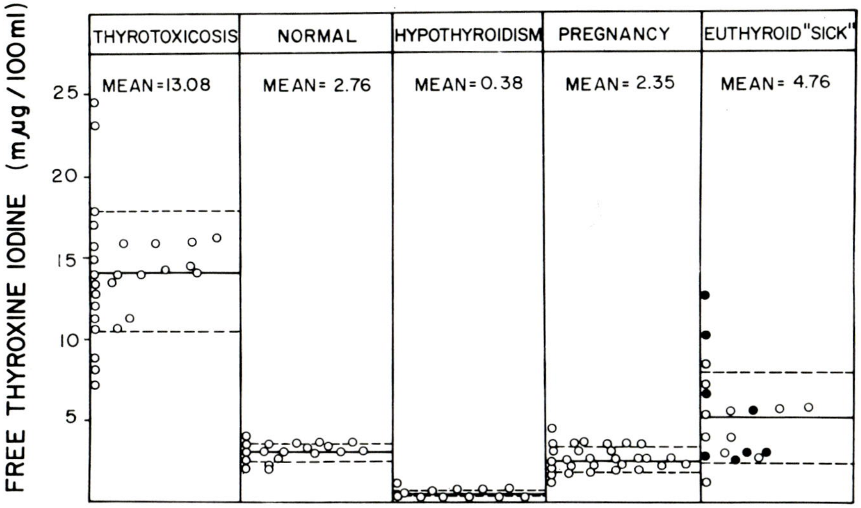

FIG. 14. Free T4 iodine in serum in health and disease. The points (○ and ●) represent the products of the free T4 percentages and the PBI or T4 iodine values. (—) mean values; (- - -) standard deviations; (●) patients with liver disease (from Sterling and Brenner, 1966, p. 157).

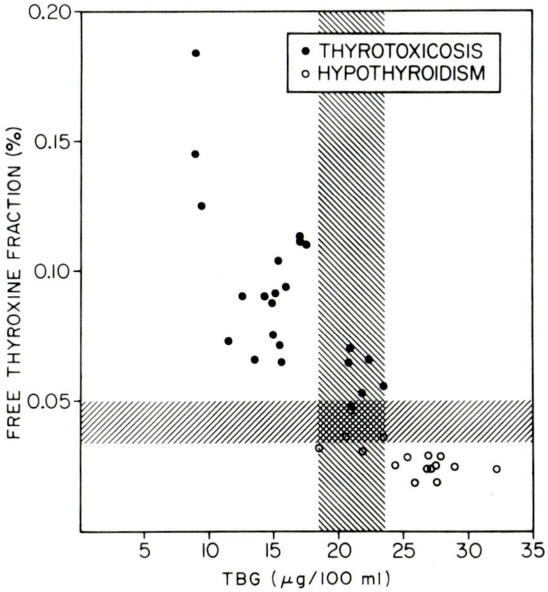

FIG. 15. Relation between free T4 fraction and TBG in thyrotoxicosis and hypothyroidism. The cross-hatched areas indicate the normal ranges (from Inada and Sterling, 1967b, p. 1447).

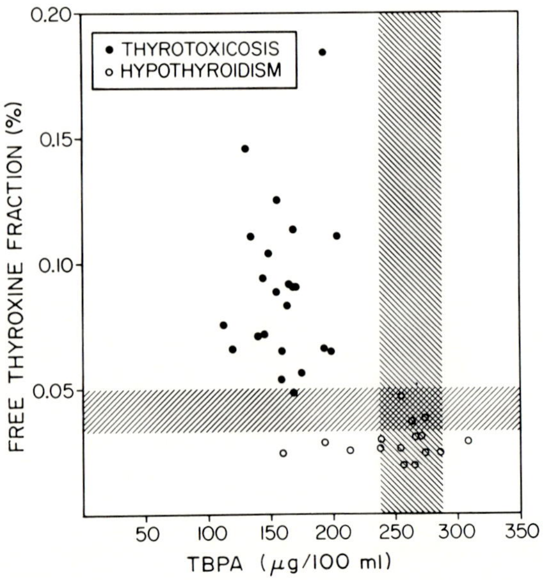

FIG. 16. Relation between free T4 fraction and TBPA in thyrotoxicosis and hypothyroidism. The cross-hatched areas indicate the normal ranges (from Inada and Sterling, 1967b, p. 1447).

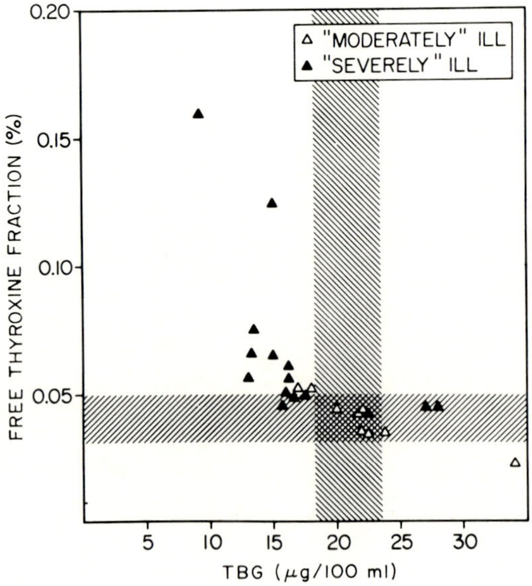

FIG. 17. Relation between free T4 fraction and TBG in major illness. The cross-hatched areas indicate the normal ranges (from Bellabarba *et al.*, 1968, p. 1026).

The inverse relation of free T4 fraction with TBPA was less striking than that with TBG (Fig. 16).

Similarly, a very good inverse relation was evident between free T4 fraction and TBG in sera from nonthyroidal illnesses, mainly neoplastic (Fig. 17). The relation to TBPA was much less striking (Fig. 18). Therefore, it was inferred that the binding capacity of TBG may often be a more important determinant of free T4 than TBPA.

As mentioned earlier, electrophoresis of serum enriched with "tracer" T4 at pH 7.4 reveals that TBPA carries much less T4 than at pH 8.6, which has been employed for most of the numerous electrophoretic studies of the binding proteins. Moreover, the work of Woeber and Ingbar (1968) employing immunoprecipitation of TBPA by a rabbit antihuman prealbumin antiserum indicated that TBPA may bind only about 15% or less of the circulating thyroxine in normal serum (in contrast to much larger amounts on electrophoresis at pH 8.6) and that the removal of TBPA from 12 normal and abnormal sera by precipitation resulted in relatively small increases in the free thyroxine fractions of the sera, except in one serum which lacked detectable TBG.

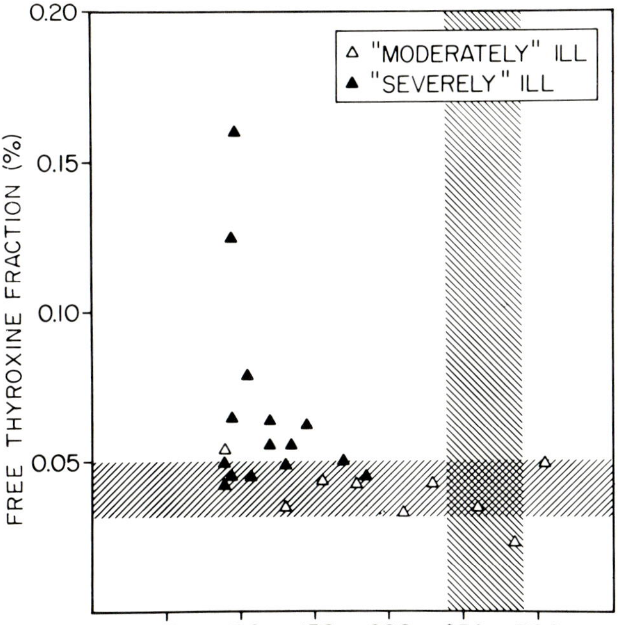

FIG. 18. Relation between free T4 fraction and TBPA in major illness. The cross-hatched areas indicate the normal ranges (from Bellabarba *et al.*, 1968, p. 1027).

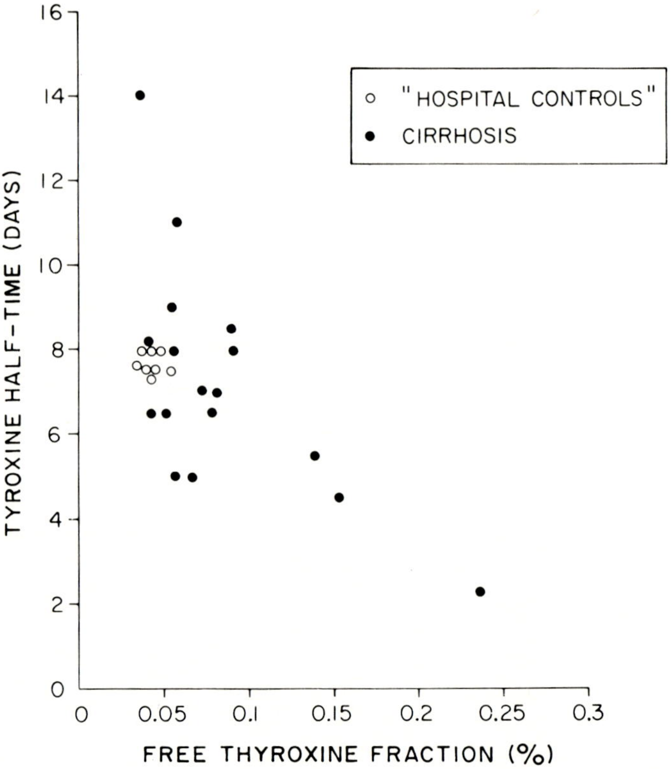

FIG. 19. Relation between T4 half-time and free T4 fraction. Correlation coefficient is -0.69 in patients with cirrhosis ($p < 0.01$) (from Inada and Sterling, 1967a, p. 1280).

The role of the free T4 fraction as a determinant of biological half-time of turnover appears plausible. In addition to indirect evidence, this inference is supported by highly significant inverse correlations in cirrhosis and in neoplastic diseases. As shown in Fig. 19, from a study in cirrhosis, high free T4 fraction is associated with short biologic half-time and low free T4 fraction with long half-time. A similar relation exists in a study of the other nonthyroidal illnesses (Fig. 20).

The relations illustrated do not exclude the possibility that additional factors may also influence hormone disposal. Considerable evidence indicates that impaired hepatic uptake and removal occurs in liver disease, and this factor may have an influence (Oppenheimer *et al.*, 1967). It is also possible that hypermetabolism per se could have an effect beyond that of reduction of binding proteins in causing the accelerated hormone disposal observed in hypermetabolic states, such as acute leukemia (Sterling, and Chodos, 1956; Bellabarba *et al.*, 1968).

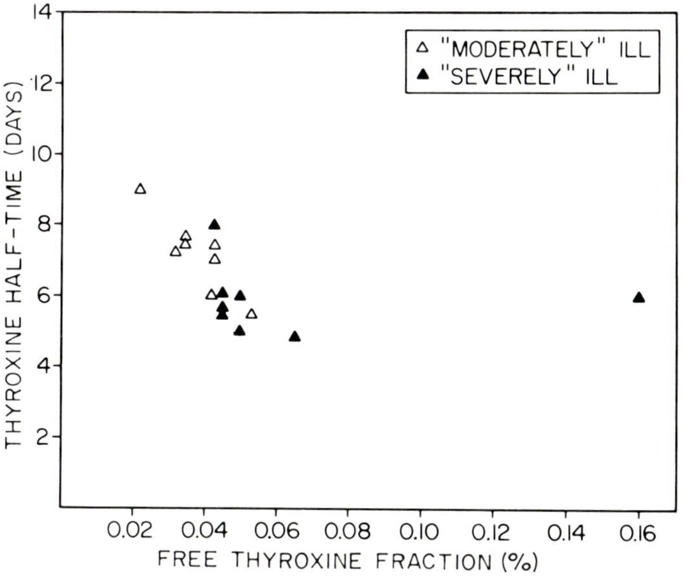

FIG. 20. Relation between T4 half-time and free T4 fraction in major illness (from Bellabarba *et al.*, 1968, p. 1028).

The administration of T3 has been reported to increase T4 turnover, perhaps by stimulation of catabolic processes (Schussler and Vance, 1968).

We cannot explain all the phenomena of T4 transport in the circulation and its turnover entirely by the physico–chemical equilibria described, as exemplified by the work of Refetoff, DeWind, and DeGroot (1967) concerning the unusual family with possible target organ refractoriness to thyroid hormone. Subsequent studies by these investigators have revealed a markedly increased extrathyroidal pool, as anticipated from elevated serum T4 concentrations. The slightly accelerated fractional turnover was, therefore, associated with more than five times normal absolute removal rate, with appropriately increased deiodination since the hormone penetrated normally into the liver and intracellular space. The sera had normal binding proteins and free T4 percentage.

Pending the outcome of these current studies, it may be appropriate to review the major currently accepted instances of abnormal T4 binding, which can produce variations of the PBI in the euthyroid state (Table IV). Pregnancy and estrogen therapy, including the ovulatory suppressants, produce elevation in TBG. The genetic trait of elevated TBG is relatively rare. Hepatic disease may raise TBG, probably by elevated circulating estrogens. It may also produce diminution of TBG as well as TBPA, presumably by failure of synthesis. Prolonged phenothiazine therapy may cause TBG

TABLE IV

ABNORMAL PROTEIN-BINDING CAPACITIES

	Increased	Decreased
TBG	Pregnancy	Androgen Rx
	Estrogen Rx	Steroids (massive)
	Ovulatory suppressants	Acromegaly
	Genetic	Nephrosis, marked hypoproteinemia
	Hepatic disease	Genetic
	Phenothiazine	Hepatic disease
	(Trilafon)	Severe illness, acidosis
	Therapy (prolonged)	Drugs: Dilantin and
	Porphyria, acute	tetrachlorthyronine
	Hypothyroidism	Thyrotoxicosis
TBPA	Steroids (massive)	Thyrotoxicosis
	Acromegaly	Severe illness
	Androgen Rx	Drugs: dinitrophenol and
	(Slight effect)	salicylates

elevation. Acute porphyria, particularly in women, causes TBG elevation (Hollander *et al.*, 1967). In hypothyroidism the elevated TBG usually present accounts for the low free T4 fraction. Diminished TBG is regularly produced (Braverman and Ingbar, 1967) by androgen therapy with some reciprocal elevation of TBPA. Massive corticosteroids and Cushing's syndrome (Oppenheimer and Werner, 1966) and acromegaly (Inada and Sterling, 1967c) have been associated with reduced TBG and reciprocal elevation of TBPA. The nephrotic syndrome as well as very marked hypoproteinemias may occasionally show diminution of TBG, less frequently than might be expected. The genetic trait of diminished or absent TBG is much more widespread than genetic elevation. Pronounced depression of TBG may exist without other phenotypic abnormalities and is compatible with high intellectual attainment; in fact there are two known instances of faculty members of medical schools who possess this interesting disorder. Severe acidosis, if uncompensated, may lessen hormone binding to TBG as well as to the other proteins. Dilantin and the salicylates interfere with binding to TBG and TBPA, respectively. Nonspecific severe illness and injury show regular diminution of TBPA and TBG (Bellabarba *et al.*, 1968). Thyrotoxicosis produces diminution of TBPA and often diminution of TBG in an increased free T4 fraction.

In a series of 23 cases of active thyrotoxicosis (Inada and Sterling, 1967b),

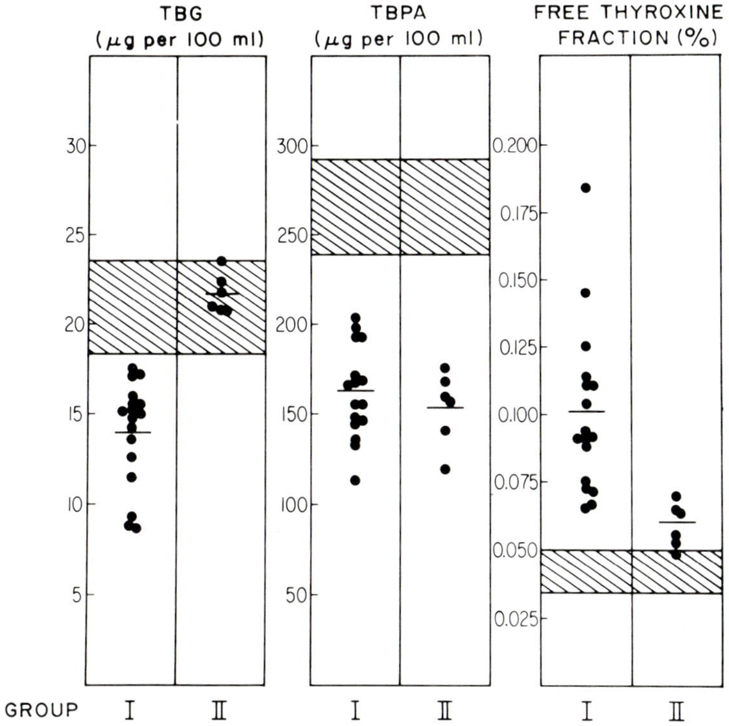

FIG. 21. TBG, TBPA, and free T4 fraction in thyrotoxicosis. The cross-hatched areas indicate the normal ranges. The horizontal lines indicate the mean values in each group of thyrotoxicosis. Group I has diminished TBG capacity; group II has normal TBG (from Inada and Sterling, 1967b, p. 1445).

the sera all showed depressions of TBPA, as shown in Fig. 21. In addition, the binding capacity of TBG was reduced in three-quarters of the sera, which have been designated Group I for purposes of illustration. The free T4 fractions in Group I with diminished TBG capacities were much higher than those in Group II with normal TBG capacities (Fig. 21). This finding suggested that in thyrotoxicosis diminution of TBG may often play a major role in causing elevation of the free T4 fraction (Inada and Sterling, 1967b).

The physiologic activity of T4 appears to depend upon the concentration of unbound hormone. It is considered likely that the free hormone diffuses across the cell membrane where it may be in equilibrium with binding sites within the cell.

ACKNOWLEDGMENTS

This work has been supported in part by the following research grants to Columbia University, Grants AM 10739-03, -04, and -05, from the U.S. Public Health Service, and Grants P-226 and P-461 from the American Cancer Society.

The author is grateful for the creative expert collaboration of Dr. Mitsuo Inada, Dr. Satoshi Hamada, Dr. Yoshihiro Takemura, Dr. Gabriel Hocman, Mr. Milton A. Brenner, and Mr. Edward S. Newman.

REFERENCES

Anbar, M., Guttmann, S., Rodan, G., and Stein, J. A. (1965). *J. Clin. Invest.* **44**, 1986.
Beierwaltes, W. H., and Robbins, J. (1959). *J. Clin. Invest.* **38**, 1683.
Bellabarba, D., Inada, M., Varsano-Aharon, N., and Sterling, K. (1968). *J. Clin. Endocrinol. Metab.* **28**, 1023.
Blackburn, C. M., McConahey, W. M., Keating, F. R., Jr., and Albert, A. (1954). *J. Clin. Invest.* **33**, 819.
Braverman, L. E., and Ingbar, S. H. (1967). *J. Clin. Endocrinol. Metab.* **27**, 389.
Braverman, L. E., Ingbar, S. H. and Sterling, K. (1970). *J. Clin. Invest.* **49**, 855.
Deiss, W. P., Albright, E. C., and Larson, F. C. (1953). *Proc. Soc. Exp. Biol. Med.* **84**, 513.
Dowling, J. T., Freinkel, N., and Ingbar, S. H. (1956a). *J. Clin. Endocrinol. Metab.* **16**, 280.
Dowling, J. T., Freinkel, N., and Ingbar, S. H. (1956b). *J. Clin. Endocrinol. Metab.* **16**, 1491.
Dowling, J. T., Freinkel, N., and Ingbar, S. H. (1956c). *J. Clin. Invest.* **35**, 1263.
Dowling, J. T., Freinkel, N., and Ingbar, S. H. (1960). *J. Clin. Endocrinol. Metab.* **20**, 1.
Engbring, N. H., and Engstrom, W. W. (1959). *J. Clin. Endocrinol. Metab.* **19**, 783.
Giorgio, N. A., and Tabachnick, M. (1968). *J. Biol. Chem.* **243**, 2247.
Hollander, C. S., Scott, R. L., Tschudy, D. P., Perlroth, M., Waxman, A., and Sterling, K. (1967). *New Engl. J. Med.* **227**, 995.
Inada, M., and Sterling, K. (1967a). *J. Clin. Invest.* **46**, 1275.
Inada, M., and Sterling, K. (1967b). *J. Clin. Invest.* **46**, 1442.
Inada, M., and Sterling, K. (1967c). *J. Clin. Endocrinol. Metab.* **27**, 1019.
Ingbar, S. H. (1958). *Endocrinology* **63**, 256.
Ingbar, S. H. (1960). *Ann. N.Y. Acad. Sci.* **86**, 440.
Ingbar, S. H. (1961). *J. Clin. Invest.* **40**, 2053.
Ingbar, S. H. (1963). *J. Clin. Invest.* **42**, 143.
Ingbar, S. H., and Freinkel, N. (1960a). *In* "Hormones in Human Plasma: Nature and Transport" (H. N. Antoniades, ed.), p. 515. Little, Brown, Boston, Massachusetts.
Ingbar, S. H., and Freinkel, N. (1960b). *Recent Progr. Horm. Res.* **16**, 353.
Ingbar, S. H., Braverman, L. E., Dawber, N. A. and Lee, G. Y. (1965). *J. Clin. Invest.* **44**, 1679.
Larson, F. C., and Albright, E. C. (1955). *Endocrinology* **56**, 737.
Lutz, J. H., and Gregerman, R. I. (1969). *J. Clin. Endocrinol. Metab.* **29**, 487.
Nicoloff, J. T., Nicoloff, R., and Dowling, J. T. (1962). *J. Clin. Invest.* **41**, 1998.
Oppenheimer, J. H., and Werner, S. C. (1966). *J. Clin. Endocrinol. Metab.* **26**, 715.
Oppenheimer, J. H., Squef, R., Surks, M. I., and Hauer, H. (1963). *J. Clin. Invest.* **42**, 1769.
Oppenheimer, J. H., Surks, M. I., Smith, J. C., and Squef, R. (1965). *J. Biol. Chem.* **240**, 173.
Oppenheimer, J. H., Bernstein, G., and Hasen, J. (1967). *J. Clin. Invest.* **46**. 762.
Purdy, R. H., Woeber, K. A., Holloway, M. T., and Ingbar, S. H. (1965). *Biochemistry* **4**, 1888.

Refetoff, S., DeWind, L. T., and DeGroot, L. J. (1967). *J. Clin. Endocrinol. Metab.* **27**, 279.
Robbins, J., and Nelson, J. H. (1958). *J. Clin. Invest.* **37**, 153.
Robbins, J., and Rall, J. E. (1955). *J. Clin. Invest.* **34**, 1331.
Robbins, J., and Rall, J. E. (1957). *Recent Progr. Horm. Res.* **13**, 161.
Robbins, J., and Rall, J. E. (1960). *Physiol. Rev.* **40**, 514.
Ross, J. E., and Tapley, D. F. (1966). *Endrocrinology* **79**, 493.
Schussler, G. S., and Vance, V. K. (1968). *J. Clin. Invest.* **47**, 720.
Seal, U. S, and Doe, R. P. (1964). *Proc. 2nd Int. Congr. Endocrinol. London* p. 325.
Sterling, K. (1958). *J. Clin. Invest.* **37**, 1348.
Sterling, K. (1964a). *Mayo Clinic Proc.* **39**, 586.
Sterling, K. (1964b). *J. Clin. Invest.* **43**, 1721.
Sterling, K. (1968). *In* "Radioisotopes in Medicine" (R. L. Hayes *et al.*, eds.), p. 293. U.S. Atomic Energy Commission.
Sterling, K., and Chodos, R. B. (1956). *J. Clin. Invest.* **35**, 806.
Sterling, K., and Brenner, M. A. (1966). *J. Clin. Invest.* **45**, 153.
Sterling, K., and Hegedus, A. (1962). *J. Clin. Invest.* **41**, 1031.
Sterling, K., and Tabachnick, M. (1961a). *Endocrinology* **68**, 1073.
Sterling, K., and Tabachnick, M. (1961b). *J. Biol. Chem.* **236**, 2241.
Sterling, K., Lashof, J. C., and Man, E. B. (1954). *J. Clin. Invest.* **33**, 1031.
Sterling, K., Rosen, P., and Tabachnick, M. (1962). *J. Clin. Invest.* **41**, 1021.
Sterling, K., Bellabarba, D., Newman, E. S., and Brenner, M. A. (1969). *J. Clin. Invest.* **48**, 1150.
Sterling, K., Brenner, M. A., and Newman, E. S. (1970). *Science*. In press.
Woeber, K. A., and Ingbar, S. H. (1968). *J. Clin. Invest.* **47**, 1710.

AUTHOR INDEX

Numbers in parentheses are reference numbers and indicate that an author's work is referred to although his name is not cited in the text.

I

J

K

N

O

R

S

W

Y

Z

SUBJECT INDEX

U

Z